Formation, Evolution, and Dynamics of Young Solar Systems

Astrophysics and Space Science Library

More information about this series at http://www.springer.com/series/5664

Martin Pessah • Oliver Gressel
Editors

Formation, Evolution, and Dynamics of Young Solar Systems

Springer

Sant Cugat Forum on Astrophysics

Editors
Martin Pessah
Niels Bohr International Academy
The Niels Bohr Institute
Copenhagen, Denmark

Oliver Gressel
Niels Bohr International Academy
The Niels Bohr Institute
Copenhagen, Denmark

ISSN 0067-0057 ISSN 2214-7985 (electronic)
Astrophysics and Space Science Library
ISBN 978-3-319-60608-8 ISBN 978-3-319-60609-5 (eBook)
DOI 10.1007/978-3-319-60609-5

Library of Congress Control Number: 2017955963

Cover illustration: Hydrodynamical simulation of a protoplanetary gaseous disk with three forming planets embedded on it and orbiting the central star. The star is located at the left margin of the image. The dark annulus is a density gap carved by a Jupiter-like planet. Two Neptune-like planets are migrating inward from the right. All the planets are gravitationally interacting and invisible in this particular region of the disk. The spiral wakes are the gravitational perturbations induced by the planets. Simulation performed by Pablo Benitez-Llambay with the code FARGO3D. Credit: Pablo Benitez-Llambay

Printed on acid-free paper

This Springer imprint is published by Springer Nature
The registered company is Springer International Publishing AG
The registered company address is: Gewerbestrasse 11, 6330 Cham, Switzerland

Foreword

We live at that extraordinary moment in history when one of the most compelling questions of humankind—are there other worlds such as our own—has been opened up to rapid scientific exploration. At the time of this writing, nearly 3500 exoplanets, with another 4800 candidates awaiting confirmation, have been discovered since the mid-1990s. This has been made possible by breakthroughs in technology that occurred in the last decade of the twentieth century and the first decade of the twenty-first century. These discoveries were made in large part by ground-based observations of the "Doppler wobble" of stars that host planets, as well as by space telescope observations of planets transiting across the disks of their host stars. Among the plethora of new insights is that the most common planets in the galaxy are super Earths that range from 1 to 10 times the mass of the Earth. Those less than about five Earth masses appear to be predominantly rocky worlds—prime candidates for the search for evidence of life.

These observational breakthroughs are driving major efforts to develop deep physical understanding of how solar systems form and evolve. Speculation on the origins of planets goes back at least as far as the early Greek philosophers, but the modern scientific quest began in the eighteenth century with the works of the famous German philosopher Immanuel Kant, and one of the greatest mathematicians and physicists of all time—the French genius, Pierre-Simon Laplace. They posited that the solar system began in a swirling disk of gas, and this became known as the Nebular Hypothesis. This insight was ultimately confirmed by the discovery of disks around all young stars through the development of infrared and millimeter astronomy in the 1990s. While those observations explored global properties of disks, they did not have the resolution to actually watch the process of planet formation. This situation changed remarkably with the opening of the Atacama Large Millimeter Array (ALMA) in 2013. The imaging of protoplanetary disks on scales of down to one Astronomical Unit (AU) using ALMA reveals that disks have a wealth of structure such as rings, gaps, and spiral waves, whose origins may be related to the action of forming planets themselves, or chemical and dust processes involving turbulence and magneto-hydrodynamic processes that are likely to be crucial for planet formation. The stage is set for one of the most remarkable eras of astronomical exploration in the history of science.

The scope of subjects that make up this volume primarily originates from a topical workshop on "Young Solar Systems" which took place in April 2016 as part of the Sant Cugat Forum on Astrophysics. The comprehensive chapters in this book invite readers to explore the great wealth and breadth of ideas in exoplanet formation. The authors cover a large range of scientific backgrounds and techniques in order to address a number of key, interlinked, multiscale problems. Questions attacked include how dust and gas evolve in disks to produce pebbles, chondrules, and planetesimals – the materials out of which both the rocky cores of gas giant planets, as well as terrestrial planets and super Earths are made. The science of young solar systems spans spatial scales ranging from the size of the smallest dust grains (a micrometer) to the scale of planets and ultimately protoplanetary disks (100 AU)—or nearly twenty orders of magnitude in spatial scale. These processes also represent an enormous range of time scales—from dust collisions within the disk body, to the lifetimes of the gaseous disk itself (3–10 million years) and thence to the time scale of terrestrial planet formation (hundreds of millions of years).

This well-crafted volume of selected comprehensive articles invites readers to explore this rich, exciting, and rapidly evolving science. It is organized into twelve chapters from leading experts that address a number of physical themes as revealed by a variety of techniques. On the observational side, the reader will find contributions from cosmochemists who study meteorites which are one of our only links to the epoch of solar system formation; and sub-millimeter astronomers who measure dust evolution and overall structure of protoplanetary disks and the end stages of these processes—the debris disks. On the theoretical side, state-of-the-art computer simulations of disks and of the angular momentum transport processes within them are presented in order to determine how disk evolution occurs. Theoretical modeling of the disk chemistry, dust evolution, pebble growth, radiative physics, and other key processes that inform planet formation theory and observational efforts alike are presented in several different chapters. The ultimate goal of all of this is to synthesize a comprehensive picture of disk evolution and planet formation which puts all of this multiscale, interdisciplinary work into a (hopefully) clear, comprehensive physical picture. This latter approach, known as population synthesis, although still in its infancy, will ultimately be the construct that allows a holistic picture of solar system formation to emerge.

A few words about the specific scientific themes and results presented in these chapters are in order. The first theme in this logical sequence focuses on the evolving environments for planet forming disks. The two observational chapters are on the latest results of the ALMA revolution mentioned above. Key issues here are the determination of disk masses (Bergin and Williams) and the gas and dust content of so-called transitional disks which mark the end phases of disk evolution which could be dominated by planet clearing and trapping (van der Marel). While disk evolution has, for decades, been largely modeled in terms of turbulence and the angular momentum transport that arises from such "turbulent viscosity," the most recent numerical simulations of non-ideal magneto-hydrodynamic processes in the very dense protoplanetary disks (chapter by Bai) demonstrate that it is magnetized disk winds that likely play the major role of transporting disk angular momentum. The irregular structure of disks, now being mapped by ALMA, has been hypothesized as sites for trapping dust and planets, and is explored in the chapter by Pinilla and Youdin.

The road from dust to pebbles and planetesimals, long considered the basic building blocks of rocky planets or the massive rocky cores of giant planets, has received a lot of attention lately as the evolution of dust in disks can begin to be measured. Does dust coagulation lead to large planetesimals? Are such bodies densely packed or quite porous, reflecting the fractal process of dust assembly (see chapter by Kataoka)? What precisely is the accretion history of terrestrial planets, and what are the basic size scales and materials involved in this process? This theme is explored in a couple of chapters. Meteoritic data, as used by cosmochemists, indicate that certain solids known as chondrites may preserve the record of these early accretion processes that lead to planetesimals ($\sim$100 km size bodies) and ultimately planets. These histories can be accessed by studying the stable isotopic compositions of individual "chondrules" (see chapter by Bizarro, Connelly, and Krot). A new development in planet formation theory is the possibility that rather than building rocky planets through the slow process of planetesimal collisions, rapid growth can occur by the accretion of centimeter-sized, enormously abundant pebbles onto a few embedded planetesimals (chapter by Ormel). It would be great if we could measure directly the composition of the construction materials out of which planets are built by accretion. One exotic regime in which this is indeed possible is in the extreme systems consisting of white dwarf stars accreting solid materials from so-called "debris disks" which are the leftovers of the planet building process (see chapter by Bonsor and Xu). The late evolution of gaseous disks is marked by the significant dynamical effects that planets have on their structure. Angular momentum exchange between the planet and disk can lead to the formation of gaps and rings as well as spiral waves. Observations and computer simulations are now poised to exploit these latest disk imaging results, as is clearly shown in the chapter by Dong, Zhu, and Fung.

At this point, the reader is well prepared to appreciate the latest research and insights in the last theme—exoplanets. As noted, arguably the most surprising of exoplanet discoveries is the vast population of super Earths. These have a very wide range in density, from that of rocky materials to nearly the density of gas giants. What accounts for this remarkable range is discussed in the chapter of Ginzburg, Inamdar, and Schlichting. Models of planetary structure depend heavily on their elemental and chemical composition. A largely unexplored aspect of planet formation is the fact that a significant portion of stars are actually members of binary stellar systems. How planets form in circumstellar or circumbinary disks in binary systems is addressed in the chapter by Kratter. Finally, pioneering population synthesis studies were first attempted in papers in the early 2000s. Several different approaches can be taken to the question of how solids are accreted onto planets—leading to different "evolutionary tracks" in the mass versus orbital radius diagram, in which planetary data is typically plotted. The emerging pebble accretion picture, as an example, presents its own challenges and results for population synthesis studies (see chapter by Bitsch and Johansen).

The editors of this volume have assembled an important and highly topical volume, providing an invaluable service to the exoplanetary and the larger astronomical communities. While the observational picture is certainly expected to change dramatically over the next few years, many of the basic, newly discovered elements of disks and planetary properties will remain. Computational and theoretical work has likewise been extremely productive and creative. The reader will enjoy and profit greatly from the study of this volume and will find it to be excellent preparation for both current research and for what lies ahead.

Hamilton, ON, Canada
March 2017

Ralph E. Pudritz

Contents

Chapter 1
The Determination of Protoplanetary Disk Masses

Edwin A. Bergin and Jonathan P. Williams

Abstract In this article we introduce methods used to determine the gas and dust masses of protoplanetary disks, with an emphasis on the lesser characterized total gas mass. Our review encompasses all the indirect tracers and the methodology that is being used to isolate the hidden H_2 via dust, CO, and HD emission. We discuss the overall calibration of gaseous tracers which is based on decades of study of the dense phases of the interstellar medium. At present, disk gas masses determined via CO and HD are (in a few instances) different by orders of magnitude, hinting at either significant evolution in total disk mass or in the CO abundance. Either of these would represent a fundamental physical or chemical process that appears to dominate the system on $\sim$million year timescales. Efforts to reconcile these differences using existing and future facilities are discussed.

1.1 Introduction

Disks are an inevitable by-product of stellar birth and the sites of planet formation. The determination of protoplanetary disk masses is fundamental for understanding almost all aspects of disk physics, from formation and evolution, to the types and compositions of planetary outcomes. As with the molecular cores from which they originate, disks consist of gas and dust. Unlike these cores and the general Interstellar Medium (ISM), however, most of the solid mass is contained in millimeter and larger sized dust grains that are not well mixed with the gas (Fig. 1.1). Measuring disk masses therefore requires accounting for each component separately and different techniques are used both for the observations and modeling.

E.A. Bergin (✉)
Department of Astronomy, University of Michigan, 1085 S. University Ave, Ann Arbor, MI 48109, USA
e-mail: ebergin@umich.edu

J.P. Williams
Institute for Astronomy, University of Hawaii at Manoa, Honolulu, HI 96822, USA
e-mail: jw@hawaii.edu

M. Pessah, O. Gressel (eds.), *Formation, Evolution, and Dynamics of Young Solar Systems*, Astrophysics and Space Science Library 445,
DOI 10.1007/978-3-319-60609-5_1

Fig. 1.1 Dust and gas in the HD 163296 protoplanetary disk. These publicly available ALMA Science Verification data show the 870 μm continuum emission from dust as contours and a color-coded first moment map of CO 3–2 emission (*blue–red* at 2–10 km s^{-1}). The elliptical $\sim 0\farcs5 \times 0\farcs7$ beam size and a physical scale bar is shown in each of the lower corners. The more compact continuum emission is due to the drift of dust grains and demonstrates the necessity to measure the gas and dust properties independently from each other

Knowledge of the dust mass, and the evolution of dust particles, is central to understand the beginnings (and endings) of planet formation. However, it is the gas, as the dominant constituent, which sets the stage for understanding key aspects of the physical and chemical evolution—and the disk *gas* mass is the least characterized quantity. Today with the operation of the Atacama Large Millimeter Array (ALMA) we are obtaining resolved images of the planet-forming zones with optically thin tracers that probe nearly the entire column of gas perhaps, in some cases, into the dust rich mid-plane where planets are born. In the coming decade the James Webb Space Telescope will offer an unprecedented view of disk systems in the mid-infrared tracing hot gas closer to the star and warmer material at greater distances. In all, a key goal is to use molecular emission and dust thermal emission to set deterministic constraints on their coupled physical and chemical evolution. In the case of molecular hydrogen, which is the most abundant species, this relates to the timescale of gas disk dissipation as a gas-rich system is needed for the formation of Jovian worlds; in contrast, super-Earths and ice giants could potentially form in a dissipating gas disk. Furthermore, knowledge of the gas mass, and its distribution, opens the ability to use the emission maps and determine the gas phase chemical composition, which has implications for the resulting composition of both gas giants (Öberg et al. 2011) and terrestrial worlds (Bergin et al. 2015).

More broadly it is the interplay between the gas disk and forming planets that can drive planetary migration and circularize planetary orbits (Baruteau et al. 2016, and references therein). Thus, the ability to connect our knowledge of planetary birth to the exploding field of detection and characterization of exoplanets is greatly aided by grounding knowledge of disk gas and dust masses.

It is the goal of this contribution to summarize the methodology of how masses, both gas and dust, are determined for protoplanetary disks. The main focus of this paper is on the lesser characterized gas mass, but briefly, in Sect. 1.2 we outline the methodology for determination of dust masses, while in Sect. 1.3 we discuss the various methods to approach estimation of the mass of hidden H_2. Thus we first show why H_2 cannot be detected in the bulk of the mass and how other tracers, such as CO or HD, can be used as calibrated tracers of the overall mass and even its distribution within young planet-forming disk systems. However, at present these disparate methods are finding significant (orders of magnitude) differences in the estimated mass. This hints that we have discovered a fundamental effect which is either related to fast (million year) gas disk dissipation or the formation of icy bodies that lock volatiles, such as water and CO, into the dust rich midplane, or perhaps both. We end with a discussion of future prospects.

1.2 The Determination of Dust Masses

Even though the dust is a minor component, at least initially, in protoplanetary disks, it is the easiest to detect because it emits over a continuum rather than the narrow spectral lines from gaseous species. Consequently, there have been more studies of the dust content in disks and there are several recent reviews that describe the current status in the field (Williams and Cieza 2011; Testi et al. 2014; Andrews 2015).

The intensity of the dust emission depends on its temperature and optical depth. The latter is determined by multiplying the projected surface density, $\Sigma_{\rm dust}$, times an opacity, κ, that quantifies the absorption cross-section per unit mass. Mie theory tells us that particles interact most strongly with radiation at wavelengths comparable to their size. Interstellar reddening shows that dust grains range from sub-micron to micron sizes in the ISM and, at wavelengths greater than the maximum grain size, the combined opacity can be approximated by a power law,

$$\kappa_\nu = \kappa_0 \left(\frac{\nu}{\nu_0} \right)^\beta \tag{1.1}$$

(Draine 2006). The normalization and index depend on the grain composition and size distribution (e.g., Pollack et al. 1994; Ossenkopf and Henning 1994). As we will see, there are many other uncertainties in determining disk masses. In derivations of disk masses from dust continuum emission, often a basic parameterization from Beckwith et al. (1990), $\kappa_0 = 10\,\mathrm{cm^2\,g^{-1}}$, $\nu_0 = 1000\,\mathrm{GHz}$, $\beta = 1$, is adopted.

Because of the rising dependence of κ_ν on frequency, disks radiate most strongly in the mid- and far-infrared. The most sensitive surveys for disks, and therefore the best statistics and constraints on their lifetime, have been made with the Spitzer and Herschel telescopes (Evans et al. 2009). However, at these high frequencies, the emission is saturated (optically thick) and largely independent of the surface density. At millimeter wavelengths, $\nu \sim 300\,\text{GHz}$, the emission becomes optically thin for surface densities $\Sigma_{\text{dust}} < 1/\kappa_\nu \sim 3\,\text{g}\,\text{cm}^{-2}$. This condition holds for most disks, although the inner tens of AU may exceed this in massive disks. The dust mass is therefore directly proportional to the millimeter flux, F_ν. For a constant temperature, T_{dust}, it is straightforward to derive

$$M_{\text{dust}} = \frac{F_\nu d^2}{\kappa_\nu B_\nu(T_{\text{dust}})}, \tag{1.2}$$

where d is the distance to the source and B_ν is the Planck function. More detailed models can be made that consider the radiative transfer and temperature structure, but the results do not greatly differ from this simple prescription. An additional uncertainty is the distance, d, but GAIA provides much needed precision, at least for optically bright sources.

It is also important to keep in mind that this prescription for the dust mass is really only a measure of the amount of particles with sizes up to a few times the observing wavelength (Draine 2006). Millimeter and larger-sized particles settle toward the midplane and drift toward the center at different rates such that they size-segregate in a disk. Observations at different wavelengths probe not only a different range of grain sizes but also different regions in the disk. This can be exploited to study disk physics, but it can also complicate the measurement of disk dust masses and must be taken into account when comparing results in the literature.

The process of planet formation begins with the agglomeration of small dust grains, from sub-micron sizes in the ISM to meters and beyond. This process appears to be remarkably fast, as evidenced from young ages for differentiated asteroids in the Solar System (Kleine et al. 2002). At least until self-gravity becomes important, the number distribution of particle sizes, a, that evolve through agglomeration and collisions is expected (and observed over a limited range of sizes) to follow a power law, $N(a) \propto a^{-p}$, with index $p \simeq 3.5$ (Testi et al. 2014). This has the interesting property that the total surface (emitting) area, $\propto \int a^2 N(a) da$, is dominated by the smallest grains but the total mass, $\propto \int a^3 N(a) da$, is mostly within the largest particles. This is the reason why disks are easier to detect than planets and it means that the millimeter derived dust mass is a lower limit to the total mass of solids in a disk.

Nevertheless, despite these caveats, two clear trends emerge from millimeter wavelength surveys of dust masses. First, for a given age, higher mass stars tend to have more massive disks than low mass stars. Second, disk masses decline rapidly with age such that few disks are detectable in the millimeter beyond a few Myr. We elaborate on each of these points below.

1.2.1 Dust Mass Dependence on Stellar Mass

To study the dependence of disk masses on stellar or other properties requires surveys that are minimally biased. Disks radiate most strongly in the infrared and Spitzer carried out complete surveys of all star-forming regions within 200 pc (Evans et al. 2009). These showed that disk lifetimes are longer around lower mass stars. Mass measurements require detecting the optically thin, and therefore weaker, millimeter emission. Before ALMA, disk mass surveys were painstakingly slow and only the nearby Taurus star-forming cloud had been systematically surveyed to high sensitivity (Andrews et al. 2013).

The Taurus survey, mainly carried out with the SMA at 1.3 mm to an rms of about 1 mJy, included 227 protostars (infrared Class II objects) ranging in mass from $\sim 0.01\,M_\odot$ to $\sim 3.0\,M_\odot$. Even though there were many non-detections, the large sample source and high sensitivity showed the disk mass scales approximately linearly with the host stellar mass with a median ratio, $M_{\rm dust}/M_{\rm star} \simeq 3 \times 10^{-5}$. For sun-like stars, this corresponds to about $10\,M_\oplus$ which is about a factor of 3 lower than the total mass of elements heavier than Hydrogen and Helium in the Solar System and suggests that our planetary system is more massive than most. However, there is a large intrinsic dispersion in dust masses for any given stellar mass of about ± 0.7 dex (i.e., ~ 5 times higher or lower). Some of this may be attributed to uncertainties in the stellar mass measurements, some to the dispersion in distance to the sources (the mean distance to Taurus is 140 pc but the cloud has a depth of about 30 pc Loinard 2013), and some may be disk–disk grain opacity variations, but it is clear that disks around stars with similar masses and ages can have substantially different amounts of dust, probably due to variations in their initial conditions. The general trend that higher mass stars have more massive disks is a key assumption in population synthesis models that predict planet mass—stellar mass correlations (Howard et al. 2012).

The advent of ALMA promises complete millimeter wavelength surveys of disks in many of the same regions surveyed in the infrared by Spitzer and Herschel. The first large, systematic survey of the dust and gas in a nearby, young star-forming region was carried out by Ansdell et al. (2016) in Lupus. This 0.9 mm survey had an rms of about 0.3 mJy and included 89 protostars. The survey was 96% complete for all protostars with stellar masses $>0.1\,M_\odot$. 62 sources were detected in the continuum and the dust mass was found to correlate with stellar mass, but with a steeper dependence, $M_{\rm dust} \propto M_{\rm star}^{1.8}$. They also found considerable dispersion about this relation, although the range was smaller, ± 0.5 dex, than that seen in Taurus.

A similarly steep dust mass—stellar mass scaling was found in a recent ALMA survey of 93 disks in the ~ 2 Myr Chamaeleon I star-forming region (Pascucci et al. 2016). They suggest the underlying reason is not faster growth of grains beyond millimeter sizes but rather shorter drift times in disks around low mass stars.

1.2.2 Dust Mass Evolution

Dust masses decline rapidly with time. Prior to ALMA, relatively few disks had been detected at millimeter wavelengths in regions older than $\sim$3 Myr (Carpenter et al. 2005; Mathews et al. 2012). Infrared excesses indicate that some dust was still present around many stars but the weakness of the millimeter emission implies that either the mass or the surface area has decreased (Williams 2012).

To follow the evolution of the dust in disks therefore requires sensitive observations. Moreover, large surveys are required to distinguish trends from the inherent dispersion in initial conditions. ALMA provides the ability to observe many faint sources quickly and recent results reveal disk evolution in far greater detail than before. Barenfeld et al. (2016) surveyed 75 sources un the 5–11 Myr old Upper Scorpius region and found that very little dust remains (median $\sim 1\,M_{\rm moon}$) remains, indicating that planetesimal formation is essentially complete by this point. The closest intermediate aged region, $\sim$3–5 Myr, is σ Orionis but it is about three times more distant than Lupus and Upper Scorpius so detection limits are higher. Williams et al. (2013) imaged the cluster with the SCUBA-2 submillimeter camera and found very few individual sources but, through a stacking analysis, determined a median dust mass $\sim 1.5\,M_{\oplus}$. During the course of writing this paper, we received new data from an ALMA survey of 92 disks that confirm this result and show that the distribution of dust masses in σ Orionis lie in between those of Lupus and Upper Scorpius. This suggests that dust masses decline steadily with time rather than through some abrupt cutoff. As the analysis of these datasets proceed, we will learn about the evolutionary dependence on stellar mass.

The ubiquity of extrasolar planets tells us that the disks are not dispersing but rather than we are witnessing the first steps in the formation of planetesimals. The implication is that most of the solid mass aggregates into millimeter and larger sized particles within a few Myr. This is consistent with cosmochemical studies of the Solar System that show chondrules form from 0–3 Myr after the first metallic flakes (Calcium-Aluminum Inclusions; CAIs) condensed from the hot nebular gas (Connelly et al. 2008).

1.2.3 Dust Density Profiles

Up to now, we have discussed global dust masses obtained from total continuum disk fluxes. Millimeter interferometry routinely achieves sub-arcsecond resolution which allows disks to be resolved in nearby star-forming regions. Consequently, we can measure the amount of dust at different radii in a disk or, equivalently, its surface density profile. The basic idea is to apply Eq. (1.2) for each pixel in the image with a radial variation in dust temperature that can be matched to the infrared SED (Lay et al. 1997; Andrews and Williams 2007).

As the sensitivity of observations improved, Hughes et al. (2008) showed that an accretion disk profile with form,

$$\Sigma(R) = (2-\gamma)\frac{M}{2\pi R_c^2}\left(\frac{R}{R_c}\right)^{-\gamma} \exp\left[-\left(\frac{R}{R_c}\right)^{2-\gamma}\right], \tag{1.3}$$

matches the continuum visibilities better than a truncated power law, $\Sigma(R) \propto R^{-\gamma}$ for $R \leq R_{put}$ and $\Sigma(R) = 0$ for $R > R_{out}$. The characteristic radii, R_c, range from a few tens to a few hundred AU and larger disks tend to be more massive (Andrews 2015). The typical resolution of pre-ALMA observations was about 50 AU in nearby star-forming regions. The dust surface densities generally rose toward the center but with a wide range of peak values in the innermost beam of a few hundredths to a few grams per square centimeter (which, a posteriori, justifies our assumption that the millimeter emission is optically thin). Guillot (1999) shows that Jupiter and Saturn contain at least 30 $M_\oplus$ of elements heavier than Hydrogen and Helium. The interpolated surface density values at their current orbital radii, 5–10 AU, match the extrapolated dust surface densities of the more massive disks.

Some disks are cleared of material within their inner few tens of AU. Indications for this were first seen with IRAS as a marked mid-infrared dip in a protostellar SED (Strom et al. 1989) and thought to signal an inside-out clearing in the late stages of planet formation. The large, sensitive Spitzer surveys showed that between about 10–20% of disks in any given region are in this "transition" phase (Williams and Cieza 2011, and the references therein). As millimeter interferometers expanded their baselines and improved their resolution below an arcsecond, the purported cavities were directly resolved and the depletion in the dust was determined to be at least two orders of magnitude within the inner tens of AU (Andrews et al. 2011). In part because the inner holes may be created by massive protoplanets, transition disks are the subject of considerable current research (see Chapter by van der Marel in these proceedings).

Resolved observations at multiple wavelengths from the sub-millimeter to centimeter reveal the surface density profiles of grains with different sizes. Disk sizes are smaller at longer wavelengths indicating that larger grains are more concentrated than smaller grains (Pérez et al. 2015; Tazzari et al. 2016). This matches theoretical expectations that the drift speed at 100 AU should increase with grain size up to about a centimeter (Armitage 2010). It should then slow down for larger grains as they have longer stopping times, but to detect and map their distribution would require high resolution meter wavelength observations at sensitivities that will only be achieved with the Square Kilometer Array (Wilner 2004).

Drift speeds are proportional to the Keplerian rotation rate and drift timescales are therefore dependent on the stellar mass. This is testable with the combination of the expanded VLA and ALMA and high resolution imaging disk imaging surveys at multiple wavelengths.

ALMA has just begun to transform our view of the dust distribution in disks. Its tremendous leap not only in collecting area but also in baseline length allows ultra-high resolution (∼20–30 milli-arcsecond) imaging of the continuum emission. This has led to instantly iconic images of multiple, narrow dust rings in HL Tau (ALMA Partnership et al. 2015) and TW Hya (Andrews et al. 2016). The structure in these two systems is predominantly radial but at least one disk shows spiral features that may be tidal signatures of an embedded planet or density waves from a massive, self-gravitating disk (Pérez et al. 2016).

Azimuthal asymmetries have been found in some large, massive transition disks (Casassus et al. 2013; van der Marel et al. 2013). Possible causes are planets, molecular snowlines, pressure confinement, or vortices. As these each rely on the gas properties to a varying extent, to understand the features in the high resolution continuum images requires understanding the amount and distribution of the gas.

1.3 The Determination of Gas Masses

1.3.1 H_2 as a Direct Probe

In the interstellar medium the gas to dust mass ratio is measured to be ∼100 (Goldsmith et al. 1997). Thus, at birth, the majority of the disk mass is found in the gas. Measuring the amount of gas is complicated, however, because the emission from the dominant constituent, molecular hydrogen (H_2), suffers from several effects related to its particular molecular physics (Field et al. 1966) that lead to its emission being weak and/or undetectable in most regions of a disk (Carmona et al. 2008). H_2 is a light molecule with large energy spacings between its rotational levels in the ground vibrational state. Furthermore, as a homo-nuclear molecule it has no dipole moment and only weaker quadrupole transitions with $\Delta J = \pm 2$ are allowed, where J is the quantized total rotational angular momentum. The total nuclear spin of H_2 is either $S = 0$ or $S = 1$ which gives rise to a singlet state ($2S + 1 = 1$, antisymmetric) or triplet ($2S + 1 = 3$, symmetric) state. The total wave function including electronic, vibrational, rotational, and nuclear spin must be antisymmetric with respect to the exchange of any two particles (Fermi-Dirac statistics). Since the ground electronic ($X^1\Sigma_g^+$) and vibrational states are symmetric, this means the combined rotational/nuclear spin wavefunction must be antisymmetric. Thus there are two distinct forms of H_2: para-H_2 ($S = 1$; singlet) where J must be even/symmetric (0, 2, 4, etc.) and ortho-H_2 ($S = 3$; triplet) with J odd/anti-symmetric ($1, 3, 5$, etc.). These two species cannot be exchanged by normal inelastic collisions, nor are they connected via radiative transitions (Dennison 1927).

Thus the fundamental ground state transition of molecular hydrogen is the $J = 2 \rightarrow 0$ or S(0) with an energy spacing of 510 K at 28.2 μm. It is this energy spacing combined with the weaker quadrupole, that makes it difficult for H_2 to emit appreciably in cold (T ∼ 20 K) molecule-dominated regions such as protoplanetary disks. This requires H_2 emission to be found, if at all, in the inner regions of the disk closer to the star where temperatures are much higher (>100 K). However, the large dust column densities in the inner tens of AU of a typical protoplanetary disk imply high dust optical depths at 28 μm (Pascucci et al. 2006). Thus, the detection of this line is more difficult because the warm gas closer to the star where the ground state line of H_2 can be excited co-exists with the optically thick dust layers.

This is essentially why stable and abundant molecules, particularly CO, have been used as gas-phase proxies to determine the physical properties of H_2 (density, temperature, mass, velocity field) in regions of star and planet formation. This can be demonstrated quite readily. For optically thin emission the line intensity is:

$$I_{\nu} = \frac{f_u N_{tot} A_{ul} h\nu}{4\pi}, \tag{1.4}$$

Where f_u is the fractional population, A_{ul} is the spontaneous emission coefficient, ν the frequency, and N_{tot} the total column density. In LTE this can be approximated as $f_u = g_u \exp(-\Delta E/kT)/Q(T)$; g_u is the degeneracy and $Q(T)$ is the partition function. The ratio of the line intensity of the ground rotational state of CO to that of H_2 can be expressed as:

$$\frac{I_{\mathrm{CO}\,1-0}}{I_{\mathrm{H_2}\,2-0}} = \frac{f_{\mathrm{CO},J=1}(T) N_{\mathrm{CO}} A_{\mathrm{CO}\,1-0} \nu_{\mathrm{CO}\,1-0}}{f_{\mathrm{H_2},J=2}(T) N_{\mathrm{H_2}} A_{\mathrm{H_2}\,2-0} \nu_{\mathrm{H_2}\,2-0}}. \tag{1.5}$$

For reference $A_{\mathrm{H_2}\,2-0} = 2.94 \times 10^{-11}\,\mathrm{s^{-1}}$, $\nu_{\mathrm{H_2}\,2-0} = 10.6\,\mathrm{THz}$, $A_{\mathrm{CO}\,1-0} = 7.24 \times 10^{-8}\,\mathrm{s^{-1}}$, $\nu_{\mathrm{CO}\,1-0} = 115.27\,\mathrm{GHz}$, and $N_{\mathrm{CO}}/N_{\mathrm{H_2}}$ is generally 10^{-4} (Sect. 1.3.2). Putting in these constants and assuming 20 K ($f_{\mathrm{CO},J=1} = 0.55; f_{\mathrm{H_2},J=2} = 4.2 \times 10^{-11}$) we find:

$$\frac{I_{\mathrm{CO}\,1-0}}{I_{\mathrm{H_2}\,2-0}} = 3.5 \times 10^{7}. \tag{1.6}$$

This quite readily brings the point home. For cold gas at 20 K, which is typical of much of the mass, CO rotational lines will be ten million times more emissive. Combined with issues regarding dust optical depth in the infrared, we must use other *calibrated* tracers.

1.3.2 Calibration of Tracers

1.3.2.1 Carbon Monoxide

For proper calibration of an H_2 proxy, a critical point is the need for a high precision estimate of the chemical abundance. Furthermore the molecule must be chemically stable such that there are strong reasons to expect its abundance will be robust under a wide range of conditions. Both of these factors lead towards carbon monoxide. It is widely abundant in the interstellar medium (ISM) and it is expected to be provided to the disk at or near its ISM abundance. In this section we explore this calibration.

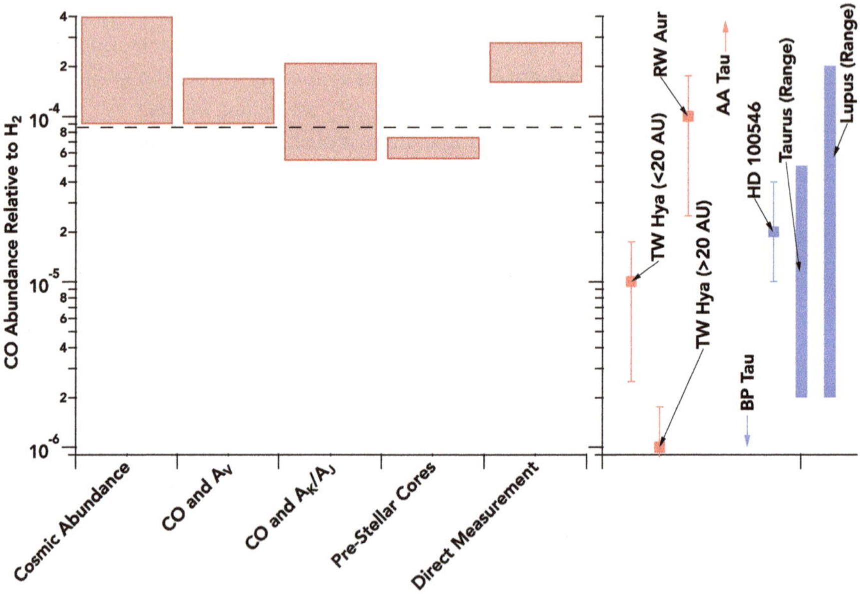

Fig. 1.2 *Left:* Abundance estimates of carbon monoxide in the interstellar medium with references given in Sect. 1.3.2. The *dashed line* shows the commonly assumed reference value from Taurus (Frerking et al. 1982). *Right:* CO abundance measurements in protoplanetary disks with references similarly given in Sect. 1.3.2. Values given in *red* derive the H_2 column from gas phase H_2 or HD, while the *blue* points determine the H_2 mass from the dust assuming a dust mass opacity coefficient and a gas/dust mass ratio of 100

Thus in Fig. 1.2 we provide various estimates of the ^{12}CO abundance in the ISM (on the left-hand side) which is the motivation for its use as a calibrated mass tracer. First we should state that the emission of ^{12}CO is generally optically thick in the ISM; instead observations of isotopologues at lower abundance are used (e.g., ^{13}CO or C^{18}O). Here we have placed all values on the same scale by multiplying by an isotopic ratio.

We start with the cosmic abundance of carbon. Very early on chemical models of dense gas demonstrated that the a primary product of ion-molecule chemistry would be carbon monoxide (Herbst and Klemperer 1973). Thus any available carbon atoms/ions would, in short order ($<10^5$ year at $n_{H_2} = 10^4\,cm^{-3}$), find their way into CO, particularly given the fact that oxygen has greater cosmic abundance when compared to carbon (Asplund et al. 2009). In the diffuse ISM, CO is (mostly) dissociated, and all cosmic carbon resides in ionized form due to the 11.2 eV ionization potential of the carbon atom. Furthermore, in this gas one can measure hydrogen directly via Ly α for H I or the Lyman-Werner bands of H_2. Thus the abundance of carbon available to form CO can be measured. Parvathi et al. (2012) provide a comprehensive look into the carbon abundance over numerous sight lines, and this provides the range in the "cosmic abundance" shown in Fig. 1.2 We note that a fraction of carbon is found to be present in the solid state (Mishra and Li 2015). Thus one cannot use the solar abundance for this calculation without subtracting the abundance of carbon locked in dust.

The next sets of measurements come from molecular clouds. Here the dust is optically thick in the ultraviolet and one must approach the question of the H_2 column density via other means to determine the H_2 abundance. This is traditionally done via the separate calibration of H I and H_2 vs extinction (A_V). This was first performed with *Copernicus* by Bohlin et al. (1978), and reconfirmed by FUSE (Rachford et al. 2009), to be $N(\mathrm{H\,I} + 2\mathrm{H}_2) \simeq 1.9 \times 10^{21}\ \mathrm{cm}^{-2}\, A_V$.[1] Thus Dickman (1978) observed several lines of sight with bright background stars in the vicinity of molecular clouds and explored the correlation of ^{13}CO against extinction (and by proxy H_2; meaning that the slope of the line is the molecular abundance). This was extended to studies using $C^{18}O$ (Frerking et al. 1982), giving the range shown as CO and A_V. The value from the latter study is often used as a canonical CO abundance (relative to H_2) of 8.5×10^{-5}, which is shown as the dashed line in Fig. 1.2.

More recently, with the advent of wide field near-infrared imaging, numerous studies have been able to make high fidelity extinction maps of a range of nearby clouds (Kramer et al. 1999; Lada et al. 1994; Harjunpää et al. 2004; Pineda et al. 2010; Ripple et al. 2013) that have been used to determine the CO abundance (CO and A_J/A_K in Fig. 1.2). Here there is slightly greater fidelity in comparison to the molecular observations. The earlier measurements discussed in the previous paragraph compared (for the most part) pencil beam observations towards a background star to much larger angular scale observations of molecular emission. In the larger scale near-infrared imaging surveys, there are observations of numerous background sources, each with a measured extinction, within a molecular beam of 10–30″. These can be convolved with a Gaussian beam of comparable dimensions to the molecular observations giving a direct comparison of A_J or A_K to the molecular emission. The calibration requires an additional step as the measurement determines A_J or A_K, which is scaled to A_V based on the assumed extinction curve (e.g., Fitzpatrick 2004) and then to the hydrogen column. Attempts have also been made to calibrate this more directly (Kramer et al. 1998; Bianchi et al. 2003; Suutarinen et al. 2013). These studies cover both quiescent and star-forming gas with an abundance range from $\sim 5 \times 10^{-5}$ to 2×10^{-4}. The highest CO abundance is associated with regions of active star formation (Ripple et al. 2013). This must hint at some potential loss of CO in regions in colder regions unassociated with stellar birth, with return in the warmer gas near young stars. The likely culprit in the freeze-out of CO onto ice-coated dust grains (e.g., Bergin and Tafalla 2007).

There is certain evidence for CO freezeout in measurements of the CO abundance in centrally concentrated pre-stellar cores, such as Barnard 68 or Lynds 1544 (Caselli et al. 1999; Bergin et al. 2002). These cores have gas density profiles estimated either via extinction or dust thermal continuum emission and have been subject to detailed chemical study. This, along with the detection of CO ice (Boogert

[1]This assumes a standard galactic extinction curve as the calibration is strictly using E(B-V).

et al. 2015) provide detailed evidence of the pervasive presence of molecular freeze-out towards dense core centers (Bergin and Tafalla 2007, and the references therein). In particular, with the physical conditions constrained as a function of position, models are able to extract chemical abundance profiles as a function of depth (e.g., Bergin et al. 2002; Hotzel et al. 2002). In general, the CO abundance profiles rise from low values at the edge, peak at visual extinctions of 1–2^m, and decline inwards as the increasing density leads to faster gas-grain collisions and subsequent freeze-out. The values shown in Fig. 1.2 are taken at the peak CO abundance from models where this information can be extracted. The fact that these are at the low end of the distribution suggests that some CO resides on grains even near the core edges, reducing the overall CO abundance in cold (<20 K) gas.

Finally, in two instances the CO abundance has been measured directly using absorption of CO and H_2 from their respective ground vibrational state (Lacy et al. 1994; Goto et al. 2015). These measurements are have fairly low precision (1 to 2σ). However, they do represent the most direct measurement.

In general, it is fair to state that the overall range of CO abundances, excluding the pre-stellar case where the effects of freeze-out are prevalent, is quite narrow between $CO/H_2 \sim 0.5\text{–}3\times10^{-4}$. This is quite reassuring for the calibration. A canonical ^{12}CO abundance of $1\text{–}2\times10^{-4}$ is entirely justified by the available information. Of course, observations of lesser abundant isotopologues are required to lower the emission line optical depth. Thus some additional assumptions regarding the atomic isotopic ratio are needed to transfer this calibration to molecular line observations.

1.3.2.2 Hydrogen Deuteride

The abundance of HD can be characterized in similar fashion as above and has additional importance due to the use of the atomic deuterium abundance to constrain cosmological models (Sarkar 1996). The abundance of primordial deuterium was set in the Big Bang and has decreased in time only due to processing through stars or astration (Epstein et al. 1976). In atomic form there are ultraviolet absorption line studies of both atomic D and H towards bright stars. These have most recently been performed by FUSE satellite with the result that the *atomic* D/H ratio is $1.56\pm0.04\times 10^{-5}$ towards sources with $\log(N_{H_2}) < 19.2$ or distance $\leq$100 pc (inside the local bubble; Wood et al. 2004; Oliveira and Hébrard 2006). Beyond this column there is greater dispersion with some lines of sight having lower values ($\sim 0.7 \times 10^{-5}$) and others with 40% higher ratios above the local value. On the low side this difference could be due to stellar astration or incorporation of deuterium in solids.

There are some measurements of HD and H_2 in both the UV (electronic states) and infrared (rotational states). Snow et al. (2008) demonstrate the clear influence of photodissociation on the abundance of hydrogen dueteride in a FUSE survey. HD, like H_2, can self-shield itself from the destructive effects of UV photons, but at reduced efficiency (Wolcott-Green and Haiman 2011). Thus, the HD/H_2 ratio approaches 2× the atomic ratio in denser shielded gas. In addition, in dense photodissociation regions (PDR) or in shocks both HD and H_2 rotational lines have been detected, allowing for a direct abundance determination. There are abundance estimates of $HD/H_2 \sim 10^{-5}$ in the Orion Bar PDR (Bertoldi et al. 1999) and in shocks (Yuan et al. 2012). Both of these are consistent with the atomic lines of sight, beyond the local bubble, where the D/H ratio is lower. However, Yuan et al. (2012) note that uncertainties in the excitation conditions encompass the primordial values.

In sum, the standard assumption for the HD/H_2 abundance is to assume the local value, but account for the fact that two hydrogen atoms are present in H_2; thus, $HD/H_2 \sim 3 \times 10^{-5}$. This is not only for the so far limited use of HD as a mass tracer in disks, but also for models of deuterium chemistry (e.g., Ceccarelli et al. 2014). However, with the exception of TW Hya, most disk systems are found at distances outside the local bubble. Hence there is clear uncertainty in this calibration with the possibility that the abundance is actually lower, meaning the mass measurement would be increased. It is worth noting that the D/H ratio of HD in the atmosphere of Jupiter is near the estimated proto-solar value (Lellouch et al. 2001) suggesting in our solar system at least that the D was in HD (for the most part) and not sequestered within other molecular forms.

1.3.3 Gas Mass Measurements Using HD

Observations of HD offer an additional indirect approach to the H_2 mass in disk systems. The general methodology is outlined by Bergin et al. (2013) and McClure et al. (2016). The TW Hya detection is shown in Fig. 1.3. Since HD is an isotopologue of H_2 it has the same chemistry in the sense that like H_2, HD will not freeze onto grains as ice. Its abundance relative to H_2 is well calibrated and is discussed in Sect. 1.3.2. Below we outline a general approach that illustrates how HD emission can be used as a probe of the gas mass, but also a more systematic look at dependencies.

We can derive the mass implied by an observed flux of an optically thin unresolved source as follows. The total number of HD molecules ($\mathcal{N}_{\rm HD}$) can be related to the line flux,

$$F_l = \frac{\mathcal{N}_{\rm HD} A_{10} h\nu f_u}{4\pi D^2}. \tag{1.7}$$

In this equation, D is the distance, ν the line center frequency, and f_u is the fractional population in the upper state:

$$f_u = 3.0 * \exp(-128.5\ {\rm K}/T)/Q(T). \tag{1.8}$$

The total gas mass of H_2 is then: $M_{\rm gas\ disk} = 2.37 * m_{\rm H}\mathcal{N}_{\rm HD}/x_{\rm HD}$. In this expression $x_{\rm HD}$ is the abundance of HD relative to H_2, $m_{\rm H}$ is the mass of a hydrogen atom, and 2.37 is the mean molecular weight per particle, including helium and heavy elements

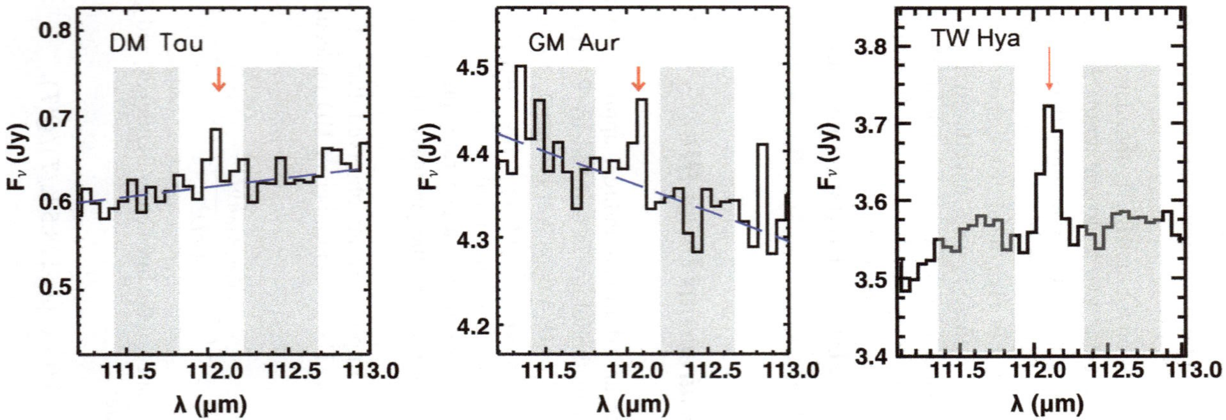

Fig. 1.3 Detection of HD $J = 1 \rightarrow 0$ towards TW Hya by Bergin et al. (2013) in the *right-hand panel*, along with GM Aur and DM Tau by McClure et al. (2016) shown in the *middle and left-hand panels*, respectively

(Kauffmann et al. 2008). Combining these two expressions gives a relation between observed line flux and mass,

$$M_{\text{gas disk}} = \frac{2.37 m_H 4\pi D^2 F_l}{A_{10} h\nu x_{\text{HD}} f_u}. \tag{1.9}$$

If we then include physical constants and note that the partition function, $Q(T)$, is near unity below $\sim$50 K (Müller et al. 2005), we obtain this simple relation that shows key dependencies where $F_{\text{TWH}} = 6.3 \times 10^{-18}\,\text{W m}^{-2}$ is the observed flux density toward TW Hya,

$$M_{\text{gas disk}} > 5.21 \times 10^{-5} \left(\frac{F_l}{F_{\text{TWH}}}\right) \left(\frac{3 \times 10^{-5}}{x_{\text{HD}}}\right) \left(\frac{D}{55\,\text{pc}}\right)^2 \exp\left(\frac{128.5\,\text{K}}{T_{\text{gas}}}\right) M_\odot. \tag{1.10}$$

The two most important factors in this expression are the abundance of HD and the sharp dependence on the gas temperature. The abundance of atomic D is discussed in Sect. 1.3.2. A bigger factor is the strong temperature dependence which requires a priori information on the gas temperature structure. It is important to note that this estimate is a lower limit as HD emission will not trace significant mass with temperatures below 20 K. Another factor is the dust opacity at 112 μm; an optically thick dust-rich midplane would be hidden from HD emission. Because of these two issues this equation is listed as a lower limit. As such the estimation of the total mass requires a model that accounts for the mass that might be hidden. Below we illustrate how models can be used to account for the hidden material and incorporate a more realistic approximation for the gas temperature structure.

It is well known that disks are flared and directly exposed to stellar heating which leads to sharp thermal gradients with disk *dust* temperature having significant radial (warmer closer to the star) and vertical (warmer on the exposed disk surface) structure. However this is only part of the picture. Sophisticated models that include the *gas* thermal physics and chemistry (Woitke et al. 2009; Thi et al. 2010; Gorti et al. 2011) have shown that midplane densities are high enough to thermally couple the gas to the dust grain, but on the disk surface the gas is more emissive in the HD line because it is directly heated by stellar irradiation and thus warmer than the dust. In the HD detection paper, Bergin et al. (2013) adopted the thermo-chemical model of TW Hya by Gorti et al. (2011) to show that the HD flux was best represented by models with gas mass $>0.05\,M_\odot$. Lower masses such as in inferred by Thi et al. (2010) of $\sim 10^{-3}\,M_\odot$ using other tracers (e.g. ^{13}CO, O I) are inconsistent with the observed HD flux. More recently, McClure et al. (2016) published two additional HD $J = 1$–0 detections at the 3σ level in DM Tau and GM Aur. This work applied a different approach using traditional models of the dust spectral energy distribution in flared disk with warm surfaces (D'Alessio et al. 1998) in concert with the constraints from the HD emission to set more stringent limits on both the gas and dust mass in these systems.

To explore general dependencies and illustrate the utility of HD emission as a mass tracer, we employ the thermochemical model of Du and Bergin (2014). The model assumes a stellar spectral type, and dust surface density distribution, along with a high energy UV radiation field motivated by observations. For the UV radiation field, B stars are dominated by the stellar field while M stars by accretion luminosity. Within this framework, the vertical structure is determined assuming hydrostatic equilibrium where we calculated the dust temperature and then iterated on the heating-cooling balance with the extensive chemistry to determine the gas temperature. The models presented here were used to explore the ground state water emission (Du et al. 2017) and are run over a range of total disk dust mass (0.25–$2.0 \times 10^{-4}\,M_{\odot}$) that is grounded to the low end of the observed distribution of dust mass in Taurus (Williams and Cieza 2011). The results are shown in Fig. 1.4, and provide the predicted HD flux at 112 μm as a function of disk dust mass (labeled in each figure), the stellar type, and the dust-to-gas mass ratio. For a fixed dust gas of the disk, a lower dust-to-gas mass ratio means a higher gas mass.

Two effects are readily apparent in Fig. 1.4. First, the effect of direct gas heating is evident as the earlier type stars (A-F) have stronger flux with greater spacing

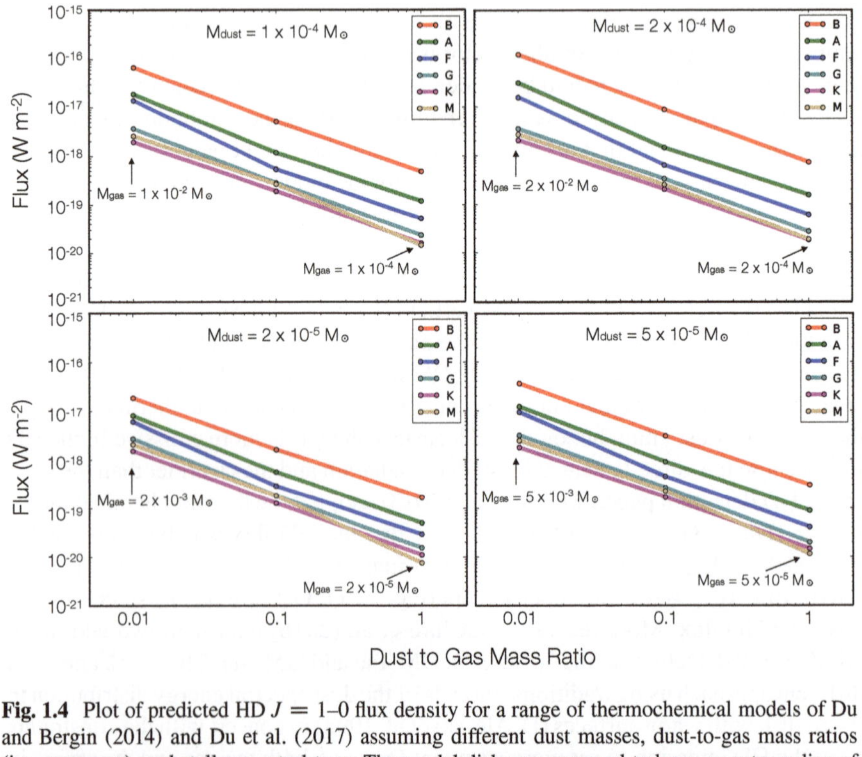

Fig. 1.4 Plot of predicted HD $J = 1$–0 flux density for a range of thermochemical models of Du and Bergin (2014) and Du et al. (2017) assuming different dust masses, dust-to-gas mass ratios (i.e., gas mass), and stellar spectral type. The model disks are assumed to have an outer radius of 400 AU and are uniformly placed at 100 pc to allow for ease in distance scaling. Most of the HD emission arises from the inner 100 AU

between lines for a given spectral type. For cooler stars (G-M), at a given gas and dust mass, the predicted HD emission flux density is similar, as this is the region the gas and dust temperatures become coupled. Regardless of these effects, the mass dependence is clear: for a given disk dust mass and stellar spectral type, over a 2 order of magnitude range of *gas* mass there is a two order of magnitude increase in the HD line flux. Clearly there are complications in this analysis as the optical depth of the line (when detected) is not constrained and there needs to be some additional information regarding the gas temperature. However, in the latter case, resolved observations with ALMA can readily provide both radial and vertical temperature information, thereby improving our ability to use the fundamental rotational transition of HD as a mass and gas surface density probe (as demonstrated by Schwarz et al. 2016).

Finally, an additional issue regards the origin of the HD line flux. All current data are spectrally unresolved. If HD is emitting from a hot wind or jet component then less mass is needed to match the overall emission. Existing models account for the entirety of the line flux as arising from the disk and, in this instance, lower masses would result. Models certainly demonstrate that the disk will be emissive in this line, but additional spectrally resolved observations are required to confirm if current assumptions are correct.

The high gas masses inferred from HD are roughly consistent with a disk gas-to-dust ratio of 100, similar to their initial conditions in the ISM. Independent evidence in support of this comes from measurements of gas accretion onto stars. Mass accretion rates typically range from 10^{-10} to $10^{-8}\,M_{\odot}\,\text{year}^{-1}$, with strong dependencies on stellar mass and age. The total accreted mass in Lupus (and other young regions with ages of 1–3 Myr) is $\sim 10^{-4}$ to $10^{-2}\,M_{\odot}$ for stellar masses ~ 0.1–$1\,M_{\odot}$, which is roughly consistent with the measured dust mass times 100 (Manara et al. 2016).

1.3.4 Gas Mass Measurements Using CO

1.3.4.1 Total Mass

An alternative approach to measuring gas masses follows the history of molecular cloud studies and uses millimeter wavelength observations of CO and its isotopologues. Interferometry, beginning with pioneering millimeter arrays at Hat Creek and Owens Valley observatories in California and continuing through the present day with ALMA, allows sensitive sub-arcsecond observations and is ideally suited to studying the gas content and distribution in protoplanetary disks. The first measurement of disk rotation was made through CO observations by Sargent and Beckwith (1987). The emission from the primary species, ^{12}CO, is so abundant that it is optically thick and therefore its emission is more sensitive to temperature than column density. As with molecular clouds and cores, the lines of the rarer isotopologues ^{13}CO and C^{18}O, are less saturated and provide a simple pathway to quantifying the gas mass. However, prior to ALMA, there were surprisingly few observations of ^{13}CO and almost none of $C^{18}O$. The reason can be found in the line survey of TW Hya by Kastner et al. (1997) who detect strong emission from CO, HCO^{+}, HCN, and CN, but very weak ^{13}CO. The bright lines are characteristic of a photon-dominated region (PDR) but the weak ^{13}CO suggests a low total gas content (assuming the calibrated abundance discussed in Sect. 1.3.2). Subsequent observations understandably tended to focus on the brighter, more easily detectable lines.

As our knowledge of protoplanetary disk structure and chemistry grew, Aikawa (2002) postulated that CO and other molecules should be found in a *warm molecular layer* with a lower boundary set by the freeze-out of molecules onto dust grains in the cold disk midplane and an upper boundary set by photo-dissociation from the central star or external radiation field (Fig. 1.5). Disk rotation allows us to tomographically map disk structure and such a chevron pattern of emission has been directly observed in the CO 3–2 line toward the HD 163296 disk (Rosenfeld et al. 2013). For a very cold or low-density disk, the CO emitting region could be so small as to explain the weak ^{13}CO emission. However, over the range of parameters expected for the temperature structure and disk size based on theory and observations, Williams and Best (2014) showed that most of the gas mass does

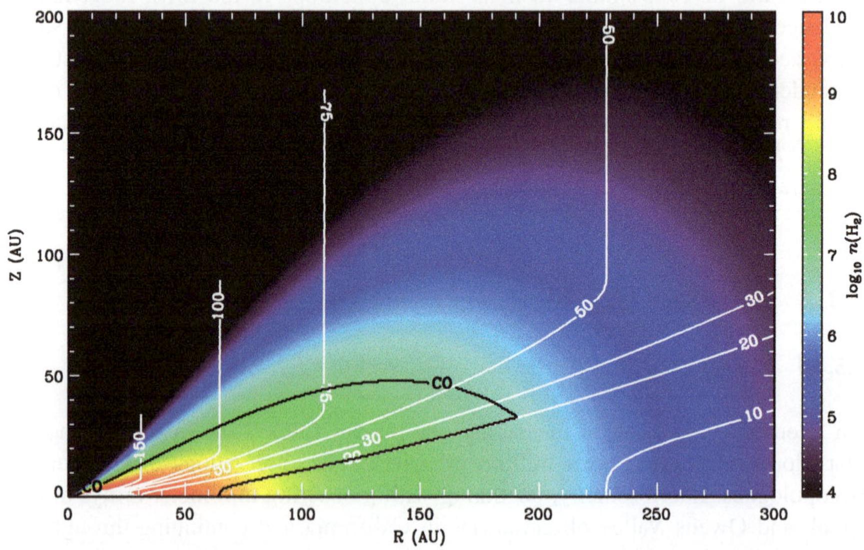

Fig. 1.5 Density and temperature distribution of a model disk. The star is at the origin and the disk is radially symmetric with mirror symmetry about the midplane at $Z = 0$. The *color scale* represents the H_2 gas density on a logarithmic scale. The gas temperature is shown and labeled in the *white* contours. The *black* contours, labeled "CO", represent the boundary of a warm molecular layer within which CO is expected to be in the gas phase and emit millimeter wavelength rotational lines

indeed lie in the warm molecular layer and therefore that the CO isotopologue lines should be a reliable measure of the total disk gas content.

With the basic assumptions of Keplerian rotation, hydrostatic equilibrium, a constant gas-phase CO abundance above the freeze-out dominated midplane, and azimuthal asymmetry, the spectral line profiles of CO and its isotopologues can be calculated with radiative transfer models for any given temperature structure, Williams and Best (2014) ran a large grid of models over a range of temperatures and densities and found a simple way to measure disk masses by plotting the ^{13}CO and C^{18}O line luminosities against each other as in Fig. 1.6.

The luminosities of the two isotopologues closely correlate of course, but disks of a fixed mass can have a wide range of values depending on molecular excitation, line optical depth, and the amount of CO freeze-out and photodissociation. The constraints from two lines help to remove the ambiguities and, despite the intrinsic

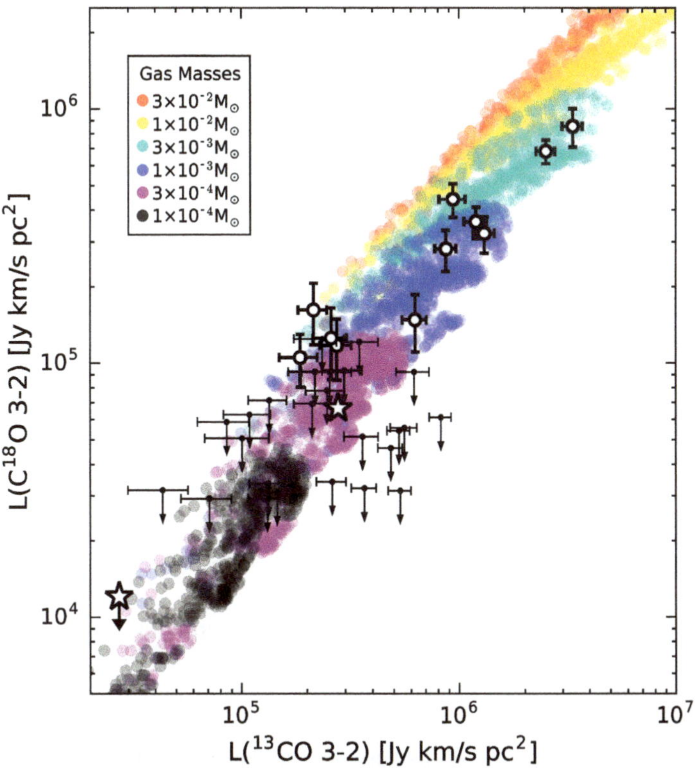

Fig. 1.6 C^{18}O 3–2 vs. ^{13}CO 3–2 line luminosity for models from the WB14 grid, color-coded by disk gas mass. The spread for each mass is due to optical depth, excitation, and variation in the gas fraction within the CO-emitting warm molecular layer over the range of disk density and temperature profiles in the models. Nevertheless, there is sufficient separation in this isotopologue plot that the total disk gas mass can generally be determined to within a factor of 3–10. Observations from the Ansdell et al. (2016) Lupus survey are overplotted

scatter, the spatially and velocity integrated line luminosities provide a simple and reliable measure of the total gas mass, to within a factor of 3–10. This simple procedure is ideally suited for the statistical analysis of large datasets. Figure 1.6 shows observed values from the Ansdell ALMA Lupus survey and the good match with this simple model description. For each data point, the mass or upper limit can be simply read off this plot.

Some data points lie below the model locus shown here indicating lower $C^{18}O$ line luminosities than predicted. This is probably due to selective photo-dissociation of this rare isotopologue, which cannot self-shield as effectively as CO (van Dishoeck and Black 1988). More detailed modeling that takes this into account confirms the basic findings here (Miotello et al. 2016).

The range of Lupus disk masses is plotted in Fig. 1.7 where the dust mass is derived following the procedure described in Sect. 1.2 and the gas masses from the CO isotopologues. The most massive dusty disks, $M_{\rm dust} > 10\,M_\oplus$, are shown here (the full sample is shown in Ansdell et al. 2016). All are detected in ^{13}CO and many in $C^{18}O$, but in general the lines are quite weak and the implied gas masses are low, typically $M_{\rm gas} \sim 1\,M_{\rm Jup}$. This mirrors the Williams and Best (2014) results in Taurus which itself has precedent in the low ^{13}CO line flux found by Kastner et al. (1997) in TW Hya. The inferred gas-to-dust mass ratios are plotted in the lower panel. The high dust and low gas masses imply ratios that are generally much lower than the fiducial ISM value of 100.

Just as molecular cloud and core masses derived from CO observations depend inversely on the molecular abundance, so the disk gas masses shown in this section are scaled to an assumed value, $[^{12}CO]/[H_2] = 10^{-4}$. This value has been calibrated in clouds and cores but there are few direct measurements in disks and they disagree. The discrepancy is most significant with the HD measurements described in Sect. 1.3.3. Ultimately, the CO abundance is the largest source of uncertainty in gas mass measurements from CO and, because of the relative ease of measurement of these lines, this is a critical issue for disk studies. We discuss this further in Sect. 1.4.

1.3.4.2 Gas Surface Density Profiles

The modeling methodology used in Fig. 1.6 to measure bulk gas masses can be extended to determine gas surface density profiles. The basic idea, described in detail in Williams and McPartland (2016), is to compare a library of model images with a resolved line map.

Figure 1.8 shows that this works very well in the case of the large, bright disk around the Herbig Ae star, HD 163296. In this case, the ^{13}CO map is well resolved, has a high signal-to-noise ratio, and the temperature structure of the disk was well constrained by modeling the CO data independently (Rosenfeld et al. 2013).

It will be interesting to carry out similar modeling of other resolved ^{13}CO line images with different gas-to-dust (or CO line to continuum) ratios. This requires a similar dynamic range in both spatial and intensity scales. Williams and McPartland

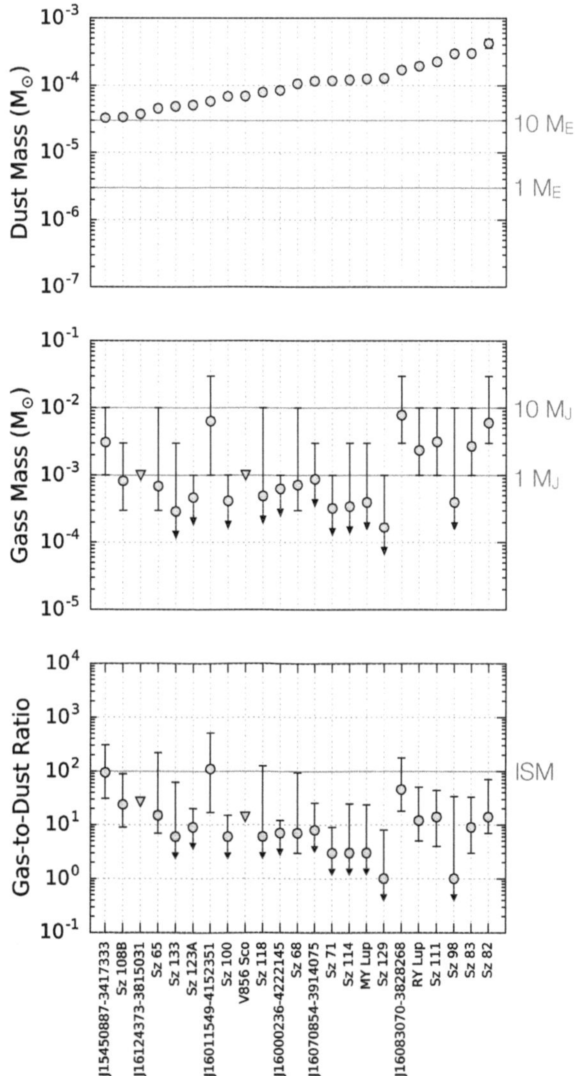

Fig. 1.7 Disk masses derived from the Lupus ALMA survey. The *top panel* shows the dust masses as derived from the 890 μm continuum equation 1.2. The *middle panel* shows the gas masses as derived from the CO isotopologue emission as described in Sect. 1.3.4 and plotted in Fig. 1.6. The *lower panel* shows the gas-to-dust ratio, and how it appears to be significantly lower than the ISM value of 100 for almost all disks here. This figure is adapted from Ansdell et al. (2016) to only show 23 most massive disks from the full sample of 90 as these were all detected in ^{13}CO and most in C^{18}O, thereby permitting the best constraints on the gas mass

(2016) show that moderate resolution (0.″2–0.″3) ALMA observations of a few to tens of minutes (mass dependent) suffice to measure the gas profiles of typical disks found around T Tauri stars.

The $J = 2$–1 transition is ideally suited for this analysis as the CO, ^{13}CO, and C^{18}O lines can all be observed simultaneously with ALMA and at the same resolution and sensitivity. A combined modeling of the CO and ^{13}CO data allows the temperature and density structure to be derived in tandem. By then comparing to the C^{18}O map, the isotopologue ratio can be determined and any radial variation thereof. This provides an interesting test of selective photo-dissociation

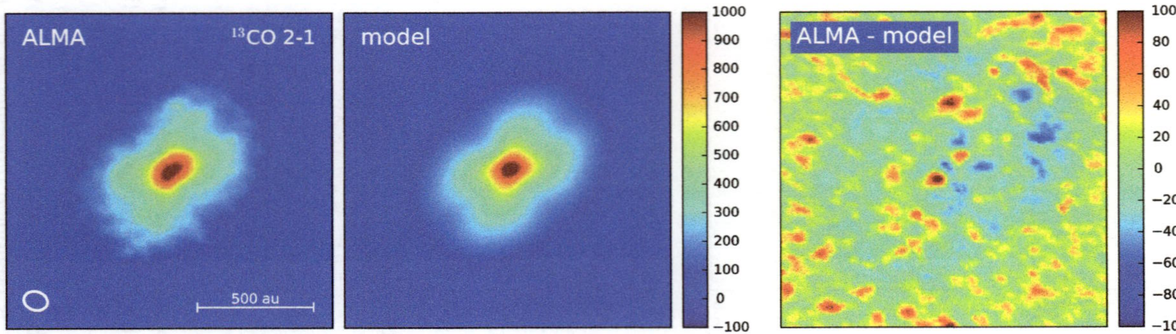

Fig. 1.8 Comparison of ^{13}CO 2–1 integrated intensity images for the HD 163296 disk. The *left panel* shows the ALMA map, the *central panel* the model fit for the best fit model parameters, and the *right panel* shows the difference between the two. The *colorbars* show the range of intensities for each map; the ALMA and model images on the same scale from −100 to 1000 mJy km s^{-1} $beam^{-1}$, and the difference image from −100 to 100 mJy km s^{-1} $beam^{-1}$

models which have been invoked for explaining the oxygen isotope variation in the Solar System (McKeegan et al. 2011). As the 1.3 mm dust continuum image is also obtained, this is a powerful combination for studying the distribution of the solid and gas components.

This can readily be extended to other molecules. Even if absolute abundances are uncertain, resolved maps can show the radial dependence of relative abundances very well. As the dust processing and chemical timescales vary with radius, we can test models that predict different abundance patterns.

1.3.5 Comparison of CO and HD

In some instances there have been attempts to measure the CO abundance in disks which provide information on its viability as a mass tracer. This is complicated by the fact that CO is known to be frozen as ice in the dense midplane as discussed in Sect. 1.3.4 (Dutrey et al. 1997; Aikawa 2002; Bergin et al. 2007; Semenov 2012). This facet is spectacularly confirmed by recent imaging of the HD 163296 disk (Rosenfeld et al. 2013). In the right-hand panel of Fig. 1.2 we provide the limited set of CO abundance measurements in disk systems, which illustrate the complexity of this problem. Thus Schwarz et al. (2016) combine the detection of HD with mulitransitional ^{13}CO and $C^{18}O$ ALMA observations of TW Hya to resolve the CO abundance structure in TW Hya (see also Nomura et al. 2016); a disk average value presented in Favre et al. (2013). The assumption in both of these cases is that the HD emission is strongly curtailed for $T < 20$ K where CO is frozen out as ice. Hence the CO/HD column density ratio is not affected by freeze-out and traces the abundance in the emissive layers. This is borne out by excitation calculations (e.g., Favre et al. 2013). With ALMA the disk thermal structure is resolved and the effect of temperature on the HD emission can be mitigated. In all cases, both within the CO snowline (near 20 AU) and beyond the CO snowline the derived CO abundance is well below the ISM value. This is argued to be due to incorporation of CO into solids either as CO_2 or perhaps processed into more chemically complex forms (Bergin et al. 2014; Furuya and Aikawa 2014; Reboussin et al. 2015; Eistrup et al. 2016).

In two cases both CO and H_2 have been observed in absorption in the UV towards RW Aur (France et al. 2014) and AA Tau (France et al. 2012). In one case the derived abundance is very close to interstellar (RW Aur), but in the case of AA Tau the CO abundance measurement is 0.4 relative to H (i.e., four orders of magnitude *above* interstellar). UV observations generally probe gas very close to the star and/or at high altitudes in the disk (with low column and UV exposure, e.g., $A_V \ll 1$). Thus it is difficult to extrapolate these numbers to the overall disk mass; regardless, they do represent useful information. These few measurements represent the only abundance estimates that are done independent of the dust. The remaining values come from a range of studies that infer the H_2 mass/column from dust observations via spectral energy distribution (SED) modeling or through resolved thermal continuum emission maps. In one study towards BP Tau, Dutrey et al. (2003) observed CO isotopologue emission with the IRAM Plateu du Bure interferometer and the 30 m telescope. They argue that the overall disk is warm and above the CO sublimation temperature. Assuming a normal gas/dust mass ratio of 100 they derive a CO abundance (relative to H_2) more than two orders of magnitude below the ISM value of $\sim 9 \times 10^{-5}$ (dashed line in Fig. 1.2). In HD 100546, Kama et al. (2016a) and Bruderer et al. (2012) use numerous molecular and atomic lines to explore the chemistry in this disk in the framework of a 2D thermochemical model. In all, if the gas/dust mass ratio is 100, they find a reduced overall amount of elemental carbon and therefore a reduced CO abundance.

In Sect. 1.3.4, we discussed the comparison of $C^{18}O$ emission and dust in the Williams and Best (2014) and Ansdell et al. (2016) surveys. Here the problem of abundance vs mass is clearly revealed. If the gas-to-dust mass ratio is 100, there is a range of CO abundances that span nearly 2 orders of magnitude. If the CO abundance is $\sim 10^{-4}$ as in the ISM, then the gas-to-dust mass ratio ranges from ~ 1 to 10. Thus the issue is revealed—is this evolution of the CO abundance or is this gas dissipation (or some combination of both). Either issue represents a fundamental statement about the overall physical and chemical evolution of disk systems.

1.4 Reconciliation?

In sum we have two disparate sets of measurements. Observations of a statistically significant sample of disks studied in the emission of CO isotopologues find that they are missing CO emission relative to measured dust masses. Recall that the dust mass measurements are a lower limit to the total solid component as grains much larger than the observing wavelength have little emission. Thus the low CO emission is not due to an overestimate of the solids. It can be interpreted as either missing carbon monoxide assuming that H_2 is still present or as very low gas/dust mass ratios (i.e. missing gas mass). The magnitude of these effects (at least two orders of magnitude) is effectively summarized in the right hand panel of Fig. 1.2. We discuss the chemical and then the physical possibilities in turn.

There are three disks with gas mass measurements from HD (Bergin et al. 2013; McClure et al. 2016) and, in the case of TW Hya and GM Aur (to a lesser extent DM Tau), this suggests that CO is missing while hydrogen gas is abundant (Favre et al. 2013; Schwarz et al. 2016; Nomura et al. 2016; Kama et al. 2016b). It is important to stress that these inferences are not limited to CO and its isotopologues alone. Surveys of [O I] 63 μm emission in Taurus find that, for normal gas/dust mass ratios (~ 100), models tend to over-predict line emission (Aresu et al. 2014). Similarly, the ground state lines of water vapor at 557 and 1113 GHz have been detected in only one T Tauri disk system (TW Hya; Hogerheijde et al. 2011), with nondetections in over 13 potentially gas-rich disk systems (Du et al. 2017). Detailed models predict that these lines would be too strong by large factors and require a reduced water abundance by two orders of magnitude under the assumption of a "normal" gas/dust ratio (Bergin et al. 2010; Hogerheijde et al. 2011; Du et al. 2017), but see also

Kamp et al. (2013). In the case of TW Hya, Du et al. (2015) performed detailed thermochemical modeling of CO isotopologue, [O I], OH, HD, and water vapor emission demonstrating that volatiles appear to be depleted in the upper atmosphere of the disk. Thus there appears to be a systematic effect that affects the main volatile carriers of at least carbon and oxygen.

One interesting chemical effect that would aid in reconciliation, at least for CO, is isotopic self-shielding. It is well known that the photodissociation of carbon monoxide proceeds via a line process which means that the presence of CO molecules along the line of sight toward the radiation source can shield other molecules downstream. This "self-shielding" effect has an innate column density dependence; hence ^{12}CO is more efficiently shielded than less abundant ^{13}CO and so on. Since we are mainly using $C^{18}O$ as our preferred H_2 mass tracer it could probe less mass in the disk than say HD, which would result in a reduced abundance measurement. Miotello et al. (2014, 2016) explore these effects in detail and show that a CO abundance reduction of nearly an order of magnitude could be accounted via this mechanism. However, HD also self-shields and hence it traces a similarly reduced column of H_2 (see Bergin et al. 2014). Furthermore, Schwarz et al. (2016) use ALMA observations of CO isotopologues towards TW Hya to constrain the thermal structure and, with HD, the CO abundance. They find that the abundance is reduced by nearly two orders of magnitude. While, self-shielding works in the right direction, it cannot account for the observed magnitude of the abundance depletion in TW Hya at least, nor does it account for the presence of similar effects seen in oxygen-bearing molecules.

An additional chemical scenario is that we are witnessing the formation of icy planetesimals or at least pebbles that are trapped in the midplane (Dutrey et al. 2003; Chapillon et al. 2010; Bergin et al. 2010; Hogerheijde et al. 2011; Du et al. 2015; Kama et al. 2016b). A wide variety of observations have now shown that the *dust* mass in disks is radially and vertically stratified due to the combined effects of coagulation, settling, and drift (see discussion in Andrews 2015). Molecular ices will preferentially form or freeze-out onto small dust grains (as they carry the surface area) as gas dynamically mixes to cold layers. Through sticking and grain growth these ice coated small grains will incorporate volatiles into larger grains that become trapped in the dust-rich midplane. This so-called "vertical cold finger effect" (Meijerink et al. 2009; Kama et al. 2016b) can deplete the upper emissive layers of volatiles, particularly water with its relatively high sublimation temperature (Krijt et al. 2016). Due to its low sublimation temperature (~20–30 K) the sequestration of CO in ice is less certain. However, a number of authors have independently characterized how the carbon in CO can be reprocessed into less volatile forms such as CO_2 and organics (Bergin et al. 2014; Furuya and Aikawa 2014; Reboussin et al. 2015; Eistrup et al. 2016). For CO the reprocessing is linked to the presence of X-rays or UV photons along with active surface chemistry and hence will be most active on surface layers and in the outer disk.

Alternately, the results to date show that most, but not all, disks have weak CO isotopologue emission, while HD (due to the limited lifetime of Herschel) has only been detected in three systems. Crucially the HD line is yet to be spectrally

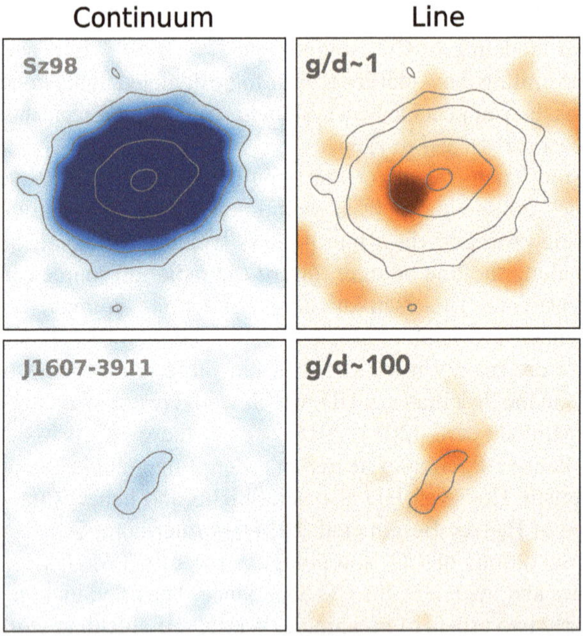

Fig. 1.9 The two most extreme gas-to-dust ratios in the Lupus disk survey. The *left column* shows the continuum emission on the same scale (1–10 mJy beam^{-1}) for two different disks, the large, bright Sz 98 on the top and the low mass J1607-3911 on the bottom. The *right column* shows the integrated ^{13}CO 3–2 emission for each disk on the same scale (40–400 mJy km s^{-1} beam^{-1}) and with the dust contours overlaid. It is clear that the ^{13}CO line-to-continuum ratio is very different in the two sources which suggests either very divergent chemistry or strikingly different gas-to-dust ratios

resolved and thus might contain contributions from unrecognized components such as outflows (jets or winds). Furthermore, the statistics show that, whatever the cause, the CO line-to-continuum ratio varies widely from disk to disk within a single star-forming region (Fig. 1.9). This is a challenge for chemical sequestration models as it is unclear why they should be 99% efficient in some cases but 0% in others with the same age. In a sense this mirrors the same issues for disks with respect to the fact that, for a given age, disks surrounding similar mass stars exhibit a range of infrared excesses and dust masses (Sect. 1.2.1), which must relate to dust evolution.

If chemistry is not the predominant cause and the [CO]/[H_2] abundance in the warm molecular layer is indeed similar to the ISM, then the conclusion would be that the gas is indeed preferentially lost relative to dust. In the HD 163296 example shown in Fig. 1.1 the gas at large radii and scale heights is dust poor due to the effects of settling and drift. These are the same regions in a disk that are less strongly bound to the star, and which can be lost through photo-evaporation, possibly assisted by magnetic fields (Alexander et al. 2014; Gressel et al. 2015).

Photoevaporative flows from disk surfaces depend strongly on the radiation field, whether EUV, FUV, or X-ray (Alexander et al. 2014). These purely hydrodynamic flows are generally strongest at small radii but the effect on the gas-to-dust has not yet been modeled. There is a direct measurement of a photoevaporative flow in one case (TW Hydra again) through mid-infrared spectroscopy of the [NeII], but the mass loss rate, $\sim 10^{-10} M_{\odot}$ year^{-1}, is too low to significantly affect the disk composition on Myr timescales (Pascucci et al. 2011).

There is much new work on angular momentum transport in disks, recently summarized by Turner et al. (2014). Just as advances in instrumentation allow new observational discoveries, so faster computers allow more detailed modeling and better treatment of non-ideal magneto-hydrodynamical effects. The ionization fraction, and therefore the coupling between the gas and dust with the magnetic field, vary widely within a disk. The dense disk mid-plane may be so shielded from radiation and cosmic rays (Cleeves et al. 2013) that the coupling may be too weak for the magneto-rotational instability (MRI; Balbus and Hawley 1991) to drive accretion onto the star. The surface layers are more highly ionized and tied to the magnetic field. The MRI may still operate in these layers and drive laminar accretion flows but disk winds that are launched at large radii may be more effective in re-distributing the angular momentum.

Such global disk winds are predicted to have mass loss rates $\sim 10^{-8} M_{\odot}$ year^{-1} and can therefore remove $10 M_{\rm Jup}$ in $\sim$1 Myr from the disk surfaces at large radii (Bai 2016). This is greater than the median disk mass around a solar mass star in Taurus for an initial gas-to-dust ratio of 100. Although there are as yet no explicit calculations of the effect on the gas-to-dust ratio, it seems that disk winds can potentially rapidly remove disk gas atmospheres but leave millimeter grains in the midplane largely unaffected. As with accretion, the winds could be variable and vary greatly from disk to disk due to the intricacies of the initial magnetic field geometry including its polarity with respect to the rotation axis.

ALMA observations show kinematically coherent CO structures around disks on $\sim$2000 AU scales that may be large scale winds (Klaassen et al. 2013, 2016). The flows have low speeds similar to Keplerian rotation rates at tens of AU, and do not appear to be swept up material at the edges of powerful jets created at the star-disk interface. Mass loss rates are estimated to be a few $10^{-8} M_{\odot}$ year^{-1}. Similarly large scale features have also been seen around FU Orionis objects. These are very young, low mass protostars that have undergone a recent (within a few decades) extreme luminosity outburst that is thought to be due to a large, stochastic accretion event. Zurlo et al. (2016) and Ruíz-Rodríguez et al. (2017) find rings and bow-shock shaped features in CO images that appear to be the lit-up edges of very wide outflow cavities. In one case, HBC 484, the cavity opening angle is so wide, $\sim 150°$, that it would intersect the disk surface. The observations are more suggestive than definitive at this stage but, if most stars undergo similarly eruptive events early in their history, such outbursts could potentially blow away disk atmospheres and drastically alter their composition. The outbursts and their effects are stochastic so we might expect a wide range in gas-to-dust ratios by the time that the stars become optically visible Class II protostars. The lost disk gas would be distributed on large angular scales and ultimately photo-dissociated. Finally, as a sign of reconciliation between the two authors, we note that we may be witnessing both of these effects, chemical sequestration and gas loss, simultaneously.

1.5 Conclusion and Future Prospects

Reconciliation between these disparate mass measurements remains possible. In the near term, within the next 2 years, the HIRMES (High Resolution Mid-infrared Spectrometer) will fly on the Stratospheric Observatory For Infrared Astronomy. HIRMES will be able to detect HD towards TW Hya and spectrally resolve the line. While TW Hya is nearly edge on with an inclination $\sim 5°$–$7°$, with little velocity structure beyond 20 AU, HIRMES will certainly be capable of determining if the HD line originates in whole or partially from a jet, wind, or from the disk. In the more distant future (late 2020s, 2030s) concepts such as SPICA (Space Infrared Telescope for Cosmology and Astrophysics; JAXA) or the Origins Space Telescope (NASA) could survey HD emission towards hundreds of disk systems and detect lines at very low levels of flux density ($\sim 10^{-21}$ to 10^{-20} W/m^2). This would provide a statistically significant sample that can be compared to the extensive archive of ALMA data that will exist by that time. In this regard, JWST has the ability to detect the pure rotational lines of H_2, that is provided there is a significantly warmer H_2 layer above the optically thick dust concentrated towards the midplane. This might prove useful in constraining the H_2 surface density in the very innermost disk radii.

Direct mass measurements, or at least useful constraints, may also be possible through ultra-high resolution ALMA imaging of dust structure. If the ratio of disk to stellar mass exceeds $\sim 20\%$, the gravity of the disk is strong enough to produce detectable features (Cossins et al. 2010; Dipierro et al. 2014). Azimuthally symmetric rings are found in HL Tau and TW Hya (ALMA Partnership et al. 2015; Andrews et al. 2016, respectively), but spirals that may indicate self-gravity have also been found (Pérez et al. 2016). A large imaging survey is underway that will show the range and occurrence of various morphologies, and that can be compared to mass determinations from spectral lines.

More generally, both suggested scenarios, gas dissipation and chemical sequestration, likely have attendant chemical effects that might be revealed via ALMA molecular line emission studies. As an example, it has been suggested that strong hydrocarbon emission seen in several disks might be related to dust evolution and the chemical sequestration of volatiles (Kama et al. 2016a; Bergin et al. 2016). At face value, if true, then these chemical effects (in this case hydrocarbon production) should track with the overall amount of CO present which may have strong radial dependencies. Along these lines, additional chemical signatures might be found in models of the gas chemistry in dissipating gaseous disk systems. Another method would be the inference of the gas density via excitation analyses using a molecule that traces close to the dense midplane in a specific location. One example could be N_2H^+ (Qi et al. 2013) which is known to show a ring structure. The inference of an H_2 gas density in a specific location can be used to pin down and constrain the overall gas density structure—such modeling would at least set some limits on the overall gas mass.

The determination of disk gas mass has been a long-term problem for decades and solving this issue will require efforts from multiple directions. However, the effects we are discussing in this manuscript encompass nearly two orders of magnitude either in gas mass or molecular abundance; or less if it is both. Such a large effect will have attendant signatures and in the coming years we are confident that this issue will be resolved, thereby making a fundamental contribution to our understanding of the physics and chemistry of planet formation.

Acknowledgements EAB was supported by funding from the National Science Foundation grant AST-1514670 and AST-1344133 (INSPIRE) along with NASA XRP grant NNX16AB48G. Reflective of taking the minority stance, JPW is supported only by NASA grant NNX15AC92G and Hawaiian aloha. We thank the editors, Oliver and Martin, for their hard work in making this book happen and gently but constantly pushing us to write this manuscript.

References

Aikawa, Y., van Zadelhoff, G.J., van Dishoeck, E.F., Herbst, E.: Warm molecular layers in protoplanetary disks. Astron. Astrophys. **386**, 622–632 (2002)

Alexander, R., Pascucci, I., Andrews, S., Armitage, P., Cieza, L..: The dispersal of protoplanetary disks. Protostars and Planets VI, pp. 475–496. University of Arizona Press, Tucson (2014) doi:10.2458/azu_uapress_9780816531240-ch021, 1311.1819

ALMA Partnership, Brogan, C.L., Pérez, L.M., Hunter, T.R., Dent, W.R.F., Hales, A.S., Hills, R.E., Corder, S., Fomalont, E.B., Vlahakis, C., Asaki, Y., Barkats, D., Hirota, A., Hodge, J.A., Impellizzeri, C.M.V., Kneissl, R., Liuzzo, E., Lucas, R., Marcelino, N., Matsushita, S., Nakanishi, K., Phillips, N., Richards, A.M.S., Toledo, I., Aladro, R., Broguiere, D., Cortes, J.R., Cortes, P.C., Espada, D., Galarza, F., Garcia-Appadoo, D., Guzman-Ramirez, L., Humphreys, E.M., Jung, T., Kameno, S., Laing, R.A., Leon, S., Marconi, G., Mignano, A., Nikolic, B., Nyman, L.A., Radiszcz, M., Remijan, A., Rodón, J.A., Sawada, T., Takahashi, S., Tilanus, R.P.J., Vila Vilaro, B., Watson, L.C., Wiklind, T., Akiyama, E., Chapillon, E., de Gregorio-Monsalvo, I., Di Francesco, J., Gueth, F., Kawamura, A., Lee, C.F., Nguyen Luong, Q., Mangum, J., Pietu, V., Sanhueza, P., Saigo, K., Takakuwa, S., Ubach, C., van Kempen, T., Wootten, A., Castro-Carrizo, A., Francke, H., Gallardo, J., Garcia, J., Gonzalez, S., Hill, T., Kaminski, T., Kurono, Y., Liu, H.Y., Lopez, C., Morales, F., Plarre, K., Schieven, G., Testi, L., Videla, L., Villard, E., Andreani, P., Hibbard, J.E., Tatematsu, K.: The 2014 ALMA long baseline campaign: first results from high angular resolution observations toward the HL Tau region. Astrophys. J. Lett. 808, L3 (2015). doi:10.1088/2041-8205/808/1/L3, 1503.02649

Andrews, S.M.: Observations of solids in protoplanetary disks. Publ. Astron. Soc. Pac. **127**, 961–993 (2015). doi:10.1086/683178, 1507.04758

Andrews, S.M., Williams, J.P.: High-resolution submillimeter constraints on circumstellar disk structure. Astrophys. J. **659**, 705–728 (2007). doi:10.1086/511741

Andrews, S.M., Wilner, D.J., Espaillat, C., Hughes, A.M., Dullemond, C.P., McClure, M.K., Qi, C., Brown, J.M.: Resolved images of large cavities in protoplanetary transition disks. Astrophys. J. **732**, 42 (2011). doi:10.1088/0004-637X/732/1/42, 1103.0284

Andrews, S.M., Rosenfeld, K.A., Kraus, A.L., Wilner, D.J.: The mass dependence between protoplanetary disks and their stellar hosts. Astrophys. J. **771**, 129 (2013). doi:10.1088/0004-637X/771/2/129, 1305.5262

Andrews, S.M., Wilner, D.J., Zhu, Z., Birnstiel, T., Carpenter, J., Pérez, L.M., Bai, X.N., Öberg, K.I., Hughes, A.M., Isella, A., Ricci, L.: Ringed substructure and a gap at 1 AU in the nearest protoplanetary disk. Astrophys. J. **820**, L40 (2016). doi:10.3847/2041-8205/820/2/L40

Ansdell, M., Williams, J.P., van der Marel, N., Carpenter, J.M., Guidi, G., Hogerheijde, M., Mathews, G.S., Manara, C.F., Miotello, A., Natta, A., Oliveira, I., Tazzari, M., Testi, L., van Dishoeck, E.F., van Terwisga, S.E.: ALMA survey of lupus protoplanetary disks. I. Dust and gas masses. Astrophys. J. **828**, 46 (2016). doi:10.3847/0004-637X/828/1/46, 1604.05719

Aresu, G., Kamp, I., Meijerink, R., Spaans, M., Vicente, S., Podio, L., Woitke, P., Menard, F., Thi, W.F., Güdel, M., Liebhart, A.: [O I] disk emission in the Taurus star-forming region. Astron. Astrophys. **566**, A14 (2014). doi:10.1051/0004-6361/201322455, 1402.2488

Armitage, P.J.: Astrophysics of Planet Formation. Cambridge University Press, Cambridge (2010)

Asplund, M., Grevesse, N., Sauval, A.J., Scott, P.: The chemical composition of the sun. Annu. Rev. Astron. Astrophys. **47**, 481–522 (2009). doi:10.1146/annurev.astro.46.060407.145222

Bai, X.N.: Towards a global evolutionary model of protoplanetary disks. Astrophys. J. **821**, 80 (2016). doi:10.3847/0004-637X/821/2/80, 1603.00484

Balbus, S.A., Hawley, J.F.: A powerful local shear instability in weakly magnetized disks. I - linear analysis. II - nonlinear evolution. Astrophys. J. **376**, 214–233 (1991). doi:10.1086/170270

Barenfeld, S.A., Carpenter, J.M., Ricci, L., Isella, A.: ALMA observations of circumstellar disks in the upper scorpius OB association. Astrophys. J. **827**, 142 (2016). doi:10.3847/0004-637X/827/2/142, 1605.05772

Baruteau, C., Bai, X., Mordasini, C., Mollière, P.: Formation, orbital and internal evolutions of Young planetary systems. Space Sci. Rev. **205**, 77–124 (2016). doi:10.1007/s11214-016-0258-z, 1604.07558

Beckwith, S.V.W., Sargent, A.I., Chini, R.S., Guesten, R.: A survey for circumstellar disks around young stellar objects. Astron. J. **99**, 924–945 (1990). doi:10.1086/115385

Bergin, E.A., Tafalla, M.: Cold dark clouds: the initial conditions for star formation. Annu. Rev. Astron. Astrophys. **45**, 339–396 (2007). doi:10.1146/annurev.astro.45.071206.100404

Bergin, E.A., Alves, J., Huard, T., Lada, C.J.: N_2H^+ and $C^{18}O$ depletion in a cold dark cloud. Astrophys. J. Lett. **570**, L101–L104 (2002)

Bergin, E.A., Aikawa, Y., Blake, G.A., van Dishoeck, E.F.: The chemical evolution of protoplanetary disks. In: Protostars and Planets V, p 751. University of Arizona Press, Tucson (2007)

Bergin, E.A., Hogerheijde, M.R., Brinch, C., et al.: Sensitive limits on the abundance of cold water vapor in the DM Tauri protoplanetary disk. Astron. Astrophys. **521**, L33 (2010). doi:10.1051/0004-6361/201015104

Bergin, E.A., Cleeves, L.I., Gorti, U., Zhang, K., Blake, G.A., Green, J.D., Andrews, S.M., Evans, N.J. II, Henning, T., Öberg, K., Pontoppidan, K., Qi, C., Salyk, C., van Dishoeck, E.F.: An old disk still capable of forming a planetary system. Nature **493**, 644–646 (2013). doi:10.1038/nature11805

Bergin, E.A., Cleeves, L.I., Crockett, N., Blake, G.A.: Exploring the origins of carbon in terrestrial worlds. Faraday Discuss. **168**, 61–79 (2014). doi:10.1039/C4FD00003J, http://dx.doi.org/10.1039/C4FD00003J

Bergin, E.A., Blake, G.A., Ciesla, F., Hirschmann, M.M., Li, J.: Tracing the ingredients for a habitable earth from interstellar space through planet formation. Proc. Natl. Acad. Sci. **112**, 8965–8970 (2015). doi:10.1073/pnas.1500954112, 1507.04756

Bergin, E.A., Du, F., Cleeves, L.I., Blake, G.A., Schwarz, K., Visser, R., Zhang, K.: Hydrocarbon emission rings in protoplanetary disks induced by dust evolution. Astrophys. J. **831**, 101 (2016). doi:10.3847/0004-637X/831/1/101, 1609.06337

Bertoldi, F., Timmermann, R., Rosenthal, D., Drapatz, S., Wright, C.M.: Detection of HD in the Orion molecular outflow. Astron. Astrophys. **346**, 267–277 (1999)

Bianchi, S., Goncalves, J., Albrecht, M., Caselli, P., Chini, R., Galli, D., Walmsley, M.: Dust emissivity in the submm/mm. SCUBA and SIMBA observations of Barnard 68. Astron. Astrophys. **399**, L43–L46 (2003)

Bohlin, R.C., Savage, B.D., Drake, J.F.: A survey of interstellar H I from L-alpha absorption measurements. II. Astrophys. J. **224**, 132–142 (1978). doi:10.1086/156357

Boogert, A.C.A., Gerakines, P.A., Whittet, D.C.B.: Observations of the icy universe. Annu. Rev. Astron. Astrophys. **53**, 541–581 (2015). doi:10.1146/annurev-astro-082214-122348, 1501.05317

Bruderer, S., van Dishoeck, E.F., Doty, S.D., Herczeg, G.J.: The warm gas atmosphere of the HD 100546 disk seen by Herschel. Evidence of a gas-rich, carbon-poor atmosphere? Astron. Astrophys. **541**, A91 (2012). doi:10.1051/0004-6361/201118218, 1201.4860

Carmona, A., van den Ancker, M.E., Henning, T., Pavlyuchenkov, Y., Dullemond, C.P., Goto, M., Thi, W.F., Bouwman, J., Waters, L.B.F.M.: A search for mid-infrared molecular hydrogen emission from protoplanetary disks. Astron. Astrophys. **477**, 839–852 (2008). doi:10.1051/0004-6361:20077846

Carpenter, J.M., Wolf, S., Schreyer, K., Launhardt, R., Henning, T.: Evolution of cold circumstellar dust around solar-type stars. Astron. J. **129**, 1049–1062 (2005). doi:10.1086/427131, astro-ph/0411020

Casassus, S., van der Plas, G., M, S.P., Dent, W.R.F., Fomalont, E., Hagelberg, J., Hales, A., Jordán, A., Mawet, D., Ménard, F., Wootten, A., Wilner, D., Hughes, A.M., Schreiber, M.R., Girard, J.H., Ercolano, B., Canovas, H., Román, P.E., Salinas, V.: Flows of gas through a protoplanetary gap. Nature **493**, 191–194 (2013). doi:10.1038/nature11769

Caselli, P., Walmsley, C.M., Tafalla, M., Dore, L., Myers, P.C.: CO depletion in the starless cloud core L1544. Astrophys. J. Lett. **523**, L165–L169 (1999). doi:10.1086/312280

Ceccarelli, C., Caselli, P., Bockelée-Morvan, D., Mousis, O., Pizzarello, S., Robert, F., Semenov, D.: Deuterium fractionation: the Ariadne's thread from the precollapse phase to meteorites and comets today. Protostars and Planets VI, pp 859–882. University of Arizona Press, Tucson (2014), 1403.7143

Chapillon, E., Parise, B., Guilloteau, S., Dutrey, A., Wakelam, V.: C I observations in the CQ Tauri proto-planetary disk: evidence of a very low gas-to-dust ratio ? Astron. Astrophys. **520**, A61 (2010). doi:10.1051/0004-6361/201014841, 1006.5244

Cleeves, L.I., Adams, F.C., Bergin, E.A.: Exclusion of cosmic rays in protoplanetary disks: stellar and magnetic effects. Astrophys. J. **772**, 5 (2013). doi:10.1088/0004-637X/772/1/5, 1306.0902

Connelly, J.N., Amelin, Y., Krot, A.N., Bizzarro, M.: Chronology of the solar system's oldest solids. Astrophys. J. **675**, L121 (2008). doi:10.1086/533586

Cossins, P., Lodato, G., Testi, L.: Resolved images of self-gravitating circumstellar discs with ALMA. Mon. Not. R. Astron. Soc. **407**, 181–188 (2010). doi:10.1111/j.1365-2966.2010.16934.x, 1004.5389

D'Alessio, P., Canto, J., Calvet, N., Lizano, S.: Accretion disks around young objects. I. The detailed vertical structure. Astrophys. J. **500**, 411 (1998). doi:10.1086/305702

Dennison, D.M.: A note on the specific heat of the hydrogen molecule. R. Soc. London Proc. Ser. A **115**, 483–486 (1927)

Dickman, R.L.: The ratio of carbon monoxide to molecular hydrogen in interstellar dark clouds. Astrophys. J. Suppl. Ser. **37**, 407–427 (1978). doi:10.1086/190535

Dipierro, G., Lodato, G., Testi, L., de Gregorio Monsalvo, I.: How to detect the signatures of self-gravitating circumstellar discs with the Atacama Large Millimeter/sub-millimeter Array. Mon. Not. R. Astron. Soc. **444**, 1919–1929 (2014). doi:10.1093/mnras/stu1584, 1409.2243

Draine, B.T.: On the submillimeter opacity of protoplanetary disks. Astrophys. J. **636**, 1114–1120 (2006). doi:10.1086/498130, astro-ph/0507292

Du, F., Bergin, E.A.: Water vapor distribution in protoplanetary disks. Astrophys. J. **792**, 2 (2014). doi:10.1088/0004-637X/792/1/2, 1408.2026

Du, F., Bergin, E.A., Hogerheijde, M.R.: Volatile depletion in the TW Hydrae disk atmosphere. Astrophys. J. Lett. **807**, L32 (2015). doi:10.1088/2041-8205/807/2/L32, 1506.03510

Du, F., Bergin, E.A., Hogerheijde, M.R., van Dishoeck, E.F., Blake, G.A., Bruderer, S., Cleeves, L.I., Dominik, C., Fedele, D., Lis, D.C., Melnick, G.J., Neufeld, D.A., Pearson, J.C., Yildiz, U.: Survey of water lines in protoplanetary disk: indications of systematic volatile depletion. Astrophys. J. **842**, 98 (2017). doi:10.3847/1538-4357/aa70ee

Dutrey, A., Guilloteau, S., Guelin, M.: Chemistry of protosolar-like nebulae: the molecular content of the DM Tau and GG Tau disks. Astron. Astrophys. **317**, L55–L58 (1997)

Dutrey, A., Guilloteau, S., Simon, M.: The BP Tau disk: a missing link between Class II and III objects? Astron. Astrophys. **402**, 1003–1011 (2003). doi:10.1051/0004-6361:20030317

Eistrup, C., Walsh, C., van Dishoeck, E.F.: Setting the volatile composition of (exo)planet-building material. Does chemical evolution in disk midplanes matter? Astron. Astrophys. **595**, 1–15 (2016)

Epstein, R.I., Lattimer, J.M., Schramm, D.N.: The origin of deuterium. Nature **263**, 198–202 (1976). doi:10.1038/263198a0

Evans, N.J., et al.: The Spitzer c2d legacy results: star-formation rates and efficiencies; evolution and lifetimes. Astrophys. J. Suppl. Ser. **181**, 321–350 (2009). doi:10.1088/0067-0049/181/2/321

Favre, C., Cleeves, L.I., Bergin, E.A., Qi, C., Blake, G.A.: A significantly low CO abundance toward the TW Hya protoplanetary disk: a path to active carbon chemistry? Astrophys. J. Lett. **776**, L38 (2013). doi:10.1088/2041-8205/776/2/L38, 1309.5370

Field, G.B., Somerville, W.B., Dressler, K.: Hydrogen molecules in astronomy. Annu. Rev. Astron. Astrophys. **4**, 207 (1966). doi:10.1146/annurev.aa.04.090166.001231

Fitzpatrick, E.L.: Interstellar extinction in the Milky Way galaxy. In: ASP Conference Series, 309: Astrophysics of Dust, pp. 33. ASP, San Francisco (2004)

France, K., Burgh, E.B., Herczeg, G.J., Schindhelm, E., Yang, H., Abgrall, H., Roueff, E., Brown, A., Brown, J.M., Linsky, J.L.: CO and H_2 absorption in the AA Tauri circumstellar disk. Astrophys. J. **744**, 22 (2012). doi:10.1088/0004-637X/744/1/22, 1109.1831

France, K., Herczeg, G.J., McJunkin, M., Penton, S.V.: CO/H_2 abundance ratio $\sim 10^{-4}$ in a protoplanetary disk. Astrophys. J. **794**, 160 (2014). doi:10.1088/0004-637X/794/2/160, 1409.0861

Frerking, M.A., Langer, W.D., Wilson, R.W.: The relationship between carbon monoxide abundance and visual extinction in interstellar clouds. Astrophys. J. **262**, 590–605 (1982). doi:10.1086/160451

Furuya, K., Aikawa, Y.: (2014) Reprocessing of ices in turbulent protoplanetary disks: carbon and nitrogen chemistry. Astrophys. J. **790**, 97. doi:10.1088/0004-637X/790/2/97, 1406.3507

Goldsmith, P.F., Bergin, E.A., Lis, D.C.: Carbon monoxide and dust column densities: the dust-to-gas ratio and structure of three giant molecular cloud cores. Astrophys. J. **491**, 615–637 (1997)

Gorti, U., Hollenbach, D., Najita, J., Pascucci, I.: Emission lines from the gas disk around TW Hydra and the origin of the inner hole. Astrophys. J. **735**, 90 (2011) doi:10.1088/0004-637X/735/2/90

Goto, M., Geballe, T.R., Usuda, T.: Infrared absorption lines toward NGC 7538 IRS 1: abundances of H2, H3+, and CO. Astrophys. J. **806**, 57 (2015). doi:10.1088/0004-637X/806/1/57

Gressel, O., Turner, N.J., Nelson, R.P., McNally, C.P.: Global simulations of protoplanetary disks with ohmic resistivity and ambipolar diffusion. Astrophys. J. **801**, 84 (2015). doi:10.1088/0004-637X/801/2/84, 1501.05431

Guillot, T.: A comparison of the interiors of Jupiter and Saturn. Planet. Space Sci. **47**, 1183–1200 (1999). arXiv:astro-ph/9907402

Harjunpää, P., Lehtinen, K., Haikala, L.K.: The relationship of CO abundance to extinction and N(H_2): observations of globules and the dependence on star formation activity. Astron. Astrophys. **421**, 1087–1099 (2004). doi:10.1051/0004-6361:20035752

Herbst, E., Klemperer, W.: The formation and depletion of molecules in dense interstellar clouds. Astrophys. J. **185**, 505–534 (1973)

Hogerheijde, M.R., Bergin, E.A., Brinch, C., et al.: Detection of the water reservoir in a forming planetary system. Science **334**, 338–340 (2011). doi:10.1126/science.1208931

Hotzel, S., Harju, J., Juvela, M., Mattila, K., Haikala, L.K.: C18O abundance in the nearby globule Barnard 68. Astron. Astrophys. **391**, 275–285 (2002). doi:10.1051/0004-6361:20020786

Howard, A.W., Marcy, G.W., Bryson, S.T., Jenkins, J.M., Rowe, J.F., Batalha, N.M., Borucki, W.J., Koch, D.G., Dunham, E.W., Gautier, T.N. III, Van Cleve, J., Cochran, W.D., Latham, D.W., Lissauer, J.J., Torres, G., Brown, T.M., Gilliland, R.L., Buchhave, L.A., Caldwell, D.A., Christensen-Dalsgaard, J., Ciardi, D., Fressin, F., Haas, M.R., Howell, S.B., Kjeldsen, H., Seager, S., Rogers, L., Sasselov, D.D., Steffen, J.H., Basri, G.S., Charbonneau, D., Christiansen,

J., Clarke, B., Dupree, A., Fabrycky, D.C., Fischer, D.A., Ford, E.B., Fortney, J.J., Tarter, J., Girouard, F.R., Holman, M.J., Johnson, J.A., Klaus, T.C., Machalek, P., Moorhead, A.V., Morehead, R.C., Ragozzine, D., Tenenbaum, P., Twicken, J.D., Quinn, S.N., Isaacson, H., Shporer, A., Lucas, P.W., Walkowicz, L.M., Welsh, W.F., Boss, A., Devore, E., Gould, A., Smith, J.C., Morris, R.L., Prsa, A., Morton, T.D., Still, M., Thompson, S.E., Mullally, F., Endl, M., MacQueen, P.J.: Planet occurrence within 0.25 AU of solar-type stars from Kepler. Astrophys. J. Suppl. Ser. **201**, 15 (2012). doi:10.1088/0067-0049/201/2/15, 1103.2541

Hughes, A.M., Wilner, D.J., Qi, C., Hogerheijde, M.R.: Gas and dust emission at the outer edge of protoplanetary disks. Astrophys. J. **678**, 1119–1126 (2008). doi:10.1086/586730, 0801.4763

Kama, M., Bruderer, S., Carney, M., Hogerheijde, M., van Dishoeck, E.F., Fedele, D., Baryshev, A., Boland, W., Güsten, R., Aikutalp, A., Choi, Y., Endo, A., Frieswijk, W., Karska, A., Klaassen, P., Koumpia, E., Kristensen, L., Leurini, S., Nagy, Z., Perez Beaupuits, J.P., Risacher, C., van der Marel, N., van Kempen, T.A., van Weeren, R.J., Wyrowski, F., Yıldız, U.A.: Observations and modelling of CO and [CI] in disks. First detections of [CI] and constraints on the carbon abundance. Astron. Astrophys. **588**, A108 (2016a)

Kama, M., Bruderer, S., van Dishoeck, E.F., Hogerheijde, M., Folsom, C.P., Miotello, A., Fedele, D., Belloche, A., Güsten, R., Wyrowski, F.: Volatile carbon locking and release in protoplanetary disks. A study of TW Hya and HD 100546. Astron. Astrophys. **592**, A83 (2016b)

Kamp, I., Thi, W.F., Meeus, G., Woitke, P., Pinte, C., Meijerink, R., Spaans M., Pascucci, I., Aresu, G., Dent, W.R.F.: Uncertainties in water chemistry in disks: an application to TW Hydrae. Astron. Astrophys. 559:A24 (2013). doi:10.1051/0004-6361/201220621, 1308.1772

Kastner, J.H., Zuckerman, B., Weintraub, D.A., Forveille, T.: X-ray and molecular emission from the nearest region of recent star formation. Science **277**, 67–71 (1997). doi:10.1126/science.277.5322.67

Kauffmann, J., Bertoldi, F., Bourke, T.L., Evans, N.J. II, Lee, C.W.: MAMBO mapping of Spitzer c2d small clouds and cores. Astron. Astrophys. **487**, 993–1017 (2008). doi:10.1051/0004-6361:200809481

Klaassen, P.D., Juhasz, A., Mathews, G.S., Mottram, J.C., De Gregorio-Monsalvo, I., van Dishoeck, E.F., Takahashi, S., Akiyama, E., Chapillon, E., Espada, D., Hales, A., Hogerheijde, M.R., Rawlings, M., Schmalzl, M., Testi, L.: ALMA detection of the rotating molecular disk wind from the young star HD 163296. Astron. Astrophys. **555**, A73 (2013). doi:10.1051/0004-6361/201321129, 1304.5436

Klaassen, P.D., Mottram, J.C., Maud, L.T., Juhasz, A.: The winds from HL Tau. Mon. Not. R. Astron. Soc. **460**, 627–633 (2016). doi:10.1093/mnras/stw989, 1604.07285

Kleine, T., Münker, C., Mezger, K., Palme, H.: Rapid accretion and early core formation on asteroids and the terrestrial planets from Hf-W chronometry. Nature **418**, 952–955 (2002). doi:10.1038/nature00982

Kramer, C., Alves, J., Lada, C., Lada, E., Sievers, A., Ungerechts, H., Walmsley, M.: The millimeter wavelength emissivity in IC5146. Astron. Astrophys. **329**, L33–L36 (1998)

Kramer, C., Alves, J., Lada, C.J., Lada, E.A., Sievers, A., Ungerechts, H., Walmsley, C.M.: Depletion of CO in a cold dense cloud core of IC 5146. Astron. Astrophys. **342**, 257–270 (1999)

Krijt, S., Ciesla, F.J., Bergin, E.A.: Tracing water vapor and ice during dust growth. Astrophys. J. **833**, 285 (2016). doi:10.3847/1538-4357/833/2/285

Lacy, J.H., Knacke, R., Geballe, T.R., Tokunaga, A.T.: Detection of absorption by H2 in molecular clouds: a direct measurement of the H2:CO ratio. Astrophys. J. **428**, L69–L72 (1994). doi:10.1086/187395

Lada, C.J., Lada, E.A., Clemens, D.P., Bally, J.: Dust extinction and molecular gas in the dark cloud IC 5146. Astrophys. J. **429**, 694–709 (1994). doi:10.1086/174354

Lay, O.P., Carlstrom, J.E., Hills, R.E.: Constraints on the HL Tauri protostellar disk from millimeter- and submillimeter-wave interferometry. Astrophys. J. **489**, 917–927 (1997)

Lellouch, E., Bézard, B., Fouchet, T., Feuchtgruber, H., Encrenaz, T., de Graauw, T.: The deuterium abundance in Jupiter and Saturn from ISO-SWS observations. Astron. Astrophys. **370**, 610–622 (2001). doi:10.1051/0004-6361:20010259

Loinard, L.: The Gould's belt distances survey. In: de Grijs, R. (ed.) Advancing the Physics of Cosmic Distances, IAU Symposium, vol. 289, pp. 36–43 (2013). doi:10.1017/S1743921312021072, 1211.1742

Manara, C.F., Rosotti, G., Testi, L., Natta, A., Alcalá, J.M., Williams, J.P., Ansdell, M., Miotello, A., van der Marel, N., Tazzari, M., Carpenter, J., Guidi, G., Mathews, G.S., Oliveira, I., Prusti, T., van Dishoeck, E.F.: Evidence for a correlation between mass accretion rates onto young stars and the mass of their protoplanetary disks. Astron. Astrophys. **591**, L3 (2016). doi:10.1051/0004-6361/201628549, 1605.03050

Mathews, G.S., Williams, J.P., Ménard, F., Phillips, N., Duchêne, G., Pinte, C.: The late stages of protoplanetary disk evolution: a millimeter survey of upper scorpius. Astrophys. J. **745**, 23 (2012). doi:10.1088/0004-637X/745/1/23, 1111.0101

McClure, M.K., Bergin, E.A., Cleeves, L.I., van Dishoeck, E., Blake, G., Evans, N.J., Green, J.D., Henning, T.K., Öberg, K.I., Pontoppidan, K.M., Salyk, C.: Mass measurements in protoplanetary disks from hydroden deuteride. Astrophys. J. **831**, 167 (2016). doi:10.3847/0004-637X/831/2/167

McKeegan, K.D., Kallio, A.P.A., Heber, V.S., Jarzebinski, G., Mao, P.H., Coath, C.D., Kunihiro, T., Wiens, R.C., Nordholt, J.E., Moses, R.W., Reisenfeld, D.B., Jurewicz, A.J.G., Burnett, D.S.: The oxygen isotopic composition of the sun inferred from captured solar wind. Science **332**, 1528 (2011). doi:10.1126/science.1204636

Meijerink, R., Pontoppidan, K.M., Blake, G.A., Poelman, D.R., Dullemond, C.P.: Radiative transfer models of mid-infrared H_2O lines in the planet-forming region of circumstellar disks. Astrophys. J. **704**, 1471–1481 (2009). doi:10.1088/0004-637X/704/2/1471, 0909.0975

Miotello, A., Bruderer, S., van Dishoeck, E.F.: Protoplanetary disk masses from CO isotopologue line emission. Astron. Astrophys. **572**, A96 (2014). doi:10.1051/0004-6361/201424712, 1410.2093

Miotello, A., van Dishoeck, E.F., Kama, M., Bruderer, S.: Determining protoplanetary disk gas masses from CO isotopologues line observations. Astron. Astrophys. **594**, A85 (2016). doi:10.1051/0004-6361/201628159, 1605.07780

Mishra, A., Li, A.: Probing the role of carbon in the interstellar ultraviolet extinction. Astrophys. J. **809**, 120 (2015). doi:10.1088/0004-637X/809/2/120, 1507.06599

Müller, H.S.P., Schlöder, F., Stutzki, J., Winnewisser, G.: The Cologne database for molecular spectroscopy, CDMS: a useful tool for astronomers and spectroscopists. J. Mol. Struct. **742**, 215–227 (2005). doi:10.1016/j.molstruc.2005.01.027

Nomura, H., Tsukagoshi, T., Kawabe, R., Ishimoto, D., Okuzumi, S., Muto, T., Kanagawa, K.D., Ida, S., Walsh, C., Millar, T.J., Bai, X.N.: ALMA observations of a gap and a ring in the protoplanetary disk around TW Hya. Astrophys. J. Lett. **819**, L7 (2016). doi:10.3847/2041-8205/819/1/L7, 1512.05440

Öberg, K.I., Murray-Clay, R., Bergin, E.A.: The effects of snowlines on C/O in planetary atmospheres. Astrophys. J. Lett. **743**, L16 (2011). doi:10.1088/2041-8205/743/1/L16, 1110.5567

Oliveira, C.M., Hébrard, G.: Variations in the D/H ratio of extended sight lines from far ultraviolet spectroscopic explorer observations. Astrophys. J. **653**, 345–360 (2006). doi:10.1086/508611, astro-ph/0609236

Ossenkopf, V., Henning, T.: Dust opacities for protostellar cores. Astron. Astrophys. **291**, 943–959 (1994)

Parvathi, V.S., Sofia, U.J., Murthy, J., Babu, B.R.S.: Probing the role of carbon in ultraviolet extinction along galactic sight lines. Astrophys. J. **760**, 36 (2012). doi:10.1088/0004-637X/760/1/36

Pascucci, I., et al.: Formation and evolution of planetary systems: upper limits to the gas mass in disks around sun-like stars. Astrophys. J. **651**, 1177 (2006)

Pascucci, I., Sterzik, M., Alexander, R.D., Alencar, S.H.P., Gorti, U., Hollenbach, D., Owen, J., Ercolano, B., Edwards, S.: The photoevaporative wind from the disk of TW Hya. Astrophys. J. **736**, 13 (2011). doi:10.1088/0004-637X/736/1/13, 1105.0045

Pascucci, I., Testi, L., Herczeg, G.J., Long, F., Manara, C.F., Hendler, N., Mulders, G.D., Krijt, S., Ciesla, F., Henning, T., Mohanty, S., Drabek-Maunder, E., Apai, D., Szűcs, L., Sacco, G., Olofsson, J.: A steeper than linear disk mass – stellar mass scaling relation. Astrophys. J. **831**, 125 (2016). doi:10.3847/0004-637X/831/2/125, 1608.03621

Pérez, L.M., Chandler, C.J., Isella, A., Carpenter, J.M., Andrews, S.M., Calvet, N., Corder, S.A., Deller, A.T., Dullemond, C.P., Greaves, J.S., Harris, R.J., Henning, T., Kwon, W., Lazio, J., Linz, H., Mundy, L.G., Ricci, L., Sargent, A.I., Storm, S., Tazzari, M., Testi, L., Wilner, D.J.: Grain growth in the circumstellar disks of the young stars CY Tau and DoAr 25. Astrophys. J. **813**, 41 (2015). doi:10.1088/0004-637X/813/1/41, 1509.07520

Pérez, L.M., Carpenter, J.M., Andrews, S.M., Ricci, L., Isella, A., Linz, H., Sargent, A.I., Wilner, D.J., Henning, T., Deller, A.T., Chandler, C.J., Dullemond, C.P., Lazio, J., Menten, K.M., Corder, S.A., Storm, S., Testi, L., Tazzari, M., Kwon, W., Calvet, N., Greaves, J.S., Harris, R.J., Mundy, L.G.: Spiral density waves in a young protoplanetary disk. Science **353**, 1519–1521 (2016). doi:10.1126/science.aaf8296, 1610.05139

Pineda, J.L., Goldsmith, P.F., Chapman, N., Snell, R.L., Li, D., Cambrésy, L., Brunt, C.: The relation between gas and dust in the Taurus molecular cloud. Astrophys. J. **721**, 686–708 (2010). doi:10.1088/0004-637X/721/1/686, 1007.5060

Pollack, J.B., Hollenbach, D., Beckwith, S., Simonelli, D.P., Roush, T., Fong, W.: Composition and radiative properties of grains in molecular clouds and accretion disks. Astrophys. J. **421**, 615–639 (1994). doi:10.1086/173677

Qi, C., Öberg, K.I., Wilner, D.J., D'Alessio, P., Bergin, E., Andrews, S.M., Blake, G.A., Hogerheijde, M.R., van Dishoeck, E.F.: Imaging of the CO snow line in a solar nebula analog. Science **341**, 630–632 (2013). doi:10.1126/science.1239560, 1307.7439

Rachford, B.L., Snow, T.P., Destree, J.D., Ross, T.L., Ferlet, R., Friedman, S.D., Gry, C., Jenkins, E.B., Morton, D.C., Savage, B.D., Shull, J.M., Sonnentrucker, P., Tumlinson, J., Vidal-Madjar, A., Welty, D.E., York, D.G.: Molecular hydrogen in the far ultraviolet spectroscopic explorer translucent lines of sight: the full sample. Astrophys. J. Suppl. Ser. **180**, 125–137 (2009). doi:10.1088/0067-0049/180/1/125

Reboussin, L., Wakelam, V., Guilloteau, S., Hersant, F., Dutrey, A.: Chemistry in protoplanetary disks: the gas-phase CO/H_2 ratio and the carbon reservoir. Astron. Astrophys. **579**, A82 (2015). doi:10.1051/0004-6361/201525885, 1505.01309

Ripple, F., Heyer, M.H., Gutermuth, R., Snell, R.L., Brunt, C.M.: CO abundance variations in the Orion molecular cloud. Mon. Not. R. Astron. Soc. **431**, 1296–1313 (2013). doi:10.1093/mnras/stt247

Rosenfeld, K.A., Andrews, S.M., Hughes, A.M., Wilner, D.J., Qi, C.: A spatially resolved vertical temperature gradient in the HD 163296 disk. Astrophys. J. **774**, 16 (2013). doi:10.1088/0004-637X/774/1/16, 1306.6475

Ruíz-Rodríguez, D., Cieza, L.A., Williams, J.P., Principe, D., Tobin, J.J., Zhu, Z., Zurlo, A.: The ALMA early science view of FUor/EXor objects - III. The slow and wide outflow of V883 Ori. Mon. Not. R. Astron. Soc. **468**, 3266–3276 (2017). doi:10.1093/mnras/stx703

Sargent, A.I., Beckwith, S.: Kinematics of the circumstellar gas of HL Tauri and R Monocerotis. Astrophys. J. **323**, 294–305 (1987). doi:10.1086/165827

Sarkar, S.: Big bang nucleosynthesis and physics beyond the standard model. Rep. Prog. Phys. **59**, 1493–1609 (1996). doi:10.1088/0034-4885/59/12/001, hep-ph/9602260

Schwarz, K.R., Bergin, E.A., Cleeves, L.I., Blake, G.A., Zhang, K., Öberg, K.I., van Dishoeck, E.F., Qi, C.: The radial distribution of H_2 and CO in TW Hya as revealed by resolved ALMA observations of CO isotopologues. Astrophys. J. **823**, 91 (2016). doi:10.3847/0004-637X/823/2/91, 1603.08520

Semenov, D.A., Chemistry in protoplanetary disks. In: Zakharova, P.E., Kuznetsov, E.D., Ostrovskii, A.B., Salii, S.V., Sobolev, A.M., Kholshevnikov, K.V., Shustov, B.M. (eds.) Physics of Space: The 41st Annual Student Scientific Conference, pp. 130–155 (2012)

Snow, T.P., Ross, T.L., Destree, J.D., Drosback, M.M., Jensen, A.G., Rachford, B.L., Sonnentrucker, P., Ferlet, R.: A new FUSE survey of interstellar HD. Astrophys. J. **688**, 1124–1136 (2008). doi:10.1086/592288, 0808.0926

Strom, K.M., Strom, S.E., Edwards, S., Cabrit, S., Skrutskie, M.F.: Circumstellar material associated with solar-type pre-main-sequence stars - a possible constraint on the timescale for planet building. Astron. J. **97**, 1451–1470 (1989). doi:10.1086/115085

Suutarinen, A., Haikala, L.K., Harju, J., Juvela, M., André, P., Kirk, J.M., Könyves, V., White, G.J.: Determination of the far-infrared dust opacity in a prestellar core. Astron. Astrophys. **555**, A140 (2013). doi:10.1051/0004-6361/201219103, 1306.3156

Tazzari, M., Testi, L., Ercolano, B., Natta, A., Isella, A., Chandler, C.J., Pérez, L.M., Andrews, S., Wilner, D.J., Ricci, L., Henning, T., Linz, H., Kwon, W., Corder, S.A., Dullemond, C.P., Carpenter, J.M., Sargent, A.I., Mundy, L., Storm, S., Calvet, N., Greaves, J.A., Lazio, J., Deller, A.T.: Multiwavelength analysis for interferometric (sub-)mm observations of protoplanetary disks. Radial constraints on the dust properties and the disk structure. Astron. Astrophys. **588**, A53 (2016). doi:10.1051/0004-6361/201527423, 1512.05679

Testi, L., Birnstiel, T., Ricci, L., Andrews, S., Blum, J., Carpenter, J., Dominik, C., Isella, A., Natta, A., Williams, J.P., Wilner, D.J.: Dust evolution in protoplanetary disks. Protostars and Planets VI, pp. 339–361. University of Arizona Press, Tucson (2014). doi:10.2458/azu_uapress_9780816531240-ch015

Thi, W.F., Mathews, G., Ménard, F., Woitke, P., Meeus, G., Riviere-Marichalar, P., Pinte, C., Howard, C.D., Roberge, A., Sandell, G., Pascucci, I., Riaz, B., Grady, C.A., Dent, W.R.F., Kamp, I., Duchêne, G., Augereau, J.C., Pantin, E., Vandenbussche, B., Tilling, I., Williams, J.P., Eiroa, C., Barrado, D., Alacid, J.M., Andrews, S., Ardila, D.R., Aresu, G., Brittain, S., Ciardi, D.R., Danchi, W., Fedele, D., de Gregorio-Monsalvo, I., Heras, A., Huelamo, N., Krivov, A., Lebreton, J., Liseau, R., Martin-Zaidi, C., Mendigutía, I., Montesinos, B., Mora, A., Morales-Calderon, M., Nomura, H., Phillips, N., Podio, L., Poelman, D.R., Ramsay, S., Rice, K., Solano, E., Walker, H., White, G.J., Wright, G.: Herschel-PACS observation of the 10 Myr old T Tauri disk TW Hya. Constraining the disk gas mass. Astron. Astrophys. **518**, L125 (2010). doi:10.1051/0004-6361/201014578

Turner, N.J., Fromang, S., Gammie, C., Klahr, H., Lesur, G., Wardle, M., Bai, X.N.: Transport and accretion in planet-forming disks. Protostars and Planets VI, pp. 411–432. University of Arizona Press, Tucson (2014). doi:10.2458/azu_uapress_9780816531240-ch018, 1401.7306

van der Marel, N., van Dishoeck, E.F., Bruderer, S., Birnstiel, T., Pinilla, P., Dullemond, C.P., van Kempen, T.A., Schmalzl, M., Brown, J.M., Herczeg, G.J., Mathews, G.S., Geers, V.: A major asymmetric dust trap in a transition disk. Science **340**, 1199–1202 (2013). doi:10.1126/science.1236770, 1306.1768

van Dishoeck, E.F., Black, J.H.: The photodissociation and chemistry of interstellar CO. Astrophys. J. **334**, 771–802 (1988)

Williams, J.P.: Astronomical evidence for the rapid growth of millimeter-sized particles in protoplanetary disks. Meteorit. Planet. Sci. **47**, 1915–1921 (2012). doi:10.1111/maps.12028, 1205.2461

Williams, J.P., Best, W.M.J.: A parametric modeling approach to measuring the gas masses of circumstellar disks. Astrophys. J. **788**, 59 (2014). doi:10.1088/0004-637X/788/1/59, 1312.0151

Williams, J.P., Cieza, L.A.: Protoplanetary disks and their evolution. Annu. Rev. Astron. Astrophys. **49**, 67–117 (2011). doi:10.1146/annurev-astro-081710-102548

Williams, J.P., McPartland, C.: Measuring protoplanetary disk gas surface density profiles with ALMA. Astrophys. J. **830**, 32 (2016). doi:10.3847/0004-637X/830/1/32, 1606.05646

Williams, J.P., Cieza, L.A., Andrews, S.M., Coulson, I.M., Barger, A.J., Casey, C.M., Chen, C.C., Cowie, L.L., Koss, M., Lee, N., Sanders, D.B.: A SCUBA-2 850-μm survey of protoplanetary discs in the σ Orionis cluster. Mon. Not. R. Astron. Soc. 435:1671–1679 (2013). doi:10.1093/mnras/stt1407, 1307.7174

Wilner, D.J.: Imaging protoplanetary disks with a square kilometer array. New Astron. Rev. **48**, 1363–1375 (2004). doi:10.1016/j.newar.2004.09.041, astro-ph/0412336

Woitke, P., Kamp, I., Thi, W.F.: Radiation thermo-chemical models of protoplanetary disks. I. Hydrostatic disk structure and inner rim. Astron. Astrophys. **501**, 383–406 (2009). doi:10.1051/0004-6361/200911821

Wolcott-Green, J., Haiman, Z.: Suppression of HD cooling in protogalactic gas clouds by Lyman-Werner radiation. Mon. Not. R. Astron. Soc. **412**, 2603–2616 (2011). doi:10.1111/j.1365-2966.2010.18080.x, 1009.1087

Wood, B.E., Linsky, J.L., Hébrard, G., Williger, G.M., Moos, H.W., Blair, W.P.: Two new low galactic D/H measurements from the far ultraviolet spectroscopic explorer. Astrophys. J. **609**, 838–853 (2004). doi:10.1086/421325, astro-ph/0403606

Yuan, Y., Neufeld, D.A., Sonnentrucker, P., Melnick, G.J., Watson, D.M.: Spitzer observations of shock-excited hydrogen deuteride in IC 443C, HH 7, and HH 54: probing the gas-phase deuterium abundance in the dense interstellar medium. Astrophys. J. **753**, 126 (2012). doi:10.1088/0004-637X/753/2/126

Zurlo, A., Cieza, L.A., Williams, J.P., Canovas, H., Perez, S., Hales, A., Mužić, K., Principe, D.A., Ruíz-Rodríguez, D., Tobin, J., Zhang, Y., Zhu, Z., Casassus, S., Prieto, J.L.: The ALMA early science view of FUor/EXor objects. I. Through the looking-glass of V2775 Ori (2016). ArXiv e-prints 1611.00765

Chapter 2
The ALMA Revolution: Gas and Dust in Transitional Disks

Nienke van der Marel

Abstract The study of protoplanetary disks was for a long time based on indirect measurements of gas and dust, limiting our understanding of planet formation. The arrival of ALMA has revolutionized our view of the structure of these disks. The Early Science ALMA observations in the last few years have revealed rings, asymmetries, dust/gas segregation, gas dynamics, evidence for dust trapping and vortices, and many more exciting phenomena that have been predicted for decades in physical models of disks. In this chapter I review the most important ALMA discoveries and the next steps in planet formation studies.

2.1 Introduction

Protoplanetary disks of gas and dust around young stars form the birth cradles of planets (reviews by, e.g., Williams and Cieza 2011; Armitage 2011; Testi et al. 2014). T Tauri and Herbig stars are usually surrounded by these disks. The large number of discovered exoplanets in the last two decades implies that planets are ubiquitous, and thus planet formation a common process in the evolution of the disk (Batalha 2014). However, the exact way in which planets form remains poorly understood. Disks evolve through viscous accretion and have a typical lifetime of a few million years (Williams and Cieza 2011). At the distance of nearby star forming regions, such as Ophiuchus, Taurus or Lupus, these disks are at most a few arcseconds in angular size, making the study of their structure particularly challenging.

Before subarcsecond imaging was technically possible, studies of protoplanetary disks were limited to spatially unresolved observations, often biased towards the brightest objects that are not necessarily representative for deriving general disk properties. Disks were identified through the measurement of infrared excess above the stellar photosphere, in particular through surveys of infrared space telescopes such as *ISO*, *IRAS*, and *Spitzer*, in combination with observed ultraviolet (UV)

N. van der Marel (✉)
Institute for Astronomy, University of Hawaii, 2680 Woodlawn dr, Honolulu, HI 96822, USA
e-mail: marel@hawaii.edu

M. Pessah, O. Gressel (eds.), *Formation, Evolution, and Dynamics of Young Solar Systems*, Astrophysics and Space Science Library 445,
DOI 10.1007/978-3-319-60609-5_2

excess, a sign of accretion. By combining photometry from optical to millimeter wavelengths, it was possible to create Spectral Energy Distributions (SEDs), which can be used to analyze the dust disk structure. Millimeter surveys were used to constrain disk masses and dust growth within specific assumptions regarding temperature and dust opacity (e.g., Beckwith and Sargent 1996; Andrews and Williams 2007; Ricci et al. 2010). Studying gas in protoplanetary disks was even more challenging: although the Keplerian rotation was seen in early line observations of a handful of bright disks (e.g., Sargent and Beckwith 1987; Koerner et al. 1993) molecular line studies were limited by both sensitivity and cloud confusion, and the density distribution of gas remained largely unknown.

Of particular interest in the studies of planet formation are the so-called transitional disks: disks with a gap or cavity in their dust distribution. This type of disk was originally identified by Strom et al. (1989) through a dip in the mid infrared emission in the SED, indicating a lack of warm dust in the system or a gap in the dust distribution (Fig. 2.1). It was proposed that these disks were in transition between a full, primordial, gas-rich disk and an older, gas-poor debris disk, although a real evolutionary sequence (implying that every disk goes through a "transition" phase) has not been proven until today. Many more transition disks have been identified since then (see Espaillat et al. 2014, for a review), especially

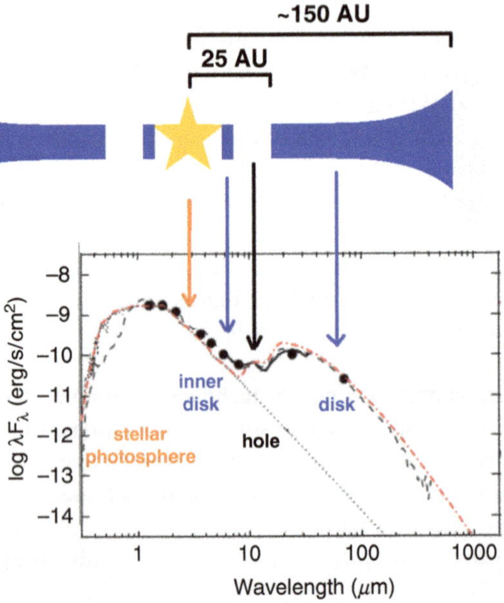

Fig. 2.1 Spectral Energy Distribution of a transition disk, showing the different components of the system and their emitting wavelengths. The stellar photosphere emits as a black-body in optical to near-infrared wavelengths, an inner disk is recognized as near-infrared excess above the photosphere due to hot dust grains, the hole is seen as a dip in the mid-infrared, while the bulk of the dust in the outer disk emits in far-infrared up to mm-wavelengths. Figure from van der Marel (2015)

by *Spitzer* and its InfraRed Spectrograph (IRS), observing the 5–35 μm wavelength range in which the characteristic dip is seen. Several of these identifications were followed up by submillimeter imaging of pioneering interferometers such as the SubMillimeterArray (SMA), Plateau de Bure Interferometer (PdBI), and Combined Array for Research in Millimeter Astronomy (CARMA) (e.g., Brown et al. 2009; Isella et al. 2010b; Andrews et al. 2011). These observations revealed ring-like structures in the submillimeter continuum at $\sim$0.3$''$ resolution, confirming the presence of large gaps in the dust distribution. However, the origin of the gaps remains unclear.

Several mechanisms have been proposed to explain the presence of dust cavities (see Fig. 2.2): (1) grain growth, where dust grains inside the cavity have grown to larger sizes that do not emit sufficiently at the observed wavelength; (2) photoevaporative clearing, where the stellar UV heats gas to temperatures above the escape speed, resulting in a photoevaporative wind that clears the disk from the inside out; (3) certain instabilities (e.g., dead zones, regions of low ionization) leading to dust concentrations in the outer disk; (4) companion clearing, where

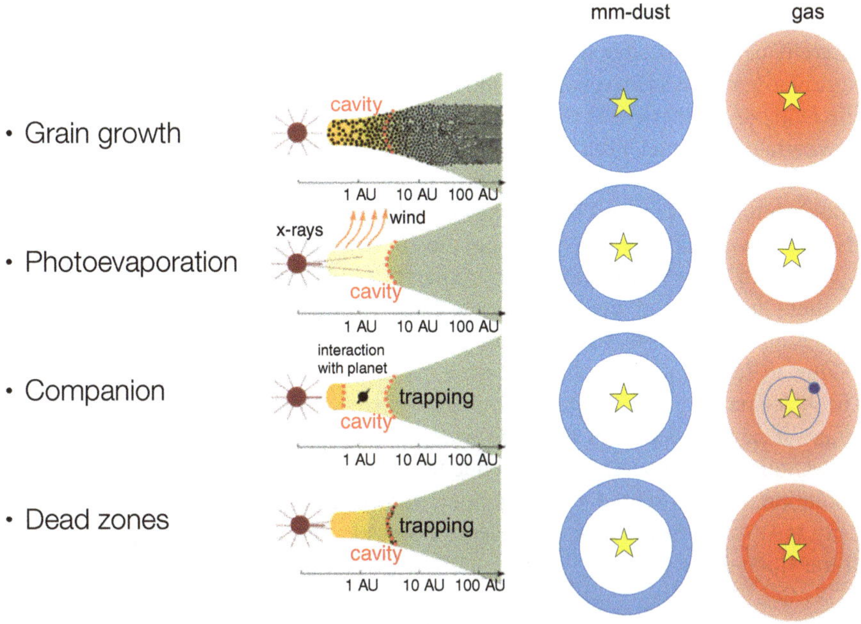

Fig. 2.2 Cavity clearing mechanisms in dust (*black dots*) and gas (*yellow*) and their expected signatures in millimeter dust (*blue*) and gas (*red*). *Top:* increased grain growth inside the cavity, while the gas density remains unchanged compared with the original disk. *Second:* photoevaporative clearing, with both the gas and dust cleared simultaneously from the inside out. *Third:* clearing by a companion, in which the gas density inside the cavity drops and the large dust particles get trapped at the edge. *Bottom:* instabilities due to, e.g., viscosity or entropy gradients create a dust trap, the gas density remains unchanged. Figure taken from van der Marel (2015)

either a planet or a star carves a gap in the dust and gas in their orbit. The grain growth mechanism has been demonstrated to be consistent with a dip in the SED, but cannot result in millimeter dust cavities according to dust evolution models (Birnstiel et al. 2012). Photoevaporation is unlikely to explain the larger observed dust cavities of tens of AU that are still accreting, but may be able to explain smaller cavities in systems with low accretion rates. For this reason, it has been proposed that there are two different types of transition disks, where only one type can be explained by photoevaporative clearing (Owen and Clarke 2012). However, the structure of the gas is key in distinguishing between the mechanisms, as demonstrated in Fig. 2.2: gas and dust are expected to be cleared simultaneously in case of photoevaporation; a companion would lower the gas density inside the gap and "trap" the millimeter dust at the gap edge in a concentrated ring (e.g., Pinilla et al. 2012); dead zones alter the ionization, but not significantly the gas density structure, and trap the millimeter dust at their edge similar to the companion scenario (e.g., Varnière and Tagger 2006; Regály et al. 2012).

The concept of dust trapping is discussed in detail in the chapter by Pinilla. In summary, dust trapping is the result of drag forces between larger dust particles and gas as a result of a pressure gradient in the disk, leading to a concentration of dust in the region of maximum pressure (a "pressure bump"). Such a pressure bump will naturally exist at the edge of a gas gap created by a recently formed planet or at the edge of a dead zone. Dust evolution models have demonstrated that dust gets trapped efficiently and experiences increased dust growth (Pinilla et al. 2012), resulting in concentrated millimeter dust rings. Although dust trapping was originally proposed as a way to solve the so-called radial drift problem in full disks, where dust particles would simply drift inward before they could grow beyond millimeter sizes, it can also explain the structures seen in transition disks.

Although early millimeter interferometry could confirm the presence of dust cavities, the origin of the cavities remained inconclusive due to the low spatial resolution and signal-to-noise of these data. Furthermore, the sensitivity is generally too low to detect molecular line emission, required to map the gas density structure. The Atacama Large Millimeter/submillimeter Array (ALMA) has opened up a large range of new possibilities to study transition disks, unravel their structure, and answer questions about their origin, due to its unprecedented sensitivity and spatial resolution. ALMA started scientific operations in 2011 with a partially completed array, but even these Early Science observations have dramatically changed our view of transition disks.

2.2 Dust Observations

2.2.1 Rings and Asymmetries

ALMA has revealed a large variety of dust structures in transition disks (see compilation of observations in Table 2.1). Figure 2.3 displays a selection of disks that were observed between Cycle 0 and 3 from various programs in continuum

Table 2.1 Transition disks observed by ALMA by Fall 2016

Name	Program	PI	Band	References
DoAr44	2012.1.00158.S[a]	van Dishoeck	7	van der Marel et al. (2016b)
HD135344B[b]	2012.1.00158.S[a]	van Dishoeck	7	van der Marel et al. (2016b)
	2011.0.00724.S	Perez	9	Pérez et al. (2014), van der Marel et al. (2015b), Pinilla et al. (2015b)
HD142527	2011.0.00465.S[a,c]	Casassus	6,7,9	Casassus et al. (2013, 2015a,b), Perez et al. (2015)
	2011.0.00318.S	Fukagawa	7	Fukagawa et al. (2013)
HD97048	2013.1.00658.S[a]	van der Plas	7	van der Plas et al. (2017)
IRS48	2013.1.00100.S[a]	van der Marel	4,7,9	van der Marel et al. (2015a, 2016b)
	2011.0.00635.S	van Dishoeck	9	van der Marel et al. (2013, 2014), Bruderer et al. (2014)
J1604-2130	2013.1.01020.S[a]	Tsukagoshi	6	Dong et al. (2017)
	2011.0.00526.S	Carpenter	7	Carpenter et al. (2014), Zhang et al. (2014), van der Marel et al. (2015b)
J16083070	2013.1.00220.S[a]	Williams	7	Ansdell et al. (2016), van der Marel et al. (in preparation)
LkCa15	2011.0.00724.S[a]	Perez	9	van der Marel et al. (2015b)
RXJ1615-3255	2011.0.00724.S[a]	Perez	9	van der Marel et al. (2015b)
RYLup	2013.1.00220.S[a]	Williams	7	Ansdell et al. (2016), van der Marel et al. (in preparation)
SR21	2012.1.00158.S[a]	van Dishoeck	7	van der Marel et al. (2016b)
	2011.0.00724.S	Perez	9	Pérez et al. (2014), van der Marel et al. (2015b), Pinilla et al. (2015b)
SR24S	2013.1.00091.S[a]	Pinilla	6	Pinilla et al. (2017)
	2011.0.00724.S	Perez	9	van der Marel et al. (2015b)
SYCha	2013.1.00437.S[a]	Pascucci	7	Pascucci et al. (2016)
SZCha	2013.1.00437.S[a]	Pascucci	7	Pascucci et al. (2016)
Sz111	2013.1.00220.S[a]	Williams	7	Ansdell et al. (2016), van der Marel et al. (in preparation)
Sz91	2011.0.00733[a]	Schreiber	3,6	Canovas et al. (2015)
	2013.1.00663.S	Canovas	7	Canovas et al. (2016)
TWHya	2015.1.00686.S[a,d]	Andrews	7	Andrews et al. (2016)
	2015.A.00005.S	Tsukagoshi	7	Tsukagoshi et al. (2016)

[a]Data used in Figs. 2.3 and/or 2.8
[b]HD135344B is also known as SAO 206462
[c]HD142527 has been observed in many ALMA programs: only the programs with published data are listed here
[d]TW Hya has been observed in many ALMA programs: only the programs at high spatial resolution are listed here

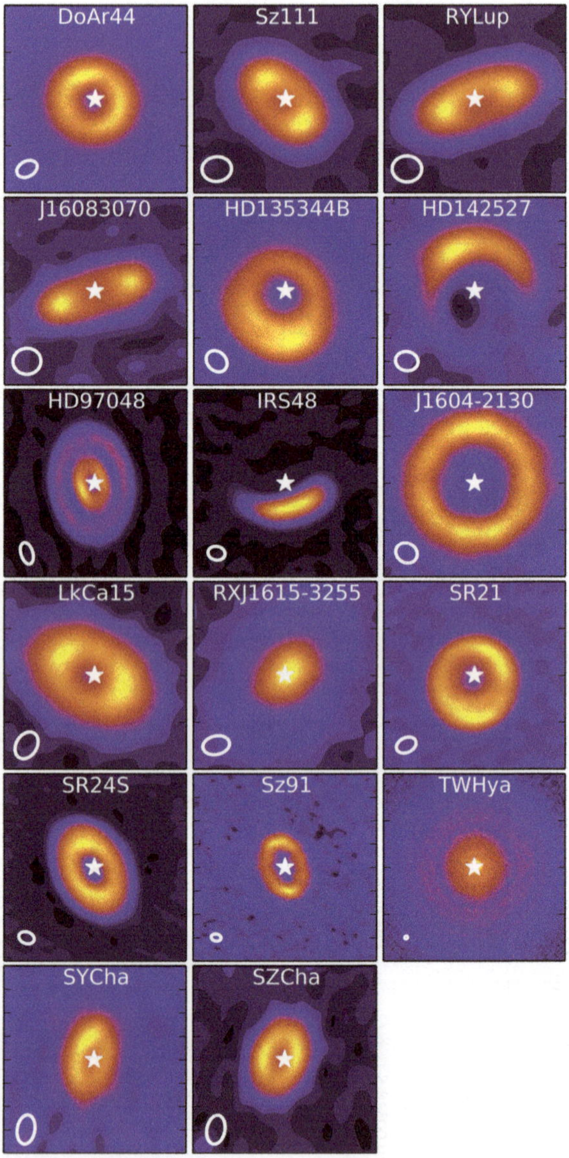

Fig. 2.3 Gallery of dust continuum observations of transition disks with ALMA from various programs. Figure available at http://www.ifa.hawaii.edu/~nmarel

emission at spatial resolutions between 0.02″ and 0.4″. Some transition disks show clear ring-like structures (e.g., J1604-2130, LkCa15, SR 24S, DoAr 44, Sz 91, Sz 111, etc. Zhang et al. 2014; van der Marel et al. 2015b, 2016b; Canovas et al. 2016; Ansdell et al. 2016), while others are azimuthally asymmetric, with contrasts of only a few up to more than 100. The most extreme examples of asymmetric disk rings are Oph IRS 48 (van der Marel et al. 2013) and HD 142527 (Casassus et al. 2013; Fukagawa et al. 2013), while others (SR 21, HD 135344B, a.k.a. SAO 206462) show minor asymmetric contrasts (Pérez et al. 2014; Pinilla et al. 2015b). These extreme asymmetries have been interpreted as dust trapping in vortices, as the result of Rossby Wave Instability of a radial pressure bump (Barge and Sommeria 1995; Klahr and Henning 1997; Lyra and Lin 2013). Trapping in a vortex results in efficient azimuthal concentration of dust (Birnstiel et al. 2013), such as seen in Oph IRS 48 and HD 142527. On the other hand, the minor asymmetries may be caused by eccentric disks rather than vortices, where the dust is not azimuthally trapped but only experiences a "traffic jam" (Ataiee et al. 2013; Pinilla et al. 2016).

Oph IRS 48 was the first disk that was indisputably identified as a dust trap (van der Marel et al. 2013), due to the clear difference in distribution of the millimeter dust, the gas (traced through ^{12}CO 6–5 emission) and small dust grains that are generally following the gas (traced by VLT-VISIR 19 μm emission). However, other disks can be explained by the trapping mechanism as well, albeit often only showing the effects of radial trapping. Although the sample of spatially resolved transition disks is still far from complete and clearly non-uniform, only a small fraction appears to show asymmetric structures. This could be a consequence of the special conditions in which a radial pressure bump becomes Rossby unstable, i.e. low viscosity and a sharp pressure gradient (de Val-Borro et al. 2007; Lyra et al. 2009), which may be intrinsically rare.

Another interesting phenomenon is revealed in the very long baseline observations, resulting in a spatial resolution of $\sim$20 mas or a few AU. Both TW Hya and HD 163296 (Isella et al. 2016; Andrews et al. 2016; Tsukagoshi et al. 2016) show distinct multiple dust rings rather than a smooth continuous disk, similar to what was found in the much younger HL Tau disk that was imaged as part of the Very Long Baseline Science Verification campaign (ALMA Partnership et al. 2015). Although gaps carved by planets with corresponding dust traps are a tempting explanation for these dust rings, at least HL Tau appears to be too young ($<$1 Myr) to have full-grown planets in its system. Another explanation that has been put forward to explain narrow rings are condensation fronts at snowline radii (e.g., Zhang et al. 2015) where the dust stickiness is increased and therefore the dust growth more efficient. As the other observed transition disk observations have lower spatial resolution, narrow rings cannot be excluded for any of them, although indications of gaps in the outer disk have been seen in HD 100546 (Walsh et al. 2014) and HD 135344B (van der Marel et al. 2016a) as well.

2.2.2 *Trapping Efficiency*

One of the properties of dust trapping is a strong grain size dependence: larger dust grains (with larger Stokes number) are more efficiently trapped (Brauer et al. 2008; Birnstiel et al. 2010, 2013) and therefore, continuum observations at different wavelengths (tracing different grain sizes) are expected to show a different structure. More specifically, at longer wavelengths, the emission is expected to be more

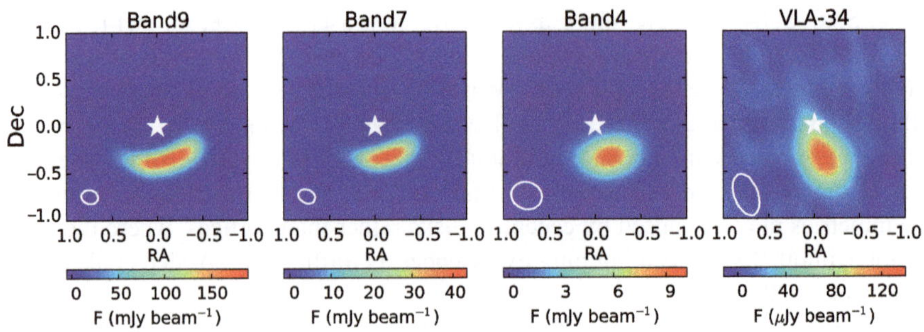

Fig. 2.4 Continuum emission of the Oph IRS 48 transition disk at four wavelengths, using combined ALMA and VLA observations (van der Marel et al. in preparation)

concentrated. Alternatively, it is also possible to measure the spectral index $\alpha_{\rm mm}$ as a function of position in the disk. Due to dust opacity effects, α is expected to decrease for larger dust grains (Draine 2006) for optically thin emission.

ALMA and VLA observations of Oph IRS 48 between 690 and 34 GHz show that the azimuthal width is $\sim$2 times narrower at the longer wavelengths (van der Marel et al. 2015a, and Fig. 2.4) and the spectral index drops significantly in the center of the dust feature (van der Marel et al. in preparation), consistent with azimuthal dust trapping. Multi-wavelength analysis of the HD 142527 and MWC 758 asymmetric disks shows the same phenomenon (Casassus et al. 2015b; Marino et al. 2015). This confirms the predictions of dust evolution models.

Radial variations in spectral index are reported in SR 21 through ALMA observations at 345 and 690 GHz (Pinilla et al. 2015b), consistent with radial trapping, while in the same work for HD 135344B no significant variations were found. The latter can be explained by the low spatial resolution of these data: if the dust ring is narrow, spatial variations in spectral index will not be detectable. Also dust optical depth can play a role: if the dust emission is optically thick, the spectral index is not a proper tracer of dust growth. Multi-wavelength ALMA observations of SR 24S also show that the dust ring is narrower at longer wavelengths, consistent with radial trapping (Pinilla et al. 2017). Multi-wavelength continuum observations at high spatial resolution are crucial to verify the dust trapping scenario in other transition disks. A synergy between ALMA and VLA/ATCA will be desirable to measure the extent of the emission over a long wavelength range.

2.3 Gas Observations

Whereas the dust distribution in disks has been studied to great detail, observational constraints on the gas distribution are rather limited, while the gas is considered to be dominating the disk properties such as mass and dynamical processes. In particular, the clearing mechanisms for transition disks described above have similar effects on the dust distribution, while the gas is affected in completely different ways and provides thus a way to distinguish between them (Fig. 2.2). Measuring the gas density inside and outside the dust cavity is a crucial step in understanding the transition disks.

The presence of accretion in the majority of the transition disks suggests that some gas must still be present inside the dust cavities. This gas is hotter than in the outer disk and can be traced by ro-vibrational and atomic lines, which can reveal the presence and kinematics of molecules, but no global density information. Ro-vibrational CO line observations at 4.7 μm of several bright transition disks have revealed Keplerian gas down to radii smaller than the dust cavity radius using the spectroastrometry technique, e.g. Oph IRS 48, SR 21 and HD 135344B (Pontoppidan et al. 2008; Brown et al. 2012), which are key targets in this thesis. For other disks, the CO line observations are often single-peaked, consistent with a photoevaporative wind (Bast et al. 2011; Brown et al. 2013), also seen in [Ne II] line emission (Pascucci et al. 2007; Lahuis et al. 2007). However, although these results are promising, they do not indisputably reveal the gas surface density structure and thus the mechanism responsible for the transition disk.

2.3.1 Gas Inside Dust Cavities

While the bulk of the gas mass is in H_2, its molecular lines cannot be used as mass tracer due to its lack of dipole moment. CO (with its isotopologues ^{13}CO and $C^{18}O$) is the second most abundant molecule in the gas and can be readily detected and spatially resolved by ALMA. Early ALMA observations revealed the presence of gas inside the dust cavity for several transition disks through ^{12}CO emission at $\sim$0.25″ resolution (van der Marel et al. 2013; Casassus et al. 2013; Bruderer et al. 2014; Pérez et al. 2014; van der Marel et al. 2015b; Canovas et al. 2015), hinting at the presence of planets as probable explanation for the transition disk cavities, with the millimeter dust being trapped in the outer part of the disk. Figure 2.5 shows a few examples, with the ^{12}CO emission peaking inside the dust cavity.

However, the proper interpretation of CO as column density tracer is challenging. CO rotational emission cannot be translated directly into mass for several reasons. First, the ^{12}CO lines are usually optically thick. Second, the gas temperature is decoupled from the dust temperature, especially in the higher layers in the disk. Third, the CO abundance is not constant with respect to H_2: in the surface layers CO is photodissociated by stellar radiation; in the mid-plane the CO freezes out onto the dust grains due to the low temperatures; and the chemical formation/destruction of CO depends strongly on the local density, temperature, and radiation conditions (van Zadelhoff et al. 2001; Woitke et al. 2009).

In order to tackle these problems, a physical-chemical modeling tool can be used. The DALI code (Bruderer et al. 2012; Bruderer 2013) first solves the dust radiative transfer in 2D for a given density structure (Fig. 2.7) and stellar radiation field. With the derived dust temperatures as starting point, the heating-cooling balance of

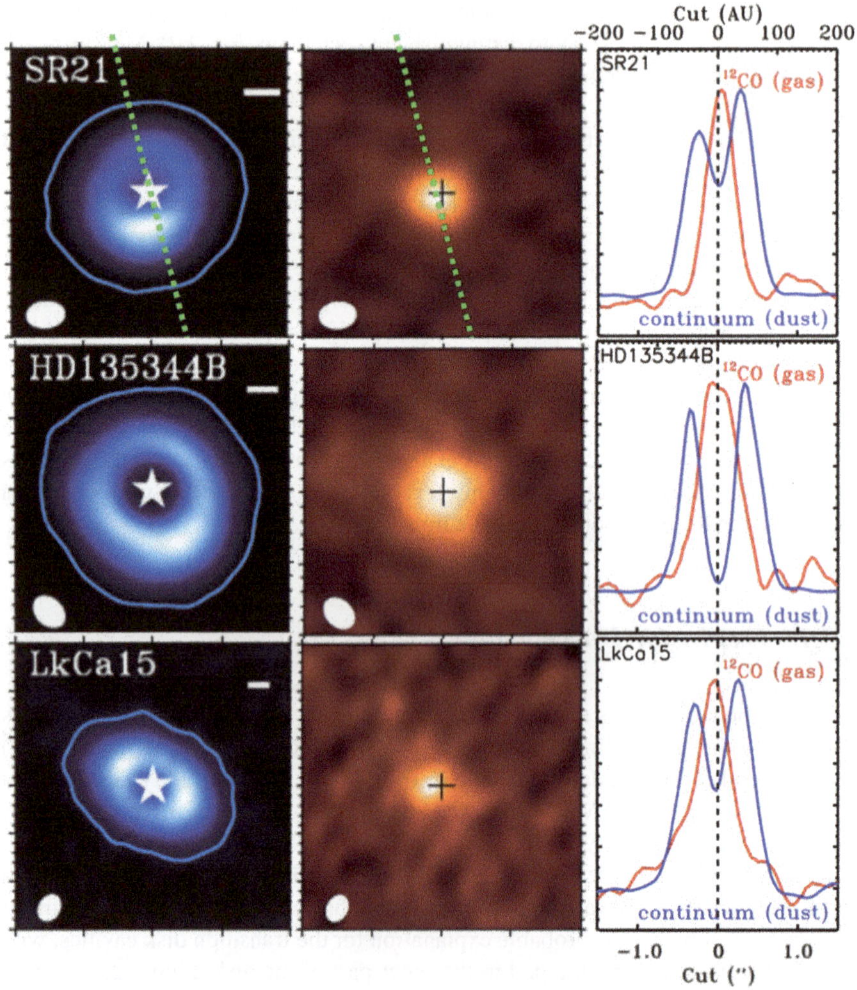

Fig. 2.5 Examples of spatially resolved ^{12}CO emission in transition disks. Figure adapted from van der Marel et al. (2015b)

the gas and chemistry are solved simultaneously to determine the gas temperature, molecular abundances and molecular excitation. All effects discussed above are thus taken into account. Special attention is paid to the chemistry and heating inside the dust-free cavities (Jonkheid et al. 2006). The resulting CO abundance in the disk in combination with the gas temperature determine the intensity and profile of the CO line emission in the final step, the ray tracer. ProDiMo (Protoplanetary Disk Modeling) (Woitke et al. 2009, 2016), that was also developed to analyze disk molecular line emission, has a comparable approach as DALI. The DALI code structure is summarized in Fig. 2.6.

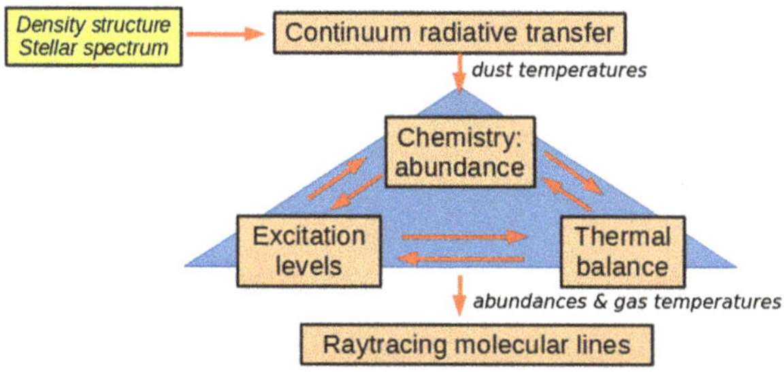

Fig. 2.6 DALI modeling structure. The input of the model is a 2D density structure and a stellar spectrum. Using dust radiative transfer (based on RADMC), the dust temperatures are calculated. DALI then solves the heating-cooling balance of the gas to determine molecular abundances, temperatures and excitation before raytracing spectra and spectral line cubes. Figure taken from van der Marel (2015)

As the ^{12}CO emission is optically thick, the absolute gas surface density and mass remain uncertain, but it is possible to constrain the depth of the gas gap within certain assumptions. The ^{12}CO emission in transition disks is dominated by the emission of the directly heated cavity wall (Bruderer 2013) and by temperature (vertical structure) rather than density. Therefore, the starting point of the modeling procedure is a dust density profile that fits the SED. Assuming an ISM gas-to-dust ratio of 100 for the outer disk, the ^{12}CO emission is subsequently calculated. Inside the cavity, the gas surface density is lowered by a factor $\delta_{\rm gas}$, as shown in Fig. 2.7, in order to fit the observations. With this approach, the gas density profiles of several transition disks were derived (Bruderer et al. 2014; van der Marel et al. 2015b). Comparison with the observations is done through cuts through the zero-moment map and the spectrum integrated over the disk region. The millimeter continuum emission is fit to the same model through the visibility curves, to constrain the dust cavity size. In two cases, Oph IRS 48 and J1604-2130, the ^{12}CO emission becomes optically thin in the center, revealing an actual cavity in the gas, with a significantly smaller radius than the millimeter dust cavity, which is fit accordingly. For the other disks, the gas density inside the dust cavity has to be lowered by a factor 10–100 in order to match the observed ^{12}CO emission in the central part of the disk. Resolving the CO emission spatially is thus crucial to compare the density structure and outside the dust cavity radius. The reason that optically thick ^{12}CO emission can still constrain density variations relies on the vertical gas temperature gradient in the disk: as the density decreases, the $\tau = 1$-surface lies closer to the midplane and is tracing lower temperatures, hence lower emission (see, e.g., Fig. 9 in Bruderer 2013). Furthermore, the procedure relies heavily on the assumptions that the SED and a gas-to-dust ratio of 100 are reasonable approaches for the gas surface density profile. In contrast to the depletion in the gas, the dust density depletion inside the cavity was found to be at least a factor 1000 in all cases. This difference in depletion between dust and gas suggests that planets are responsible for the observed structure: according to planet–disk interaction models, planets lower the gas surface

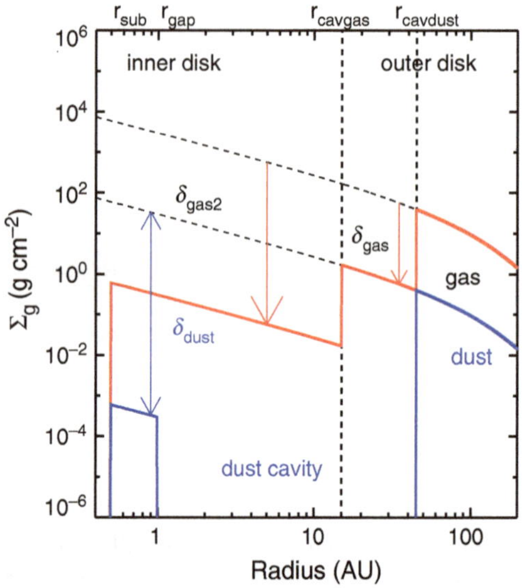

Fig. 2.7 DALI input density structure of dust and gas (adapted from van der Marel et al. 2015b)

density along their orbit, trapping the millimeter dust at the edge, resulting in a much deeper depletion of dust than gas inside the cavity (Pinilla et al. 2012).

2.3.2 *Gas Cavities*

Whereas spatially resolved optically thick ^{12}CO emission can provide some information about the gas structure, optically thin isotopologues are required to constrain the gas density properly. ALMA observations of spatially resolved ^{13}CO and $C^{18}O$ emission reveals the presence of gas cavities for *all* transition disks observed to date, where the gas cavity is found to be smaller than the dust cavity radius. This phenomenon was predicted in planet–disk interaction models (Pinilla et al. 2012), suggesting that these gaps are indeed caused by planets. A gallery of continuum and ^{13}CO zero-moment maps of various transition disks is shown in Fig. 2.8.

Several of these disks with CO isotopologue data (SR 21, HD 135344B, DoAr 44, IRS 48, and several Lupus transition disks) have been analyzed with DALI as well, in order to constrain the size and depth of the gaps (van der Marel et al. 2016b, and in preparation). Although detections of planet candidates in disks have been claimed (Huélamo et al. 2011; Kraus and Ireland 2012; Sallum et al. 2015; Currie et al. 2015), several of these have been demonstrated to originate from inner disk emission (Olofsson et al. 2013; Thalmann et al. 2016), and direct imaging of planets in disks thus remains challenging. The depth and shape of the gas gap depend primarily on the planet mass and the disk viscosity (Zhu et al. 2011; Pinilla et al. 2012; Fung et al. 2014). Quantifying these can thus provide important clues on the properties of embedded unseen planets. The models derived from the ^{12}CO data were used as a

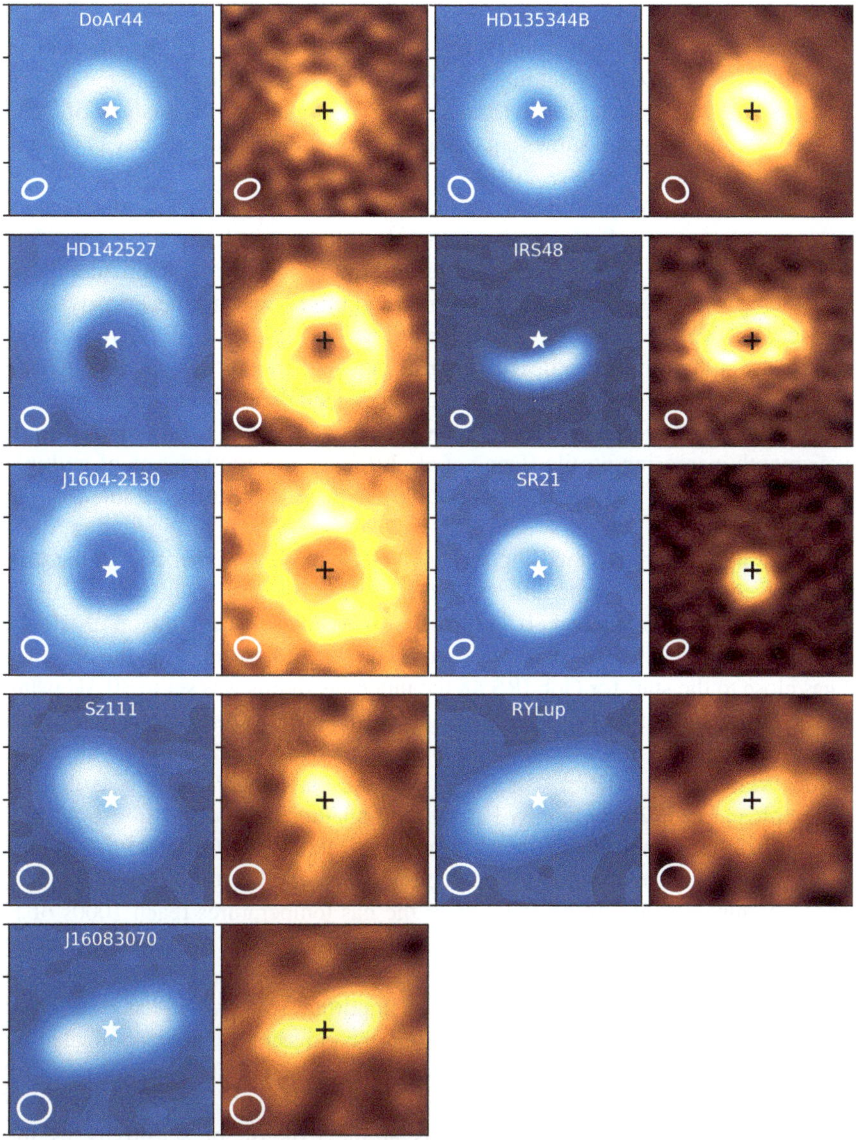

Fig. 2.8 Gallery of continuum versus integrated ^{13}CO emission of spatially resolved observations of transition disks with ALMA from various programs. Figure available at http://www.ifa.hawaii.edu/~nmarel

starting point, but as the gas cavity radius is now an additional free parameter, it was found that the gas gaps were in fact smaller, but also deeper, with a depletion factor of up to several orders of magnitude. This is not in contradiction with the previous results, where the gas cavity was set to be equal to the dust cavity, consistent with a shallower gap depth. The isotopologue data are however a more direct tracer and thus more reliable estimator of the gas density. The deep gaps are consistent with clearing by Jupiter-mass planets and low disk viscosities (van der Marel et al. 2016b). A parametrized model that was used to analyze CO isotopologue data of HD 142527 found a similar result (Perez et al. 2015), and also the analysis of ^{12}CO data of J1604-2130 which shows a gas gap implies a deep, small gas cavity (Zhang et al. 2014; van der Marel et al. 2015b).

The surface density profiles that are used to fit these data contain steep cavity edges, which are not physical: planet–disk interaction models predict a gradual radial increase over several AUs. However, smooth profiles contain more free parameters, which makes the modeling more complex and not necessarily more reliable. However, for the sample analyzed by van der Marel et al. (2016b) it was found that simple smooth profiles were consistent with the data as well, although the spatial resolution did not allow to distinguish between the different model profiles. On the other hand, new CO 2–1 isotopologue data on J1604-2130 can only be fit by a smooth increasing profile (Dong et al. 2017), so the combination of multiple CO isotopologues at high spatial resolution is crucial to constrain the shape of the profile properly.

Interestingly, a radial increasing smooth profile in the hot inner part of the disk was also required to fit near infrared rovibrational CO data taken with CRIRES of both the HD 135344B and HD 139614 transition disks (Carmona et al. 2014). Spectroastrometric CRIRES data was used before ALMA to constrain the presence of gas close to the star: for Oph IRS 48 the inner gas radius was set to 30 AU using the CRIRES data (Brown et al. 2012), which is more or less consistent with the 25 AU radius found in the spatially resolved ALMA data (Bruderer et al. 2014; van der Marel et al. 2016b). For SR 21, a 7 AU gas cavity radius was derived (Pontoppidan et al. 2008), which has so far remained unresolved in the ALMA observations but is not inconsistent. In general, rotational CO line emission that is measured with submillimeter telescopes is not suitable to constrain the gas density within $\sim$10 AU to the star, where the gas temperatures reach 1000s of K and the excitation conditions are uncertain. The derived gas density profiles in the inner 10 AU of the studies mentioned above are merely an extrapolation from radii further out. The spatial resolution plays an important role as way: at $\sim 0.25''$, at which most of these observations were taken, the emission remains dominated by the cavity edge, whereas higher resolution can reveal density variations further in (see Fig. 2.9 for a comparison between $0.25''$ and $0.1''$ data). On the other hand, increasing the spatial resolution is at the expense of sensitivity as the surface brightness decreases, and longer integration times are required to detect the emission. Ultimately a combination of spatially resolved submillimeter observations and spectrally resolved near infrared data will be necessary to constrain the gas density over the full range of the disk (Banzatti and Pontoppidan 2015).

Another interesting aspect of spatially resolved CO emission inside the cavity is the possibility of observing accretion flows from the outer disk onto the star. These gas streamers were initially claimed to be found in the HD 142527 disk (Casassus et al. 2013), but later found to be consistent with a warp: a radially variable inclination in the disk (Rosenfeld et al. 2014; Casassus et al. 2015a), which

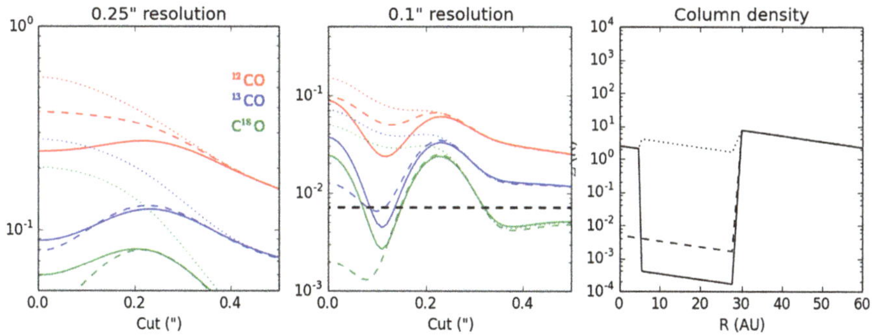

Fig. 2.9 The importance of spatial resolution for observing spatially resolved CO emission. The *left* and *middle panels* show the radial cut of the expected emission of the zero moment map of the three CO isotopologues (*red*, *blue*, *green* for ^{12}CO, ^{13}CO and C^{18}O) convolved with a 0.25″ and 0.1″ beam. The different line styles represent different types of gap, as shown in the *right panel*: *dashed* represents a disk with a deep hole (no inner disk), *solid* represents a disk with a deep gap (inner gas disk present), and *dotted* represents a disk with a minor density depletion in the inner disk. This substructure remains unresolved in the 0.25″ beam, but becomes visible at 0.1″ resolution

is also found by radiative transfer analysis of the scattered light image of this disk, which shows dips along the dust ring that are caused by shadows of the inclined inner disk (Marino et al. 2015).

2.3.3 Small Dust Grains

In addition to the gas structure, the distribution of the smaller dust grains compared to the millimeter dust grains can also potentially provide clues about the origin of the transition disk cavities. Small dust grains (∼micrometer-sized) have much lower Stokes numbers and are thus more coupled to the gas. As a result, smaller dust grains are expected to be present inside dust cavities caused by planets, as a result of dust filtration: larger particles remain trapped at the edge while smaller dust grains flow through the gap (Rice et al. 2006; Zhu et al. 2011). In recent years, near infrared imaging (scattered/polarized light) of disks has been greatly improved due to the developments in coronography and adaptive optics, with new instruments such as NACO and SPHERE at the VLT and HiCIAO and SCExAO at the Subaru telescopes. In particular the SEEDS survey (Strategic Explorations of Exoplanets and Disks with Subaru) and the SPHERE GTO programs have resulted in a large number of high quality near infrared images of small dust grains, revealing a range of structures, such as rings, spiral arms, and asymmetries (e.g., Ginski et al. 2016; de Boer et al. 2016; Stolker et al. 2016). As the dust emission is highly optically thick at this wavelength, the structures are not tracing density variations, but can still provide some information on the location of the small dust grains. The lack of large dust cavities in scattered light images was already proposed to be related to the planet clearing and trapping scenario by Dong et al. (2012). de Juan Ovelar et al. (2013) showed predictions for continuum images for ALMA and SPHERE using dust evolution models in combination with planet–disk interaction models, and later presented a method to constrain turbulence using a combination of continuum observations (de Juan Ovelar et al. 2016). The spatial segregation between scattered light and submillimeter data was observationally confirmed, e.g. HD 135344B and J1604-2130 (Garufi et al. 2013; Pinilla et al. 2015a). Also for the asymmetric disks Oph IRS 48 and HD 142527 infrared observations have revealed a segregation between the milllimeter and micrometer-grains, where the infrared images tracing the latter are showing ring-like structures (Geers et al. 2007; van der Marel et al. 2013; Avenhaus et al. 2014).

Scattered light imaging can thus provide constraints on the origin of the dust cavities, but due to the high optical depth and inner working angle at these wavelengths it remains difficult to obtain quantitative information about the density structure inside the disk cavity.

2.4 Large Surveys

The power of ALMA does not lie only in the possibility of deep line studies, but also in snapshot surveys of dozens of sources in only a fraction of the time that it would take with one of the smaller (sub)millimeter arrays. Complete samples of protoplanetary disks in star forming regions are crucial for statistical analysis and linking the disk properties to questions on disk evolution, planet formation, time scales, and exoplanet population synthesis models. Especially the possibility of detecting CO emission and thus constraining both dust *and* gas masses make these surveys very valuable. Existing ALMA surveys of Lupus (Ansdell et al. 2016), Chamaeleon (Pascucci et al. 2016), and Upper Sco (Carpenter et al. 2014; Barenfeld et al. 2016) have resulted in dozens of new submillimeter detections of protoplanetary disks in both dust and gas. The main results of these surveys for protoplanetary disks are discussed in the chapter by Williams.

Complete disk surveys can also help to answer questions about transition disks: how common are they? What does this imply about their lifetime and origin? What role do they have in the disk evolution process? In the Lupus survey (Ansdell et al. 2016) it was noted that 10–20% of the disks (depending on the definition) are transition disks, and that they are generally the brightest disks in the entire sample in both continuum and ^{13}CO emission. This implies that either these disks are intrinsically brighter due to the irradiated cavity wall, or that only the most massive disks go through a transition disk phase with a large dust cavity (van der Marel et al. in preparation)

Also, disk surveys can result in unexpected discoveries of new transition disks: most of the known transition disks with millimeter dust cavities were follow-ups on *Spitzer*-selected disks with a clear dip in the infrared part of their SED, but a handful of imaged disks (MWC 758, AB Aur, RY Tau and WSB 60) do not show this signature in their SED, while their millimeter image reveals a clear cavity (e.g. Isella et al. 2010a; Andrews et al. 2011). Transition disks do not form a well-defined sample with similar properties, and for a better understanding of their role in disk evolution the sample needs to be expanded to include all disks with signatures of gaps and cavities. The results of the Very Long Baseline ALMA observations, showing narrow dust rings in every disk that has been observed at $\sim$20 mas resolution, may complicate this even further: if every disk that appears to be smooth at $\sim 0.2''$ scales in fact shows substructure at higher resolution, many more processes in disks, including active planet formation, may be happening simultaneously, implying that in fact every disk is "transitional", i.e. going through an evolutionary process where the structure is actively changing.

Expansion of the number of transition disks will provide an unbiased view on their properties, as the current sample is biased towards the brightest disks with largest cavities, and around generally earlier type stars, due to their easier observability. Current survey studies of transition disks based on their SED show a large variety of both disk and stellar properties of transition disks (Najita et al. 2007; Cieza et al. 2010, 2012; Romero et al. 2012; Currie and Sicilia-Aguilar 2011; Espaillat et al. 2012; van der Marel et al. 2016c), which could possibly divide them up in different categories and different evolutionary paths (e.g. Owen 2016).

Overall, larger ALMA sample studies in uniform settings are crucial for a better understanding of transition disks.

Acknowledgements The author would like to thank Jonathan Williams for useful comments on the manuscript. The author is supported by the Beatrice Watson Parrent Fellowship at the University of Hawaii.

References

ALMA Partnership, Brogan, C.L., Pérez, L.M., Hunter, T.R., Dent, W.R.F., Hales, A.S., Hills, R.E., Corder, S., Fomalont, E.B., Vlahakis, C., Asaki, Y., Barkats, D., Hirota, A., Hodge, J.A., Impellizzeri, C.M.V., Kneissl, R., Liuzzo, E., Lucas, R., Marcelino, N., Matsushita, S., Nakanishi, K., Phillips, N., Richards, A.M.S., Toledo, I., Aladro, R., Broguiere, D., Cortes, J.R., Cortes, P.C., Espada, D., Galarza, F., Garcia-Appadoo, D., Guzman-Ramirez, L., Humphreys, E.M., Jung, T., Kameno, S., Laing, R.A., Leon, S., Marconi, G., Mignano, A., Nikolic, B., Nyman, L.A., Radiszcz, M., Remijan, A., Rodón, J.A., Sawada, T., Takahashi, S., Tilanus, R.P.J., Vila Vilaro, B., Watson, L.C., Wiklind, T., Akiyama, E., Chapillon, E., de Gregorio-Monsalvo, I., Di Francesco, J., Gueth, F., Kawamura, A., Lee, C.F., Nguyen Luong, Q., Mangum, J., Pietu, V., Sanhueza, P., Saigo, K., Takakuwa, S., Ubach, C., van Kempen, T., Wootten, A., Castro-Carrizo, A., Francke, H., Gallardo, J., Garcia, J., Gonzalez, S., Hill, T., Kaminski, T., Kurono, Y., Liu, H.Y., Lopez, C., Morales, F., Plarre, K., Schieven, G., Testi, L., Videla, L., Villard, E., Andreani, P., Hibbard, J.E., Tatematsu, K.: The 2014 ALMA long baseline campaign: first results from high angular resolution observations toward the HL Tau region. Astrophys. J. Lett. **808**, L3 (2015). doi:10.1088/2041-8205/808/1/L3, 1503.02649

Andrews, S.M., Williams, J.P.: A submillimeter view of circumstellar dust disks in ρ Ophiuchi. Astrophys. J. **671**, 1800–1812 (2007). doi:10.1086/522885, 0708.4185

Andrews, S.M., Wilner, D.J., Espaillat, C., Hughes, A.M., Dullemond, C.P., McClure, M.K., Qi, C., Brown, J.M.: Resolved images of large cavities in protoplanetary transition disks. Astrophys. J. **732**, 42 (2011). doi:10.1088/0004-637X/732/1/42, 1103.0284

Andrews, S.M., Wilner, D.J., Zhu, Z., Birnstiel, T., Carpenter, J.M., Pérez, L.M., Bai, X.N., Öberg, K.I., Hughes, A.M., Isella, A., Ricci, L.: Ringed substructure and a gap at 1 au in the nearest protoplanetary disk. Astrophys. J. **820**, L40 (2016). doi:10.3847/2041-8205/820/2/L40, 1603.09352

Ansdell, M., Williams, J.P., van der Marel, N., Carpenter, J.M., Guidi, G., Hogerheijde, M., Mathews, G.S., Manara, C.F., Miotello, A., Natta, A., Oliveira, I., Tazzari, M., Testi, L., van Dishoeck, E.F., van Terwisga, S.E.: ALMA survey of lupus protoplanetary disks. I. Dust and gas masses. Astrophys. J. **828**, 46 (2016). doi:10.3847/0004-637X/828/1/46, 1604.05719

Armitage, P.J.: Dynamics of protoplanetary disks. Annu. Rev. Astron. Astrophys. **49**, 195–236 (2011). doi:10.1146/annurev-astro-081710-102521, 1011.1496

Ataiee, S., Pinilla, P., Zsom, A., Dullemond, C.P., Dominik, C., Ghanbari, J.: Asymmetric transition disks: vorticity or eccentricity? Astron. Astrophys. **553**, L3 (2013). doi:10.1051/0004-6361/201321125, 1304.1736

Avenhaus, H., Quanz, S.P., Schmid, H.M., Meyer, M.R., Garufi, A., Wolf, S., Dominik, C.: Structures in the protoplanetary disk of HD142527 seen in polarized scattered light. Astrophys. J. **781**, 87 (2014). doi:10.1088/0004-637X/781/2/87, 1311.7088

Banzatti, A., Pontoppidan, K.M.: An empirical sequence of disk gap opening revealed by rovibrational CO. Astrophys. J. **809**, 167 (2015). doi:10.1088/0004-637X/809/2/167, 1507.06650

Barenfeld, S.A., Carpenter, J.M., Ricci, L., Isella, A.: ALMA observations of circumstellar disks in the upper Scorpius OB Association. Astrophys. J. **827**, 142 (2016). doi:10.3847/0004-637X/827/2/142, 1605.05772

Barge, P., Sommeria, J.: Did planet formation begin inside persistent gaseous vortices? Astron. Astrophys. **295**, L1–L4 (1995). astro-ph/9501050

Bast, J.E., Brown, J.M., Herczeg, G.J., van Dishoeck, E.F., Pontoppidan, K.M.: Single peaked CO emission line profiles from the inner regions of protoplanetary disks. Astron. Astrophys. **527**, A119 (2011). doi:10.1051/0004-6361/201015225, 1101.1091

Batalha, N.M.: Exploring exoplanet populations with NASA's Kepler mission. Proc. Natl. Acad. Sci. **111**, 12647–12654 (2014). doi:10.1073/pnas.1304196111, 1409.1904

Beckwith, S.V.W., Sargent, A.I.: Circumstellar disks and the search for neighbouring planetary systems. Nature **383**, 139–144 (1996). doi:10.1038/383139a0

Birnstiel, T., Dullemond, C.P., Brauer, F.: Gas- and dust evolution in protoplanetary disks. Astron. Astrophys. **513**, A79 (2010). doi:10.1051/0004-6361/200913731, 1002.0335

Birnstiel, T., Andrews, S.M., Ercolano, B.: Can grain growth explain transition disks? Astron. Astrophys. **544**, A79 (2012). doi:10.1051/0004-6361/201219262, 1206.5802

Birnstiel, T., Dullemond, C.P., Pinilla, P.: Lopsided dust rings in transition disks. Astron. Astrophys. **550**, L8 (2013). doi:10.1051/0004-6361/201220847, 1301.1976

Brauer, F., Dullemond, C.P., Henning, T.: Coagulation, fragmentation and radial motion of solid particles in protoplanetary disks. Astron. Astrophys. **480**, 859–877 (2008). doi:10.1051/0004-6361:20077759, 0711.2192

Brown, J.M., Blake, G.A., Qi, C., Dullemond, C.P., Wilner, D.J., Williams, J.P.: Evidence for dust clearing through resolved submillimeter imaging. Astrophys. J. **704**, 496–502 (2009). doi:10.1088/0004-637X/704/1/496

Brown, J.M., Herczeg, G.J., Pontoppidan, K.M., van Dishoeck, E.F.: A 30 AU radius CO gas hole in the disk around the Herbig Ae Star Oph IRS 48. Astrophys. J. **744**, 116 (2012). doi:10.1088/0004-637X/744/2/116, 1110.2095

Brown, J.M., Pontoppidan, K.M., van Dishoeck, E.F., Herczeg, G.J., Blake, G.A., Smette, A.: VLT-CRIRES survey of rovibrational CO emission from protoplanetary disks. Astrophys. J. 770:94 (2013). doi:10.1088/0004-637X/770/2/94, 1304.4961

Bruderer, S.: Survival of molecular gas in cavities of transition disks. I. CO. Astron. Astrophys. **559**, A46 (2013). doi:10.1051/0004-6361/201321171, 1308.2966

Bruderer, S., van Dishoeck, E.F., Doty, S.D., Herczeg, G.J.: The warm gas atmosphere of the HD 100546 disk seen by Herschel. Evidence of a gas-rich, carbon-poor atmosphere? Astron. Astrophys. **541**, A91 (2012). doi:10.1051/0004-6361/201118218, 1201.4860

Bruderer, S., van der Marel, N., van Dishoeck, E.F., van Kempen, T.A.: Gas structure inside dust cavities of transition disks: Ophiuchus IRS 48 observed by ALMA. Astron. Astrophys. **562**, A26 (2014). doi:10.1051/0004-6361/201322857, 1312.2756

Canovas, H., Caceres, C., Schreiber, M.R., Hardy, A., Cieza, L., Ménard, F., Hales, A.: A ring-like concentration of mm-sized particles in Sz 91. Mon. Not. R. Astron. Soc. **458**, L29–L33 (2016). doi:10.1093/mnrasl/slw006, 1601.06801

Canovas, H., Schreiber, M.R., Cáceres, C., et al.: Astrophys. J. **805**, 21 (2015)

Carmona, A., Pinte, C., Thi, W.F., Benisty, M., Ménard, F., Grady, C., Kamp, I., Woitke, P., Olofsson, J., Roberge, A., Brittain, S., Duchêne, G., Meeus, G., Martin-Zaïdi, C., Dent, B., Le Bouquin, J.B., Berger, J.P.: Constraining the structure of the transition disk HD 135344B (SAO 206462) by simultaneous modeling of multiwavelength gas and dust observations. Astron. Astrophys. **567**, A51 (2014). doi:10.1051/0004-6361/201322534, 1403.6193

Carmona, A., Thi, W.F., Kamp, I., et al.: Astron. Astrophys. **598**, A118 (2017)

Carpenter, J.M., Ricci, L., Isella, A.: An ALMA continuum survey of circumstellar disks in the upper Scorpius OB Association. Astrophys. J. **787**, 42 (2014). doi:10.1088/0004-637X/787/1/42, 1404.0387

Casassus, S., van der Plas, G., M, S.P., Dent, W.R.F., Fomalont, E., Hagelberg, J., Hales, A., Jordán, A., Mawet, D., Ménard, F., Wootten, A., Wilner, D., Hughes, A.M., Schreiber, M.R., Girard, J.H., Ercolano, B., Canovas, H., Román, P.E., Salinas, V.: Flows of gas through a protoplanetary gap. Nature **493**, 191–194 (2013). doi:10.1038/nature11769, 1305.6062

Casassus, S., Marino, S., Pérez, S., Roman, P., Dunhill, A., Armitage, P.J., Cuadra, J., Wootten, A., van der Plas, G., Cieza, L., Moral, V., Christiaens, V., Montesinos, M.: Accretion kinematics through the warped transition disk in HD142527 from resolved CO(6-5) observations. Astrophys. J. **811**, 92 (2015a). doi:10.1088/0004-637X/811/2/92, 1505.07732

Casassus, S., Wright, C.M., Marino, S., Maddison, S.T., Wootten, A., Roman, P., Pérez, S., Pinilla, P., Wyatt, M., Moral, V., Ménard, F., Christiaens, V., Cieza, L., van der Plas, G.: (2015b) A compact concentration of large grains in the HD 142527 protoplanetary dust trap. Astrophys. J. **812**, 126. doi:10.1088/0004-637X/812/2/126, 1505.07743

Cieza, L.A., Schreiber, M.R., Romero, G.A., Mora, M.D., Merin, B., Swift, J.J., Orellana, M., Williams, J.P., Harvey, P.M., Evans, N.J. II The nature of transition circumstellar disks. I. The Ophiuchus molecular cloud. Astrophys. J. **712**, 925-941 (2010). doi:10.1088/0004-637X/712/2/925, 1001.4825

Cieza, L.A., Schreiber, M.R., Romero, G.A., Williams, J.P., Rebassa-Mansergas, A., Merín, B.: The nature of transition circumstellar disks. III. Perseus, Taurus, and Auriga. Astrophys. J. **750**, 157 (2012). doi:10.1088/0004-637X/750/2/157, 1203.6849

Currie, T., Cloutier, R., Brittain, S., Grady, C., Burrows, A., Muto, T., Kenyon, S.J., Kuchner, M.J.: Resolving the HD 100546 protoplanetary system with the Gemini Planet Imager: evidence for multiple forming, accreting planets. Astrophys. J. Lett. **814**, L27 (2015). doi:10.1088/2041-8205/814/2/L27, 1511.02526

Currie, T., Sicilia-Aguilar, A.: Astrophys. J. **732**, 24 (2011)

de Boer, J., Salter, G., Benisty, M., Vigan, A., Boccaletti, A., Pinilla, P., Ginski, C., Juhasz, A., Maire, A.L., Messina, S., Desidera, S., Cheetham, A., Girard, J.H., Wahhaj, Z., Langlois, M., Bonnefoy, M., Beuzit, J.L., Buenzli, E., Chauvin, G., Dominik, C., Feldt, M., Gratton, R., Hagelberg, J., Isella, A., Janson, M., Keller, C.U., Lagrange, A.M., Lannier, J., Menard, F., Mesa, D., Mouillet, D., Mugrauer, M., Peretti, S., Perrot, C., Sissa, E., Snik, F., Vogt, N., Zurlo, A., SPHERE Consortium: Multiple rings in the transition disk and companion candidates around RX J1615.3-3255. High contrast imaging with VLT/SPHERE. Astron. Astrophys. **595**, A114 (2016). doi:10.1051/0004-6361/201629267, 1610.04038

de Juan Ovelar, M., Min, M., Dominik, C., Thalmann, C., Pinilla, P., Benisty, M., Birnstiel, T.: Imaging diagnostics for transitional discs. Astron. Astrophys. **560**, A111 (2013). doi:10.1051/0004-6361/201322218, 1309.1039 de Juan Ovelar, M., Pinilla, P., Min, M., Dominik, C., Birnstiel, T.: Constraining turbulence mixing strength in transitional discs with planets using SPHERE and ALMA. Mon. Not. R. Astron. Soc. **459**, L85–L89 (2016). doi:10.1093/mnrasl/slw051, 1603.09357

de Val-Borro, M., Artymowicz, P., D'Angelo, G., Peplinski, A.: Vortex generation in protoplanetary disks with an embedded giant planet. Astron. Astrophys. **471**, 1043–1055 (2007). doi:10.1051/0004-6361:20077169, 0706.3200

Dong, R., Rafikov, R., Zhu, Z., Hartmann, L., Whitney, B., Brandt, T., Muto, T., Hashimoto, J., Grady, C., Follette, K., Kuzuhara, M., Tanii, R., Itoh, Y., Thalmann, C., Wisniewski, J., Mayama, S., Janson, M., Abe, L., Brandner, W., Carson, J., Egner, S., Feldt, M., Goto, M., Guyon, O., Hayano, Y., Hayashi, M., Hayashi, S., Henning, T., Hodapp, K.W., Honda, M., Inutsuka, S., Ishii, M., Iye, M., Kandori, R., Knapp, G.R., Kudo, T., Kusakabe, N., Matsuo, T., McElwain, M.W., Miyama, S., Morino, J.I., Moro-Martin, A., Nishimura, T., Pyo, T.S., Suto, H., Suzuki, R., Takami, M., Takato, N., Terada, H., Tomono, D., Turner, E.L., Watanabe, M., Yamada, T., Takami, H., Usuda, T., Tamura, M.: The missing cavities in the SEEDS polarized scattered light images of transitional protoplanetary disks: a generic disk model. Astrophys. J. **750**, 161 (2012). doi:10.1088/0004-637X/750/2/161, 1203.1612

Dong, R., van der Marel, N., Hashimoto, J., Chiang, E., Akiyama, E., Liu, H.B., Muto, T., Knapp, G.R., Tsukagoshi, T., Brown, J., Bruderer, S., Koyamatsu, S., Kudo, T., Ohashi, N., Rich, E., Satoshi, M., Takami, M., Wisniewski, J., Yang, Y., Zhu, Z., Tamura, M.: The sizes and depletions of the dust and gas cavities in the transitional disk J160421.7-213028. Astrophys. J. **836**, 201 (2017). doi:10.3847/1538-4357/aa5abf, 1701.05189

Draine, B.T.: On the submillimeter opacity of protoplanetary disks. Astrophys. J. **636**, 1114–1120 (2006). doi:10.1086/498130, astro-ph/0507292

Espaillat, C., Ingleby, L., Hernández, J., Furlan, E., D'Alessio, P., Calvet, N., Andrews, S., Muzerolle, J., Qi, C., Wilner, D.: On the transitional disk class: linking observations of T Tauri stars and physical disk models. Astrophys. J. **747**, 103 (2012). doi:10.1088/0004-637X/747/2/103, 1201.1518

Espaillat, C., Muzerolle, J., Najita, J., et al.: Protostars and Planets VI, p. 497 (2014)

Fukagawa, M., Tsukagoshi, T., Momose, M., Saigo, K., Ohashi, N., Kitamura, Y., Inutsuka, Si., Muto, T., Nomura, H., Takeuchi, T., Kobayashi, H., Hanawa, T., Akiyama, E., Honda, M., Fujiwara, H., Kataoka, A., Takahashi, S.Z., Shibai, H.: Local enhancement of the surface density in the protoplanetary ring surrounding HD 142527. Publ. Astron. Soc. Jpn. 65, L14 (2013). doi:10.1093/pasj/65.6.L14, 1309.7400

Fung, J., Shi, J.M., Chiang, E.: How empty are disk gaps opened by giant planets? Astrophys. J. **782**, 88 (2014). doi:10.1088/0004-637X/782/2/88, 1310.0156

Garufi, A., Quanz, S.P., Avenhaus, H., Buenzli, E., Dominik, C., Meru, F., Meyer, M.R., Pinilla, P., Schmid, H.M., Wolf, S.: Small vs. large dust grains in transitional disks: do different cavity sizes indicate a planet? SAO 206462 (HD 135344B) in polarized light with VLT/NACO. Astron. Astrophys. **560**, A105 (2013). doi:10.1051/0004-6361/201322429, 1311.4195

Geers, V.C., Pontoppidan, K.M., van Dishoeck, E.F., Dullemond, C.P., Augereau, J.C., Merín, B., Oliveira, I., Pel, J.W.: Spatial separation of small and large grains in the transitional disk around the young star <ASTROBJ>IRS 48</ASTROBJ>. Astron. Astrophys. **469**, L35–L38 (2007). doi:10.1051/0004-6361:20077524, 0705.2969

Ginski, C., Stolker, T., Pinilla, P., Dominik, C., Boccaletti, A., de Boer, J., Benisty, M., Biller, B., Feldt, M., Garufi, A., Keller, C.U., Kenworthy, M., Maire, A.L., Ménard, F., Mesa, D., Milli, J., Min, M., Pinte, C., Quanz, S.P., van Boekel, R., Bonnefoy, M., Chauvin, G., Desidera, S., Gratton, R., Girard, J.H.V., Keppler, M., Kopytova, T., Lagrange, A.M., Langlois, M., Rouan, D., Vigan, A.: Direct detection of scattered light gaps in the transitional disk around HD 97048 with VLT/SPHERE. Astron. Astrophys. **595**, A112 (2016). doi:10.1051/0004-6361/201629265, 1609.04027

Huélamo, N., Lacour, S., Tuthill, P., Ireland, M., Kraus, A., Chauvin, G.: A companion candidate in the gap of the T Chamaeleontis transitional disk. Astron. Astrophys. **528**, L7 (2011). doi:10.1051/0004-6361/201016395, 1102.4982

Isella, A., Carpenter, J.M., Sargent, A.I.: Investigating planet formation in circumstellar disks: CARMA observations of Ry Tau and Dg Tau. Astrophys. J. **714**, 1746–1761 (2010a). doi:10.1088/0004-637X/714/2/1746, 1003.4318

Isella, A., Guidi, G., Testi, L., et al.: Phys. Rev. Lett. **117**, 251101 (2016)

Isella, A., Natta, A., Wilner, D., Carpenter, J.M., Testi, L.: Millimeter imaging of MWC 758: probing the disk structure and kinematics. Astrophys. J. **725**, 1735–1741 (2010b). doi:10.1088/0004-637X/725/2/1735, 1010.3016

Jonkheid, B., Kamp, I., Augereau, J.C., van Dishoeck, E.F.: Modeling the gas-phase chemistry of the transitional disk around HD 141569A. Astron. Astrophys. **453**, 163–171 (2006). doi:10.1051/0004-6361:20054769, astro-ph/0603515

Klahr, H.H., Henning, T.: Particle-trapping eddies in protoplanetary accretion disks. Icarus **128**, 213–229 (1997). doi:10.1006/icar.1997.5720

Koerner, D.W., Sargent, A.I., Beckwith, S.V.W.: A rotating gaseous disk around the T Tauri star GM Aurigae. Icarus **106**, 2 (1993). doi:10.1006/icar.1993.1154

Kraus, A.L., Ireland, M.J.: Astrophys. J. **745**, 5 (2012)

Lahuis, F., van Dishoeck, E.F., Blake, G.A., Evans, N.J. II, Kessler-Silacci, J.E., Pontoppidan, K.M.: c2d Spitzer IRS spectra of disks around T Tauri stars. III. [Ne II], [Fe I], and H_2 gas-phase lines. Astrophys. J. **665**, 492–511 (2007). doi:10.1086/518931, 0704.2305

Lyra, W., Johansen, A., Klahr, H., Piskunov, N.: Standing on the shoulders of giants. Trojan earths and vortex trapping in low mass self-gravitating protoplanetary disks of gas and solids. Astron. Astrophys. **493**, 1125–1139 (2009). doi:10.1051/0004-6361:200810797, 0810.3192

Lyra, W., Lin, M.-K.: Astrophys. J. **775**, 17 (2013)

Marino, S., Perez, S., Casassus, S.: Shadows cast by a warp in the HD 142527 protoplanetary disk. Astrophys. J. Lett. **798**, L44 (2015) doi:10.1088/2041-8205/798/2/L44, 1412.4632

Najita, J.R., Strom, S.E., Muzerolle, J.: Demographics of transition objects. Mon. Not. R. Astron. Soc. **378**, 369–378 (2007). doi:10.1111/j.1365-2966.2007.11793.x, 0704.1681

Olofsson, J., Szűcs, L., Henning, T., Linz, H., Pascucci, I., Joergens, V.: The Herschel/PACS view of disks around low-mass stars in Chamaleon-I. Astron. Astrophys. **560**, A100 (2013). doi:10.1051/0004-6361/201321967, 1310.0834

Owen, J.E.: The origin and evolution of transition discs: successes, problems, and open questions. Publ. Astron. Soc. Aust. **33**, e005 (2016). doi:10.1017/pasa.2016.2, 1512.06873

Owen, J.E., Clarke, C.J.: Two populations of transition discs? Mon. Not. R. Astron. Soc. **426**, L96–L100 (2012). doi:10.1111/j.1745-3933.2012.01334.x, 1205.5564

Pascucci, I., Hollenbach, D., Najita, J., Muzerolle, J., Gorti, U., Herczeg, G.J., Hillenbrand, L.A., Kim, J.S., Carpenter, J.M., Meyer, M.R., Mamajek, E.E., Bouwman, J.: Detection of [Ne II] emission from Young circumstellar disks. Astrophys. J. **663**, 383–393 (2007). doi:10.1086/518535, astro-ph/0703616

Pascucci, I., Testi, L., Herczeg, G.J., Long, F., Manara, C.F., Hendler, N., Mulders, G.D., Krijt, S., Ciesla, F., Henning, T., Mohanty, S., Drabek-Maunder, E., Apai, D., Szűcs, L., Sacco, G., Olofsson, J.: A steeper than linear disk mass-stellar mass scaling relation. Astrophys. J. **831**, 125 (2016). doi:10.3847/0004-637X/831/2/125, 1608.03621

Pérez, L.M., Isella, A., Carpenter, J.M., Chandler, C.J.: Large-scale asymmetries in the transitional disks of SAO 206462 and SR 21. Astrophys. J. Lett. **783**, L13 (2014). doi:10.1088/2041-8205/783/1/L13, 1402.0832

Perez, S., Casassus, S., Ménard, F., Roman, P., van der Plas, G., Cieza, L., Pinte, C., Christiaens, V., Hales, A.S.: CO gas inside the protoplanetary disk cavity in HD 142527: disk structure from ALMA. Astrophys. J. **798**, 85 (2015). doi:10.1088/0004-637X/798/2/85, 1410.8168

Pinilla, P., Benisty, M., Birnstiel, T.: Ring shaped dust accumulation in transition disks. Astron. Astrophys. **545**, A81 (2012). doi:10.1051/0004-6361/201219315, 1207.6485

Pinilla, P., de Boer, J., Benisty, M., Juhász, A., de Juan Ovelar, M., Dominik, C., Avenhaus, H., Birnstiel, T., Girard, J.H., Huelamo, N., Isella, A., Milli, J.: Variability and dust filtration in the transition disk J160421.7-213028 observed in optical scattered light. Astron. Astrophys. **584**, L4 (2015a). doi:10.1051/0004-6361/201526981, 1510.00412

Pinilla, P., van der Marel, N., Pérez, L.M., van Dishoeck, E.F., Andrews, S., Birnstiel, T., Herczeg, G., Pontoppidan, K.M., van Kempen, T.: Testing particle trapping in transition disks with ALMA. Astron. Astrophys. **584**, A16 (2015b). doi:10.1051/0004-6361/201526655, 1509.03040

Pinilla, P., Klarmann, L., Birnstiel, T., Benisty, M., Dominik, C., Dullemond, C.P.: A tunnel and a traffic jam: How transition disks maintain a detectable warm dust component despite the presence of a large planet-carved gap. Astron. Astrophys. **585**, A35 (2016). doi:10.1051/0004-6361/201527131, 1511.04105

Pinilla, P., Pérez, L.M., Andrews, S., et al.: Astrophys. J. **839**, 99 (2017)

Pontoppidan, K.M., Blake, G.A., van Dishoeck, E.F., Smette, A., Ireland, M.J., Brown, J.: Spectroastrometric imaging of molecular gas within protoplanetary disk gaps. Astrophys. J. **684**, 1323–1329 (2008). doi:10.1086/590400, 0805.3314

Regály, Z., Juhász, A., Sándor, Z., Dullemond, C.P.: Possible planet-forming regions on submillimetre images. Mon. Not. R. Astron. Soc. 419:1701–1712 (2012). doi:10.1111/j.1365-2966.2011.19834.x, 1109.6177

Ricci, L., Testi, L., Natta, A., Neri, R., Cabrit, S., Herczeg, G.J.: Dust properties of protoplanetary disks in the Taurus-Auriga star forming region from millimeter wavelengths. Astron. Astrophys. **512**, A15 (2010). doi:10.1051/0004-6361/200913403, 0912.3356

Rice, W.K.M., Armitage, P.J., Wood, K., Lodato, G.: Dust filtration at gap edges: implications for the spectral energy distributions of discs with embedded planets. Mon. Not. R. Astron. Soc. **373**, 1619–1626 (2006). doi:10.1111/j.1365-2966.2006.11113.x, astro-ph/0609808

Romero, G.A., Schreiber, M.R., Cieza, L.A., Rebassa-Mansergas, A., Merín, B., Smith Castelli, A.V., Allen, L.E., Morrell, N.: The nature of transition circumstellar disks. II. Southern molecular clouds. Astrophys. J. **749**, 79 (2012). doi:10.1088/0004-637X/749/1/79, 1203.6816

Rosenfeld, K.A., Chiang, E., Andrews, S.M.: Fast radial flows in transition disk holes. Astrophys. J. **782**, 62 (2014). doi:10.1088/0004-637X/782/2/62, 1312.3817

Sallum, S., Follette, K.B., Eisner, J.A., Close, L.M., Hinz, P., Kratter, K., Males, J., Skemer, A., Macintosh, B., Tuthill, P., Bailey, V., Defrère, D., Morzinski, K., Rodigas, T., Spalding, E., Vaz, A., Weinberger, A.J.: Accreting protoplanets in the LkCa 15 transition disk. Nature **527**, 342–344 (2015). doi:10.1038/nature15761, 1511.07456

Sargent, A.I., Beckwith, S.: Kinematics of the circumstellar gas of HL Tauri and R Monocerotis. Astrophys. J. **323**, 294–305 (1987). doi:10.1086/165827

Stolker, T., Dominik, C., Avenhaus, H., et al.: Astron. Astrophys. **595**, A113 (2016)

Strom, K.M., Strom, S.E., Edwards, S., Cabrit, S., Skrutskie, M.F.: Circumstellar material associated with solar-type pre-main-sequence stars - a possible constraint on the timescale for planet building. Astron. J. **97**, 1451–1470 (1989). doi:10.1086/115085

Testi, L., Birnstiel, T., Ricci, L., et al.: Protostars and Planets VI, p. 339 (2014)

Thalmann, C., Janson, M., Garufi, A., Boccaletti, A., Quanz, S.P., Sissa, E., Gratton, R., Salter, G., Benisty, M., Bonnefoy, M., Chauvin, G., Daemgen, S., Desidera, S., Dominik, C., Engler, N., Feldt, M., Henning, T., Lagrange, A.M., Langlois, M., Lannier, J., Le Coroller, H., Ligi, R., Ménard, F., Mesa, D., Meyer, M.R., Mulders, G.D., Olofsson, J., Pinte, C., Schmid, H.M., Vigan, A., Zurlo, A.: Resolving the planet-hosting inner regions of the LkCa 15 disk. Astrophys. J. Lett. **828**, L17 (2016). doi:10.3847/2041-8205/828/2/L17, 1608.08642

Tsukagoshi, T., Nomura, H., Muto, T., Kawabe, R., Ishimoto, D., Kanagawa, K.D., Okuzumi, S., Ida, S., Walsh, C., Millar, T.J.: A gap with a deficit of large grains in the protoplanetary disk around TW Hya. Astrophys. J. Lett. **829**, L35 (2016). doi:10.3847/2041-8205/829/2/L35, 1605.00289

van der Marel, N.: Mind the gap: gas and dust in planet-forming disks. PhD thesis, Leiden University (2015)

van der Marel, N., van Dishoeck, E.F., Bruderer, S., Birnstiel, T., Pinilla, P., Dullemond, C.P., van Kempen, T.A., Schmalzl, M., Brown, J.M., Herczeg, G.J., Mathews, G.S., Geers, V.: A major asymmetric dust trap in a transition disk. Science **340**, 1199–1202 (2013). doi:10.1126/science.1236770, 1306.1768 van der Marel, N., van Dishoeck, E.F., Bruderer, S., van Kempen, T.A.: Warm formaldehyde in the Ophiuchus IRS 48 transitional disk. Astron. Astrophys. **563**, A113 (2014). doi:10.1051/0004-6361/201322960, 1402.0392

van der Marel, N., Pinilla, P., Tobin, J., van Kempen, T., Andrews, S., Ricci, L., Birnstiel, T.: A concentration of centimeter-sized grains in the Ophiuchus IRS 48 dust trap. Astrophys. J. Lett. **810**, L7 (2015a). doi:10.1088/2041-8205/810/1/L7, 1508.01003

van der Marel, N., van Dishoeck, E.F., Bruderer, S., Pérez, L., Isella, A.: Gas density drops inside dust cavities of transitional disks around young stars observed with ALMA. Astron. Astrophys. **579**, A106 (2015b). doi:10.1051/0004-6361/201525658, 1504.03927

van der Marel, N., Cazzoletti, P., Pinilla, P., Garufi, A.: Vortices and spirals in the HD135344B transition disk. Astrophys. J. **832**, 178 (2016a). doi:10.3847/0004-637X/832/2/178, 1607.05775

van der Marel, N., van Dishoeck, E.F., Bruderer, S., Andrews, S.M., Pontoppidan, K.M., Herczeg, G.J., van Kempen, T., Miotello, A.: Resolved gas cavities in transitional disks inferred from CO isotopologs with ALMA. Astron. Astrophys. **585**, A58 (2016b). doi:10.1051/0004-6361/201526988, 1511.07149

van der Marel, N., Verhaar, B.W., van Terwisga, S., Merín, B., Herczeg, G., Ligterink, N.F.W., van Dishoeck, E.F.: The (w)hole survey: an unbiased sample study of transition disk candidates based on Spitzer catalogs. Astron. Astrophys. **592**, A126 (2016c). doi:10.1051/0004-6361/201628075, 1603.07255

van der Plas, G., Wright, C.M., Ménard, F., et al.: Astron. Astrophys. **597**, A32 (2017)

van Zadelhoff, G.J., van Dishoeck, E.F., Thi, W.F., Blake, G.A.: Submillimeter lines from circumstellar disks around pre-main sequence stars. Astron. Astrophys. **377**, 566–580 (2001). doi:10.1051/0004-6361:20011137, astro-ph/0108375

Varnière, P., Tagger, M.: Reviving dead zones in accretion disks by Rossby vortices at their boundaries. Astron. Astrophys. **446**, L13–L16 (2006). doi:10.1051/0004-6361:200500226, astro-ph/0511684

Walsh, C., Juhász, A., Pinilla, P., Harsono, D., Mathews, G.S., Dent, W.R.F., Hogerheijde, M.R., Birnstiel, T., Meeus, G., Nomura, H., Aikawa, Y., Millar, T.J., Sandell, G.: ALMA hints at the presence of two companions in the disk around HD 100546. Astrophys. J. Lett. **791**, L6 (2014). doi:10.1088/2041-8205/791/1/L6, 1405.6542

Williams, J.P., Cieza, L.A.: Protoplanetary disks and their evolution. Annu. Rev. Astron. Astrophys. **49**, 67–117 (2011). doi:10.1146/annurev-astro-081710-102548, 1103.0556

Woitke, P., Kamp, I., Thi, W.F.: Radiation thermo-chemical models of protoplanetary disks. I. Hydrostatic disk structure and inner rim. Astron. Astrophys. **501**, 383–406 (2009). doi:10.1051/0004-6361/200911821, 0904.0334

Woitke, P., Min, M., Pinte, C., Thi, W.F., Kamp, I., Rab, C., Anthonioz, F., Antonellini, S., Baldovin-Saavedra, C., Carmona, A., Dominik, C., Dionatos, O., Greaves, J., Güdel, M., Ilee, J.D., Liebhart, A., Ménard, F., Rigon, L., Waters, L.B.F.M., Aresu, G., Meijerink, R., Spaans, M.: Consistent dust and gas models for protoplanetary disks. I. Disk shape, dust settling, opacities, and PAHs. Astron. Astrophys. **586**, A103 (2016). doi:10.1051/0004-6361/201526538, 1511.03431

Zhang, K., Blake, G.A., Bergin, E.A.: Evidence of fast pebble growth near condensation fronts in the HL Tau protoplanetary disk. Astrophys. J. Lett. **806**, L7 (2015). doi:10.1088/2041-8205/806/1/L7, 1505.00882

Zhang, K., Isella, A., Carpenter, J.M., Blake, G.A., et al.: Astrophys. J. **791**, 42 (2014)

Zhu, Z., Nelson, R.P., Hartmann, L., Espaillat, C., Calvet, N.: Transitional and pre-transitional disks: gap opening by multiple planets? Astrophys. J. **729**, 47 (2011). doi:10.1088/0004-637X/729/1/47, 1012.4395

[illegible]

Varnière, P., Tagger, M.: Reviving dead zones in accretion disks by Rossby vortices at their boundaries. Astron. Astrophys. 446, L13–L16 (2006). doi:10.1051/0004-6361:200600226, astro-ph/0511064

Walsh, C., Juhász, A., Pinilla, P., Harsono, D., Mathews, G.S., Dent, W.R.F., Hogerheijde, M.R., Birnstiel, T., Meeus, G., Nomura, H., Aikawa, Y., Millar, T.J., Sandell, G.: ALMA hints at the presence of two companions in the disk around HD 100546. Astrophys. J. Lett. 791, L6 (2014) [illegible]

[illegible]

Woitke, P., Min, M., Pinte, C., Thi, W.-F., Kamp, I., Rab, C., Anthonioz, F., Antonellini, S., Baldovin-Saavedra, C., Carmona, A., Dominik, C., Dionatos, O., Greaves, J., Güdel, M., Ilee, J.D., Liebhart, A., Ménard, F., Rigon, L., Waters, L.B.F.M., Aresu, G., Meijerink, R., Spaans, M.: Consistent dust and gas models for protoplanetary disks. I. Disk shape, dust settling, opacities, and PAHs. Astron. Astrophys. 586, A103 (2016). doi:10.1051/0004-6361/201526538, 1511.03431

Zhang, K., Blake, G.A., Bergin, E.A.: Evidence of fast pebble growth near condensation fronts in the HL Tau protoplanetary disk. Astrophys. J. Lett. 806, L7 (2015). doi:10.1088/2041-8205/806/1/L7, 1505.00882

Zhang, K., Isella, A., Carpenter, J.M., Blake, G.A., et al.: Astrophys. J. 791, 42 (2014)

Zhu, Z., Nelson, R.P., Hartmann, L., Espaillat, C., Calvet, N.: Transitional and pre-transitional disks: gap opening by multiple planets? Astrophys. J. 729, 47 (2011). doi:10.1088/0004-637X/729/1/47, 1012.4395

Chapter 3
Wind-Driven Global Evolution of Protoplanetary Disks

Xue-Ning Bai

Abstract It has been realized in the recent years that magnetized disk winds likely play a decisive role in the global evolution of protoplanetary disks (PPDs). Motivated by recent local simulations, we first describe a global magnetized disk wind model, from which wind-driven accretion rate and wind mass loss rate can be reliably estimated. Both rates are shown to strongly depend on the amount of magnetic flux threading the disk. Wind kinematics is also affected by thermodynamics in the wind zone (particularly far UV heating/ionization), and the mass loss process can be better termed as "magneto-photoevaporation." We then construct a framework of PPD global evolution that incorporates wind-driven and viscously driven accretion as well as wind mass loss. For typical PPD accretion rates, the required field strength would lead to wind mass loss rate at least comparable to disk accretion rate, and mass loss is most significant in the outer disk (beyond $\sim$10 AU). Finally, we discuss the transport of magnetic flux in PPDs, which largely governs the long-term evolution of PPDs.

3.1 Introduction

Global evolution of protoplanetary disks (PPDs) is governed by the processes of angular momentum transport and outflow mass loss. These processes directly control disk structure and evolution, which set the timescales of disk dispersal and hence planet formation (e.g., see Armitage 2011; Turner et al. 2014 and Alexander et al. 2014 for reviews). They also strongly affect the evolution of dust grains, which are building blocks of planets, and feedback to disk thermal and chemical structures (e.g., see Testi et al. 2014 and Henning and Semenov 2013 for reviews). For planets

X.-N. Bai (✉)
Institute for Advanced Study, Tsinghua University, Beijing 100084, China

Tsinghua Center for Astrophysics, Tsinghua University, Beijing 100084, China

Institute for Theory and Computation, Harvard-Smithsonian Center for Astrophysics, 60 Garden St., Cambridge, MA 02138, USA
e-mail: xueningbai@gmail.com

M. Pessah, O. Gressel (eds.), *Formation, Evolution, and Dynamics of Young Solar Systems*, Astrophysics and Space Science Library 445,
DOI 10.1007/978-3-319-60609-5_3

formed within disk lifetime, planet–disk interaction leads to planet migration, which is also sensitive to global disk structure and evolution (e.g., see Baruteau et al. 2014 for a review). In brief, a reliable global evolutionary picture of PPD is key to understanding most processes of planet formation.

Angular momentum transport in most existing models of global PPD evolution is based on the viscous α-disk model (Shakura and Sunyaev 1973), where the underlying assumption is that the disk is turbulent presumably due to the magneto-rotational instability (MRI, Balbus and Hawley 1991). However, the extremely weak level of ionization introduces strong non-ideal magnetohydrodynamic (MHD) effects that suppress or damp the MRI in most regions in PPDs, while pure hydrodynamic mechanisms appear unable to provide sufficiently level of turbulent viscosity (e.g., Turner et al. 2014). Angular momentum transport in PPDs is thus most likely governed by magnetized disk winds, as demonstrated in disk simulations that properly taking into account these non-ideal MHD effects (e.g., Bai and Stone 2013b; Bai 2013, 2014, 2015; Lesur et al. 2014; Gressel et al. 2015; Simon et al. 2015). In the meantime, the MRI may operate in the surface layer of the outer disk, owing to strong far-UV (FUV) ionization at disk surface (Perez-Becker and Chiang 2011), which can lead to vigorous turbulence and mediate a certain level of viscous transport of angular momentum (Simon et al. 2013b,a; Bai 2015).

The theory of magnetized disk winds has been developed for decades, following the seminal work of Blandford and Payne (1982). The wind exerts a torque on the disk surface that efficiently extracts angular momentum and drives accretion. The analytical theory was subsequently generalized and refined to better account for disk structure and boundary conditions, and has been applied to jets and outflows from general accretion disks and young stellar objects (e.g., Pudritz and Norman 1983; Konigl 1989; Lovelace et al. 1991; Pelletier and Pudritz 1992; Ferreira and Pelletier 1995; Li 1995; Ostriker 1997; Vlahakis et al. 2000). Global MHD simulations, less restricted by simplifying assumptions such as self-similarity and time-independence, have been widely adopted for studying global wind structure and evolution, as well as the wind launching process and disk-wind connection (e.g., Ouyed et al. 1997; Krasnopolsky et al. 1999; Kato et al. 2002; Casse and Keppens 2002; Pudritz et al. 2006; Zanni et al. 2007; Tzeferacos et al. 2009; Porth and Fendt 2010; Ramsey and Clarke 2011). It is now well known that the wind properties depend mainly on the strength and distribution of the magnetic flux threading the disk, and on the mass loading, with the latter mainly controlled by disk physics and thermodynamics.

However, none of the aforementioned global wind studies are directly applicable to PPDs. First, they generally fail to take into account realistic disk microphysics, particularly the non-ideal MHD effects, whose diffusivities are determined by ionization-recombination chemistry. Second, poloidal field strength considered in these works is typically close to equipartition with disk midplane gas pressure, which is orders of magnitude too strong and would deplete the disk in $\sim 10^3$ years instead of typical disk lifetime of a few Myrs (Haisch et al. 2001). Thirdly, most previous works considered the wind to be "cold" (with notable exceptions by Casse and Ferreira 2000 and Tzeferacos et al. 2013), whereas the wind zone in PPDs is strongly heated by high-energy stellar radiation, including FUV.[1] Therefore, it is highly desirable to develop realistic MHD wind models for PPDs.

Mass loss and dispersal of PPDs have conventionally been attributed to "photoevaporation," which is a pure thermal (hydrodynamic) wind launched from the disk surface heated by external X-ray and UV radiation to high temperatures (Alexander et al. 2014). However, this is incompatible with the realization that disk angular momentum has to rely on MHD disk wind. So far, photoevaporation models have been developed in parallel with MHD disk wind models without much overlap. In reality, the nature of PPD wind is likely a marriage of both magnetic and thermal effects: PPD wind is both hot and magnetically dominated.

In sum, the paradigm shift towards MHD wind-driven PPD evolution calls for a model framework in replacement of the conventional α-disk models. The new framework must incorporate prescriptions of wind-driven accretion and mass loss, that properly reflect disk physics and thermodynamics. In this chapter, we present a first effort for developing such a framework. In this chapter, we review and synthesize the key results from the very recent works of Bai et al. (2016), Bai (2016), and Bai and Stone (2017), which represent a first effort for developing such a framework.

In Sect. 3.2, we describe the basic ingredients of our wind-driven PPD evolution model. In Sect. 3.3, we construct a physically motivated model of PPD winds and discuss wind properties within the relevant parameter range for PPDs. Applying this model, we describe the new framework of PPD evolution in Sect. 3.4, and show that disk evolution is largely controlled by the amount of poloidal magnetic flux threading the disk and how this flux evolves with time. We conclude and outline future directions in Sect. 3.5, highlighting the importance of better understanding magnetic flux transport in PPDs.

3.2 Basic Model Ingredients

We construct a 1 D disk model on the evolution of the disk surface density $\Sigma(R)$ as a function of disk cylindrical radius R in the presence of turbulence, wind torque, and mass loss. In the meantime, we adopt a simple treatment of disk vertical structure to estimate the vertical location z_a where external far-UV (FUV) radiation can penetrate, and z_b ($\geq z_a$) where the wind is launched (wind base), based on which we calculate the transport properties.

[1]While the FUV luminosity from young stars is only a very small fraction ($\sim 10^{-3}$) of the stellar luminosity, it is primarily absorbed in the disk surface and the energy deposited plays a dominant role in heating and ionizing the surface layer (e.g., Bergin et al. 2007).

3.2.1 Disk Evolutionary Equations

We begin by writing down the equations governing the surface density evolution. Let $\dot{M}_{\rm acc}(R)$ be the net accretion rate at cylindrical radius R, where positive values correspond to net accretion. Let $\dot{M}_{\rm loss}(R)$ be the cumulative wind mass loss rate enclosed within radius R. The equation for surface density evolution reads

$$2\pi R\frac{\partial\Sigma}{\partial t}=\frac{\partial\dot{M}_{\rm acc}}{\partial R}-\frac{\partial\dot{M}_{\rm loss}}{\partial R}=\frac{\partial}{\partial R}\left[\frac{dR}{dj_K}\left(\frac{\partial W_r}{\partial R}+\frac{\partial W_w}{\partial R}\right)\right]-\frac{\partial\dot{M}_{\rm loss}}{\partial R}\ , \tag{3.1}$$

where $j_K(R)$ is Keplerian specific angular momentum (and v_K, Ω_K are the corresponding Keplerian velocity and angular frequency),

$$\begin{aligned}W_r(R)=&2\pi R^2\int_{-z_b}^{z_b}dz\left(\overline{\rho\delta v_R\delta v_\phi}-\frac{\overline{B_RB_\phi}}{4\pi}\right)\\ \equiv&2\pi R^2\int_{-z_b}^{z_b}dz\alpha(z)\rho c_{s,d}^2\equiv 2\pi R^2\tilde{\alpha}\Sigma c_{s,d}^2\end{aligned} \tag{3.2}$$

accounts for angular momentum flux generated from internal stresses, including Reynolds stress ($\rho\delta v_R\delta v_\phi$), Maxwell stress ($-B_RB_\phi/4\pi$), and potentially stress from disk self-gravity (not included), with overline representing averaging over the local volume. Here, we follow the α convention, with $c_{s,d}$ being disk sound speed. Note that α can be position dependent, and we use $\tilde{\alpha}$ to represent an effective, (pressure-weighted) vertically averaged value, with the integral carried in between the upper and lower side of the wind base z_b. The differential torque from the MHD wind at radius R is given by

$$\frac{\partial W_w(R)}{\partial R}=2\pi R^2\left(-\frac{\overline{B_zB_\phi}}{4\pi}\right)\Bigg|_{-z_b}^{z_b}=(\lambda-1)\frac{\partial\dot{M}_{\rm loss}}{\partial R}j_K\ , \tag{3.3}$$

where the torque is exerted at the wind base z_b, and the torque can be conveniently connected to the mass loss rate via the wind lever arm λ (dimensionless). For wind launched from radius R_0, the physical meaning of $\lambda(R_0)$ is that it is the ratio of specific angular momenta in the wind flow to that in the Keplerian disk at $R=R_0$, and it is related to wind Alfvén radius R_A by $\lambda\equiv(R_A/R_0)^2>1$. The $\partial W_w/\partial R$ term represents the *excess* angular momentum extracted from the disk, related to the wind torque. In the above, we have dropped the hydrodynamic term in the definition of W_w, which corresponds to the standard definition of the wind base z_b (Wardle and Koenigl 1993): at the wind base, $v_\phi(z_b)=v_K$.

Combining the above, we arrive at the master equation that we will solve to study global disk evolution

$$2\pi R\frac{\partial\Sigma}{\partial t}=\frac{\partial}{\partial R}\left[\frac{2}{v_K}\frac{\partial}{\partial R}(2\pi R^2\tilde{\alpha}\Sigma c_s^2)\right]+\frac{\partial}{\partial R}\left[2(\lambda-1)R\left(\frac{\partial\dot{M}_{\rm loss}}{\partial R}\right)\right]-\frac{\partial\dot{M}_{\rm loss}}{\partial R}\ . \tag{3.4}$$

It clearly generalizes the viscous evolution equation to include wind-driven accretion and mass loss terms (the 2nd and 3rd terms on the right-hand side). To evaluate these terms, we must determine $\tilde{\alpha}$, λ, and $\partial \dot{M}_{\rm loss}/\partial R$ at each radius.

3.2.2 *Disk Structure*

We now describe our model disk structure that will be employed to evaluate $\tilde{\alpha}$, λ, and $\partial \dot{M}_{\rm loss}/\partial R$. The basic picture is illustrated in Fig. 3.1. Fiducially, as initial condition, we consider a disk model with a power-law radial profile and an exponential cutoff motivated from sub-millimeter observations (Andrews et al. 2009, 2010)

$$\Sigma_0(R) = 500\,{\rm g\,cm^{-2}} R_{\rm AU}^{-1} \cdot \exp\left(-R/R_d\right) , \tag{3.5}$$

where $R_{\rm AU}$ is radius measured in AU. We choose the cutoff radius to be $R_d = 100$ AU throughout this work. The total disk mass is about 0.035 $M_\odot$.

Thermodynamically, we assume the disk to follow a two-temperature profile with $T = T_d$ in the disk interior and $T = T_a$ at disk surface (atmosphere), and transitions at vertical height $z = z_a$. Both layers are treated as vertically isothermal. Disk temperature T_d is largely determined by stellar irradiation, given by

$$T_d = 280\,{\rm K}\, {R_{\rm AU}}^{-1/2} , \tag{3.6}$$

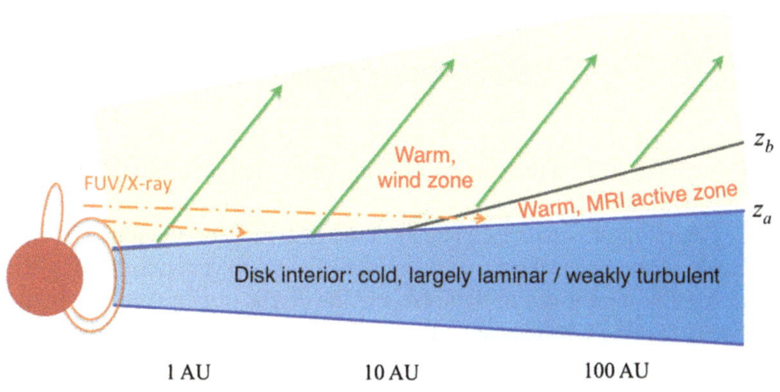

Fig. 3.1 Cartoon illustration of our global PPD evolution model. The bulk of the disk is cold (thin), where magnetic fields are poorly coupled to the gas with very weak level of turbulence. The surface layer is much warmer and much better ionized due to strong external far-UV/X-ray heating/ionization, and the gas is well coupled to magnetic fields, enabling MHD wind launching. The cold-warm interface is marked by z_a, and we use z_b to denote the wind base where the wind is launched. In general, $z_b \approx z_a$ unless the net vertical magnetic field is not sufficiently strong. In this case, the warm surface layer can become MRI active (achieved at the outer disk), and $z_b > z_a$, sandwiching a warm MRI layer. Not reflected in this cartoon picture is that the disk is flared, making FUV radiation easier to penetrate into the outer disk. Figure taken from Bai (2016)

as in the standard minimum-mass solar nebular (MMSN) model (Weidenschilling 1977; Hayashi 1981), where $R_{\rm AU}$ is radius measured in AU. The corresponding disk isothermal sound speed $c_{s,d}$ and disk scale height H_d satisfy $c_{s,d}/v_K = (H_d/R) \approx 0.034R_{\rm AU}^{1/4}$. With H_d/R increasing with R, the disk is flared, as is well known from SED modeling of T Tauri disks (Kenyon and Hartmann 1987; Chiang and Goldreich 1997).

Disk surface (atmosphere) is hotter as a result of strong heating from stellar far-UV (FUV) radiation and X-rays. For simplicity, we assume its temperature is a constant factor higher than the disk

$$T_a = f\,T_d\ . \tag{3.7}$$

In practice, we take $f = 3 - 8$ in our calculations, consistent with typical results from more detailed thermo-chemical calculations (e.g., Walsh et al. 2010), as well as observational constraints from the HD 163296 disk (Rosenfeld et al. 2013). Associated with T_a there is the atmosphere sound speed $c_{s,a}$ and scale height $H_a = c_{s,a}/\Omega_K$.

The density structure in the disk interior can be obtained directly from hydrostatic equilibrium. In the disk atmosphere/wind zone, the density structure deviates from hydrostatic, and will be discussed separately in the next section.

The transition height z_a is determined by how deep FUV/X-rays can penetrate. By default, we assume that the penetration depth of FUV and X-rays are comparable, denoted by $\Sigma_{\rm FUV}$.[2] Its value is uncertain and depends on the FUV luminosity from the protostar, as well as the abundance of very small grains. Here we quote from Perez-Becker and Chiang (2011)

$$\Sigma_{\rm FUV} = 0.01\text{–}0.1\,{\rm g\,cm^{-2}}\ , \tag{3.8}$$

and treat it as a constant at all radii.[3] The location of the FUV front (z_a) is determined by tracing radial rays from the central star, until the column density traversed by the rays equals to $\Sigma_{\rm FUV}$.

The key element of our picture is that in the disk interior, the ionization level is sufficiently low, and hence the gas and magnetic fields are very poorly coupled. As a result, the MRI is suppressed or damped, and the disk is largely laminar. In the disk surface, we note that FUV does not only provide heating, but is also capable of fully ionizing carbon and sulfur species, leading to a substantial increase in ionization fraction to $\sim 10^{-(4-5)}$. This is generally sufficient to make gas and magnetic fields

[2]In reality, X-rays may be able to heat deeper layers. The situation is discussed in Bai (2016) and it was concluded that this effect does not strongly affect our model results.

[3]The value quoted from Perez-Becker and Chiang (2011) corresponds to a vertical column, whereas in our calculation it corresponds to the radial column towards the star. The two numbers differ by a factor of a few ($\sim$3) depending on the level of disk flaring, though here we have ignored the difference.

well coupled (Perez-Becker and Chiang 2011), which is essential for efficient wind launching.

Another important location is the wind base z_b, from which the wind is launched. Numerical simulations of the inner regions of PPDs have found that in general, z_b coincides with the FUV ionization front (Bai and Stone 2013b; Gressel et al. 2015), and hence $z_b = z_a$. Towards the outer disk, however, FUV radiation can effectively penetrate deeper into the disk (in terms of z_a/H_d) as the disk is flared, and this can make the surface layer subject to the MRI (Perez-Becker and Chiang 2011). A disk wind can also be launched from the surface MRI turbulent layer in the presence of net vertical field (Suzuki and Inutsuka 2009; Fromang et al. 2013; Bai and Stone 2013a; Suzuki and Inutsuka 2014), and local simulations applicable to the outer PPDs found evidence that disk outflow is launched from a higher position than the FUV front (Simon et al. 2013a).

Whether a surface MRI layer exist depends on whether vertical magnetic field strength at $z = z_a$ is strong enough to suppress the MRI. Given the net vertical field strength B_z, we can estimate the plasma β of the net vertical field at $z = z_a$: $\beta_0(z_a) = 8\pi\rho_a c_{s,a}^2/B_z^2$, where ρ_a is gas density at z_a. We note that in the ideal MHD regime, the most unstable MRI wavelength is given by (e.g., Hawley et al. 1995) $\lambda_m/H_a \approx 9.18\beta_0^{-1/2}$. Due to the rapid density drop with height, we expect the MRI to be able to operate when $\lambda_m \leq H_a$. In practice, we assume that the disk surface becomes MRI turbulent when $\beta_0(z_a) > 50$. In this case, we determine the wind base location z_b by setting $\beta_0(z_b) = 50$. This treatment is by no means rigorous, but at least local simulations lend support to this treatment (Simon et al. 2013a). Without better knowledge of the wind launching process from turbulent disk surfaces and its global kinematics, we expect it to be a reasonable first approximation.

3.2.3 *Effective Viscosity*

Overall, we expect very low, but non-zero level of turbulence in regions where the MRI is inactive. Contribution to turbulent transport may include pure hydrodynamic instabilities, such as the vertical shear instability (Nelson et al. 2013; Stoll and Kley 2014), convective overstability and baroclinic vortex amplification (Klahr and Hubbard 2014; Lyra 2014; Raettig et al. 2013), and the zombie vortex instability (Marcus et al. 2015). Whether they can operate largely depends on disk thermodynamics, but in general, they are found to produce very limited viscous transport with $\alpha < 10^{-3}$. In this work, we adopt $\alpha = \alpha_0 = 2 \times 10^{-4}$ at locations $z < z_a$ to represent residual "viscosity" from such hydrodynamic instabilities. Because of the dominant role played by disk winds, we have found that the exact value of α_0 adopted here is unimportant.

If an MRI active zone is present, we set $\alpha = \alpha_1 = 0.2$ in this region ($z_a < z < z_b$). We choose a relatively large α because the net vertical field at this surface layer is effectively strong (with β_0 of the order 100), and it is well known from MRI simulations (e.g., Hawley et al. 1995; Simon et al. 2013a) that the resulting α is larger than zero net flux case and is of the order 0.1 or higher.

Additionally, while transport by the MRI is suppressed or strongly damped by non-ideal MHD effects at $z < z_a$, there is still Maxwell stress resulting from either weak MRI turbulence or large-scale fields (e.g., Okuzumi and Hirose 2011; Simon et al. 2013a; Lesur et al. 2014; Bai 2014, 2015), and its strength increases with net vertical magnetic flux. Incorporating these considerations, we adopt the following form of α in the disk zone at $z < z_a$:

$$\alpha_d = \text{Min}[\text{Max}(10.0\beta_0^{-1}, \alpha_0), \alpha_1] \,. \tag{3.9}$$

This way, the α_d value ranges from α_0 for the weak field case to α_1 for the strong field case which joins the α value in the surface MRI active zone. We further comment that the overall disk evolution is insensitive to the exact prescription of α values because again, as we will see, disk evolution is largely wind-driven.

We apply the α values in the disk and surface regions discussed above to Eq. (3.2), from which an effective value, $\tilde{\alpha}$, can be determined. Note that we quote the $\tilde{\alpha}$ value based on sound speed in the disk interior $c_{s,d}$.

3.3 Model of Disk Wind

Wind torque (lever arm) and wind mass loss rates are crucial ingredients of our disk evolution framework. In this section, we construct a wind model to calculate these quantities given the physical parameters. Our model assumes axisymmetry and steady state, and is constructed based on standard conservation laws along poloidal field lines in axisymmetric MHD (e.g., Spruit 1996).

We prescribe the geometry of a typical poloidal field line anchored to the disk at the wind base (R_0, z_b). For simplicity, the field lines are taken to be straight, with a constant inclination angle θ relative to the equatorial plane at $z \geq z_b$. The poloidal field strength is prescribed as

$$B_p(R) = B_{p0}\frac{1+q}{(R/R_0) + q(R/R_0)^2} \,, \tag{3.10}$$

where B_{p0} is the poloidal field strength at the wind base, and the parameter q controls the radius at which the poloidal field lines transition from being parallel (so that $B_p \propto R^{-1}$) to diverging (so that $B_p \propto R^{-2}$). The former applies if the wind is launched from a radially extended region in an approximately self-similar manner, while the latter applies towards large R with the solenoidal constraint demanding $B_p(R) \propto R^{-2}$.

Along the field line, let ρ, $\boldsymbol{v}_p$, and $v_\phi = \Omega R$ be gas density, poloidal, and toroidal velocities, with Ω being gas angular velocity. In the disk wind zone, the gas is largely in the ideal MHD regime owing to much high levels of ionization from external UV/X-rays (e.g., see Fig. 9 of thermo-chemical calculations of Walsh et al. 2012, where ideal MHD applies when the ambipolar Elsasser number $Am \gg 1$). The ideal MHD equations for a steady axisymmetric wind yield four constants of the flow. The first three are

$$k \equiv \frac{4\pi\rho v_p}{B_p}, \quad \omega \equiv \Omega - \frac{kB_\phi}{4\pi\rho R}, \quad l = \Omega R^2 - \frac{RB_\phi}{k}. \tag{3.11}$$

The first is related to wind mass flux, and the other two correspond to angular velocity of the field line, and specific angular momentum of the wind flow. With the above relation, it is straightforward to show that

$$l = \omega R_A^2 , \tag{3.12}$$

where R_A is the Alfvén radius, at which the poloidal velocity v_p equals to the poloidal component of the Alfvén velocity $v_{Ap} = B_p/\sqrt{4\pi\rho}$.

Thermodynamics is reflected in the energy equation via the enthalpy $h \equiv \int dP/\rho$. We assume that the gas in the wind zone can be approximately described by an isothermal equation of state $P = \rho c_{s,a}^2$, so that we simply have $h(\rho) = c_{s,a}^2 \log(\rho/\rho_0)$, where ρ_0 is the density at the wind base. Energy conservation is then expressed via the Bernoulli constant

$$E = \frac{v^2}{2} + h + \Phi - \omega R v_\phi = \frac{v_p^2 + (v_\phi - \omega R)^2}{2} + h + \Phi_{\rm eff} , \tag{3.13}$$

where $v = \sqrt{v_p^2 + v_\phi^2}$ is the total flow speed, $\Phi = -GM_*/\sqrt{R^2+z^2}$ is the gravitational potential, and $\Phi_{\rm eff} = \Phi - \frac{1}{2}\omega^2 R^2$.

Although constant along each poloidal line, the four flow quantities $\{k, \omega, l, E\}$ vary between lines. Once we have prescribed the shape of a poloidal field line and its strength B_p along the line, as well as the enthalpy $h(\rho)$, the wind solution is fully specified by the flow constants. However, not all of the flow constants can be chosen independently if the wind is to reach infinity with a terminal speed exceeding all of its characteristic velocities. The Bernoulli constant E is essentially determined by physical conditions at the wind base. Also, one expects $\omega \approx \Omega_K(R_0)$, where $\Omega_K(R_0)$ is the Keplerian angular frequency at the wind base, and so we provisionally take $\omega = \Omega_K$. The remaining constants k and l are found by imposing regularity conditions at the two critical points (slow and fast magnetosonic points, reached when poloidal velocity v_p equals to the poloidal component of the slow and fast magnetosonic speeds).

For a given disk model, therefore, the wind solution is determined by the following parameters:

- Physical parameters: B_{p0} [Eq. (3.10)] and surface sound speed $c_{s,a}$.
- Geometric parameters: field divergence parameter q [Eq. (3.10)], base height z_b, and field inclination angle θ.

Needless to say, our wind model involves substantial simplifications, especially with regard to the poloidal field and thermodynamics. In reality, the poloidal field geometry and strength are determined by force balance perpendicular to flux surfaces, as described by the Grad-Shafranov equation (GSE), a nonlinear partial differential equation in R and z. However, besides the numerical difficulties involved in finding the solution to the GSE (which is intrinsically global and generally requires MHD simulations), the solution depends on the global distribution of magnetic flux, which is largely unconstrained a priori. As a result, even when self-similarity is imposed, a diverse set of wind solutions is possible with vastly different field geometries. Therefore, rather than solve the GSE, we have chosen to parametrize the poloidal field in terms of B_{p0} (its strength at the wind base), θ (inclination to the disk), and q (divergence radius $q^{-1}R_0$). We expect the effects of more realistic field geometries to be roughly captured by systematically varying θ and q. Similarly, the role of external heating is encapsulated by an isothermal sound speed $c_{s,a}$. In brief, we have traded the accuracy and fidelity of individual solutions to the power to survey a broad range of parameters. The simplicity and transparency of our model also greatly facilitate interpretation of wind properties.

For all calculations in this section, we normalize lengths, velocities, and gas densities by their values at the footpoint of the field-line on the disk (wind base), so that $R_0 = v_K = \rho_0 = 1$. Correspondingly, time is in units of $1/\Omega_K$. Poloidal field strength is conveniently parameterized by $v_{A0} \equiv B_{p0}/\sqrt{4\pi\rho_0}$, the poloidal Alfvén velocity at the footpoint. Based on the discussion in Appendix A of Bai et al. (2016), we choose the following as fiducial parameters:

$$v_{A0} = c_{s,a} = 0.1 v_K \ , \quad z_b = 0.15 R_0 \ , \quad \theta = 45^\circ \ , \tag{3.14}$$

with $\omega = \Omega_K$. In particular, the value of v_{A0} is chosen based on the required field strength that can drive disk accretion rates consistent with observations. While the choice of inclination angle is somewhat arbitrary, we have found in Bai et al. (2016) that the wind kinematics is not very sensitive to θ for θ in between 30° and 60°. We take $q = 0.25$ as our fiducial choice for the functional form of B_p in Eq. (3.10), which means that transition from R^{-1} (parallel) to R^{-2} (diverging) occurs at around $R \sim 2R_0$.

We find that v_{A0}, $c_{s,a}$ are the more important model parameters. We vary v_{A0}/v_K in the range between 0.01 and 10, and $c_{s,a}/v_K$ between 0.03 and 0.3. The wind properties turn out to be also modestly sensitive to the geometric parameter q. Nevertheless, given the limited space for this chapter, we primarily explore the role of v_{A0} and $c_{s,a}$ in this chapter, and the readers are encouraged to consult Bai et al. (2016) for further details.

3.3.1 Wind Kinematics

In Fig. 3.2, we show the radial profiles of various wind diagnostic quantities for our fiducial set of parameters, with different lines representing wind solutions from different v_{A0}. We see that all solutions accelerate monotonically from the wind-launching region, smoothly pass the critical points, and approach terminal velocities comparable to or beyond the Keplerian velocity at the wind base. Typically, near the wind base, the poloidal velocity is very small, and the gas is pressure supported, with significant contribution from magnetic pressure. In all solutions, the gas density decreases with R (and hence z) as if in (magneto-)hydrostatic equilibrium in this region. At larger radii, the wind achieves its asymptotic velocity $v_p^{\rm inf}$, and hence the gas density scales in the same way as B_p, i.e. R^{-2}, while the scalings $B_\phi \propto R^{-1}$ and $v_\phi \propto R^{-1}$ follow directly from the conservation laws.

The most important wind diagnostic is the Alfvén radius R_A (i.e. R_A/R_0): it characterizes the efficiency of the wind for extracting disk angular momentum. Since $\omega \approx \Omega_K$, the excess angular momentum per unit mass in the wind is

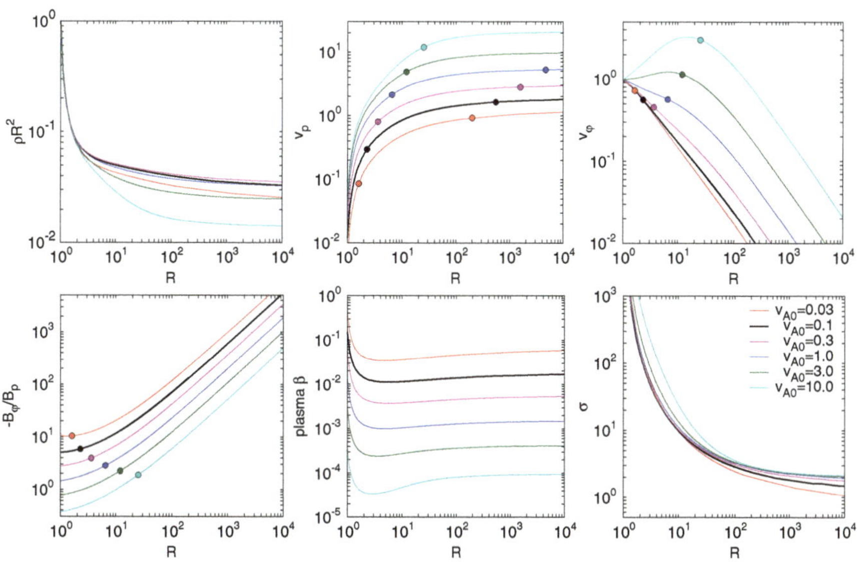

Fig. 3.2 Profiles of various physical quantities along poloidal field lines as functions of cylindrical radius R. We vary poloidal field strength (given in v_{A0}, see legend) while fixing $c_{s,a} = 0.1 v_K$, $q = 0.25$. *Bold black line* corresponds to our fiducial set of parameters. Alfvén points of the solutions are marked by *filled circles* in three of the panels. Fast magnetosonic points are marked in *top middle panel* (but lie outside the box for two top curves). *Bottom middle panel:* β is the ratio of gas pressure to magnetic pressure, computed from the total field. *Bottom right panel:* σ is the ratio of Poynting flux to kinetic energy flux. Figure taken from Bai et al. (2016)

$\Omega_K(R_A^2 - R_0^2)$. The removal of this excess from the disk leads to a direct relation between the wind mass loss rate and wind-driven accretion rate:

$$\dot{M}_{\rm acc}\frac{dj_K}{dR} = \frac{d\dot{M}_{\rm loss}}{dR}\Omega_K(R_A^2 - R_0^2) = (\lambda - 1)\frac{d\dot{M}_{\rm loss}}{dR}j_K\ , \tag{3.15}$$

which is related to Eq. (3.3). Noting that $d\ln j_K/d\ln R = 1/2$ in Keplerian disks, we obtain

$$\xi \equiv \left.\frac{d\dot{M}_{\rm loss}/d\ln R}{\dot{M}_{\rm acc}}\right|_{R=R_0} = \frac{1}{2}\frac{1}{(R_A/R_0)^2 - 1} = \frac{1}{2(\lambda - 1)}\ . \tag{3.16}$$

Ferreira and Pelletier (1995) call ξ the "ejection index."

The Alfvén radius R_A is a rapidly increasing function of v_{A0}. Using a small $v_{A0} = 0.03v_K$, we obtain $R_A \approx 1.61R_0$. Such small R_A would result in extremely efficient mass loss, with ejection index $\xi \approx 0.4$. If the wind were launched over an extended region of the disk—say 0.3–15 AU as suggested by local disk simulations (Bai 2013)—the integrated wind mass loss rate (assuming R_A/R_0 roughly constant) would even exceed the accretion rate: $\dot{M}_{\rm loss} \approx 1.5\dot{M}_{\rm acc}$! On the other hand, increasing v_{A0} to v_K moves the Alfvén radius to $\sim 5R_0$, and mass loss is almost negligible compared with accretion with $\xi \sim 0.02$. Our fiducial solution ($v_{A0} = c_{s,a} = 0.1v_K$) has $R_A/R_0 \approx 2.29$, and gives a modest ejection index $\xi \approx 0.12$ which, if approximately constant over an extended radial range (e.g., again 0.3–15 AU), still yields a total mass loss rate that is a considerable fraction (47%) of the wind-driven accretion rate.

Another important wind diagnostic concerns with the wind launching mechanism. MHD disk winds come in two flavors. The magnetocentrifugal mechanism (Blandford and Payne 1982) operates when poloidal field is sufficiently strong to enforce the wind flow to corotate with the field line footpoint at the launching radius R_0, allowing them to be centrifugally accelerated. On the other hand, if poloidal field is not sufficiently strong, disk rotation would wind up poloidal field to produce strong toroidal field near the wind launching region. As toroidal field builds up, the wind is driven by its magnetic pressure. This is sometimes referred to as magnetic tower outflow (Lynden-Bell 2003). We see from Fig. 3.2 that corotation is not enforced unless $v_{A0} \geq 10c_{s,a}$. We also see that only when $v_{A0} \geq 3c_{s,a}$, the ratio of $|B_\phi/B_p|$ near the wind launching region falls below 1, leading to a situation favorable for centrifugal acceleration. However, such field strength is more than a factor $\sim 10^3$ stronger than the expected field strength in PPDs (to match accretion rates). Therefore, for realistic field strength in PPDs, the disk wind must be largely driven by magnetic pressure gradient.

For the purpose of locating the wind base (see previous section), we also need the density profile in the vicinity of the wind base so as to calculate the column densities. We see from the first panel of Fig. 3.2 that due to magnetic forces, density can deviate substantially from hydrostatic equilibrium. This deviation provides additional gas column that can partially block the stellar FUV radiation (discussed in more detail in Bai (2016), see also Panoglou et al. (2012) for a calculation with very different parameters). Interestingly, the wind density profile is almost independent of wind magnetization up to a substantial distance ($\sim$2 times wind launching radius) from the wind base. We have also verified this trend with other choices of wind temperature. A reasonable fitting formula for gas density profile along the wind streamline is found to be

$$\rho(R) \approx \rho_b(R_0) \exp\left[-\left(\frac{c_{s,a}}{v_K}\right)^{-0.6} \sqrt{\frac{R-R_0}{R_0}}\right], \tag{3.17}$$

where ρ_b is gas density at the wind base, and R is the cylindrical radius along the wind streamline. This formula is accurate to within order unity (most cases within $\sim$20%) up to $R = 2R_0$ for wind temperature range $c_{s,a}/v_K = 0.05$–0.3. Because this density profile is solely used for finding the wind base location, we may convert it into a vertical density profile at radius R_0 and $z > z_b$ for convenience by replacing $R - R_0$ by $z - z_b$ (which is because FUV photons penetrate through approximately the same column density in these two cases to reach the next wind streamline).

3.3.2 Parameter Dependence

In this subsection, we focus on important wind diagnostics and discuss how they vary with main physical parameters, $c_{s,a}$ and v_{A0}. We vary these parameters in two ways. First, we can vary v_{A0} at fixed $c_{s,a}$. It can be endowed with two interpretations. The most straightforward is changing field strength without changes in other parameters. Another interpretation is the change in FUV penetration depth $\Sigma_{\rm FUV}$ without changing field strength and temperature: deeper penetration allows the wind to be launched from regions with higher density, hence smaller v_{A0}, and vice versa. This interpretation reflects the role of thermodynamics from external heating. Second, we can vary $c_{s,a}$ and v_{A0} simultaneously, keeping their ratio $c_{s,a}/v_{A0}$ constant. This is equivalent to changing the wind temperature without changing field strength: at the wind base (where pressure is fixed), $c_{s,a}$ and v_{A0} scale the same way with density.

Besides R_A, another important diagnostic quantity is the dimensionless "mass loading" parameter:

$$\mu \equiv \frac{\omega R_0}{B_{p0}} k = \frac{\omega R_0 v_{p0}}{v_{A0}^2} . \tag{3.18}$$

A wind is considered "heavily loaded" if $\mu > 1$, and "lightly loaded" if $\mu \ll 1$. The mass loss rate per logarithmic radius is thus given by

$$\frac{d\dot{M}_{\rm loss}}{d\ln R} = 2\pi R^2 (\rho v_p)_{z_b} = \frac{2\pi R_0}{\omega} \mu \rho_0 v_{A0}^2 = \frac{R_0}{2\omega} \mu B_{p0}^2 . \tag{3.19}$$

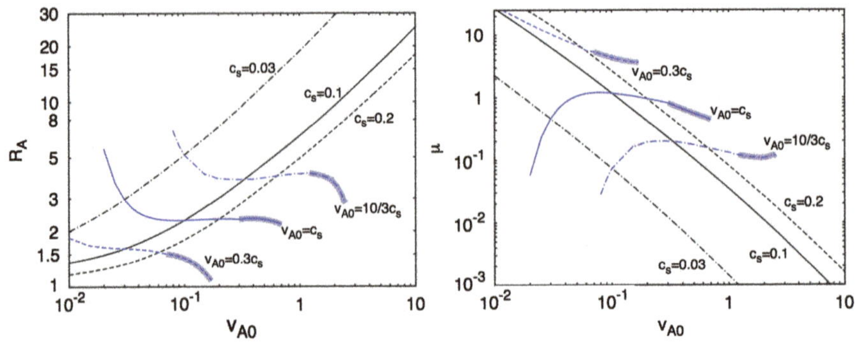

Fig. 3.3 Dependence of Alfvén radius R_A (*left*) and mass loading μ (*right*) on v_{A0} for different choices of $c_{s,a}$ (fixed or $\propto v_A$), as marked, with $q = 0.25$ throughout. Our solutions are less reliable in shaded regions where the slow magnetosonic points fall below the wind base. Figures taken from Bai et al. (2016)

In Fig. 3.3 we show the dependence of R_A/R_0 and μ on v_A and $c_{s,a}$. Black lines correspond to varying v_{A0} at fixed $c_{s,a}$. Clearly, enhanced field strength or reduced FUV penetration depth leads to a stronger wind higher terminal velocities, larger Alfvén radius, smaller mass loading, and less toroidally dominated (more loosely wound) magnetic field.

The blue lines correspond to varying the wind temperature at fixed penetration depth and field strength. We see that over a large range, these lines are relatively flat, suggesting that the location of R_A and mass loading μ are to a large extent controlled by the ratio of $v_{A0}/c_{s,a}$ at the wind launching region. The contrast among these lines is very significant. Varying v_{A0}/c_s by a factor of $\sim$3 above and below unity makes R_A/R_0 vary from $\sim$2.3 to $\sim$4 and $\sim$1.4, respectively. Because the ratio $\dot{M}_{\rm loss}/\dot{M}_{\rm acc}$ depends quadratically on R_A/R_0, such variations in the latter would easily give variations in fractional wind mass loss by factors $\sim$10. This discussion best demonstrates the sensitive dependence of the wind mass loss rate on the thermodynamics.

In reality, stronger external radiation is likely to lead to both higher wind temperature and deeper penetration. Therefore, enhancing external radiation may correspond to moving in between rightward (larger $c_{s,a}$) and to the bottom left (deeper penetration) in the left panel of Fig. 3.3, along directions between tangents to the blue and black curves.

The results obtained above can be condensed into a fitting formula for the mass loading parameter μ

$$\mu = \mu_0 \left(\frac{v_{A0}}{v_K} \right)^{-1.48+0.17\log_{10}(c_{s,a}/v_K)-0.1\log_{10}(v_{A0}/v_K)} , \tag{3.20}$$

where $\mu_0 = 0.5(c_{s,a}/v_K) - 0.015$. We have also examined additional wind solutions and confirm that this fitting formula is accurate as long as $c_{s,a}/v_K \geq 0.033$ (note that it would lead to $\mu < 0$ if $c_{s,a}/v_K \leq 0.03$). For all models considered in this work, the disk surface is sufficiently warm and satisfies this condition. Similarly for the Alfvén radius, we find

$$\frac{R_A}{R_0} \approx \frac{\sqrt{1.5[1 + (0.2\mu)^{-2/3}]}}{[1 + (0.2\mu)^{1/4}(c_{s,a}/v_K)^2]^2} , \tag{3.21}$$

which is accurate for μ up to 100 for the range of $c_{s,a}/v_K$ considered in this work (≤ 0.2). Note that this fitting relation becomes invalid and would predict $R_A/R_0 < 1$ for very large μ and when the wind is warm. In reality, the wind would transition to become pure thermally driven towards higher temperature, and no longer extracts disk angular momentum. In practice, because we have imposed the condition $\beta_0 \leq 50$ at the wind base (relatively strong field), we never encounter a situation with $\mu > 100$. These fitting relations encapsulate the necessary physics and will provide estimates for λ and $\partial \dot{M}_{\rm loss}/\partial R$ for the disk evolution equation (3.4).

3.4 Global Evolution

With preparations above, we are ready to present calculations and results for our global disk evolution model in this section.

Our calculations are carried out on a logarithmic grid with inner boundary at $R_i = 0.1$ AU, and outer boundary at $R_o = 1000$ AU using 200 grid points. We do not attempt to model the dynamics of the innermost disk region, where the physics is more complex with thermal ionization of Alkali species and thermionic emission from dust grains (e.g., Desch and Turner 2015). While this region is expected to be fully turbulent due to the MRI, it contains only a very small fraction of total disk mass and angular momentum and should not affect the bulk of disk evolution. The viscous term in Eq. (3.4) is integrated with standard zero torque boundary conditions. The wind-driven accretion (advection) term is integrated using the standard upwind method.

Our model is mainly specified by three sets of parameters. Most important of all, we have seen that essentially all aspects of angular momentum transport and mass loss depend on the strength of net vertical field strength threading the disk B_z. It is associated with the poloidal field at the wind base B_{p0} by $B_z \approx B_{p0}\cos\theta$ (we take $\theta = 45°$), and reflected in the wind base Alfvén speed $v_{A0} = B_{p0}/\sqrt{4\pi\rho_b}$. The radial profile of $B_z(R)$, and its time evolution, however, are largely unconstrained. Before having a reliable understanding of magnetic flux transport, we here simply consider a phenomenological approach, as adopted in Armitage et al. (2013), and assume that magnetic flux is distributed in such a way that midplane plasma β of the net vertical field is constant. We consider three different values for initial magnetic field strength, corresponding to midplane $\beta_0(z = 0) = 10^3$, 10^4, 10^5 and 10^6, with $\beta_0 = 10^5$ as the fiducial value.

We further consider two scenarios on the evolution of total magnetic flux $\Phi_B = 2\pi \int_{R_i}^{R_o} B_z(R) R dR$. In the first scenario, we assume Φ_B is conserved during disk evolution. Therefore, as accretion proceeds, disk becomes more strongly magnetized. In the second scenario, we assume that the disk loses magnetic flux in a way such that Φ_B is proportional to total disk mass $M_d = 2\pi \int_{R_i}^{R_o} \Sigma(R) R dR$. This treatment avoids significant accumulation of magnetic flux and results in slower evolution.

The other two parameters are the FUV penetration depth, and the disk atmosphere temperature. We consider $\Sigma_{\rm FUV} = 0.01$ and $0.1\,{\rm g\,cm^{-1}}$, taking the former as fiducial. For atmosphere temperatures, we fix $f = 3$ [Eq. (3.7)], as we have found that the wind transport properties are insensitive to T_a (Bai et al. 2016).

3.4.1 Transport Properties in a Static Disk

In Fig. 3.4, we show the radial profiles of various diagnostic quantities from our fiducial model (solid lines). First, we can see that up to $R \sim 50$ AU, the locations z_a and z_b coincide with each other, meaning that the disk is largely laminar up to this radius, with wind being the dominant mechanism for angular momentum transport. The value of z_a/R increases with increasing R, meaning that the FUV ionization front traces a flared geometry. On the other hand, because the disk itself is flared, the value of z_a/H_d actually decreases with increasing R: from $\sim$5 at 1 AU to $\sim$3.5 at $\sim$100 AU. This means that the FUV radiation effectively penetrates deeper (geometrically) towards the outer disk. These numbers are similar to local disk simulations where the FUV penetration depth is measured in the vertical domain (Bai and Stone 2013b; Simon et al. 2013a).

The fact that FUV geometrically penetrates deeper towards the outer disk allows the MRI to operate at $R \sim 50$ AU and beyond in our fiducial model. Correspondingly, the wind base location z_b no longer coincides with z_a, and is located higher in the atmosphere. With z_a/H_d decreasing with R, the extent of the MRI zone increases towards larger R, and hence the value of $\tilde{\alpha}$ increases towards the outer disk. Nevertheless, because only a very small fraction of mass resides in the MRI active zone, the averaged value $\tilde{\alpha}$ remains relatively small ($\sim 10^{-3}$ or less).

We find that in the inner disk, the wind Alfvén radius (or lever arm) decreases with increasing R. This is a consequence of z_b/H_d getting smaller, and hence when assuming constant midplane β_0, the wind base is less strongly magnetized (larger plasma β, or smaller v_{A0}/c_s, at the wind base). As a result, the wind becomes less efficient in transporting angular momentum towards larger R, and the disk has to lose more mass to maintain its accretion rate (i.e., higher mass loading). When the MRI zone is present, because we have set fixed $\beta_0 = 50$ at the wind base location, the Alfvén radius and mass loading parameter become approximately constants.

The middle right panel of Fig. 3.4 shows the radial profiles of wind-driven accretion rate, viscous accretion rate, and wind mass loss rate per logarithmic

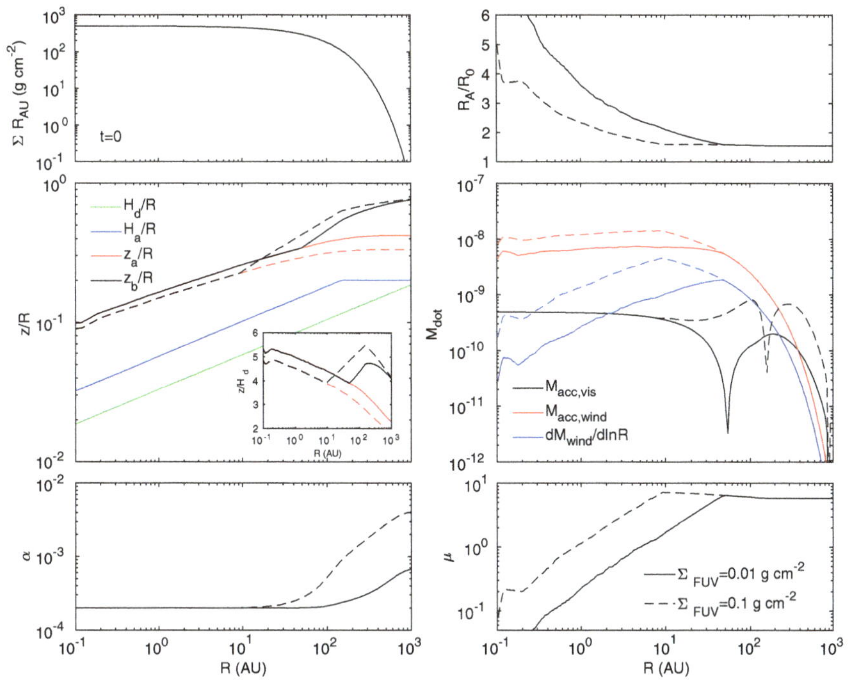

Fig. 3.4 Radial profiles of general diagnostics from our fiducial wind model (*solid lines*, with $\beta_0 = 10^5$, $\Sigma_{\rm FUV} = 0.01\,{\rm g\,cm^{-2}}$, and $T_a = 3T_d$), as well as a model with deeper FUV penetration ($\Sigma_{\rm FUV} = 0.1\,{\rm g\,cm^{-2}}$) at the beginning of evolution ($t = 0$). *Top left*: surface density profile (multiplied by *R*/AU). *Middle left*: location of the wind base z_b (*black*), FUV ionization front z_a (*red*), together with disk thickness H_d (*green*), and disk atmosphere scale height H_a (*blue*). They are normalized to *R* in the main plot (note we enforce $H_a/R \le 0.2$), and we further plot z_a/H_d and z_b/H_d in the *inset*. *Bottom left*: effective viscosity α. *Top right*: ratio of wind Alfvén radius to wind launching radius. *Middle right*: wind-driven accretion rate (*red*), viscous-driven accretion rate (*black*), and wind mass loss rate per logarithmic radii (*blue*). Note viscous accretion rate changes sign at around $R \sim 140$ AU. *Bottom right*: mass loading parameter of the disk wind. In all panels, *solid and dashed lines* correspond to calculations with FUV penetration depth of $\Sigma_{\rm FUV} = 0.01$ and $0.1\,{\rm g\,cm^{-2}}$, respectively. Figure made based on data from Bai (2016)

radii. We see that clearly, almost the entire disk relies on disk wind to transport angular momentum. The only exception is near the disk outer edge, where disk surface density exponentially falls off and viscous transport picks up. Because of the dominant role played by the wind, viscous disk spread is largely suppressed, except for regions well beyond the characteristic disk size of $R_d = 100$ AU.

The mass loss profile is best quantified by the mass loss rate per logarithmic radii $d\dot{M}_{\rm loss}/d\ln R$. As the lever arm $\lambda \equiv [R_A(R)/R]^2$ decreases towards large *R*, we see that wind mass loss rate rapidly increases with *R*. The mass loss rate per logarithmic radii is nearly two orders of magnitude below the accretion rate at the innermost radius, while it can become a significant fraction of the accretion rate at tens of AU. Thus, we expect that most of the mass loss in PPDs occurs through the outer disk via very slow winds, in contrast with observations which typically can only trace winds launched from the inner disk (e.g., Hartigan et al. 1995; Pascucci and Sterzik 2009; Natta et al. 2014). On the other hand, the kinetic flux of the wind (not shown in the figure) is more smoothly distributed across different disk radii, as can be envisioned by multiplying the mass loss rate profile by a radial scaling R^{-1} ($\sim v_K^2$).

We also show in dashed lines in Fig. 3.4 with $\Sigma_{\rm FUV} = 0.1$ g cm^{-2}, which mimics the case of enhanced FUV luminosity to allow its deeper penetration depth. We see that larger $\Sigma_{\rm FUV}$ makes the wind to be launched from deeper layer in the disk, and enlarges the size of the MRI active zone. As a consequence, $v_{A0}/c_{s,a}$ is smaller at the wind base, leading to smaller R_A and higher mass loading μ. Both accretion rates and mass loss rates increase, but mass loss rates increase more. This exhibits the effect of what we term "magneto-photoevaporation" (Bai et al. 2016).

3.4.2 Global Disk Evolution

We now explore the long-term disk evolution in this subsection. We here mainly focus on the global mass budget, in terms of evolution timescale, fractional mass loss through accretion and wind, etc., to minimize uncertainties associated with the magnetic flux distribution and evolution.

In Fig. 3.5, we show the time evolution of disk surface density and the associated disk mass (M_d). Two scenarios of magnetic flux evolution are considered, namely,

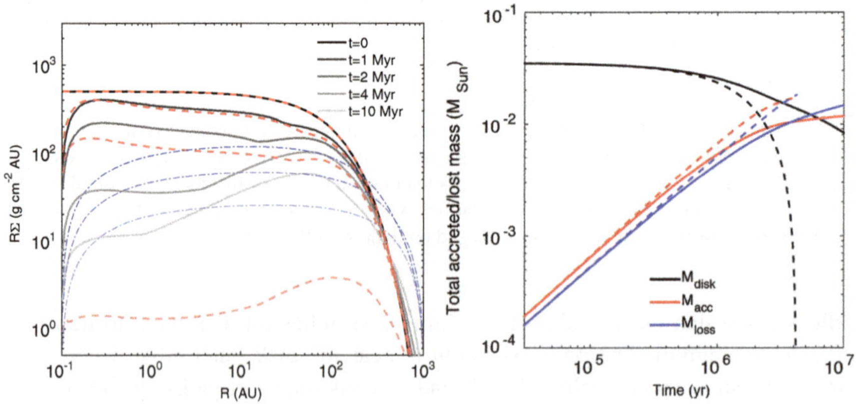

Fig. 3.5 *Left*: Time evolution of disk surface density (given by $R\Sigma$) using our fiducial model parameters. Lines with different transparencies represent different times, as indicated by the legend. *Black solid lines* correspond to the case where total magnetic flux Φ_B is set to be proportional to disk mass M_d, while *red dashed lines* represent the case where Φ_B is conserved. For comparison, we also show in *blue thin dash-dotted lines* results from a pure viscous evolution calculations assuming constant $\alpha = 0.01$ at $t = 1.0$, 2.0 and 4.0 Myrs. *Right*: Time evolution of total disk mass (*black*), mass accreted to the central protostar (*red*), and mass lost through the wind (*blue*). *Solid lines* correspond to the case with $\Phi_B \propto M_d$, while *dashed lines* represent the case with Φ_B conserved. Figures taken from Bai (2016)

either total magnetic flux $\Phi_B \propto M_d$, or Φ_B is constant. We see that at early stages, disk evolution proceeds very similarly in these two scenarios. After about 1 Myrs, the disk has lost about 30% of its mass through accretion and wind. Subsequently, the evolutionary paths in the two scenarios diverge.

When assuming $\Phi_B \propto M_d$, the accretion rate gradually decreases with time as magnetic flux is lost, and the evolution slows down. Even after 10^7 years of evolution, the disk still possesses about 23% of its initial mass. We also see that towards later evolution, wind mass loss dominates accretion. When assuming Φ_B is constant, on the other hand, we see that accretion rate and mass loss rate maintain approximately a constant level for ~2 Myrs. This is because accretion rates are largely regulated by the strength of poloidal fields, and in this case the field strength roughly stays constant. Constant accretion/mass loss rates lead to rapid depletion of disk materials, and we see that total disk mass plunges down in a runaway manner shortly after $t \sim 2$ Myrs.

The two scenarios discussed here can be considered as two extreme limits on magnetic flux evolution. Rapid loss of magnetic flux leads to slow evolution and long disk lifetime, while conservation of magnetic flux leads to a two-timescale behavior: disk depletion occurs on timescales much shorter than disk lifetime. The latter behavior was also discussed by Armitage et al. (2013) when assuming constant total magnetic flux. The reality may lie in between the two extreme scenarios, which we will use shortly as a way to constrain disk lifetime.

For comparison, we also show in dash-dotted lines in the left panel of Fig. 3.5 results from a pure viscous disk evolution model assuming constant $\alpha = 0.01$. Besides the sequential drop in surface density, disk evolution is characterized by the expansion of the outer disk (viscous spreading). Over the course of a few Myrs, the disk size (defined by the radius above a certain threshold surface density) has expanded by more than a factor of 2. This is much more significant than that in our fiducial disk evolution models, where viscous spreading is largely suppressed. Thus, we conclude that wind-dominated PPD evolution likely undergoes very little expansion as compared with viscous evolution models.

To better quantify the timescale of disk evolution and the significance of mass loss from disk winds, we have carried out a series of disk evolution calculations scanning the parameter space. We do not evolve the disk all the way to the end, but terminate the evolution when the disk has lost half of its mass, and call this time half disk lifetime $t_{\rm half}$. We do so mainly because long-term evolution calculations likely bare large uncertainties due to our ignorance on magnetic flux evolution. For each set of parameters, we perform two runs, evolving the disk either assuming magnetic flux conservation $\Phi_B = \text{const}$, or $\Phi_B \propto M_d$. As discussed earlier, we may consider these two cases as two extreme scenarios, which set the lower and upper limits of $t_{\rm half}$. Similarly, we evaluate $f_{\rm wind}$, the fractional mass loss from the wind compared to total mass loss due to both wind and accretion, over evolution time up to $t_{\rm half}$.

In Fig. 3.6, we plot $t_{\rm half}$ and $f_{\rm wind}$ from all runs in this parameter study, organized as a function of initial midplane β_0. It best summaries the main results discussed in this paper.

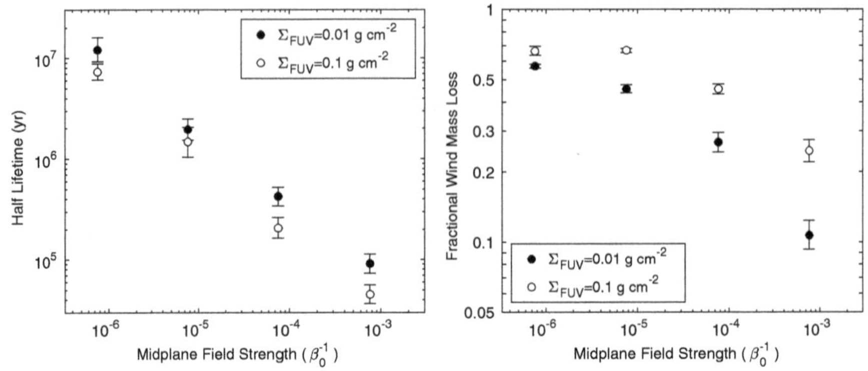

Fig. 3.6 Half disk lifetime ($t_{\rm half}$, *left*) and fractional mass loss via disk wind ($f_{\rm wind}$, *right*) from all our global disk evolution calculations. Each symbol reflects two disk evolution calculations, assuming total magnetic flux $\Phi_B \propto M_d$ (disk mass) or $\Phi_B = $ const. They are marked as upper and lower limits, with their geometric mean marked by the *circle*. *Filled and open circles* correspond to calculations assuming $\Sigma_{\rm FUV} = 0.01$ and $0.1\,{\rm g\,cm^{-2}}$, respectively. Symbols are plotted as a function of midplane field strength (characterized by β_0). Figure made based on data from Bai (2016)

Disk lifetime is determined by a combination of mass accretion and mass loss processes. It mostly depends on the amount of magnetic flux threading the disks. For our fiducial disk model, midplane plasma $\beta_0 \sim 10^5$ yields $2 \times t_{\rm half}$ that is in best agreement with observational constraints of disk lifetime (e.g., Haisch et al. 2001; Fedele et al. 2010) on a few Myr time scale. Note that increasing or reducing the field strength by a factor of only ~3 (a factor ~10 change in β_0) would yield disk lifetime that is either too short or too long.

All calculations with $\beta_0 = 10^5$ yield a fractional wind mass loss $f_{\rm wind}$ around one half (0.3–0.7). Wind mass loss becomes progressively less important towards stronger disk magnetization, but such scenario is very unlikely as constrained by disk lifetime. Therefore, we expect that wind mass loss plays a rather significant role in global disk evolution.

Stronger FUV radiation (hence larger penetration depth) leads to both enhanced accretion rate and enhanced outflow rate, with the latter being more significant (hence larger $f_{\rm wind}$). In our fiducial model, $f_{\rm wind}$ increases from 0.45 to 0.65 as $\Sigma_{\rm FUV}$ increases from 0.01 to $0.1\,{\rm g\,cm^{-2}}$, which again manifests the effect of magneto-photoevaporation. On the other hand, reduction of $t_{\rm half}$ is only modest. A factor of 10 difference in $\Sigma_{\rm FUV}$ typically leads to a factor of at most 3 difference in $t_{\rm half}$.

3.5 Summary, Discussion, and Future Directions

In this work, we have constructed a framework to study global evolution of PPDs that incorporates wind-driven accretion, wind mass loss, and viscous transport. The model is motivated by recent local simulations of PPD gas dynamics that have properly incorporated disk microphysics, which suggest that the MRI is largely suppressed in the disk interior and magnetized wind is launched in the externally heated and ionized surface layer from the FUV ionization front (Bai and Stone 2013b; Gressel et al. 2015). We further consider the fact that the well-ionized surface layer can be subject to the MRI in the outer region of PPDs (Perez-Becker and Chiang 2011; Simon et al. 2013a), which contributes to viscous angular momentum transport. Global wind kinematics is obtained from solving conservation laws along prescribed poloidal field lines, which provides reliable estimates of the wind-driven accretion rate and wind mass loss rate for given parameters. Our main findings include

- Disk evolution is largely dominated by MHD wind-driven accretion and mass loss. Contribution from the MRI can be important in the outermost disk but viscous spread is suppressed.
- The disk evolution timescale sensitively depends on the amount of magnetic flux threading PPDs, and how it co-evolves with the disk. Disk dispersal is rapid if the disk is able to retain most of its magnetic flux during evolution, otherwise, disk dispersal is gradual.
- Given typical disk lifetime of a few Myrs, the disk loses comparable amount of mass via disk wind and accretion. Most of the wind mass loss proceeds through the outer disk (beyond $\sim$10 AU). Fractional mass loss via disk wind increases with decreasing disk magnetization (increasing disk lifetime).
- The depth of FUV penetration is the main thermodynamic factor that affects disk evolution. Deeper FUV penetration depth slightly enhances wind-driven accretion rate, and more strongly enhances wind mass loss.

Mass loss from PPDs has long been considered as a result of photoevaporation. As a pure thermal wind, it does not exert a torque to the disk, and hence would evaporate the disk regardless of internal processes of disk angular momentum transport. As PPD winds are now expected to be magnetized in nature, wind mass loss and angular momentum transport are intrinsically coupled, leading to a correlation between disk lifetime and fractional wind mass loss. Our MHD wind model has taken into account both effects of magnetic field and thermodynamics, and we have shown that wind mass loss is modestly sensitive to thermodynamic effects, whereas the dominant role is played by amount of magnetic flux. We therefore suggest "magneto-photoevaporation" as a more appropriate term for a unified description of mass loss process from PPDs.

One important implication of wind mass loss is that mass is primarily removed from disk surface, while in low turbulent environment, most dust/solids settle and reside around disk midplane. The removal of largely dust-free gas, and in combination with radial drift (e.g., Youdin and Shu 2002), can directly enhance the dust-to-gas mass ratio. Similar conclusion has been reached in Gorti et al. (2015) in the context of photoevaporation on top of viscous disk evolution. If the disk loses a similar amount of mass via MHD winds compared with accretion, the enhancement factor would be a factor of $\sim$2 on average. The fact that wind mass loss is most significant towards the outer disk implies that the enhancement of dust-to-gas ratio may proceed from the outer disk inward. Using CO as gas tracer, tentative observational evidence of enhanced dust-to-(CO) gas ratio (by a factor of $\sim$10) has been inferred (Williams and Best 2014; Ansdell et al. 2016), and we expect gas removal by disk wind to be at least one important contributing factor (in addition to intrinsic CO depletion, either chemically (e.g., Yu et al. 2016) or dynamically (e.g., Xu et al. 2016). Theoretically, enhanced dust-to-gas mass ratio leads to favorable conditions for planetesimal formation (Johansen et al. 2009), and particularly allowing for smaller, more strongly coupled dust to participate in planetesimal formation (Bai and Stone 2010; Carrera et al. 2015).

3.5.1 Future Directions: Magnetic Flux Transport in PPDs

The largest uncertainty in our global disk evolution calculation lies in our ignorance on magnetic flux transport. With phenomenological treatment in this work, we have already seen the sensitive dependence of the disk evolutionary path on the specific prescriptions of magnetic flux evolution. This is why we have mainly focused on the bulk mass budget of the disk rather than the details of disk surface density evolution.

Magnetic flux transport has conventionally been studied in the advection-diffusion framework, with inward advective transport from viscously driven accretion and outward diffusive transport by turbulent or physical resistivity (Lubow et al. 1994). More recent works have taken into account disk vertical structure (Guilet and Ogilvie 2012, 2014), or radial resistivity profile (Okuzumi et al. 2014; Takeuchi and Okuzumi 2014). However, they all fall into the same framework, and have ignored additional disk microphysics, particularly the Hall effect and ambipolar diffusion, as well as the wind-driven accretion process.

In Fig. 3.7, we show our numerical experiments of magnetic flux transport in PPDs using the new Athena++ MHD code. We have implemented all three non-ideal MHD effects, and the simulations have included the Hall effect and ambipolar diffusion (AD). As a proof-of-concept, the Hall and AD diffusivities are not self-consistently obtained, but are prescribed based on chemistry calculations. More details on the simulation setup and analysis of the results are described in Bai and Stone (2017). The two non-ideal MHD effects introduce additional mechanisms for magnetic flux transport. In particular, the direction of transport due to the Hall term depends on the polarity of the large-scale poloidal field.

We have found that for poloidal field aligned with disk rotation axis, the Hall effect very efficiently drags magnetic flux radially inward in the disk midplane region as a result of the Hall drift (fed by the vertical gradient of toroidal magnetic field), and pushes magnetic flux outward above/below the midplane. This leads to a radially stretched field configuration, which in the meantime produces very strong, and oppositely directed toroidal field around the midplane. The overall evolution is a global manifestation of the Hall-shear instability (Kunz 2008), previously discussed in local shearing-box simulations (Lesur et al. 2014; Bai 2014). The inward transport

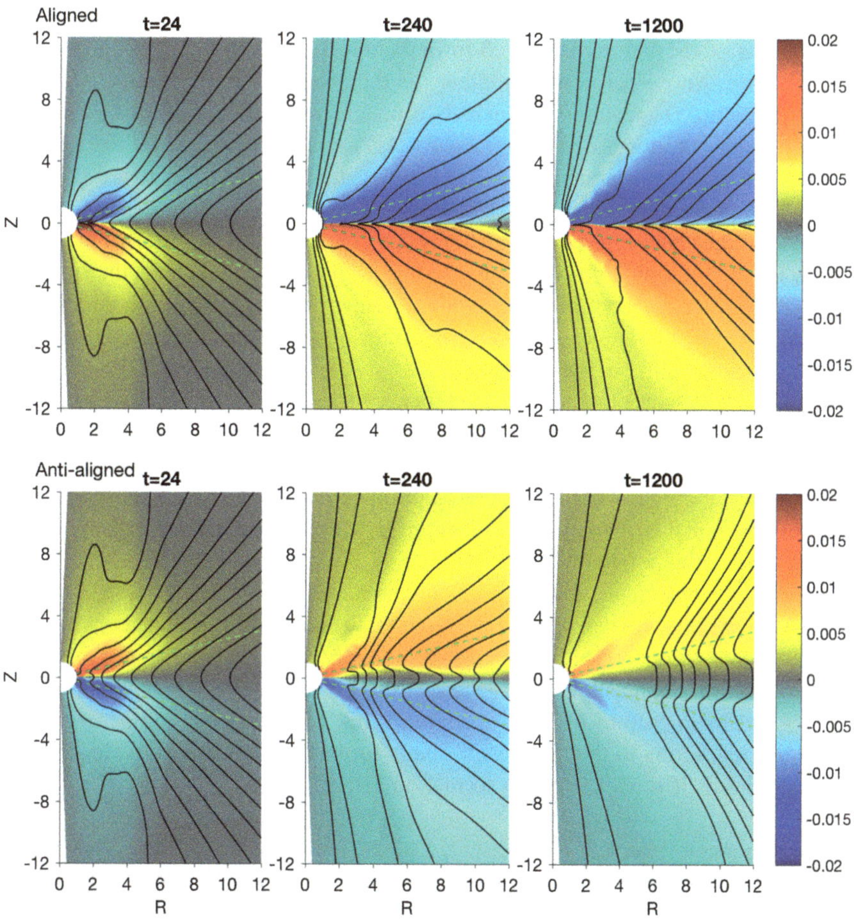

Fig. 3.7 Snapshots of magnetic field configuration represented by equally spaced contours (poloidal field) and color (toroidal field RB_ϕ) at $t = 24$, 240 and $1200\Omega_0^{-1}$ from our simulations of preliminary studies of magnetic flux, where Ω_0 is the Keplerian angular frequency of the innermost orbit. The poloidal field strength corresponds to midplane plasma $\beta_0 = 10^4$. *Top and bottom panels* are for simulations with poloidal field aligned and anti-aligned with disk rotation. *Green-dashed lines* mark 2.5 scale heights above/below the midplane (disk aspect ratio is $H/R = 0.1$), which roughly correspond to the transition from the non-ideal MHD dominated disk zone to the disk corona where ideal MHD applies as a proxy of FUV ionization. Figure taken from Bai and Stone (2017)

by Hall drift is balanced by outward transport by AD, and we have found that upon achieving a quasi-steady state, magnetic flux is slowly transported outward.

When poloidal field is anti-aligned with disk rotation, we find that the opposite occurs: the Hall drift pushes magnetic flux outward around the midplane region, driving poloidal field into a concave configuration across the midplane. Later on, poloidal field lines around the midplane straighten. This was discussed in Bai

(2014), where the situation is exactly the opposite as in the aligned case: horizontal components of the field are reduced towards zero instead of undergoing runaway amplification. Above/below the midplane, on the one hand, the Hall effect weakens compared with AD because of the density drop, and in the meantime, the concavely bent field configuration is more favorable for AD to transport magnetic flux outward. Overall, we find that magnetic flux is systematically transported outward at a rate about twice faster than the aligned case.

These studies have revealed the rich physics governing the evolution of magnetic flux in PPDs. It is thus of crucial importance in the future to address the problem of magnetic flux transport in PPDs in much greater detail. In particular, the distinct roles played by different non-ideal MHD effects require ionization-recombination chemistry to be properly incorporated into the studies. The MHD disk winds are also an integral part of the disk dynamics and magnetic flux evolution, which demands the simulation domain to extend to at least near the polar region to properly accommodate wind propagation. Moreover, the polarity-dependence in magnetic flux transport may already have operated in much earlier stages, since protostellar core collapse and disk formation (Tsukamoto et al. 2015; Wurster et al. 2016), which provides the initial condition for PPDs.

To conclude, we are at the beginning of a paradigm shift in our understandings of PPD evolution. The present work only represents an initial effort in the construction of a new framework. We anticipate that as more physics are incorporated, many of which are becoming feasible using global PPD simulations, we will be approaching the most realistic picture of PPD evolution. This will further allow us to couple global disk evolution to the theories of planet formation, and hopefully, lead to the most deterministic model of planet formation at last!

Acknowledgements I thank my collaborators, Jiani Ye, Jeremy Goodman, Feng Yuan, and Jim Stone for their contribution to part of the works presented here, and an anonymous referee for helpful comments. Finally, I acknowledge support from Institute for Theory and Computation at Harvard-Smithsonian Center for Astrophysics.

References

Alexander, R., Pascucci, I., Andrews, S., Armitage, P., Cieza, L.: The dispersal of protoplanetary disks. Protostars and Planets VI, pp. 475–496. University of Arizona Press, Tucson (2014), 1311.1819

Andrews, S.M., Wilner, D.J., Hughes, A.M., Qi, C., Dullemond, C.P.: Protoplanetary disk structures in Ophiuchus. Astrophys. J. **700**, 502–1523 (2009), 0906.0730

Andrews, S.M., Wilner, D.J., Hughes, A.M., Qi, C., Dullemond, C.P.: Protoplanetary disk structures in Ophiuchus. II. Extension to fainter sources. Astrophys. J. **723**, 1241–1254 (2010), 1007.5070

Ansdell, M., Williams, J.P., van der Marel, N., Carpenter, J.M., Guidi, G., Hogerheijde, M., Mathews, G.S., Manara, C.F., Miotello, A., Natta, A., Oliveira, I., Tazzari, M., Testi, L., van Dishoeck, E.F., van Terwisga, S.E.: ALMA survey of Lupus protoplanetary disks I: dust and gas masses (2016) ArXiv e-prints 1604.05719

Armitage, P.J.: Dynamics of protoplanetary disks. Annu. Rev. Astron. Astrophys. **49**, 195–236 (2011), 1011.1496
Armitage, P.J., Simon, J.B., Martin, R.G.: Two timescale dispersal of magnetized protoplanetary disks. Astrophys. J. Lett. **778**, L14 (2013), 1310.6745
Bai, X.N.: Wind-driven accretion in protoplanetary disks. II. Radial dependence and global picture. Astrophys. J. **772**, 96 (2013), 1305.7232
Bai, X.N.: Hall-effect-controlled gas dynamics in protoplanetary disks. I. Wind solutions at the inner disk. Astrophys. J. **791**, 137 (2014), 1402.7102
Bai, X.N.: Hall effect controlled gas dynamics in protoplanetary disks. II. Full 3D simulations toward the outer disk. Astrophys. J. **798**, 84 (2015), 1409.2511
Bai, X.N.: Towards a global evolutionary model of protoplanetary disks. Astrophys. J. **821**, 80 (2016), 1603.00484
Bai, X.N., Stone, J.M.: The effect of the radial pressure gradient in protoplanetary disks on planetesimal formation. Astrophys. J. Lett. **722**, L220–L223 (2010), 1005.4981
Bai, X.N., Stone, J.M.: Local study of accretion disks with a strong vertical magnetic field: magnetorotational instability and disk outflow. Astrophys. J. **767**, 30 (2013a), 1210.6661
Bai, X.N., Stone, J.M.: Wind-driven accretion in protoplanetary disks. I. Suppression of the magnetorotational instability and launching of the magnetocentrifugal wind. Astrophys. J. **769**, 76 (2013b), 1301.0318
Bai, X.N., Stone, J.M.: Hall effect-mediated magnetic flux transport in protoplanetary disks. Astrophys. J. **836**, 46 (2017). doi:10.3847/1538-4357/836/1/46, 1612.03912
Bai, X.N., Ye, J., Goodman, J., Yuan, F.: Magneto-thermal disk wind from protoplanetary disks. Astrophys. J. **818**, 152 (2016), 1511.06769
Balbus, S.A., Hawley, J.F.: A powerful local shear instability in weakly magnetized disks. I - linear analysis. II - nonlinear evolution. Astrophys. J. **376**, 214–233 (1991)
Baruteau, C., Crida, A., Paardekooper, S.J., Masset, F., Guilet, J., Bitsch, B., Nelson, R., Kley, W., Papaloizou, J.: Planet-disk interactions and early evolution of planetary systems. Protostars and Planets VI, pp. 667–689. University of Arizona Press, Tucson (2014), 1312.4293
Bergin, E.A., Aikawa, Y., Blake, G.A., van Dishoeck, E.F.: The chemical evolution of protoplanetary disks. Protostars and Planets V, pp. 751–766.University of Arizona Press, Tucson (2007) astro-ph/0603358
Blandford, R.D., Payne, D.G.: Hydromagnetic flows from accretion discs and the production of radio jets. Mon. Not. R. Astron. Soc. **199**, 883–903 (1982)
Carrera, D., Johansen, A., Davies, M.B.: How to form planetesimals from mm-sized chondrules and chondrule aggregates. Astron. Astrophys. **579**, A43 (2015), 1501.05314
Casse, F., Ferreira, J.: Magnetized accretion-ejection structures. V. Effects of entropy generation inside the disc. Astron. Astrophys. **361**, 1178–1190 (2000), astro-ph/0008244
Casse, F., Keppens, R.: Magnetized accretion-ejection structures: 2.5-dimensional magnetohydrodynamic simulations of continuous ideal jet launching from resistive accretion disks. Astrophys. J. **581**, 988–1001 (2002), arXiv:astro-ph/0208459
Chiang, E.I., Goldreich, P.: Spectral energy distributions of T Tauri stars with passive circumstellar disks. Astrophys. J. **490**, 368–376 (1997), arXiv:astro-ph/9706042
Desch, S.J., Turner, N.J.: High-temperature Ionization in protoplanetary disks. Astrophys. J. **811**, 156 (2015), 1508.07878
Fedele, D., van den Ancker, M.E., Henning, T., Jayawardhana, R., Oliveira, J.M.: Timescale of mass accretion in pre-main-sequence stars. Astron. Astrophys. **510**, A72 (2010), 0911.3320
Ferreira, J., Pelletier, G.: Magnetized accretion-ejection structures. III. Stellar and extragalactic jets as weakly dissipative disk outflows. Astron. Astrophys. **295**, 807 (1995)
Fromang, S., Latter, H., Lesur, G., Ogilvie, G.I.: Local outflows from turbulent accretion disks. Astron. Astrophys. **552**, A71 (2013), 1210.6664
Gorti, U., Hollenbach, D., Dullemond, C.P.: The impact of dust evolution and photoevaporation on disk dispersal. Astrophys. J. **804**, 29 (2015), 1502.07369
Gressel, O., Turner, N.J., Nelson, R.P., McNally, C.P.: Global Simulations of protoplanetary disks with Ohmic resistivity and ambipolar diffusion. Astrophys. J. **801**, 84 (2015), 1501.05431

Guilet, J., Ogilvie, G.I.: Transport of magnetic flux and the vertical structure of accretion discs - I. Uniform diffusion coefficients. Mon. Not. R. Astron. Soc. **424**, 2097–2117 (2012), 1205.6468
Guilet, J., Ogilvie, G.I.: Global evolution of the magnetic field in a thin disc and its consequences for protoplanetary systems. Mon. Not. R. Astron. Soc. **441**, 852–868 (2014), 1403.3732
Haisch, K.E. Jr., Lada, E.A., Lada, C.J.: Disk frequencies and lifetimes in Young clusters. Astrophys. J. Lett. **553**, L153–L156 (2001), astro-ph/0104347
Hartigan, P., Edwards, S., Ghandour, L.: Disk accretion and mass loss from Young stars. Astrophys. J. **452**, 736 (1995)
Hawley, J.F., Gammie, C.F., Balbus, S.A.: Local three-dimensional magnetohydrodynamic simulations of accretion disks. Astrophys. J. **440**, 742–763 (1995)
Hayashi, C.: Structure of the solar nebula, growth and decay of magnetic fields and effects of magnetic and turbulent viscosities on the nebula. Prog. Theor. Phys. Suppl. **70**, 35–53 (1981)
Henning, T., Semenov, D.: Chemistry in protoplanetary disks. Chem. Rev. **113**, 9016–9042 (2013), 1310.3151
Johansen, A., Youdin, A., Mac Low, M.: Particle clumping and planetesimal formation depend strongly on metallicity. Astrophys. J. Lett. **704**, L75–L79 (2009), 0909.0259
Kato, S.X., Kudoh, T., Shibata, K.: 2.5-Dimensional nonsteady magnetohydrodynamic simulations of magnetically driven jets from geometrically thin disks. Astrophys. J. **565**, 1035–1049 (2002)
Kenyon, S.J., Hartmann, L.: Spectral energy distributions of T Tauri stars - disk flaring and limits on accretion. Astrophys. J. **323**, 714–733 (1987)
Klahr, H., Hubbard, A.: Convective Overstability in radially stratified accretion disks under thermal relaxation. Astrophys. J. **788**, 21 (2014), 1403.6721
Konigl, A.: Self-similar models of magnetized accretion disks. Astrophys. J. **342**, 208–223 (1989)
Krasnopolsky, R., Li, Z.Y., Blandford, R.: Magnetocentrifugal launching of jets from accretion disks. I. Cold axisymmetric flows. Astrophys. J. **526**, 631–642 (1999), arXiv:astro-ph/9902200
Kunz, M.W.: On the linear stability of weakly ionized, magnetized planar shear flows. Mon. Not. R. Astron. Soc. **385**, 1494–1510 (2008), 0801.0974
Lesur, G., Kunz, M.W., Fromang, S.: Thanatology in protoplanetary discs. The combined influence of Ohmic, Hall, and ambipolar diffusion on dead zones. Astron. Astrophys. **566**, A56 (2014), 1402.4133
Li, Z.Y.: Magnetohydrodynamic disk-wind connection: self-similar solutions. Astrophys. J. **444**, 848–860 (1995)
Lovelace, R.V.E., Berk, H.L., Contopoulos, J.: Magnetically driven jets and winds. Astrophys. J. **379**, 696–705 (1991)
Lubow, S.H., Papaloizou, J.C.B., Pringle, J.E.: Magnetic field dragging in accretion discs. Mon. Not. R. Astron. Soc. **267**, 235–240 (1994)
Lynden-Bell, D.: On why discs generate magnetic towers and collimate jets. Mon. Not. R. Astron. Soc. **341**, 1360–1372 (2003), astro-ph/0208388
Lyra, W.: Convective overstability in accretion disks: three-dimensional linear analysis and nonlinear saturation. Astrophys. J. **789**, 77 (2014), 1405.3437
Marcus, P.S., Pei, S., Jiang, C.-H., Barranco, J.A., Hassanzadeh, P., Lecoanet, D.: Zombie vortex instability. I. A purely hydrodynamic instability to resurrect the dead zones of protoplanetary disks. Astrophys. J. **808**(1), 16, A87 (2015). http://adsabs.harvard.edu/abs/2015ApJ...808...87M
Natta, A., Testi, L., Alcalá, J.M., Rigliaco, E., Covino, E., Stelzer, B., D'Elia, V.: X-shooter spectroscopy of young stellar objects. V. Slow winds in T Tauri stars. Astron. Astrophys. **569**, A5 (2014), 1406.5630
Nelson, R.P., Gressel, O., Umurhan, O.M.: Linear and non-linear evolution of the vertical shear instability in accretion discs. Mon. Not. R. Astron. Soc. **435**, 2610–2632 (2013), 1209.2753
Okuzumi, S., Hirose, S.: Modeling magnetorotational turbulence in protoplanetary disks with dead zones. Astrophys. J. **742**, 65 (2011), 1108.4892
Okuzumi, S., Takeuchi, T., Muto, T.: Radial transport of large-scale magnetic fields in accretion disks. I. Steady solutions and an upper limit on the vertical field strength. Astrophys. J. **785**, 127 (2014), 1310.7446

Ostriker, E.C.: Self-similar Magnetocentrifugal disk winds with cylindrical asymptotics. Astrophys. J. **486**, 291–306 (1997), astro-ph/9705226

Ouyed, R., Pudritz, R.E., Stone, J.M.: Episodic jets from black holes and protostars. Nature **385**, 409–414 (1997)

Panoglou, D., Cabrit, S., Pineau Des Forêts, G., Garcia, P.J.V., Ferreira, J., Casse, F.: Molecule survival in magnetized protostellar disk winds. I. Chemical model and first results. Astron. Astrophys. **538**, A2 (2012), 1112.3248

Pascucci, I., Sterzik, M.: Evidence for disk photoevaporation driven by the central star. Astrophys. J. **702**, 724–732 (2009), 0908.2367

Pelletier, G., Pudritz, R.E.: Hydromagnetic disk winds in young stellar objects and active galactic nuclei. Astrophys. J. **394**, 117–138 (1992)

Perez-Becker, D., Chiang, E.: Surface layer accretion in conventional and transitional disks driven by far-ultraviolet ionization. Astrophys. J. **735**, 8 (2011), 1104.2320

Porth, O., Fendt, C.: Acceleration and collimation of relativistic magnetohydrodynamic disk winds. Astrophys. J. **709**, 1100–1118 (2010), 0911.3001

Pudritz, R.E., Norman, C.A.: Centrifugally driven winds from contracting molecular disks. Astrophys. J. **274**, 677–697 (1983)

Pudritz, R.E., Rogers, C.S., Ouyed, R.: Controlling the collimation and rotation of hydromagnetic disc winds. Mon. Not. R. Astron. Soc. **365**, 1131–1148 (2006), astro-ph/0508295

Raettig, N., Lyra, W., Klahr, H.: A parameter study for baroclinic vortex amplification. Astrophys. J. **765**, 115 (2013), 1212.4464

Ramsey, J.P., Clarke, D.A.: Simulating protostellar jets simultaneously at launching and observational scales. Astrophys. J. Lett. **728**, L11 (2011), 1012.3723

Rosenfeld, K.A., Andrews, S.M., Hughes, A.M., Wilner, D.J., Qi, C.: A spatially resolved vertical temperature gradient in the HD 163296 disk. Astrophys. J. 774, 16 (2013), 1306.6475

Shakura, N.I., Sunyaev, R.A.: Black holes in binary systems. Observational appearance. Astron. Astrophys. **24**, 337–355 (1973)

Simon, J.B., Bai, X.N., Armitage, P.J., Stone, J.M., Beckwith, K.: Turbulence in the outer regions of protoplanetary disks. II. Strong accretion driven by a vertical magnetic field. Astrophys. J. **775**, 73 (2013a), 1306.3222

Simon, J.B., Bai, X.N., Stone, J.M., Armitage, P.J., Beckwith, K.: Turbulence in the outer regions of protoplanetary disks. I. Weak accretion with no vertical magnetic flux. Astrophys. J. **764**, 66 (2013b), 1210.4164

Simon, J.B., Lesur, G., Kunz, M.W., Armitage, P.J.: Magnetically driven accretion in protoplanetary discs. Mon. Not. R. Astron. Soc. **454**, 1117–1131 (2015), 1508.00904

Spruit, H.C.: Magnetohydrodynamic jets and winds from accretion disks. In: Wijers, R.A.M.J., Davies, M.B., Tout, C.A. (eds.) NATO ASIC Proceedings 477: Evolutionary Processes in Binary Stars, pp. 249–286 (1996)

Stoll, M.H.R., Kley, W.: Vertical shear instability in accretion disc models with radiation transport. Astron. Astrophys. **572**, A77 (2014), 1409.8429

Suzuki, T.K., Inutsuka, S.i.: Disk winds driven by magnetorotational instability and dispersal of protoplanetary disks. Astrophys. J. Lett. **691**, L49–L54 (2009)

Suzuki, T.K., Inutsuka, S.i.: Magnetohydrodynamic simulations of global accretion disks with vertical magnetic fields. Astrophys. J. **784**, 121 (2014), 1309.6916

Takeuchi, T., Okuzumi, S.: Radial transport of large-scale magnetic fields in accretion disks. II. Relaxation to steady states. Astrophys. J. **797**, 132 (2014), 1310.7380

Testi, L., Birnstiel, T., Ricci, L., Andrews, S., Blum, J., Carpenter, J., Dominik, C., Isella, A., Natta, A., Williams, J.P., Wilner, D.J.: Dust evolution in protoplanetary disks. Protostars and Planets VI, pp. 339–361.University of Arizona Press, Tucson (2014), 1402.1354

Tsukamoto, Y., Iwasaki, K., Okuzumi, S., Machida, M.N., Inutsuka, S.: Bimodality of circumstellar disk evolution induced by the hall current. Astrophys. J. Lett. **810**, L26 (2015), 1506.07242

Turner, N.J., Fromang, S., Gammie, C., Klahr, H., Lesur, G., Wardle, M., Bai, X.N.: Transport and accretion in planet-forming disks. Protostars and Planets VI, pp. 411–432. University of Arizona Press, Tucson (2014) 1401.7306

Tzeferacos, P., Ferrari, A., Mignone, A., Zanni, C., Bodo, G., Massaglia, S.: On the magnetization of jet-launching discs. Mon. Not. R. Astron. Soc. **400**, 820–834 (2009)

Tzeferacos, P., Ferrari, A., Mignone, A., Zanni, C., Bodo, G., Massaglia, S.: Effects of entropy generation in jet-launching discs. Mon. Not. R. Astron. Soc. **428**, 3151–3163 (2013)

Vlahakis, N., Tsinganos, K., Sauty, C., Trussoni, E.: A disc-wind model with correct crossing of all magnetohydrodynamic critical surfaces. Mon. Not. R. Astron. Soc. **318**, 417–428 (2000), astro-ph/0005582

Walsh, C., Millar, T.J., Nomura, H.: Chemical processes in protoplanetary disks. Astrophys. J. **722**, 1607–1623 (2010), 1008.4305

Walsh, C., Nomura, H., Millar, T.J., Aikawa, Y.: Chemical processes in protoplanetary disks. II. On the importance of photochemistry and X-ray ionization. Astrophys. J. **747**, 114 (2012), 1201.2613

Wardle, M., Koenigl, A.: The structure of protostellar accretion disks and the origin of bipolar flows. Astrophys. J. **410**, 218–238 (1993)

Weidenschilling, S.J.: The distribution of mass in the planetary system and solar nebula. Astrophys. Space Sci. **51**, 153–158 (1977)

Williams, J.P., Best, W.M.J.: A parametric modeling approach to measuring the gas masses of circumstellar disks. Astrophys. J. **788**, 59 (2014), 1312.0151

Wurster, J., Price, D.J., Bate, M.R.: Can non-ideal magnetohydrodynamics solve the magnetic braking catastrophe? Mon. Not. R. Astron. Soc. **457**, 1037–1061 (2016), 1512.01597

Xu, R., Bai, X.N., Oberg, K.: Runaway freeze-out of volatiles in weakly turbulent protoplanetary disks (2016). ArXiv:160900796 1609.00796

Youdin, A.N., Shu, F.H.: Planetesimal formation by gravitational instability. Astrophys. J. **580**, 494–505 (2002), arXiv:astro-ph/0207536

Yu, M., Willacy, K., Dodson-Robinson, S.E., Turner, N.J., Evans II, N.J.: Probing planet forming zones with rare CO isotopologues. Astrophys. J. **822**, 53 (2016), 1603.08930

Zanni, C., Ferrari, A., Rosner, R., Bodo, G., Massaglia, S.: MHD simulations of jet acceleration from Keplerian accretion disks. The effects of disk resistivity. Astron. Astrophys. **469**, 811–828 (2007), arXiv:astro-ph/0703064

Chapter 4
Particle Trapping in Protoplanetary Disks: Models vs. Observations

Paola Pinilla and Andrew N. Youdin

Abstract With modern astronomical observatories, such as ALMA, we are now able to constrain the physical processes that govern planet formation by confronting model predictions with observations. Of particular interest is the growth and evolution of dust grains into larger solids, a crucial first step in the formation of planetesimals that shapes the diversity of exoplanetary systems. In traditional disk models, with smooth radial gradients, dust growth inevitably leads to the fast inward migration of pebbles, a challenge to understanding both planet formation and disk observations. However, recent high resolution observations at different wavelengths show that disks are not smooth, but show diverse structures, including rings, asymmetries, dips of emission, and spiral arms. In this chapter, we summarize how models of evolving gas and dust disks can explain these observed structures. We explore several phenomena that can create particle traps in gas disks, such as magneto-rotational instabilities and embedded planets. We place these models in the context of planetesimal formation theory broadly, and discuss open questions and future opportunities for both theoretical and observational work.

4.1 Introduction

Astronomers observe protoplanetary disks to better understand planet formation. In turn, disk observations constrain theories of planet formation. The goal of this chapter is to understand the modern interplay between planet formation theory and disk observations. Our particular focus will be the trapping of solids in disks. Particle trapping is especially important because it can be readily observed with modern instruments, and because it provides a theoretical basis for overcoming serious obstacles in the theory of planet formation. The idea that disk structures promote planet formation predates modern observations. Weizsäcker (1943) postulated that planets formed between large scale vortices in a turbulent disk. While prescient (and also attempting to explain Bode's Law of planetary spacing), von Weizsäcker's

P. Pinilla (✉) • A.N. Youdin
Steward Observatory, University of Arizona, 933 N. Cherry Ave., Tucson, AZ 85719, USA
e-mail: pinilla@email.arizona.edu; youdin@email.arizona.edu

M. Pessah, O. Gressel (eds.), *Formation, Evolution, and Dynamics of Young Solar Systems*, Astrophysics and Space Science Library 445,
DOI 10.1007/978-3-319-60609-5_4

theory is inconsistent with modern theories of particle concentration in vortices. Whipple (1972) realized that if a disk contains a ring with a local gas pressure maximum, then solid particles would collect in that ring. This basic mechanism remains relevant today.

The occurrence of protoplanetary disks around young stars is a consequence of angular momentum conservation during the collapse of molecular clouds. Young stars retain only a small fraction of the angular momentum of the original clouds, because their disks redistribute angular momentum into small amounts of gas and dust at large orbital distances. To explain the reservoir of angular momentum, Terebey et al. (1984) predicted that protoplanetary disks would extend a few hundred, or even a thousand, astronomical units (AU) from the host star. With a large radial extent, disks span a wide temperature range, from $\gtrsim$1000 K in the inner disk to 10–20 K in the outer parts, approaching the lower limit of 3 K set by the cosmic microwave background. Various species of gas and dust in disks span a broad range of temperatures and densities as they reprocess starlight (and also radiate the disk's own luminosity). Thus, protoplanetary disks have complex spectra, observable over a large range of wavelengths, from optical to centimeter-emission.

With the use of a coronograph, disks can be imaged out to tens of AU at optical and near-infrared wavelengths due to the scattering of light by small dust particles ($\lesssim$10 μm) in their surface layers. For the combined spectral energy distribution (SED) of the star plus disk, the emission at different wavelengths probes different regions of the disk, as shown in Fig. 4.1. In the infrared (from near to far-infrared), optically thick thermal emission dominates the disk's SED. The near-infrared (NIR, from $\sim$0.75 to 1.5 μm) emission probes the very inner regions ($\lesssim$1 AU) of disks, the mid-infrared (MidIR, from $\sim$1.5 to 15 μm) emission probes regions between $\sim$1 to $\sim$10 AU, and the far-infrared (FIR, from $\sim$15 to 100 μm) probe regions at few tens of AU in a typical T-Tauri disk. At sub-millimeter wavelengths, the (mostly) optically thin emission comes (mostly) from solids in the midplane of the outer disk, out to $\sim$100 AU. Most of the disk mass is expected to be in these outer regions, which contain most of the disk's surface area. High angular resolution observations at sub-mm wavelengthssub-can constrain the radial dust mass distribution of disks, a key input for planet formation models.

Observations of the gas content in disks are much more challenging. Molecular hydrogen, H_2, is most abundant constituent of young gas-rich disks. However, it has no permanent electric dipole moment and hence very weak rotational and vibrational lines. Other molecular lines, such as hydrogen deuteride (HD) and CO (in particular its isotopologues), can be used to trace the gas content in disks in addition to disk kinematics (e.g., rotation). However the mass fraction of these minor species relative to H_2—and thus the total disk mass—is uncertain. To combat these uncertainties, observations of both the gas and dust components give a complementary understanding of disk structure and planet formation. The complex interactions between gas and dust affect the disk dynamics and the growth of solids, as explained in this chapter.

Our observational understanding of disks has advanced dramatically over the last 30 years. The Infrared Astronomical Satellite (IRAS) showed that 60–80%

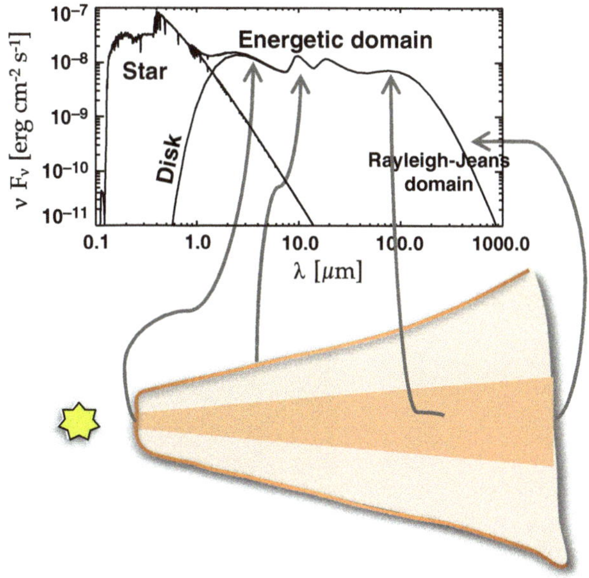

Fig. 4.1 (*Top*) The spectral energy distribution (SED) of a star plus its protplanetary disk. (*Bottom*) A cartoon indicating the different regions of the disk probed by the SED. The near-infrared (NIR, from ~0.75 to 1.5 μm) emission probes the hot inner regions ($\lesssim$1 AU) of disks, the mid-infrared (MidIR, from ~1.5 to 15 μm) emission probes warm regions between ~1 to ~10 AU, and the far-infrared (FIR, from ~15 to 100 μm) probes regions at few tens of AU in a typical T-Tauri disk. At sub-millimeter wavelengths, the (mostly) optically thin emission comes mainly from solids in the midplane of the outer disk. Figure adapted from Dullemond et al. (2007)

of young stars possesses protoplanetary disks, as measured by the infrared excess of circumstellar material (Strom et al. 1989). The shape of the spectral energy distributions (SEDs) is consistent with a dusty, flared disk (Adams et al. 1987; Kenyon and Hartmann 1987). Contemporaneously, millimeter wavelength observations of accreting T-Tauri stars revealed the presence of large grains (Weintraub et al. 1989). The Hubble Space Telecope eventually provided resolved images of flattened disks, in silhouette against the background light in the Orion Nebula as shown in Fig. 4.2 (O'dell and Wen 1994). More recently, infrared space telescopes such as *Spitzer* and *Herschel* have significantly refined our understanding of disk's inner regions, at ~0.5–20 AU. Knowledge of the outer disk has been refined by sub-mm interferometers such as the Sub-millimeter Array (SMA) and the Plateau de Bure Interferometer (PdBI). Observations have consistently shown that massive disks of gas and dust, i.e. with a mass within roughly an order of magnitude of our Solar System, persist for $\lesssim$10 Myr around young stars (e.g., Haisch et al. 2001; Hernández et al. 2007; Fedele et al. 2010; Pfalzner et al. 2014; Ribas et al. 2015).

Careful modeling of observed disk SEDs and images is required to better understand the structure and evolution of protoplanetary disks (see, e.g., reviews by Wyatt 2008; Dullemond and Monnier 2010; Williams and Cieza 2011). However

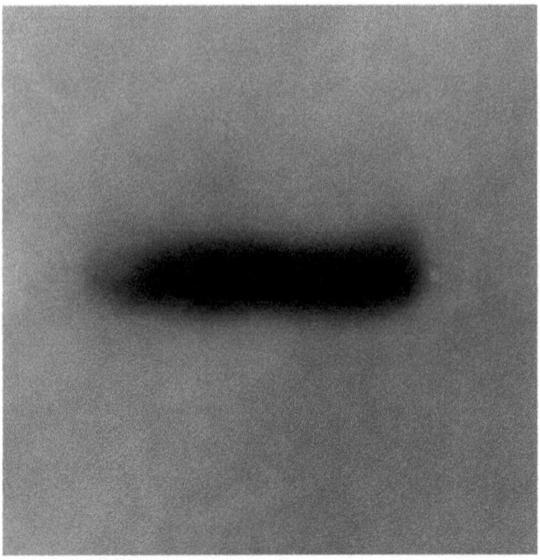

Fig. 4.2 Image of an edge-on disk around a young star in the Orion nebula taken with the Hubble Space Telescope (HST). The disk is seen in silhouette against the background light of the Orion Nebula. Image credit: Mark McCaughrean (Max-Planck-Institute for Astronomy), C. Robert O'Dell (Rice University), and NASA/ESA

certain features of the disk, and thus certain details of planet formation do not reveal themselves observationally. For instance, dust grains larger than centimeter sizes do not emit significant radiation due to their very low opacities, leaving an observational gap from centimeter pebbles to planets.

Theoretical models are required not only to interpret observations, but also to fill these gaps in observational knowledge. Modern theory is complemented by detailed computer simulations. Relevant computational methods include 3D radiative transfer calculations, 3D (magneto-) hydrodynamics, and numerical methods to follow the coagulation, fragmentation, and bouncing of dust particles in disks. The standard model for planet formation is the core accretion hypothesis, which connects the formation of terrestrial planets to gas-rich giant planets (see Youdin and Kenyon 2013, for a review of various planet formation theories). In the core accretion scenario, a massive solid core, that forms while the gas rich disk is still present, induces runaway gas accretion, leading to the formation of a giant planet (Perri and Cameron 1974; Pollack et al. 1996). The minimum core mass required for significant gas accretion is typically $\sim 10\,M_{\oplus}$, but depends on local disk conditions (Piso and Youdin 2013).

This core accretion phase is simultaneously affected by the gravitational interaction between the core and the surrounding disk of gas and planetesimals, which cause the planet to migrate. When planets interact with the disk, spiral density waves are driven, and torques exerted on both disk and planet. Theory and simulations show that the resulting planet migration is most often inwards, but can be outwards for certain planet masses and disk conditions (see, e.g., reviews by Kley and Nelson 2012; Baruteau et al. 2014). The torques exerted by massive planets can eventually open gaps in the disk, which aid in disk clearing, as covered in Sect. 4.6.1.

In this chapter, we primarily focus on the aspects of planet formation that are closely tied to disk observations, specifically to the trapping of roughly mm-size dust grains in resolved disk structures, as seen in different protoplanetary disks

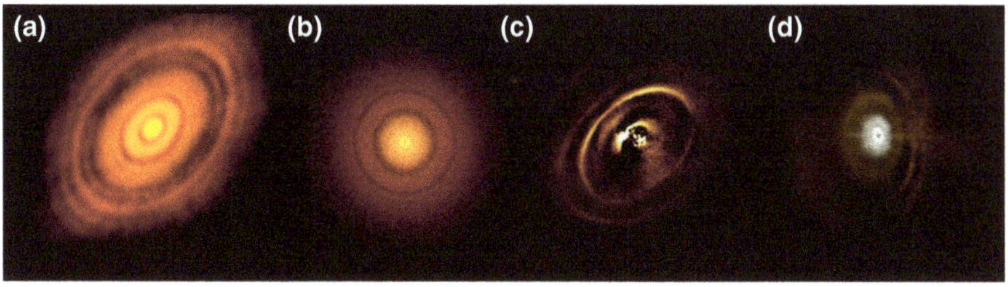

Fig. 4.3 Multiple rings have been observed in high resolution images of protoplanetary disks.The above rings were imaged by ALMA at sub-mm wavelengths ((**a**) HL Tau, ALMA Partnership et al. 2015), ((**b**) TW Hya, Andrews et al. 2016) or by VLT/SPHERE in scattered starlight ((**c**) de Boer et al. 2016, RX J1615), and ((**d**) Ginski et al. 2016, HD 97048). A variety of theories have been proposed to explain these rich structures, see text

with modern powerful telescopes such as ALMA and VLT/SPHERE (Fig. 4.3). We review a range of theories for these particle traps. Some mechanisms do not involve planets, in which case thus these particle traps can be connected to the first generation of planetesimal formation. Other scenarios involve planet-induced disk structures, which probes a later evolutionary stage of planet formation. Nevertheless, planet-induced traps are likely sites of subsequent generations of planet formation. In Sect. 4.8, we complement our review of observable structures with a summary of planetesimal formation theory, some aspects of which remain unobservable.

4.2 Radial Drift and Particle Trapping

One of the most challenging problems of the growth from small grains to planetesimals is to understand how the "radial drift barrier" is overcome (e.g., Whipple 1972; Weidenschilling 1977). In standard gas disks, aerodynamic drag causes significant inward migration of solids between $\sim 10\,\mu\text{m}$ and ~ 1 km in size, as explained in detail below. If solids do not grow through this size range fast enough, planet-forming solids will be lost to the inner disk or host star. In addition to being an obstacle to planet formation, this radial drift is a challenge for understanding observations of protoplanetary disks, which indicate that many mm-sized solids persist in the outer disk for millions of years, despite radial drift.

In this section, we will quantify how drag forces in disks produce radial drift. We will then show how long-lived particle traps can impede radial drift. Both phenomena exemplify a general tendency of particles to seek regions of higher gas pressure.

Aerodynamical drag forces depend on the properties of solid grains, such as mass and cross-sectional area, and of the gas disk, such as density and molecular viscosity. Though many drag laws exist, small solids in a gas disk typically fall in the *Epstein regime*. Epstein drag holds for subsonic relative motion (easily satisfied for the nearly circular orbits of interest) and for particle radii, a, which satisfy $a \lesssim (9/4)\lambda_{\rm mfp}$. The mean free path of gas molecules is given by

$$\lambda_{\rm mfp} = \frac{\mu m_{\rm H}}{\sqrt{2}\rho\sigma_m}, \quad (4.1)$$

where μ is the mean molecular mass, $m_{\rm H}$ the proton mass, ρ the gas mass density, and $\sigma_m = 1.58 \times 10^{-15}\,{\rm cm}^2$ is the molecular cross section for molecular hydrogen.

For properties typical of the solar nebula near 1 AU, $\rho \sim 5 \times 10^{-10}\,{\rm g\,cm}^{-3}$ and $(9/4)\lambda_{\rm mfp} \sim 5\,{\rm cm}$. At larger disk radii, the gas density drops sharply and $\lambda_{\rm mfp}$ increases further. Thus Epstein drag is the most relevant case for the grain sizes and disks conditions that are treated in this chapter. However other drag regimes are sometimes relevant. For larger grain sizes, first the viscous Stokes law applies. Then, for yet larger sizes, various turbulent drag laws—the ones we experience on a daily basis—apply (Adachi et al. 1976). For the numerical calculations presented in this chapter, the appropriate drag regime (Epstein or otherwise) is always used.

The details of the drag laws can be captured in a characteristic *stopping time*, t_s, which can then be compared to other relevant timescales. To define t_s, we express the drag force for a particle of mass m, moving relative to the gas with a velocity $\Delta\mathbf{v}$, as

$$\mathbf{F}_{\rm drag} = -\frac{m}{t_s}\Delta\mathbf{v}\,. \quad (4.2)$$

Physically, t_s gives the exponential decay time of $\Delta\mathbf{v}$, ignoring the other forces that generate the relative motion. In the Epstein drag regime, the stopping time is given by

$$t_s = \frac{\rho_s}{\rho}\frac{a}{v_{\rm TH}}, \quad (4.3)$$

with ρ_s being the material volume density of the solid and $v_{\rm TH} = \sqrt{8/\pi}\,c_s$ the mean thermal velocity of the gas molecules, closely related to the isothermal sound speed $c_s = \sqrt{P/\rho}$, which in turn depends on the thermal pressure P.

The dimensionless Stokes number ${\rm St} \equiv \Omega t_s$ measures the importance of drag forces by comparing t_s to the Keplerian frequency at radius, r,

$$\Omega = \sqrt{\frac{GM_\star}{r^3}}, \quad (4.4)$$

where G is the gravitational constant and $M_\star$ is the stellar mass. Small (large) values of St correspond to strong (weak, respectively) drag forces. For optimal coupling at ${\rm St} \simeq 1$, the effects of drag forces are usually most pronounced. For Epstein drag near the disk midplane

$$\mathrm{St} = \frac{a\rho_s}{\Sigma_g}\frac{\pi}{2}, \tag{4.5}$$

where the column density of gas, $\Sigma_g = \sqrt{2\pi}\rho c_s/\Omega$, follows from vertical hydrostatic balance, and our exact value assumes a vertically isothermal disk.

Gas drag modifies the orbits of particles, making them slightly non-Keplerian. Our derivation of the orbital motion of particles in gas disks mirrors that of Takeuchi and Lin (2002) and Youdin (2010). We point the interested reader to Nakagawa et al. (1986) for an alternate derivation which (a) treats the collection of solids as a pressureless fluid and thereby (b) accounts for the collective feedback of the total particle mass on the coupled drift motion of solids and gas. Here we consider the "test particle" limit, valid when the distributed mass density of particles is below the gas density. This limit is appropriate for moderately turbulent disks, which limit the sedimentation of solids to a dense midplane layer.

We consider the motion of particles and gas near the disk midplane, where the height $z \ll r$. The radial and azimuthal speeds are $v_{r,\mathrm{dust\ or\ gas}}$, $v_{\phi,\mathrm{dust\ or\ gas}}$ for the dust (i.e., solid) or gas components as indexed. Gas drag damps motions to nearly circular, Keplerian motion with $v_{r,\mathrm{dust\ or\ gas}} \sim 0$ and $v_{\phi,\mathrm{dust\ or\ gas}} \sim v_\mathrm{K} \equiv \Omega r$ to leading order. We derive the leading order deviations from this Keplerian motion.

For the gas, orbital motion is determined by the centrifugal balance of radial gravity and pressure forces as

$$\frac{v_{\phi,\mathrm{gas}}^2}{r} = \frac{GM_\star}{r^2} + \frac{1}{\rho}\frac{dP}{dr}. \tag{4.6}$$

We define the pressure support parameter as

$$\eta = -\frac{1}{\rho r\Omega^2}\frac{dP}{dr}. \tag{4.7}$$

In regions with smooth pressure gradients, i.e. $|dP/dr| \sim P/r$, pressure support is weak and $\eta \sim (c_s/v_\mathrm{K})^2 \sim 10^{-3} \ll 1$. We can thus solve Eq. (4.6) as[1]

$$v_{\phi,\mathrm{gas}} = v_\mathrm{K}\sqrt{1-\eta} \simeq v_\mathrm{K}\left(1-\eta/2\right), \tag{4.8}$$

showing that gas orbital motion is slightly sub-Keplerian in a smooth disk where the pressure decreases monotonically with radius (as the density and temperature also do). The radial motion of the gas is determined by the disk's angular momentum transport, often approximated as a viscous accretion process. Here $v_{r,\mathrm{gas}}$ is treated as a free parameter.

[1]Many works use the original (Adachi et al. 1976), here primed, definition of $\eta' = \eta/2$ so that $v_{\phi,\mathrm{gas}} - v_\mathrm{K} \simeq \eta v_\mathrm{K}$. The lack of a uniform convention requires attention for this factor of 2.

The response of a solid particle to this non-Keplerian gas motion is given by the radial and angular momentum equations

$$\frac{dv_{r,\mathrm{dust}}}{dt} = \frac{v_{\phi,\mathrm{dust}}^2}{r} - \frac{v_\mathrm{K}^2}{r} - \frac{1}{t_s}(v_{r,\mathrm{dust}} - v_{r,\mathrm{gas}}), \quad (4.9)$$

$$\frac{d(rv_{\phi,\mathrm{dust}})}{dt} = -\frac{r}{t_s}(v_{\phi,\mathrm{dust}} - v_{\phi,\mathrm{gas}}), \quad (4.10)$$

where the right-hand side of Eq. (4.9) includes the relevant radial accelerations: centrifugal, gravity, and drag, respectively. The right-hand side of Eq. (4.10) gives the torque from azimuthal drag forces. For a simple algebraic solution of the particle motion, we must eliminate time derivatives and linearize. The particle radial acceleration, $dv_{r,\mathrm{dust}}/dt \sim v_{r,\mathrm{dust}}^2/r$ is a higher order term that can be neglected in the radial force balance. Since the leading contribution to angular momentum is Keplerian motion, the rate of angular momentum change is given, to good approximation, by the migration rate as

$$\frac{d(rv_{\phi,\mathrm{dust}})}{dt} \simeq \frac{drv_\mathrm{K}}{dr} v_{r,\mathrm{dust}} = \frac{1}{2} v_k v_{r,\mathrm{dust}}, \quad (4.11)$$

The nonlinearity of $v_{\phi,\mathrm{dust}}$ in Eq. (4.9) is eliminated to leading order as

$$\begin{aligned} v_{\phi,\mathrm{dust}}^2 - v_k^2 &= (v_{\phi,\mathrm{dust}} + v_k)(v_{\phi,\mathrm{dust}} - v_k) \\ &\simeq 2v_\mathrm{K}\left[(v_{\phi,\mathrm{dust}} - v_{\phi,\mathrm{gas}}) - (v_k - v_{\phi,\mathrm{gas}})\right] \\ &\simeq 2v_\mathrm{K}\left[v_{\phi,\mathrm{dust}} - v_{\phi,\mathrm{gas}} - \eta v_\mathrm{K}/2\right]. \end{aligned} \quad (4.12)$$

With these simplifications, the leading order solution of Eqs. (4.9) and (4.10) is

$$v_{\phi,\mathrm{dust}} - v_\mathrm{K} = -\frac{\eta v_\mathrm{K} + v_{r,\mathrm{gas}}\mathrm{St}}{2(1+\mathrm{St}^2)}, \quad (4.13)$$

$$v_{r,\mathrm{dust}} = \frac{v_{r,\mathrm{gas}}}{1+\mathrm{St}^2} - \frac{\eta v_\mathrm{K}\mathrm{St}}{1+\mathrm{St}^2}. \quad (4.14)$$

Equation (4.13) shows that large solids ($\mathrm{St} \gg 1$) exhibit Keplerian motion with $v_{\phi,\mathrm{dust}} \to v_\mathrm{K}$, while small dust (with $\mathrm{St} \ll 1$) has the same sub-Keplerian orbital speed as the gas with $v_{\phi,\mathrm{dust}} \to v_\mathrm{K}(1-\eta/2)$.

The radial drift of dust particles in Eq. (4.14) is plotted in the left panel of Fig. 4.4 at 3 AU from the star, for two different models of the gas disk, the standard minimum mass solar nebula (MMSN, which is an estimation of the minimum amount of material necessary to build the planets of our Solar System), and the minimum mass extrasolar nebula (MMEN, which is the minimum amount of material to build super-Earths, i.e. $M \sin i \lesssim M\oplus$ and with periods lower than 100 days, found in multiple-planet systems), from Chiang and Laughlin (2013). The smallest particles

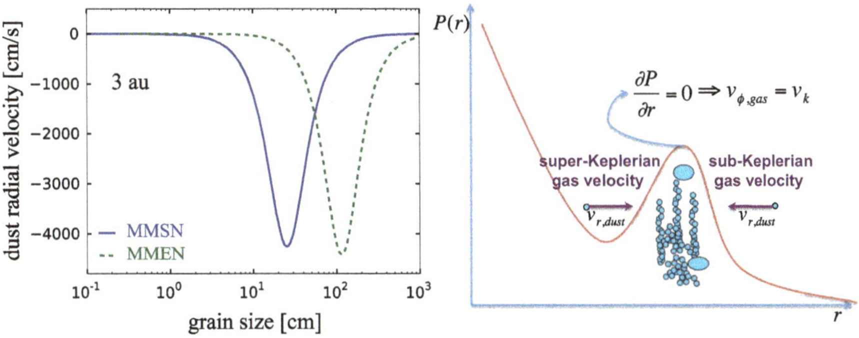

Fig. 4.4 *Left panel*: total radial dust velocity as a function of grain size [Eq. (4.14)] calculated at 3 AU from the star, assuming two different profiles for the gas surface density, the minimum mass solar nebula (MMSN), and the minimum mass extrasolar nebula (MMEN). *Right panel*: sketch of particle trapping in a pressure maximum

with St $\lesssim v_{r,\text{gas}}/(\eta v_\text{K}) \ll 1$, move with the gas accretion flow at $v_{r,\text{dust}} \sim v_{r,\text{gas}}$, as given by the first term in Eq. (4.14). This motion is not evident in the plot since $v_{r,\text{gas}} \sim$ a few cm/s (for a gas accretion rate $\dot{M} \sim 10^{-8} M_\odot$/yr) is small compared to the fastest drift speeds of $\eta v_\text{K}/2 \simeq 40$ m/s.

Thus the second term of Eq. (4.14), due to the gas headwind, explains the fastest particle drift speeds. This term becomes small for both large and small particles, St $\ll 1$ and $\gg$1, giving the fastest inward drift for optimal coupling at St $= 1$. In the case of the MMSN, the particles with St $= 1$ correspond to $\sim$30 cm sized particles, but for the denser MMEN this critical size is $\sim$1 m. The maximum inward drift speed for St $= 1$ is $\eta v_\text{K}/2 \sim 4000\,\text{cm}\,\text{s}^{-1}$. At these speeds, optimally coupled solids drift towards the star on timescales of $\sim$300 yr.

These rapid timescales are a serious challenge to planetesimal formation by coagulation, which requires particle to grow gradually through St ~ 1. Indeed, the infamous "meter-sized barrier" refers in part to these rapid drift speeds, and also to the fact that particles in this size range are prone to destructive collisions, as we discuss shortly. The association of the drift barrier with "meter sizes" is most appropriate near 1 AU in gas rich disks. Moving to the outer disk, particles with St $= 1$ are smaller, because Σ_g decreases, see Eq. (4.5). At $\sim$50 AU, the fastest drifting solids are only $\sim$ millimeter sized in MMSN type models.

The drift speed estimates above assume a smooth disk, with $dP/dr < 0$ [and thus $\eta > 0$ from Eq. (4.7)]. This assumption should hold for the *average* behavior of disks. However disks could develop pressure bumps for a variety of reasons, as detailed below. Regions where the pressure gradient is positive, $dP/dr > 0$ (and $\eta < 0$), will drive the *outward* drift of solids. Thus particles are driven towards a pressure maximum, from either side of the maximum. At the pressure maximum, $dP/dr = 0$ and drift stops. The right panel of Fig. 4.4 depicts this particle collection mechanism. While pressure minima also have $dP/dr = 0$, they are an unstable equilibrium; particles will drift away from particle minimum if they are nudged in any direction.

So far, we have assumed an axisymmetric disk in our discussion of drift and particle trapping. It turns out that non-axisymmetric pressure maxima can also act as effective particle traps. For example, anti-cyclonic vortices have a pressure maximum at their center, and particles can collect near the centers of these vortices (e.g., Barge and Sommeria 1995; Li et al. 2000, 2005; Klahr and Bodenheimer 2003; Varnière and Tagger 2006; Lin 2014). Spiral arms in self-gravitating disks are another form of non-axisymmetric pressure maximum that can collect solids (e.g., Rice et al. 2004; Dipierro et al. 2015a). Predominantly axisymmetric particle traps could be generated at the outer edge of gap carved by a massive planet (e.g., Rice et al. 2006; Paardekooper and Mellema 2006; Gonzalez et al. 2012; Zhu et al. 2012). Magnetohydrodynamic (MHD) effects could also cause ring-like pressure bumps, either to zonal flows (e.g., Johansen et al. 2009a; Uribe et al. 2011; Dittrich et al. 2013; Simon and Armitage 2014) or at the boundary of MHD active and dead zones (e.g., Kretke and Lin 2007; Dzyurkevich et al. 2010; Drążkowska et al. 2013; Ruge et al. 2016).

We will explore the details of these mechanisms in this chapter. Here, we note some general features of all of these quasi-static, i.e. slowly evolving, particle traps. The first is that particles with $\mathrm{St} = 1$ will drift most rapidly into particle traps, and thus should concentrate most effectively. Very small solids will be coupled to the accretion flow of the gas, $v_{r,\mathrm{gas}}$, as explained above, and will not be trapped as effectively. Moreover disk turbulence will preferentially diffuse smaller grains away from particle traps. To better understand this effect, we note that the diffusion rate of solids is fairly similar for $\mathrm{St} \ll 1$ up to $\mathrm{St} \simeq 1$ (Youdin and Lithwick 2007). However, the slower drift rate of $\mathrm{St} \ll 1$ solids means that they return more slowly to the pressure maximum and are preferentially lost.

Our goal is to understand how particle traps affect multi-wavelength observations of protoplanetary disks. For this goal, we must understand how grain sizes and St values evolve due to coagulation and fragmentation. Of course this size evolution may differ once particles are caught in dust traps, and moreover these traps may favor the gravitational collapse into planetesimals as a solution to the meter-size barrier.

4.3 Overview of Dust Evolution

The evolution of dust properties in protoplanetary disks—such as mass and size—arises from complex physical processes with several unresolved questions. For example: what is the fragmentation velocity for grain collisions and how does it depend on the material properties? Do the dust particles grow as fluffy aggregates or compact solids, and can we observationally distinguish these possibilities? Many of these questions have been addressed in several reviews as, e.g., Dominik et al. (2007), Blum and Wurm (2008), Testi et al. (2014), and Birnstiel et al. (2016). In this section, we present a short summary of the most relevant issues for the models presented in this chapter.

The initial conditions of the dust distribution in protoplanetary disks is not well understood. Usually, it is assumed that dust particles in protoplanetary disks are inherited from the interstellar medium as sub-micron-sized particles that are well coupled to the gas (Mathis et al. 1977). These particles are expected to evolve via collisions and transport. Dust collisions can lead to different outcomes: sticking, due to electrostatic van der Waals forces; bouncing, which does not alter the mass of the grains after the impact; mass transfer, where mass is transferred between grains (to the larger grain for growth); and erosion and/or fragmentation where grains lose mass into smaller collisional products. Van der Waals forces between molecules arise from dipole–dipole interactions. For dust particles with mantles with electric dipoles (e.g., particles with water ice mantels), the attractive force is strong. The collision outcome varies greatly with particle size and impact speed, with other factors such as porosity, impact angle, and grain composition also playing important roles (e.g., Blum and Wurm 2008; Windmark et al. 2012).

Laboratory experiments and numerical simulations of dust collisions have been probing the vast parameter space of collision outcomes (e.g., Kempf et al. 1999; Blum and Wurm 2000, 2008; Krause and Blum 2004; Paszun and Dominik 2006; Wada et al. 2009, 2011; Gundlach and Blum 2015; Musiolik et al. 2016). Overall, these experiments and models agree with the fact that collisions between dust aggregates without water ice mantles and with an impact velocity of $>1\,\mathrm{m\,s^{-1}}$ do not result in sticking collisions. However, grains with water ice mantles can impact at relatively high velocities, i.e. $\gtrsim 10$–$50\,\mathrm{m\,s^{-1}}$, and potentially avoid fragmentation.

The relative velocities between dust grains arises from several effects, including thermal Brownian motion, settling to the midplane, turbulent mixing, and radial drift. Brownian motion dominates the collision rate for very small particles (sub-micron sized dust particles). For larger particles (micron to sub-millimeter dust particles) settling to the midplane is an important source for dust collisions. Ignoring for the moment the stirring by gas turbulence, the vertical velocity of grains settling to the midplane $v_{z,\mathrm{settling}}$ is calculated by equating F_{drag} [Eq. (4.2)] with the vertical gravitational force, $F_{\mathrm{grav}} = -m\Omega^2 z$, giving $v_{z,\mathrm{settling}} = -z\Omega\mathrm{St}$ (Dullemond and Dominik 2004). Hence, the settling speed decreases close to the midplane where the gas density increases, and it also increases for particles with higher Stokes number. Note that here and elsewhere in this section we assume $\mathrm{St} \lesssim 1$. For $\mathrm{St} > 1$ orbital inclination oscillations dominate the vertical motion (Youdin 2010).

When particles start to decouple from the gas, turbulence and radial drift are the dominant sources for their relative velocities. As with radial drift, relative velocities due to turbulence, v_{turb}, depend on the Stokes number (Ormel and Cuzzi 2007), roughly as

$$v_{\mathrm{turb}}^2 \sim 3\alpha\mathrm{St}\, c_s^2\,, \tag{4.15}$$

where the larger St in the collisional pair is the relevant value, with c_s being the sound speed and α as the dimensionless parameter in the effective turbulent viscosity of Shakura and Sunyaev (1973),

$$\nu = \alpha c_s h \quad \text{with} \quad h = \frac{c_s}{\Omega}. \tag{4.16}$$

An estimate of the maximum grain size produced by collisional growth follows from the assumption that sticking occurs below the fragmentation velocity $v_{\rm frag}$ and turbulence dominates collision speeds. In this case, Eq. (4.15) and the definition of the Stokes number [Eq. (4.5)] give the maximum grain size as

$$a_{\rm frag} = \frac{2}{3\pi}\frac{\Sigma}{\rho_s \alpha}\frac{v_{\rm frag}^2}{c_s^2}, \tag{4.17}$$

known as the fragmentation barrier. Growth to St $\simeq 1$ requires weak turbulence and/or high $v_{\rm frag}$ as

$$\alpha \lesssim \frac{1}{3}\frac{v_{\rm frag}^2}{c_s^2} \sim 3 \times 10^{-7} \left(\frac{v_{\rm frag}}{1\ \rm m/s}\right)^2 \left(\frac{c_s}{1\ \rm km/s}\right)^{-2}. \tag{4.18}$$

When particles stick efficiently, growth can be limited due to the rapid timescale of radial drift. A simple estimate of the growth times scale is $t_{\rm growth} = \Sigma_g/(\Omega \Sigma_d)$, with Σ_g and Σ_d the gas and dust surface density, respectively (Birnstiel et al. 2012). Comparing to the drift timescales [using Eq. (4.14)] gives the drift barrier,

$$a_{\rm drift} = \frac{2\Sigma_d}{\pi \rho_s}\frac{v_K^2}{c_s^2}\left|\frac{d\ln P}{d\ln r}\right|^{-1}. \tag{4.19}$$

Overall, it is expected that in smooth disks, the maximum grain size decreases with radius as multi-wavelength observations at (sub-) millimeter emission show for several disks (e.g., Banzatti et al. 2011; Guilloteau et al. 2011; Pérez et al. 2012; Tazzari et al. 2016).

4.4 Observational Evidence of Millimeter Grains in Protoplanetary Disks

At (sub)-millimeter wavelengths, the total flux at a given frequency F_ν from the optically thin region is sensitive to the dust mass of the disk and it is given by (e.g., Hildebrand 1983)

$$F_\nu = \frac{M_{\rm dust}}{d^2}\kappa_\nu B_\nu(T), \tag{4.20}$$

where d is the distance to the source, κ_ν is the dust opacity at a given frequency, T is the characteristic temperature of the emitting region, and $B_\nu(T)$ is the Planck function,

$$B_\nu(T) = \frac{2h\nu^3}{c^2}\frac{1}{\exp\left(\frac{h\nu}{k_B}\right) - 1}. \tag{4.21}$$

At long wavelengths, the Rayleigh-Jeans approximation holds with

$$B_\lambda(\lambda, T) = \frac{2hc^2}{\lambda^5}\frac{1}{e^{\frac{hc}{\lambda k_B T}} - 1}. \tag{4.22}$$

The dust opacity in the (sub)-millimeter regime can be fit to a powerlaw in frequency, ν, as $\kappa_\nu \propto \nu^\beta$, giving $F_\nu \propto \nu^{\beta+2} = \nu^{\alpha_{\rm mm}}$, where $\alpha_{\rm mm}$ is known as the spectral index (i.e., $\alpha_{\rm mm} \approx \beta + 2$) and it is the slope of the spectral energy distribution (SED) at mm-emission. For small particles (sub-micron sized particles), as in the inter-stellar medium (ISM) dust, the spectral index has values of ∼3.5–4.0 (e.g., Finkbeiner et al. 1999), while for large particles, the spectral index is expected to have lower values $\alpha_{\rm mm} \lesssim 3.5$ (Draine 2006; Rodmann et al. 2006). Therefore, by observing protoplanetary disks at two optically thin wavelengths, it is possible to obtain information about the grain size.

The first large survey of millimeter observations of protoplanetary disks revealed that $\beta \approx 1$ for Taurus-Auriga (Beckwith and Sargent 1991) and Ophiuchus (Andre and Montmerle 1994) regions. Recent millimeter observations of protoplanetary disk in different star formation regions show that the spectral index has low values ($\alpha_{\rm mm} \lesssim 3.5$) for most of the disks (e.g., Ricci et al. 2010, 2011, 2012; Ubach et al. 2012; Testi et al. 2014). On the other hand, dust evolution models predict that mm-sized particles rapidly drain from the outer disk due to large radial drift speeds. At million year timescales (below typical ages) disk models are already depleted of large grains and predict a high spectral index values ($\alpha_{\rm mm} \gtrsim 3.5$), in disagreement with observations $\alpha_{\rm mm}$ (Birnstiel et al. 2010b). These standard disk models have smooth radial density profiles. We next consider the advantages of abandoning this assumption.

4.5 Global Pressure Bumps and Zonal Flows

We now consider the possibility that disks contain pressure bumps, i.e. nearly axisymmetric rings where the radial profile of the midplane gas pressure has a local maximum. These pressure bumps are synonymous with *zonal flows* in disks, because the reversal of the radial pressure gradient causes—due to geostrophic balance with Coriolis forces—a reversal in orbital speeds relative to the Keplerian flow. MHD effects, specifically the development of the magneto-rotational instability (MRI), have been shown to cause zonal flows and pressure bumps (Johansen et al. 2009a; Uribe et al. 2011; Dittrich et al. 2013; Simon and Armitage 2014). However, the real nature of zonal flows, i.e. their lifetime, width, amplitude, is subject to uncertainties about the outcome of the MRI in protoplanetary disks.

To analyze the ability of zonal flows to trap particles, Pinilla et al. (2012b) included a sinusoidal perturbation in the gas surface density with a given amplitude and width (defined as a factor of the local scale height) into dust evolution models by Birnstiel et al. (2010a). In these models, the dynamics and the growth of the dust are both considered. Both process depend on the Stokes number of the particles and for this reason these two phenomena are strongly correlated. In the models, dust particles grow, fragment, and crater depending on the relative velocities. The grain size distribution evolves via the Smoluchowski coagulation equation (Smoluchowski 1916). Relative velocities arise from various sources: Brownian motion, settling, turbulent motion, and drift velocities in the azimuthal and radial direction.

Two different factors were considered to constrain the values of the amplitude and wavelength of the sinusoidal perturbation in order to trap particles. First, a minimum amplitude of 10% with respect to the background density is required to have a local outward movement of the dust particles, assuming one scale height for the width of the sinusoidal perturbation (higher amplitude is needed if the bumps are assumed to be wider). Second a maximum amplitude of ~30% keeps the angular momentum per unit mass increasing outwards (so the disk remains Rayleigh stable).

Figure 4.5 summarizes the main results of dust evolution models assuming no pressure bumps, bumps with an amplitude of 10 and 30%, and a model where the direct output from global MHD simulations that show zonal flows is assumed. In these models, the transport, growth, and fragmentation of particles are taking into account for several grain particles, covering sizes from 1-μm size to 2 m. Specifically, Fig. 4.5 shows the vertically integrated dust density distribution after 1 Myr of evolution as a function of grain size and distance from the star. For each case, the same turbulent parameter is assumed, with a value of $\alpha = 10^{-3}$. In addition, the solid line in each panel of Fig. 4.5 represents the grain size for which the Stokes number is equal to unity [it is hence a representation of the gas surface density profile, see Eq. (4.5)].

The discrepancy between the models including radial drift and observations could be solved by including pressure bumps of ~30% of amplitude. As shown in Fig. 4.5, when the pressure bumps are only 10% amplitude, the pressure gradient is not high enough to trap the dust particles in the outer regions of the disk after million years time scale. In this case, turbulent mixing spread the dust particles from the pressure bumps without effective trapping. Contrary, in the case of pressure bumps with ~30% of amplitude, the pressure gradient is enough to trap sufficient millimeter-sized particles in the outer regions of the disks, such that there is a good agreement between the observed spectral indices and models (Pinilla et al. 2012b).

Moreover, long-lived zonal flows from MHD simulations by Uribe et al. (2011) agree with the amplitudes and wavelengths assumed for the sinusoidal perturbation in the gas surface density, with values of ~25% of amplitude and a width of one

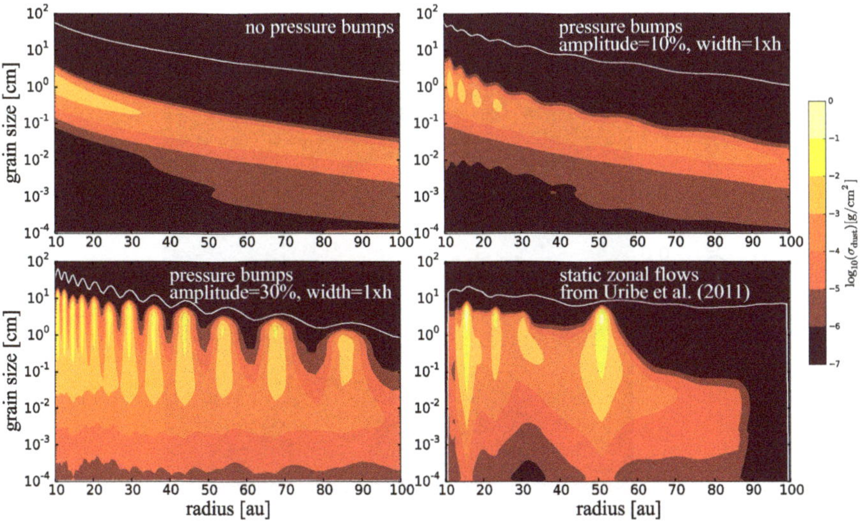

Fig. 4.5 Vertically integrated dust density distribution after 1 Myr of evolution as a function of grain size and distance from the star for four different cases: no pressure bumps (*top-left panel*), with global pressure bumps with an amplitude of 10% and one scale height for the width (*top-right panel*), as before but with an amplitude of 30% (*bottom-left panel*), and with static pressure bumps from MHD simulations from Uribe et al. (2011). The *solid white line* in each case shows the size of St = 1 solids. Figures adapted from Pinilla et al. (2012b)

scale height approximately. Considering such kind of pressure bumps, the retention of dust particles in the outer parts of the disks is possible after million years timescales (Fig. 4.5, bottom right panel).

In addition, the simulated images using the CASA ALMA simulator (including atmospheric and calibration noise) at different millimeter wavelengths show that with high angular resolution, multiple ring structures should be detectable (Fig. 4.6 illustrates the particular case at 0.85 mm), as observed already in two protoplanetary disks, HL Tau and TW Hya, with long-baseline observations from ALMA (ALMA Partnership et al. 2015; Andrews et al. 2016). Whether or not the origin of the observed structures are due to MHD effects is still an open question because of the unknown nature of zonal flows from MHD simulations. In principle, these pressure bumps have to be long-lived to trap mm-grains. For simplicity, in the simulations shown in Fig. 4.5 the gas density profile is considered static while the dust evolves, which can be a good approximation if the evolution of the bumps is longer than drift timescales, e.g. if they evolve with viscous timescales.

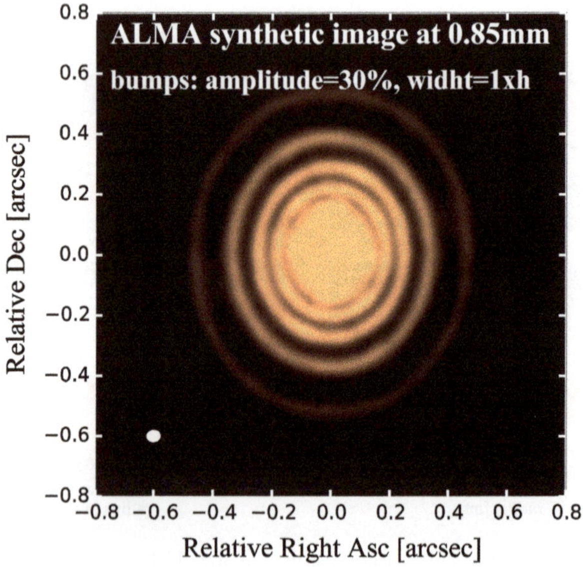

Fig. 4.6 ALMA synthetic image (including atmospheric and calibration noise) at 0.85 mm assuming a configuration of 50 antennas for an angular resolution of ~5 mas, and assuming a distance of 140 pc and stellar parameters of a typical T-Tauri star. To create the image, we assumed the model with pressure bumps of an amplitude of 30% and with a width of one scale height. Figure adapted from Pinilla et al. (2012b)

4.5.1 The Extreme Case of Brown Dwarf Disks

The drift velocity of the dust particles is determined by how much the gas azimuthal velocity differs from the Keplerian speed. This difference depends on the stellar properties, more precisely

$$v_{\text{drift, dust}} \propto v_{\phi,\text{gas}} - v_k \propto \frac{L_\star^{1/4}}{\sqrt{M_\star}}. \tag{4.23}$$

For a young Brown Dwarf, the luminosity is 3–5% of the solar luminosity, while the mass is between 1–5% of the solar mass (e.g Burrows et al. 2001). Thus radial drift speeds in young brown dwarves should increase compared to more massive stars, according to Eq. (4.23). Pinilla et al. (2013) studied the scenario of particle trapping in disks around brown dwarfs (BDs), for which observations also reveal the existence of millimeter-sized particles (e.g., Ricci et al. 2014). Such observations are a challenge for models of dust evolution in disks, and question the universality of the processes. In fact, the radial drift velocities are expected to be more than twice as fast for particles in BD disks than in disks around T-Tauri stars, depleting the disk in grains before they can grow. Since the radial drift velocities are much higher in BD disks, the pressure bumps needed to trap the particles and explain the observations

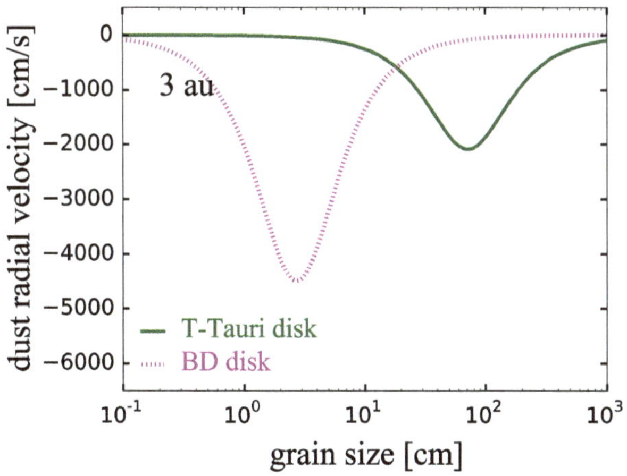

Fig. 4.7 Total radial dust velocity as a function of grain size [Eq. (4.14)] calculated at 3 AU from the star, assuming typical parameters for a disk around a T-Tauri star and a Brown Dwarf (BD, Pinilla et al. 2013)

must have a higher amplitude ($\gtrsim$40%) in a compact disk ($\sim$15–30 AU), to account for current mm-observations of BD disks. Figure 4.7 illustrates the total dust radial velocities [Eq. (4.14)] as a function of grain size for a case a disk around a T-Tauri star vs. a disk around a BD. In this case typical parameters are assumed for both cases, as, for example, a power law for the gas surface density and temperature (see Pinilla et al. 2013, for more details). Since the BD disks are less dense, smaller grains have St = 1 at the same distance from the star. At 3 AU, St = 1 corresponds to $\sim$2 cm particles for the BD disk vs. $\sim$70 cm for the T-Tauri disk, for which the radial drift velocity can be more than twice for the BD case (Fig. 4.7). Further investigation is needed to understand if pressure bumps with such high amplitude can result from MHD simulations assuming BD disk conditions.

4.6 Transition Disks

Observations done with IRAS in 1989 by Strom et al. (1989) revealed for the first time a new kind of disks, with little or no excess emission at $\lambda < 10\ \mu$m, but with a significant excess at $\lambda \geq 10\ \mu$m. This feature has been interpreted as a optically thick disk with a inner hole, where the dust has been depleted or it has become optically thin. These disks were called *transition disks*, because they may indicate a transition between a gas- and dust-rich disk to a debris disk (gas-poor disk).

Recent observations of transition disks (TDs) have provided important insights into the processes of planetary formation and circumstellar disk dissipation (e.g., Espaillat et al. 2014). Large cavities (>10 AU) have been spatially resolved in (sub-)millimeter images of TDs (e.g., Andrews et al. 2011; van Dishoeck et al.

2015), and multiple processes have been proposed to reproduce the lack of infrared (IR) emission from these disks. TDs are thought to be intermediate systems between young (few Myr) optically thick protoplanetary disks and older ($\gtrsim$10 Myr) optically thin debris disks, where significant evolution has taken place. Multiple processes can reproduce the lack of IR emission from TDs, such as photoevaporation (e.g., Owen and Clarke 2012; Alexander et al. 2014), planet–disk interaction (e.g., Zhu et al. 2012), or dead zones (Regály et al. 2012; Flock et al. 2015). Distinguishing between the different physical explanations requires spatially resolved observations since various mechanisms predict different or similar behaviors for the dust and gas. In this section, we focused on embedded planets and dead zones as potential origins of the structures seen in transition disks.

Recent millimeter observations of gas and dust in TDs suggest that particle trapping in a single and broad pressure maximum is the primary cause of the seen structures in these disks, such as rings and asymmetries (e.g., van der Marel et al. 2016). The idea of embedded planet(s) as potential origin of the pressure maximum is supported by direct imaging of planet candidates in some TDs, as, for example, in HD 169142 (Quanz et al. 2013b; Reggiani et al. 2014), HD 142527 (Biller et al. 2012), HD 100546 (Quanz et al. 2013a, 2015; Currie et al. 2015), and LkCa 15 (Kraus and Ireland 2012; Sallum et al. 2015).

4.6.1 *Trapping by Embedded Planets*

Planet disk interactions and planet migration are complex processes that have been studied in great detail (Lin and Papaloizou 1979, 1986a,b; Rafikov 2002; Masset and Papaloizou 2003; Crida et al. 2006, see reviews by Kley and Nelson 2012; Baruteau et al. 2014). The gap opening process by a planet mainly depends on the balance between the disk viscous torque, the planet gravitational torque, and the pressure torque. In a disk with average values of viscosity ($\alpha \sim 10^{-3}$), a 1 M_{Jup} mass planet opens a clear gas gap in the disk. Gap opening planets experience type II migration, with timescales typically equal to the viscous timescales of the disk.

When a planet opens a gap, a local maximum at the outer edge of the gap is formed in the gas density and pressure, which acts as a powerful dust trap. Pinilla et al. (2012a) studied particle trapping induced by embedded massive planets. The top panels of Fig. 4.8 show the vertically integrated dust density distribution after 1 Myr of evolution as predicted by combining hydrodynamical simulations of planet disk interaction with dust evolution models from Birnstiel et al. (2010a) (explained in the previous section). The planet is on a fixed circular orbit at 20 AU. For the hydrodynamical simulations, we used the two-dimensional hydrodynamical code FARGO (Masset 2000), assuming a locally isothermal disk, with a power-law surface density, a disk mass of 5% of the stellar mass, a flared disk geometry, and a constant α viscosity ($\alpha = 10^{-3}$). In these models, particles are trapped at the outer edge of the gap and the distribution of millimeter particles peak at the location of the pressure maximum. Assuming the inner edge to be the location where the gas

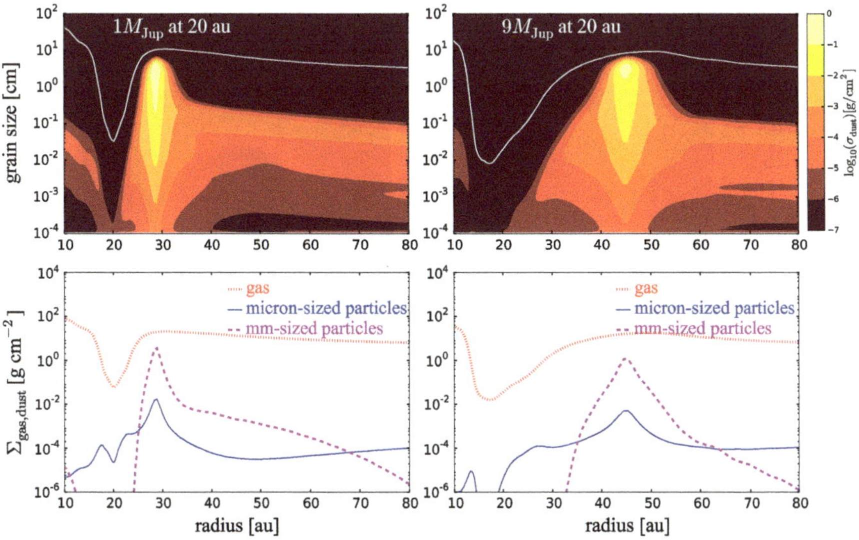

Fig. 4.8 *Top panels*: Vertically integrated dust density distribution after 1 Myr of evolution as a function of grain size and distance from the star for a disk hosting a 1 $M_{\rm Jup}$ mass planet (*left-panel*) and a 9 $M_{\rm Jup}$ mass planet (*right-panel*), with the planet at 20 AU in both cases. The *solid white line* in each case shows the size of St = 1 solids. *Bottom panels*: Comparison of the gas surface density and the dust distribution for small particles (micron-sized) and large particles (millimeter-sized) for 1 and 9 $M_{\rm Jup}$ mass planet (*left* and *right panel*, respectively). Figures adapted from models in Pinilla et al. (2012a)

surface density starts to steeply decrease, the outer edge of the gas gap is taken to be the location where the gas density reaches a similar value than the inner edge. The location of the outer edge of the gas gap is much closer in than the peak of the pressure maximum. In the cases of Fig. 4.8, the pressure maximum is at ~29 and ~48 AU, while the outer edge of the gas gap is ~26 and ~34 AU for 1 and 9 $M_{\rm Jup}$, respectively. Comparing with the Hill radius r_H, defined as $r_H = r_p(M_p/3M_\star)^{1/3}$ (being r_p and M_p the orbital radius and mass of the planet, respectively), the location where the large particles (millimeter and centimeter-sized grains) accumulate is ~7 r_H and ~10 r_H for $\lesssim$5 $M_{\rm Jup}$ and $\gtrsim$5 $M_{\rm Jup}$ planet mass, respectively. This location is further out than the outer edge of the gas gap, which can be at most at 5 R_H (Pinilla et al. 2012a). As a result, a large segregation exist between the location of the gas gap (and therefore small particles that follow the gas) and the peak of the mm-emission. In addition, it is also expected that the mm-dust is much more depleted inside the cavity than the small-grains and the gas, as it is illustrated in the bottom panels of Fig. 4.8. For instance, for the case of 1 $M_{\rm Jup}$, the gas is depleted around one order of magnitude while the millimeter dust is depleted for more than three orders of magnitude (bottom-left panel of Fig. 4.8).

This segregation between the gas (or small grains) and large mm-sized particles has been observed in several TDs (e.g., Garufi et al. 2013; Perez et al. 2015a;

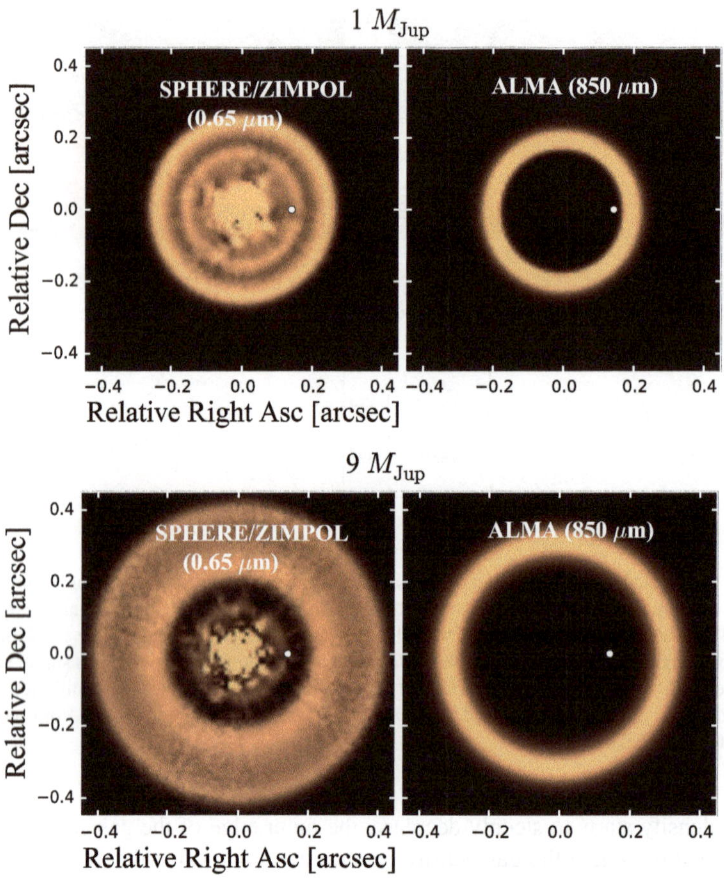

Fig. 4.9 Simulated images of the planet-induced particle trapping models shown in Fig. 4.8. *Left panels* show the polarized intensity in the R-band (0.65 μm) including instrumental effects of the ZIMPOL polarimeter with the SPHERE adaptive optics system. *Right panels* correspond to ALMA simulated images at 0.85 mm with an angular resolution of 0.1″. Figures adapted from de Juan Ovelar et al. (2016)

van der Marel et al. 2016), suggesting that particle trapping trigger by embedded planets is the origin of the observed structures in TDs (see also Dong et al. 2015b). In principle, any broad pressure bump can cause such segregation as long as dust diffusion is also included. From the models, it is expected that the gap or cavity size changes depending on the wavelength of the observation. Figure 4.9 shows the synthetic images from the models in Fig. 4.8 after radiative transfer and instrument calculations considering the dust density distributions of the dust evolution models after 1 Myr of evolution and assuming a planet located at 20 AU radius. As shown in de Juan Ovelar et al. (2013), the separation between the micron- and mm-sized dust grains is very sensitive to the mass of the embedded planet, making the segregation more significant for more massive planets.

Recently de Juan Ovelar et al. (2016) also proved that the expected structures due to trapping by a planet notably depend on viscosity. For high values of viscosity ($\alpha = 10^{-2}$), there is too much fragmentation and diffusion of particles, such that the ring-like structures are unlikely to form at any wavelength. While for low values of viscosity ($\alpha = 10^{-4}$) the growth inside the trap at the edge of the gap is so effective because the turbulent and drift relative velocities are too low, such that most of the collisions lead to growth to meter-sized objects, emptying the disk in (sub-)mm grains that are the ones traced with mm-observations. Due to the effective growth, the amount of micron-sized particles is very few and sub-structures due to the presence of the planet are very difficult to observe in scattered light images. As a result, only intermediate values of viscosity ($\alpha = 10^{-3}$) can produce smaller or no optical/NIR cavity while a clear cavity at longer wavelengths, which can be detectable with current telescopes.

For TDs, the spatially integrated spectral index ($\alpha_{\rm mm}$) is higher than for regular protoplanetary disks. Nonetheless, the values of the observed $\alpha_{\rm mm}$ still correspond to disks with mm-sized particles. In addition, the current observations of TDs show that there is a positive correlation between the spectral index and size of the cavity at mm-wavelengths (Pinilla et al. 2014). This is in agreement with models of planet–disk interaction and dust evolution, since when the particle trap is further out (hence the cavity size is larger), the size of the trapped grains is smaller [the maximum grain size due to the fragmentation and the drift limit decrease with radius, Eqs. (4.17) and (4.19)], increasing the integrated spectral index for TDs with larger cavities.

4.6.1.1 Example: The Transition Disk J160421.7-213028

An excellent example of the spatial segregation between the small and large grains is the TD J160421.7-213028 (hereafter J1604). Observations with SPHERE/ ZIMPOL in polarized differential imaging at 0.626 μm show a bright ring extending from 40 to 80 AU, with a brightness peak located at ~60 AU distance from the star. This dust ring lies inside the mm-cavity observed with ALMA at 880 μm (Zhang et al. 2014) (Fig. 4.10). This indicates dust filtration in a particle trap, which can be induced by a massive planet embedded in the disk at 20–40 AU. At this location is where observations of CO indicate that the gas cavity size is located. J1604 is a TD where most of the predictions of particle trapping by a planet fit very well.

This target has a very interesting characteristic, a dip in the annulus seen in the SPHERE image was detected at ~46°. A dip was also previously detected with HiCIAO at H-band, but at a different azimuthal location ~85° (Mayama et al. 2012). Thus, the dip must be rotating quickly, with at least an angular speed of ~12°/year, to explain the difference of the dip location between the HiCIAO and SPHERE data. The dip rotation is clearly inconsistent with the local Keplerian angular velocity of at ~0.8°/year at 60 AU. The most likely explanation is that the dip is a shadow cast by an object orbiting closer in to the central star, but future observations are needed to test this idea.

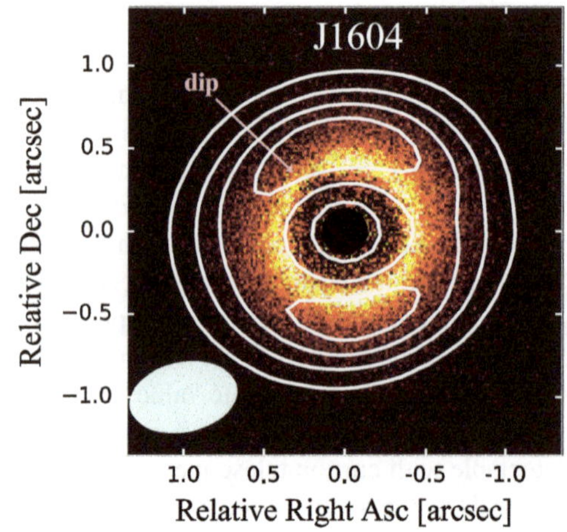

Fig. 4.10 Spatial segregation of small and large particles in J160421.7-213028 (J1604). This overlay shows both the 0.626 μm emission from SPHERE/VLT, which traces micron-sized particles (color map, Pinilla et al. 2015b), and also the 880 μm map from ALMA observations, which traces mm-sized grains (contour lines at 20, 40, 60, 80 and 100% of the peak value, from Zhang et al. 2014). Figure adapted from Pinilla et al. (2015b)

4.6.2 Pre-transition Disks: The Role of the Snow Line

Pre-transition disks have a strong near-infrared (NIR) excess that comes from a optically thick ring of dust that is in the innermost regions of the disk. The name *pre-transition disks* was given to these objects because they may be at an earlier evolutionary stage than TDs with completely empty dust cavities.

Models of planet–disk interaction and dust evolution failed to reproduce the NIR excess of some TDs independent of the mass or location of the planet (de Juan Ovelar et al. 2013). When the planet is very massive ($\gtrsim 5\,M_{\rm Jup}$) and complete filtration of dust particles of all sizes happens, and the grains that remain in the inner part of the disk (in front of the planet location) eventually grow and drift towards the star in time scales shorter than 0.1 Myr. On the other hand, if the embedded planet allows partial filtration and small grains pass through the gap, these particles in any case grow efficiently in the inner parts and quickly drift towards the star.

To solve this inconsistency between models and observations, the snow line plays a fundamental role since it can help to re-create a ring of dust in the innermost parts of the disk despite the presence of a large planet-carved gap. The process for creating pre-transition disk-like structures in the context of planet–disk interaction can be understood by the following steps:

1. Step 1-The tunnel: When partial filtration occurs and small particles pass through the planetary gap, these small particles clump and efficiently grow because they are expected to have water ice mantles beyond the snow line. Once grown, the grains drift very fast such that the gap appears empty at mm-emission.
2. Step 2-The traffic jam: Once particles have tunneled from the outer to the inner disk and reach regions where it is warm enough for the water ice to melt (at the so-called snow line), they become much less sticky and break up into small fragments. For these small particles, the drift velocities is reduced and thus they pile up close to the snow line, forming an optically thick ring and NIR excess, which remains as long as the disk is fed of dust through the tunnel.

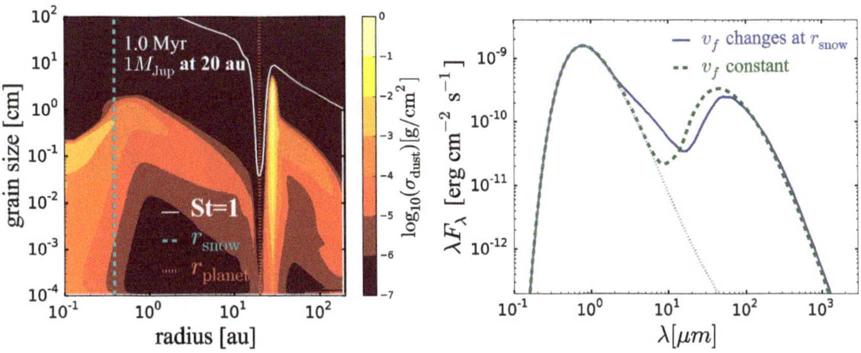

Fig. 4.11 *Left panel*: Vertically integrated dust density distribution after 1 Myr of dust evolution, when 1 $M_{\rm Jup}$ mass planet is embedded in the disk at 20 AU. At the location of the snow line (0.4 AU for a typical T-Tauri disk), the fragmentation velocity changes, leading to the re-appearance of small grains that are well coupled and have slow velocities. *Right panel*: Synthetic SEDs from models of 1 $M_{\rm Jup}$ mass planet at 20 AU, assuming and neglecting the changes of the fragmentation velocity v_f at the location of the snow line. Figures adapted from Pinilla et al. (2016b)

Pinilla et al. (2016b) combined hydrodynamical models of planet–disk interaction, dust evolution, changing the dust properties and collision velocities near the snow line. Figure 4.11 illustrates the case of 1 $M_{\rm Jup}$ mass planet embedded at 20 AU including the variation of the fragmentation velocity from 10 m s^{-1} outside the snow line (0.4 AU, assuming typical values of a T-Tauri star) to 1 m s^{-1} inside the snow line. Near the snow line, small grains are re-created due to the effective fragmentation, which are more couple to the gas and their drift velocities are lower. As a consequence, the NIR excess re-appears when the snow line effects are considered for the dynamics of the dust (right panel of Fig. 4.11).

When the planet is massive enough ($\gtrsim 5\,M_{\rm Jup}$) and total dust filtration at the outer edge of the gap happens, the NIR excess disappears after $\gtrsim$2 Myr, creating a typical SED of a TD with an empty dust cavity. In this case, the disk evolves from a pre-transition disk appearance to a TD. As a consequence, by knowing the age of the disk and its SED (whether or not it shows a NIR excess), it is possible to constrain the planet mass that carves the cavity.

4.6.3 *Multiple Planets*

In the context of embedded planetary companions that can clear a gap and hence be the origin of cavities in TDs, the wide observed dust cavities suggest that they may form by multiple planets, which individual gaps overlap to form a wider one (Zhu et al. 2011; Isella et al. 2013). For multiple planets, it is expected that shallow cavities where the gas is less depleted compared to the case where a single massive planet is embedded (Duffell and Dong 2015). In this case, the potential trapping

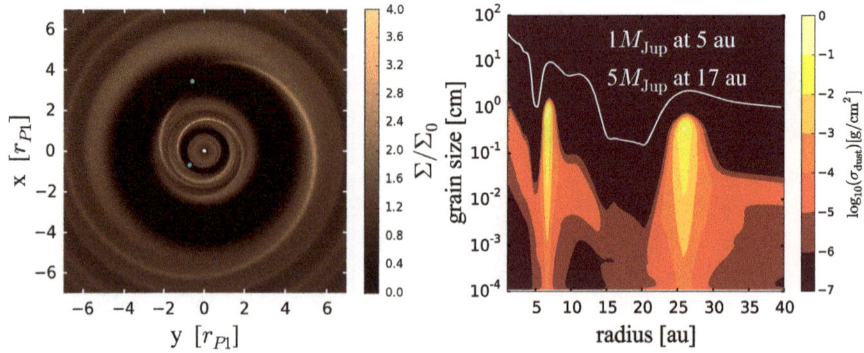

Fig. 4.12 *Left panel*: Simulated 2D gas surface density with two planets embedded in the disk, after 2000 orbits of the inner planet. With a ratio of outer to inner planet orbital radii of $r_{P2}/r_{P1} = 3.5$, the individual gaps do not overlap. The planet-to-stellar mass ratio is 4×10^{-4} (inner) and 2.0×10^{-3} (outer), i.e. 1 and 5 M_{Jup} around the 2.5 $M_{\odot}$ central star. The scaled disk viscosity is $\alpha = 10^{-3}$. *Right panel*: Vertically integrated dust density distribution after 1 Myr of evolution in the gas disk shown in the *left panel*. Figure adapted from Pinilla et al. (2015c)

would occur at the location of the pressure maximum expected at the outer edge of the resulting gap that results from overlapping the gaps from each planet.

Multiple planets whose gaps do not overlap can create multiple pressure bumps and therefore several ring-like structures. The left panel of Fig. 4.12 shows an example of a hydrodynamical simulation when two planets are embedded in the disk. The locations of the planets are such that the individual gaps do not overlap, with $r_{P2}/r_{P1} = 3.5$. The planet-to-stellar mass ratio is 4×10^{-4} and 2.0×10^{-3} for the inner and outer planet, respectively, and the disk viscosity is assumed to be $\alpha = 10^{-3}$. In the right panel of Fig. 4.12 the corresponding vertically integrated dust density distribution after 1 Myr of dust evolution is illustrated. In this case, the central star is considered to be 2.5 $M_{\odot}$, such that the embedded planets are 1 and 5 M_{Jup}. Because the gaps do not overlap, there are two distinct locations where the pressure gradient is positive and hence two pressure traps where the mm-dust particles accumulate. Pinilla et al. (2015c) explored in detail the case of multiple planets embedded in disks and the resulting gas and dust structures for different disk and planet parameters, such as disk viscosity, fragmentation velocity of particles, and planet masses.

In the case shown in Fig. 4.12, the gaps are not deep enough to have complete filtration of the small particles and therefore small grains are distributed smoothly in the entire disk. As a consequence, high angular resolution observation of the mm-emission of such kind of planet configuration would show a two-ring like structure, but none cavity at NIR or optical emission (Pinilla et al. 2015c).

HL Tau is an example where several rings at mm-emission have been resolved and several authors have invoked multiple embedded planets as an explanation for the seen structures (Dong et al. 2015b; Dipierro et al. 2015b; Akiyama et al. 2016; Jin et al. 2016). These works agree that in the case of HL Tau, the mass of the embedded planets should be sub-Jovian to explain the shape of the observed rings. When planets are not massive enough to open gaps in the disks, the process of particle trapping does not happen. However, the perturbations that a low-mass planet can have on the gas azimuthal velocities [Eq. (4.6)] can slow down the drift velocities of the particles, creating traffic jams of the particles that are slightly decoupled from the gas (e.g., Rosotti et al. 2016). As a result, it is possible to have accumulations of dust at the outer location of the orbit of a low mass planet too, without opening a gap in the gas (hence without any pressure bump). Beside planets, additional interpretations for the structures seen in HL Tau include sintering of dust particles (a similar physical process at the location of snow lines as described in Sect. 4.6.2 and introduced by Okuzumi et al. 2016), and secular gravitational instability (Takahashi and Inutsuka 2016, see Sect. 4.8).

4.6.3.1 Example of Multiple Planets: The Transition Disk HD 100546

Observations of the disk around the Herbig Ae star HD 100546 have indicated the presence of two potential companions, making HD 100546 an excellent laboratory to study particle trapping triggered by massive planets. An inner planet located at ~10 AU distance from the star has been proposed to model the [O I] 6300 Å emission line (Acke and van den Ancker 2006), and the images obtained with near and mid-infrared interferometry that show an inner cavity (Tatulli et al. 2011; Mulders et al. 2013). Detailed modeling of the cavity wall at 10 μm with MIDI/VLT demonstrated that the mass of a potential planet located at ~10 AU is a very massive companion with a mass of 20–30$M_{\rm Jup}$. On the other hand, high-contrast imaging confirmed the presence of what it seems to be a massive planet (~15$M_{\rm Jup}$) at ~50 AU, which may be its formation stage (Quanz et al. 2013a, 2015; Currie et al. 2015).

ALMA-Cycle 0 observations of HD 100546 show that the radial extent of the gas is ~400 AU, while most of the emission at millimeter wavelengths comes from a narrow ring concentrated at ~26 AU, with a width of ~21 AU. A much fainter ring of emission (a factor of ~100 lower) comes from an outer ring, which is centered at ~190 AU, with a width of ~75 AU (Walsh et al. 2014). ALMA-Cycle 0 observations showed for the first time a multiple-ring structure in HD 100546, which is consistent with the two-planet scenario. The left panel of Fig. 4.13 shows the residual ring of emission in the outer part of HD 100546 after subtracting the best ring model for the inner emission and clearly show a strong residual ring that lies outside of the outer planet candidate.

To study the effect of the two proposed planets in HD100546 on the dust density distribution of mm-particles, Pinilla et al. (2015a) combined analytical predictions of planet–disk interactions with the dust evolution models, to directly compare with ALMA data and give constraints on the disk and planet parameters, such as viscosity, and mass and age of the planet(s). Pinilla et al. (2015a) demonstrated that in the case that only an inner massive planet is embedded in the disk, the outer ring of emission observed with ALMA remains unexplained. However, if an outer planet is also embedded in the outer part of the disk, and it is as massive as

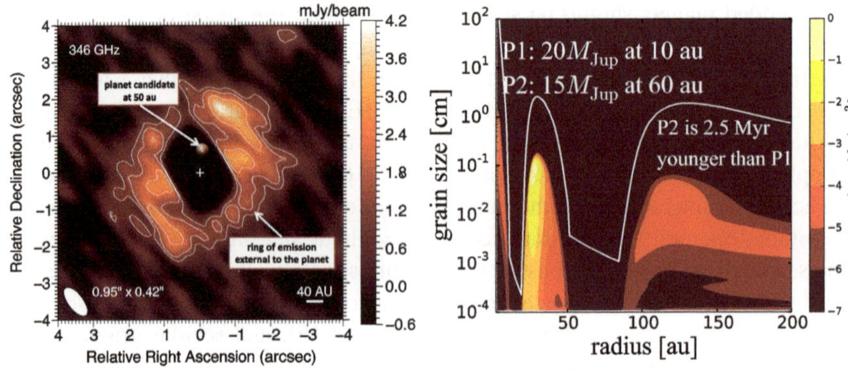

Fig. 4.13 *Left panel*: ALMA-Cycle 0 observations at 867 μm with a beam of 0.95×0.42″ HD 100546 after subtracting the best ring model for the inner emission. The residual ring lies outside the location of the outer planet candidate at 50 AU (Quanz et al. 2013a, 2015; Currie et al. 2015). Figure adapted from Walsh et al. (2014). *Right panel*: vertically integrated dust density distribution as a function of grain size and radial disk extension, when the two massive planet-candidates for HD 100546 are embedded in the disk (20$M_{\rm Jup}$ at ∼10 AU and 15$M_{\rm Jup}$ at ∼50–70 AU). In this case, the outer planet is embedded in the disk after 2.5 Myr of evolution, i.e. the outer planet is much younger than the inner planet. Figure adapted from Pinilla et al. (2015a)

suggested by Quanz et al. (2013a, 2015), and Currie et al. (2015) (i.e., ∼15 $M_{\rm Jup}$), this outer planet should be at least ∼2–3 Myr younger than the inner planet, to trap the right amount of dust particles and produce an outer ring of emission that is 100 times fainter than the inner ring of emission as observed with ALMA-Cycle 0 (right panel of Fig. 4.13). This supports the fact that the outer planet may be in the act of formation. In the case that the outer planet is very massive and it has a similar age than the inner planet, the amount of the trapped dust is too high and the models are in disagreement with current ALMA and ATCA observations of this disk (Walsh et al. 2014; Wright et al. 2015). To prove the existence of the planet and have insights on its mass and location, Perez et al. (2015b) proposed to measure the effects that a cicumplanetary disk can have on the gas kinematics. In addition, planets can also have an important effect on the chemical evolution and clear tracers such as HCN and its isotopologues can detect planetary heating (Cleeves et al. 2015). ALMA future observations will give crucial insights about embedded planets in systems as HD 100546. Recent mm-observations of other transition disks, such as HD 97048, HD 169142, and HD 163296 (Walsh et al. 2016; van der Plas et al. 2017; Isella et al. 2016; Fedele et al. 2017), also show multiple rings structure and suggest that TDs may have more sub-structures than a single and predominant pressure bump.

4.6.4 Trapping at the Outer Edge of a Dead Zone

Angular momentum is transported outward in protoplanetary disks to sustain accretion onto the central star (e.g., Armitage 2011). Different mechanisms have been proposed for angular momentum transfer, including MRI turbulence, MHD winds, and disk self-gravity (e.g., Turner et al. 2014). MRI has been the leading mechanism for transferring angular momentum throughout the disk, which arises from the magnetic tension between two adjacent fluid elements of the disk that rotate with different speed (e.g., Balbus and Hawley 1991). In the denser regions of the disk, such as the midplane, where X-rays, FUV-rays, cosmic rays cannot completely penetrate, the disk can decouple from the magnetic field and MRI is expected to be inefficient. These regions of low ionization are called *dead zones*, where the effective α-viscosity is expected to have low values ($\lesssim 10^{-4}$), generated perhaps by the hydrodynamic vertical shear instability (Lin and Youdin 2015). The boundaries of dead zones might explain gaps and asymmetries of transition disks. The shape of the dead zone in protoplanetary disks is still an open question, because it depends on different factors such as dust density and its properties, the magnetic field, chemistry, and the sources of ionization (e.g., Dzyurkevich et al. 2010).

The accretion rate in a steady-state of protoplanetary disks depends on the disk viscosity, such that $\dot{M} = 3\pi\Sigma\nu$ (ν defined in Eq. (4.16), which is directly proportional to α-viscosity). At the outer edge of a dead zone where the effective α changes from high values outside to low values inside the dead zone, a gas accumulation is expected. As a result, the surface gas density has a bump at the outer edge of a dead zone and hence a pressure bump is formed that is capable to trap particles. Pinilla et al. (2016a) studied the trapping of dust at the outer edge of dead zones and analyze if the potential observational signatures can create structures like a transition disk. For this a simplified parametrization of α changing with radius is assumed, which depends on the gas surface density to mimic the effect of MRI-driven turbulence.

Figure 4.14 shows the vertically integrated dust density distribution after 0.5 Myr of dust evolution (left panel), when a dead zone (with $\alpha = 10^{-4}$) extends up to 20 AU. At the location of the dead zone edge α increases to $\sim 10^{-2}$ and a bump in the gas surface density forms, which has amplitude of a factor of ~ 5. In this bump large grains (mm and cm-sized particles) are trapped. Inside the dead zone, there is effective growth because the relative velocities due to turbulence are low due to the low values of α. Because of this effective growth, the dead zone is depleted in small grains (right-upper panel in Fig. 4.14). As a consequence of the trapping and the effective growth inside the dead zone, the synthetic images derived from radiative transfer simulations show that the cavity and a ring-like emission at optical and mm-emission are co-located (lower panels in Fig. 4.14). Pinilla et al. (2016a) found that the cavities are expected to be observed at short (optical and NIR emission) and long wavelengths independent on the shape of the dead zone that is assumed, as observed in transition disks.

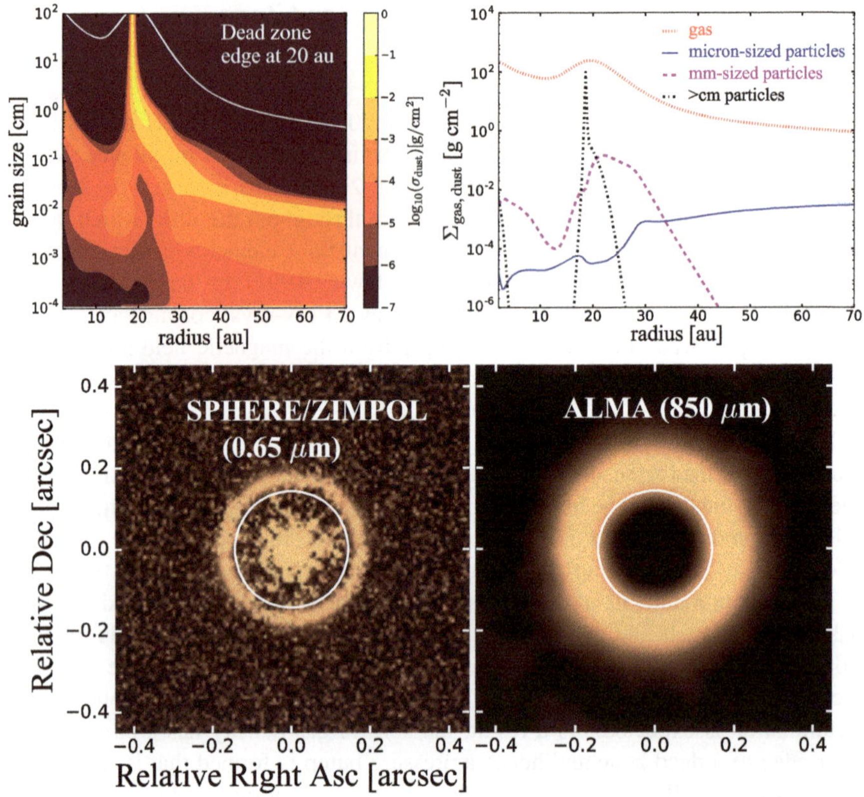

Fig. 4.14 (*Upper left*:) Vertically integrated dust density distribution after 0.5 Myr of dust evolution in a disk with a dead zone extending out to 20 AU. Outside the dead zone edge, a bump in the gas surface density forms, trapping large particles. (*Upper right*:) A comparison of the gas surface density and the dust distribution for both small particles (micron-sized) and large particles (millimeter and centimeter-sized). (*Lower panels*:) Simulated images of the dead zone (edge shown by the *solid white circle*) particle trapping models shown above. (*Lower left*:) the polarized intensity in the R-band (0.65 μm) from instrumental simulations for ZIMPOL with SPHERE. (*Lower right*:) ALMA simulated images at 0.85 mm with 0.1″ angular resolution. Adapted from Pinilla et al. (2016a)

Observationally, gas inside the mm-cavity seems to be much more depleted (by several orders of magnitude, e.g., van der Marel et al. 2015) than the dead zone predictions (less than one order of magnitude). To solve this inconsistency in the case of dead zones, Pinilla et al. (2016a) included the effect of a MHD wind that removes mass from the inner disk to the dead zone models. In these cases, the inner part of the disk is highly depleted, and a broad and distinct ring in the final gas surface density remains at million year timescales. In this case, large particles are trapped at the peak of the gas bump and a large segregation between gas (thus small grains) and large particles is created as in the case of planets. A potential method to distinguish both mechanisms as possible origins for transition disk structures is to spatially resolved broadening of molecular line profiles to measure radial changes of turbulence.

4.7 Asymmetries: Spiral Arms in Self-gravitating Disks and Vortices

In previous sections, we explain mechanisms for only radial trapping that create ring-like, nearly axisymmetric structures. However, azimuthal or non-axisymmetric trapping can also occur due to anti-cyclonic vortices and spiral arms in self-gravitating disks. In this section, we discuss both mechanisms and their observational consequences.

4.7.1 Trapping in Self-gravitating Spiral Arms

Self-gravity can be an important source of angular momentum transport, in particular in the outer parts regions where the disk is cold and MRI may not be effective to ionize the disk (Lin and Pringle 1987; Fromang et al. 2004; Vorobyov and Basu 2009). The physical processes that can lead to gravitational unstable disk and its evolution include the mass, size, and disk temperature (see Kratter and Lodato 2016, for a recent review). The criterion for a disk to be gravitationally unstable is given by the Toomre parameter, defined as

$$Q = \frac{c_s \Omega}{\pi G \Sigma} \sim \frac{M_\star}{M_{\rm disk}} \frac{h}{r} \lesssim 1 - 2, \qquad (4.24)$$

where $M_\star$ is the stellar mass and $M_{\rm disk} = \pi \Sigma r^2$ is a local estimate (at r) of the disk mass. The total disk mass is close to the maximum value of this local estimate (for finite disks). Disks which are more massive and colder and thinner (i.e., colder) are more likely to become unstable. Gravitational instability (GI) can lead to the fragmentation of the disk and the formation of clumps at very early times of evolution in Class 0 and I objects. The outcome of GI can be either the formation of stellar companions, large scale spiral arms, or smaller scale gravito-turbulence, depending on disk properties such as the mass infall rate onto the disk (Kratter et al. 2010a,b; Rafikov 2009). We focus on the parameter space of spiral arms (high mass disks with moderate infall rates) because these large scale regions of high pressure are effective particle traps (Rice et al. 2004).

Dipierro et al. (2015a) modeled the trapping of dust particles by self-gravitating spiral arms in a 1D (radial) + 1D (azimuthal) dust dynamics model, using the gas disk model from Lodato and Rice (2004). For the initial condition, it was assumed a $M_{\rm disk}/M_\star = 0.25$ with power laws for the initial density and temperature profiles, such that $\Sigma \propto r^{-1}$ and $T \propto r^{-1/2}$. In this case, for the dust evolution the processes of growth and fragmentation were neglected, but as a first estimate we assumed a

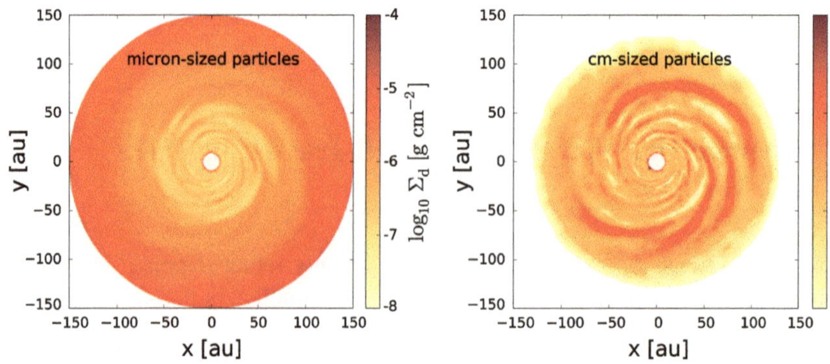

Fig. 4.15 Dust surface density for micron-sized particles [1–100] μm (*left-panel*) and centimeter-sized particles [1–100] cm (*right-panel*) in the case of trapping by self-gravitating spiral arms. Figure adapted from models in Dipierro et al. (2015a)

dust distribution from one micron-sized to $a_{\rm max} \approx a_{\rm frag}$ [Eq. (4.17)], and solve the advection-diffusion equation for each grain size. For the turbulence, it was assumed $\alpha = 0.05$ as a simplified parameter for the gravito-turbulence expected from GI simulations (Rice et al. 2005).

Figure 4.15 illustrates the effect of trapping in self-gravitating spiral arms. In particular, it shows the dust surface density for micron-sized particles [1–100] μm and centimeter-sized particles [1–100] cm. The distribution of the small grains that are well coupled to the gas (the micron-sized grains) are showing similar structures than the gas and the disk radial extension is as the gas, i.e. ~150 AU. The disk extension of the centimeter-sized particles is smaller $\lesssim$100 AU and the dust density of the large particles is enhanced inside the spiral arms due to the trapping.

As a consequence of the trapping of the large grains inside spirals arms induced by self-gravity, it is expected that strong spiral structures can be detected with high angular resolution at millimeter and centimeter wavelengths. At short wavelengths, less dominant but still detectable spiral arms are predicted (Dipierro et al. 2015a). Most detections of spiral arms in protoplanetary disks have been at scattered light in the NIR, which trace the distribution of small hot dust particles in the upper layers of the disk (e.g., Canovas et al. 2013; Garufi et al. 2013; Avenhaus et al. 2014; Benisty et al. 2015; Stolker et al. 2016). For most of these disks, the mass estimated from mm dust emission is inconsistent with self-gravity as the explanation for the observed structures. Planet–disk interaction is an alternative explanation for spiral arms seen in scattered light imaging (e.g., Pohl et al. 2015; Dong et al. 2015a). However, some work finds that embedded planets fail to create the high contrasts and opening angles observed in the spiral arms of several protoplanetary disks (Juhász et al. 2015). Other explanations for the observed spiral arms include interaction with an external binary (Dong et al. 2016), an unstable accretion shock from an external envelope (Lesur et al. 2015), MHD effects (Lyra et al. 2015), and light-travel-time effects caused by the shadow of an object orbiting close to the central star (Kama et al. 2016).

4.7.2 Trapping in Anti-cyclonic Vortices

Anti-cyclonic vortices can form in protoplanetary disks as a result of the Rossby wave instability (RWI). To trigger the RWI, disks must contain a special radial structure which produces an extremum in the vortensity (vertical vorticity per unit surface density) and/or entropy (Lovelace et al. 1999). Since certain pressure bumps can trigger the RWI, mechanisms that lead to the formation of vortices are similar to those discussed previously for pressure bumps: planet–disk interactions and transitions to dead zones. In the case of planet–disk interaction as the origin of vortices, only under suitable conditions of massive planet and low disk viscosity, the vortices are long-lived and able to trap particles (e.g., Ataiee et al. 2013; Zhu and Stone 2014; Hammer et al. 2017). However they may form, anti-cyclonic vortices are regions of high pressure, where particles with St$\lesssim$ 1 are effectively trapped, creating high contrast asymmetries in the mm- or cm-emission.

To study the trapping by an azimuthal pressure bump, such as the ones from anticyclonic vortices, Birnstiel et al. (2013) investigated the azimuthal distribution of particles taking into account turbulent mixing and azimuthal drift and assuming a steady state dust density distribution which is radially confined due to a particle trap. A weak and long-lived azimuthal gas pressure gradients can produce strong azimuthal concentrations of the dust with St$\gtrsim$ α. Figure 4.16 shows the distribution of micron-sized (1–10 μm) and millimeter-sized (1–10 mm) particles when a smooth pressure bump is assumed in the azimuthal direction (Birnstiel et al. 2013). These calculations use a hybrid 1D (radial) + 1D (azimuthal) analytical model. For the radial dust density distribution, particles accumulate in a radial pressure bump induced by a massive planet in the disk (Sect. 4.6), while for the azimuthal distribution, mixing and drift are both considered.

For Stokes numbers below unity, larger grains (traced at longer wavelengths), are concentrated more narrowly in azimuth (see also Lyra and Lin 2013). The concentration of dust particles inside the vortex can lead to dust-to-gas ratios over

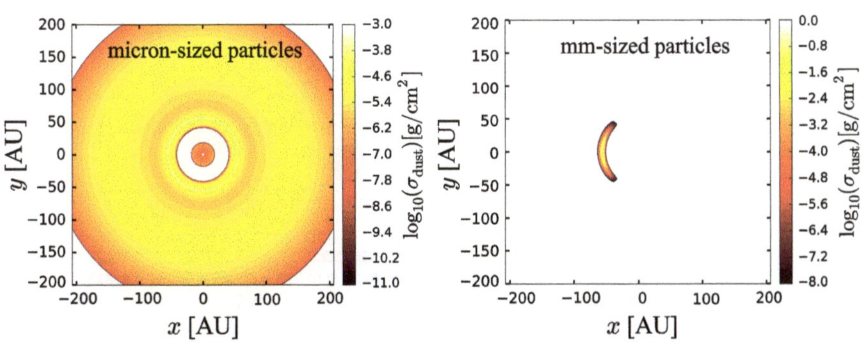

Fig. 4.16 Dust surface density for micron-sized particles 1–10 μm (*left-panel*) and millimeter-sized particles ~1–10 mm (*right-panel*) in the case of trapping by a smooth pressure bump in the azimuthal direction. Figure adapted from models in Birnstiel et al. (2013)

unity and feedback from the dust to the gas can lead to the vortex destruction (e.g., Chang and Oishi 2010).

Azimuthally asymmetric mm-emission has been observed in several transition disks, suggesting that whatever causes the inner cavity can also create a long-lived vortex that traps particles. Some of these asymmetries are low-contrast (e.g., SR 21, HD 135344B, LkHα 330; Isella et al. 2013) and others are high-contrast (e.g., HD 142527 and IRS 48; Casassus et al. 2013; Fukagawa et al. 2013; van der Marel et al. 2013). These high contrast disks, the asymmetries are azimuthally narrower in the longer wavelength emission, as expected for vortex trapping (Casassus et al. 2015; van der Marel et al. 2015).

However, this wavelength dependence is not evident in some transition disks with low contrast asymmetries. For instance, Fig. 4.17 shows ALMA Cycle 0 and I observations of SR 21 at 0.45 and 0.88 mm. The images are restored with the same beam of $0.27''$ in order to be comparable. Analyzing the disk structures at the two wavelengths, the azimuthal asymmetry does not become narrower at longer wavelengths contrary to theoretical predictions of trapping by a vortex. The Stokes numbers of the particles that are traced at these two wavelengths are lower than unity, assuming the gas surface density obtained from observations of CO and its isotopologues (Pinilla et al. 2015d; van der Marel et al. 2016).

At the moment, the origin of the asymmetries in transition disks like SR 21 and HD 135344B is unclear. Intriguingly, optical and NIR scattered light observations of the latter system reveal strong spiral structures which may be related to the asymmetry observed at mm-emission. An alternative mechanism to create asymmetries in protoplanetary disks, which has a weaker grain size dependence, is disk eccentricity (Ataiee et al. 2013). Disk can become eccentric when high-mass planets ($\gtrsim 5\,M_{\rm Jup}$) are embedded in the disk (Kley and Dirksen 2006).

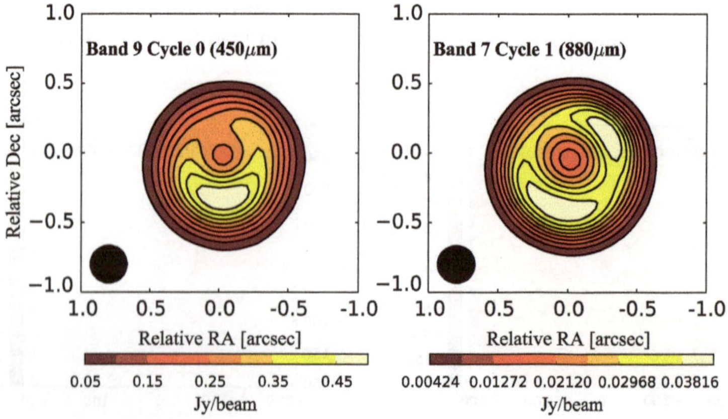

Fig. 4.17 ALMA Cycle 0 and I observations of the transition disk SR 21 at 0.45 and 0.88 mm, respectively. The two images have been restored with the same beam for comparison. Figure adapted from Pinilla et al. (2015d)

4.8 The Formation of Planetesimals

In this section we consider how modern models of planetesimal formation relate to particle traps of various types. For more complete summaries of work on planet formation, we suggest detailed reviews of the subject (Chiang and Youdin 2010; Johansen et al. 2014). This chapter has focused on long-lived, large scale particle traps, as these are the most amenable to direct observation. Such traps are also promising sites of planetesimal formation for several reasons. First, saving particles from rapid radial drift gives more time for planetesimal formation. Second, collision speeds should be lower and more amenable to growth in particle traps, due to drift speed and perhaps less turbulence. Third, the collection of particles to high densities can trigger the gravitational collapse of small solids into larger planetesimals. These theoretical advantages for planetesimal formation are especially relevant now that long lived particle traps are an observational reality.

Several caveats must be attached to the hypothesis that planetesimals primarily form in long lived particle traps. First (and for this chapter foremost) particle traps cannot be too efficient at converting dust and pebbles into planetesimals, or else the particle traps would not be observable and disk's mm-excess would be too small. For the coagulation models presented in this chapter, solids do not grow past the $\mathrm{St} = 1$ barrier of $\sim$cm- to m-sized solids (see, e.g., Figs. 4.5, 4.8, and other grain size figures in this chapter). Planetesimals could form in models of this type if additional physics, notably self-gravity, were included in a full (possibly magneto-) hydrodynamical model. Second, particle traps *caused by a planet* cannot be the primary cause of planetesimal formation, either in systems that do not form massive gap-opening planets or assuming that the core accretion hypothesis holds. Thirdly, particle traps must exist at many locations in the disk to be a leading cause of planetesimal formation. A range of locations is certainly possible given the diversity of proposed mechanisms for particle traps.

These caveats serve to motivate the two main ideas of this section. Firstly, self-gravity may ultimately be responsible for the assembly of $>$km-sized planetesimals. Secondly, the long lived particle traps discussed in this section are not the only way to trap particles and trigger planetesimal formation. We summarize the streaming instability (SI), a powerful particle concentration mechanism that arises from drag forces (Youdin and Goodman 2005; Youdin and Johansen 2007), and can trigger planetesimal formation either with or without existing pressure bumps and particle traps (Johansen et al. 2007, 2009b). We also describe related secular gravitational instabilities (SGI) which depends on the interplay of drag forces and self-gravity (Youdin 2011). While streaming instabilities are too small scale to be directly observed, SGI provide an alternate explanation for observable dust rings, which does not depend on gas pressure maxima (Takahashi and Inutsuka 2016).

4.8.1 The Gravitational Collapse of Solids into Planetesimals

Earlier in this chapter, we described the role of *gas* self-gravity. Particles can also become strongly self-gravitating, if particle densities become large due to sedimentation to the midplane and/or particle trapping. The hypothesis that gravitational instabilities of solids are responsible for planetesimal formation is long-standing (Safronov 1969; Goldreich and Ward 1973).

The main obstacle to the gravitational collapse hypothesis is that turbulent gas tends to stir particles out of the midplane, and also away from particle traps. Some level of midplane turbulence is hard to avoid in disks, due to the global pressure gradients introduced in Sect. 4.2. These pressure gradients drive a vertical shear between a particle-laden midplane—which forces gas to orbit closer to Keplerian speeds—and the more sub-Keplerian gas away from the midplane. Turbulence driven by *vertical shear instabilities* has long been appreciated as an impediment to particle sedimentation and thus the high densities needed for gravitational collapse (Cuzzi et al. 1993). Radial pressure gradients are now known to also drive streaming instabilities, as described below.

There currently exist several scenarios for planetesimals to form by gravitational collapse despite disk turbulence. These possibilities include pressure bumps as particle traps, streaming instabilities and SGI. These scenarios are not mutually exclusive, as a range of processes could enhance particle densities until collapse occurs. Particle traps have an additional advantage that radial pressure gradients at the traps' pressure maxima. Thus, the otherwise "inevitable" turbulence caused by radial pressure gradients should be much weaker in particle traps.

The promise of particle traps for planet formation was simulated in detail by Johansen et al. (2007). In this groundbreaking work, particle traps (in the form of MRI zonal flows) combined with the streaming instability to cause the clumping of $\mathrm{St} = 0.25$–1 sized solids which collapsed gravitationally into Ceres mass planetesimals. We next summarize the streaming instability, which (if nothing else) is currently the most thoroughly vetted hypothesis for the formation of 1–100 km planetesimals.

4.8.2 Streaming Instabilities and Planetesimal Formation

The streaming instability is a robust instability in protoplanetary disks, which arises from the relative drift (streaming) between solids and gas (Youdin and Goodman 2005). Under the right conditions, described below, the streaming instability causes strong particle clumping (Johansen and Youdin 2007), which can trigger gravitational collapse into planetesimals (Johansen et al. 2009b).

The key physical ingredient for the streaming instability is the back reaction of drag forces on the *gas* in protoplanetary disks. This back reaction naturally produces the streaming instability as particles drift through the disk gas. The energy for the instability comes ultimately from the disk's radial pressure gradient, specifically the pressure work done on the disk gas as it flows radially outward (Youdin and Johansen 2007). This radial outflow of gas is a consequence of angular momentum conservation with the solids that drift inwards.

The streaming instability is closely related to the secular drag instability (Goodman and Pindor 2000). While the streaming instability treats drag as a microscopic force on particles (as in Sect. 4.2 but also including the back reaction on the gas), the secular drag instability considers the macroscopic "plate" drag on the particle midplane layer. The fact that both treatments of drag forces predict particle clumping helps explain why dissipative drag forces are able to clump particles.

From the basic ingredients of the streaming instability, the optimal conditions for the streaming instability to drive strong particle clumping are readily understood. The three key parameters for the streaming instability are particle size (or St), the mass ratio of solids to gas, and the strength of the radial pressure gradient. Particle concentration is strongest for intermediate sized particles, near St $\sim$ 0.3 (Yang et al. 2016). The fact that St of order unity gives the optimum size for particle concentration is consistent with the fact that streaming instabilities depend on radial drift.

Since the streaming instability relies on the feedback of particles on gas, a high particle density is needed. To quantify this density criterion requires some care, because in real disks, the midplane particle density is affected by particle sedimentation. In the artificial case where particle sedimentation is turned off, the average density ratio of particles to gas, ϵ, can be fixed in a computer simulation, and $\epsilon = 0.2$–1 gives the strongest particle clumping, consistent with the need for strong feedback (Johansen and Youdin 2007). In the realistic case with particle sedimentation, the *surface* density ratio of particles to gas is fixed and referred to as Z, an effective "metallicity" or particle-to-gas-mass ratio in a local patch of the disk. The density ratio in the midplane then evolves according to the balance between particle sedimentation and stirring caused by the streaming instability and any other sources of midplane turbulence and related dynamical activity.

The required metallicity for strong particle clumping by the streaming instability is at least $Z \gtrsim 0.015$. This threshold is determined by self-consistent hydrodynamic simulations where streaming instabilities (i.e., particle–gas interactions) are the only source of turbulence (Johansen et al. 2009b). In these simulations St$\sim$ 0.25–1 with a headwind speed of $\eta v_{\mathrm{K}} = 0.1 c_s$. As particle sizes depart from the optimum St values, higher Z values are needed to trigger strong clumping. As headwind speeds increase (or decrease), the minimum Z needed for strong clumping increases (or decreases, respectively) (Bai and Stone 2010b). This finding highlights the delicate balance between particle stirring and particle clumping. The radial pressure gradient is the driving force for the streaming instability, but since the streaming instability also stirs particles, higher Z values are needed for particle clustering as the radial pressure gradient increases.

Note that there is no way to distinguish between particle stirring caused by "streaming instabilities" vs. that caused by "vertical shear instabilities" in any meaningful way. There is no difference between streaming instabilities with vertical stratification (on the one hand) and vertical shear instabilities with non-zero St (on the other), because the physical ingredients are identical, namely: Keplerian dynamics with vertical gravity, a radial pressure gradient and two way drag forces between particles and gas.[2]

The fact that effective metallicities near the cosmic value of 0.01 can trigger strong particle concentration via streaming instabilities is promising for planetesimal formation. The fact that modestly super-Solar abundances can lead to enhanced particle sedimentation and gravitational collapse was expected from earlier studies of vertical shear instabilities (Sekiya 1998; Youdin and Shu 2002). Enhanced Z values can be achieved several ways, including particle traps, gas removal, and also by simple radial drift in a smooth disk, which produces particle pileups as solids drift towards the inner disk (Youdin and Chiang 2004). Of course, a "metallicity" near solar in large pebbles (St near unity) requires a combination of efficient particle growth and/or more significant enhancements in the *total* particle-to-gas ratio, to account for the small dust that is not optimally sized and doesn't take part in strong clumping (Bai and Stone 2010a). Whether coagulation can produce enough of the optimally sized solids for the streaming instability (to produce planetesimals) is subject of ongoing research that depends on the disk and coagulation model (Drążkowska and Dullemond 2014).

The streaming instability can strongly concentrate solids even when self-gravity is weak, or turned off completely. However, self-gravity is required to create gravitationally bound planetesimals, to determine their masses (Johansen et al. 2009b; Simon et al. 2016; Schäfer et al. 2016) and to contract the particle clumps to solid densities. Currently simulations cannot track in detail the gravitational collapse of particle clumps into planetesimals all the way to solid densities. Initial attempts to model this final collapse phase show that binary and even triple planetesimals can result from the fragmentation of collapsing clumps, which may explain the binaries in the primordial Kuiper Belt (Nesvorný et al. 2010).

Figure 4.18 shows a planetesimal formation simulation from Simon et al. (2016) with $\mathrm{St} = 0.3$, $Z = 0.02$ and $\eta v_\mathrm{K} = 0.1 c_s$. The snapshots span ten local orbital periods. In the early snapshot at the upper left, the streaming instability is concentrating particles into azimuthally elongated rings. By the final snapshot in the lower right, self-gravity has caused these rings to contract and then fragment into multiple planetesimals. The number, N, of planetesimals of a mass m follows a roughly powerlaw distribution $dN/dm \propto m^{-1.6}$ (Simon et al. 2016) with some evidence of an exponential taper (vs. a sharp cutoff) at high masses (Schäfer et al. 2016). The masses produced in this brief formation phase should quickly evolve as the planetesimals continue to accrete (Johansen et al. 2015).

[2]It is tempting to define the strong vs. weak clumping regimes as the "streaming" vs. "vertical shear" regimes. However the fact that clumping can be weak without vertical shear for the unstratified streaming instability (e.g., with low St and ϵ values) shows that such a distinction is misleading.

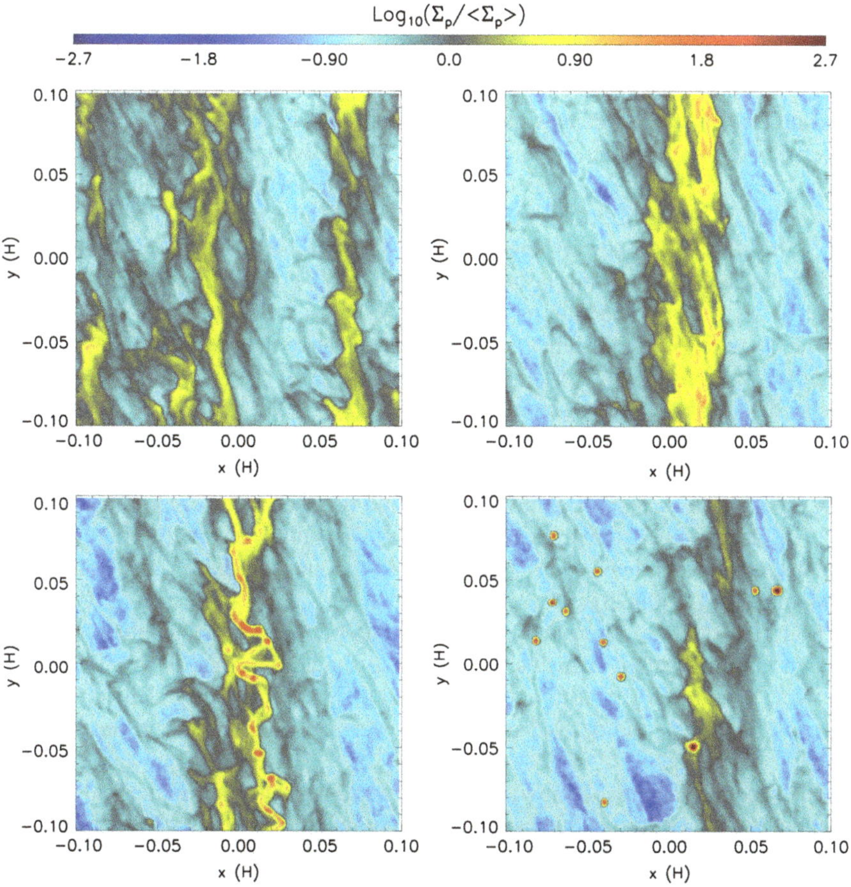

Fig. 4.18 Snapshots from a planetesimal formation simulation by Simon et al. (2016) of particle clumping by streaming instabilities and gravitational collapse. Images show the column density of solids with St $= 0.3$ versus radius and azimuth (x, y) in units of the gas scale height. Time evolves from top-left to bottom-right over ten local orbits. The circular clumps in the final panel are planetesimals with masses equivalent to $\sim$100 km radii

For this chapter, we should emphasize the small scale of streaming instability clumping is $< 0.2H \sim 0.005r$, too small even for ALMA. And if the clumping scales could be measured, the rapid concentration and collapse occurs over several orbits, only a tiny fraction of the disk's few Myr lifetime. Thus at any given moment, only a small fraction (if any) of a disk could be expected to be in a strong clumping phase due to streaming instabilities.

4.8.3 Secular Gravitational Instabilities

We now briefly discuss secular gravitational instabilities (SGI), a hypothesized mechanism for both planetesimal formation (Ward 2000; Youdin 2011) and observed dust rings (Takahashi and Inutsuka 2016). Given the apparent effectiveness of the streaming instability, it is reasonable to ask how any other instability could be relevant. SGI could be relevant for smaller particles (lower St) which are not concentrated by the hydrodynamic streaming instability. The key distinction for SGI (compared to streaming instabilities) is that self-gravity is important from the outset at creating particle clumps.

The "secular" in SGI refers to the role of dissipative drag forces in gravitational collapse. Absent frictional drag, self-gravity is usually only significant above the Roche density, when self-gravity is stronger than tidal gravity. However, with friction, radially wide particle rings can gradually contract due to self-gravity, even below the Roche density (Youdin 2005). The key to this slow mode of gravitational collapse is the exchange of angular momentum between the particle ring and disk gas. For planetesimal formation to occur, these rings must eventually fragment into planetesimals, or perhaps even larger cores as the mass in these rings can exceed Earth masses (Youdin 2011).

There are two main problems with SGI. The first is that gravitational collapse is opposed by turbulent diffusion which acts to spread the rings as they attempt to contract (Youdin 2011; Shariff and Cuzzi 2011). The effects of turbulent stirring are especially strong in the desired regime of lower St where collapse timescales are long.

The second main problem with SGI is uncertainty. Detailed numerical simulations have yet to be carried out, in part due to computational cost. Existing analytical studies of linear instability parameterize key uncertainties (namely the turbulent diffusion) and ignore important physical effects like vertical stratification. Thus more study is needed to determine the actual relevance of SGI.

The optimistic view of SGI is that its limitations as a planetesimal formation mechanism—the very slow contraction of wide rings—make it possible to observe these particle rings. Since the non-linear evolution of SGI is poorly understood, it is also possible that these rings evolve into vortex-like blobs (Takahashi and Inutsuka 2014). Particle clumps created by SGI should not be associated with a strong pressure maximum. This provides a key observational distinction between particle traps generated by pressure bumps of various types.

The physics of the SGI (particle self-gravity and drag) also applies inside pressure maxima, and could be important for plantesimal formation there. The larger questions posed by the SGI are similar to the one posed by streaming instabilities. Are particles primarily concentrated by structures in the gas disk (that are not directly caused by particle dynamics)? Or do particles spontaneously cause their own clumping, potentially over a wider range of disk locations? Coupled theoretical and observational studies are needed to understand the relative importance of these different mechanisms.

4.9 Summary and Future Perspectives

In protoplanetary disks, the millimeter-sized particles in the outer regions at $\gtrsim$50 AU (or meter-sized objects in the very first AU) are expected to experience a rapid inward drift, caused by the head-wind that particles experience in standard gas disks where pressure decreases radially. This radial drift barrier is predicted to be more severe for dust particles in disks around Brown Dwarfs than in T-Tauri or Herbig disks. As a result, millimeter/centimeter-sized particles are expected to be rapidly depleted in any smooth protoplanetary disks, before planetesimals can form. This is in disagreement with millimeter-observations of protoplanetary disks, which evidence the existence of large-sized particles in disks in different star-formation regions and around different type of stars.

Pressure bumps and/or streaming instabilities are possible solutions for this radial drift barrier. Pressure bumps have the distinct advantage of being directly observable. In this chapter we present the main results of dust evolution models when pressure bumps of different origin exist, including: zonal flows, the outer edge of a dead zone, embedded planet(s), self-gravitating spiral arms, and anti-cyclonic vortices. The results of the models touched in these chapter and the comparison with observations are summarized as follows:

1. Models of particle evolution should explain the observed values of the spectral index, i.e. the spatially integrated value when high spatial resolution is not available. The low values observed indicate mm/cm-sized particles in the outer regions of protoplanetary disks. Slowly evolving pressure bumps with amplitudes of $\sim$25–30% above the background can retain mm/cm particles and explain observed spectral indices. In this explanation, multiple ring-like structures should be revealed when disks are studied at high resolution, which has now been observed for HL Tau and TWHya.
2. For disks around Brown Dwarfs (BDs), the amplitude of pressure bumps must be higher, $\sim$40–60% to explain mm-observations of BD disks. MHD models assuming BD parameters are needed to understand if pressure bumps with this amplitude are possible.
3. Transition disks (TDs) have particle traps exterior to a central cavity that is cleared of mm-emission. Embedded planets (inside the cavity) or a dead zone boundary (at the cavity edge) could explain these TDs. Both mechanisms lead to similar structures of the gas and particles of different sizes. Future gas observations of TDs could determine the dominant mechanism, with import implications for exoplanet statistics. For instance, spatially resolved observations of molecular line widths could measure turbulence as a function of radius, revealing boundaries between dead and active regions in the disks.
4. A radially broad pressure bump can form as a result of a massive planet embedded in the disk. Such bumps cause the spatial segregation of small and large grains. Other mechanisms, such as dead zones with MHD winds, could also generate broad bumps with similar size segregation of dust particles.

5. Pre-transition disks are similar to TDs, but with a near infrared excess caused by small grains in the inner disk. To explain this behavior, we first note that even when large particles are trapped in the outer part of the disk by a pressure bump, small grains can be mixed and dragged with the gas and move towards the inner regions. In normal TDs these small grains do not build up and indeed can be depleted by efficient growth and radial migration. Pre-transition disks can be explained by the effect of the water snow line. When dust particles lose their ice mantles inside the snow line, they fragment more easily. The build-up of small dust in the inner disk can explain the near-infrared excess of pre-transition disks. Disk can evolve from a pre-transition disk to a TD when a massive planet carves a deep enough gap to stop the replenishment of dust from the outer disk.
6. In addition to rings, non-axisymmetric particle traps are caused by anti-cyclonic vortices and spiral arms in self-gravitating disks. Vortices can form by the Rossby wave instability, which develops at the outer edge of gaps opened by planets or at the outer edge of dead zones. Particle trapping in vortices leads to high contrast asymmetries at mm-emission. The azimuthal extent of emission is predicted to decrease at longer wavelengths, as larger solids collect closer to the core of the vortex. Similarly, emission from spiral arms is expected to be more compact and with higher contrast at longer wavelengths. Observations of molecular lines that the gas content in protoplanetary disks can help determine if spiral arms are caused by self-gravity vs. planetary companions or other mechanisms.
7. The trapping of particles in pressure maxima is a promising route to planetesimal formation via enhanced coagulation and/or gravitational collapse. However, planetesimal formation cannot be too efficient in particle traps, or else the mm-emission will be lost. The streaming instability is a planetesimal formation mechanism that does not require pressure maxima and is difficult to observe. The mechanism is complementary, as the streaming instability is facilitated by the trapping of particles in pressure maxima.

We are now in an era where current telescopes have the resolution and sensitivity to observe the structures expected from particle trapping by different physical mechanisms. Telescopes like ALMA and instruments at the VLT (such as SPHERE, PIONEER, and GRAVITY) are revolutionizing our understanding of disk structures and planet formation. Future facilities, such as James Webb Space Telescope (JWST) and the European Extremely Large Telescope (EELT), will continue transforming our comprehension of how, where, and when the first steps of planet formation occur.

Acknowledgements We are very grateful to Martin Pessah and Oliver Gressel for their patience with this chapter. PP acknowledges support by NASA through Hubble Fellowship grant HST-HF2-51380.001-A awarded by the Space Telescope Science Institute, which is operated by the Association of Universities for Research in Astronomy, Inc., for NASA, under contract NAS 5-26555. ANY acknowledges support from the NSF through grant AST-1616929 and support from NASA through grant 16-ATP16-0052.

References

Acke, B., van den Ancker, M.E.: Resolving the disk rotation of HD 97048 and HD 100546 in the [O I] 6300 Å line: evidence for a giant planet orbiting HD 100546. Astron. Astrophys. **449**, 267–279 (2006). doi:10.1051/0004-6361:20054330, astro-ph/0512562

Adachi, I., Hayashi, C., Nakazawa, K.: The gas drag effect on the elliptical motion of a solid body in the primordial solar nebula. Prog. Theor. Phys. **56**, 1756–1771 (1976)

Adams, F.C., Lada, C.J., Shu, F.H.: Spectral evolution of young stellar objects. Astrophys. J. **312**, 788–806 (1987). doi:10.1086/164924

Akiyama, E., Hasegawa, Y., Hayashi, M., Iguchi, S.: Planetary system formation in the protoplanetary disk around HL Tauri. Astrophys. J. **818**, 158 (2016). doi:10.3847/0004-637X/818/2/158, 1511.04822

Alexander, R., Pascucci, I., Andrews, S., Armitage, P., Cieza, L.: The dispersal of protoplanetary disks. In: Protostars and Planets VI, pp. 475–496 (2014). doi:10.2458/azu_uapress_9780816531240-ch021, 1311.1819

ALMA Partnership, Brogan, C.L., Pérez, L.M., Hunter, T.R., Dent, W.R.F., Hales, A.S., Hills, R.E., Corder, S., Fomalont, E.B., Vlahakis, C., Asaki, Y., Barkats, D., Hirota, A., Hodge, J.A., Impellizzeri, C.M.V., Kneissl, R., Liuzzo, E., Lucas, R., Marcelino, N., Matsushita, S., Nakanishi, K., Phillips, N., Richards, A.M.S., Toledo, I., Aladro, R., Broguiere, D., Cortes J.R., Cortes, P.C., Espada, D., Galarza, F., Garcia-Appadoo, D., Guzman-Ramirez, L., Humphreys, E.M., Jung, T., Kameno, S., Laing, R.A., Leon S., Marconi, G., Mignano, A., Nikolic, B., Nyman, L.A., Radiszcz, M., Remijan A., Rodón, J.A., Sawada, T., Takahashi, S., Tilanus, R.P.J., Vila Vilaro, B., Watson, L.C., Wiklind, T., Akiyama, E., Chapillon, E., de Gregorio-Monsalvo I., Di Francesco, J., Gueth, F., Kawamura, A., Lee, C.F., Nguyen Luong, Q., Mangum, J., Pietu, V., Sanhueza, P., Saigo, K., Takakuwa, S., Ubach, C., van Kempen, T., Wootten, A., Castro-Carrizo, A., Francke, H., Gallardo, J., Garcia, J., Gonzalez, S., Hill, T., Kaminski, T., Kurono, Y., Liu, H.Y., Lopez, C., Morales, F., Plarre, K., Schieven, G., Testi, L., Videla, L., Villard, E., Andreani, P., Hibbard, J.E., Tatematsu, K.: The 2014 ALMA long baseline campaign: first results from high angular resolution observations toward the HL Tau region. Astrophys. J. Lett. **808**, L3 (2015). doi:10.1088/2041-8205/808/1/L3, 1503.02649

Andre, P., Montmerle, T.: From T Tauri stars to protostars: circumstellar material and young stellar objects in the rho Ophiuchi cloud. Astrophys. J. **420**, 837–862 (1994). doi:10.1086/173608

Andrews, S.M., Wilner, D.J., Espaillat, C., Hughes, A.M., Dullemond, C.P., McClure, M.K., Qi, C., Brown, J.M.: Resolved images of large cavities in protoplanetary transition disks. Astrophys. J. **732**, 42 (2011). doi:10.1088/0004-637X/732/1/42, 1103.0284

Andrews, S.M., Wilner, D.J., Zhu, Z., Birnstiel, T., Carpenter, J.M., Pérez, L.M., Bai, X.N., Öberg, K.I., Hughes, A.M., Isella, A., Ricci, L.: Ringed substructure and a gap at 1 au in the nearest protoplanetary disk. Astrophys. J. Lett. **820**, L40 (2016). doi:10.3847/2041-8205/820/2/L40, 1603.09352

Armitage, P.J.: Dynamics of protoplanetary disks. Annu. Rev. Astron. Astrophys. **49**, 195–236 (2011). doi:10.1146/annurev-astro-081710-102521, 1011.1496

Ataiee, S., Pinilla, P., Zsom, A., Dullemond, C.P., Dominik, C., Ghanbari, J.: Asymmetric transition disks: vorticity or eccentricity? Astron. Astrophys. **553**, L3 (2013). doi:10.1051/0004-6361/201321125, 1304.1736

Avenhaus, H., Quanz, S.P., Schmid, H.M., Meyer, M.R., Garufi, A., Wolf, S., Dominik, C.: Structures in the protoplanetary disk of HD142527 seen in polarized scattered light. Astrophys. J. **781**, 87 (2014). doi:10.1088/0004-637X/781/2/87, 1311.7088

Bai, X., Stone, J.M.: Dynamics of solids in the midplane of protoplanetary disks: implications for planetesimal formation. Astrophys. J. **722**, 1437–1459 (2010a). doi:10.1088/0004-637X/722/2/1437, 1005.4982

Bai, X., Stone, J.M.: The effect of the radial pressure gradient in protoplanetary disks on planetesimal formation. Astrophys. J. Lett. **722**, L220–L223 (2010b). doi:10.1088/2041-8205/722/2/L220, 1005.4981

Balbus, S.A., Hawley, J.F.: A powerful local shear instability in weakly magnetized disks. I - Linear analysis. II - Nonlinear evolution. Astrophys. J. **376**, 214–233 (1991). doi:10.1086/170270

Banzatti, A., Testi, L., Isella, A., Natta, A., Neri, R., Wilner, D.J.: New constraints on dust grain size and distribution in CQ Tauri. Astron. Astrophys. **525**, A12 (2011). doi:10.1051/0004-6361/201015206, 1009.2697

Barge, P., Sommeria, J.: Did planet formation begin inside persistent gaseous vortices? Astron. Astrophys. **295**, L1–L4 (1995). astro-ph/9501050

Baruteau, C., Crida, A., Paardekooper, S.J., Masset, F., Guilet, J., Bitsch, B., Nelson, R., Kley, W., Papaloizou, J.: Planet-disk interactions and early evolution of planetary systems. In: Protostars and Planets VI, pp. 667–689 (2014). doi:10.2458/azu_uapress_9780816531240-ch029, 1312.4293

Beckwith, S.V.W., Sargent, A.I.: Particle emissivity in circumstellar disks. Astrophys. J. **381**, 250–258 (1991). doi:10.1086/170646

Benisty, M., Juhasz, A., Boccaletti, A., Avenhaus, H., Milli, J., Thalmann, C., Dominik, C., Pinilla, P., Buenzli, E., Pohl, A., Beuzit, J.L., Birnstiel, T., de Boer, J., Bonnefoy, M., Chauvin, G., Christiaens, V., Garufi, A., Grady C., Henning, T., Huelamo, N., Isella, A., Langlois, M., Ménard, F., Mouillet, D., Olofsson, J., Pantin, E., Pinte, C., Pueyo, L.: Asymmetric features in the protoplanetary disk MWC 758. Astron. Astrophys. **578**, L6 (2015). doi:10.1051/0004-6361/201526011, 1505.05325

Biller, B., Lacour, S., Juhász, A., Benisty, M., Chauvin, G., Olofsson, J., Pott, J.U., Müller, A., Sicilia-Aguilar, A., Bonnefoy, M., Tuthill, P., Thebault, P., Henning, T., Crida, A.: A likely close-in low-mass stellar companion to the transitional disk star HD 142527. Astrophys. J. Lett. **753**, L38 (2012). doi:10.1088/2041-8205/753/2/L38, 1206.2654

Birnstiel, T., Dullemond, C.P., Brauer, F.: Gas- and dust evolution in protoplanetary disks. Astron. Astrophys. **513**, A79 (2010a). doi:10.1051/0004-6361/200913731, 1002.0335

Birnstiel, T., Ricci, L., Trotta, F., Dullemond, C.P., Natta, A., Testi, L., Dominik, C., Henning, T., Ormel, C.W., Zsom, A.: Testing the theory of grain growth and fragmentation by millimeter observations of protoplanetary disks. Astron. Astrophys. **516**, L14 (2010b). doi:10.1051/0004-6361/201014893, 1006.0940

Birnstiel, T., Klahr, H., Ercolano, B.: A simple model for the evolution of the dust population in protoplanetary disks. Astron. Astrophys. **539**, A148 (2012). doi:10.1051/0004-6361/201118136, 1201.5781

Birnstiel, T., Dullemond, C.P., Pinilla, P.: Lopsided dust rings in transition disks. Astron. Astrophys. **550**, L8 (2013). doi:10.1051/0004-6361/201220847, 1301.1976

Birnstiel, T., Fang, M., Johansen, A.: Dust evolution and the formation of planetesimals. Space Sci. Rev. (2016). doi:10.1007/s11214-016-0256-1, 1604.02952

Blum, J., Wurm, G.: Experiments on sticking, restructuring, and fragmentation of preplanetary dust aggregates. Icarus **143**, 138–146 (2000). doi:10.1006/icar.1999.6234

Blum, J., Wurm, G.: The growth mechanisms of macroscopic bodies in protoplanetary disks. Annu. Rev. Astron. Astrophys. **46**, 21–56 (2008). doi:10.1146/annurev.astro.46.060407.145152

Burrows, A., Hubbard, W.B., Lunine, J.I., Liebert, J.: The theory of brown dwarfs and extrasolar giant planets. Rev. Mod. Phys. **73**, 719–765 (2001). doi:10.1103/RevModPhys.73.719, astro-ph/0103383

Canovas, H., Ménard, F., Hales, A., Jordán, A., Schreiber, M.R., Casassus, S., Gledhill, T.M., Pinte, C.: Near-infrared imaging polarimetry of HD 142527. Astron. Astrophys. **556**, A123 (2013). doi:10.1051/0004-6361/201321924, 1306.6379

Casassus, S., van der Plas, G., Sebastian Perez, M., Dent, W.R.F., Fomalont, E., Hagelberg, J., Hales, A., Jordán, A., Mawet, D., Ménard, F., Wootten, A., Wilner, D., Hughes, A.M., Schreiber, M.R., Girard, J.H., Ercolano, B., Canovas, H., Román, P.E., Salinas, V.: Flows of gas through a protoplanetary gap. Nature **493**, 191–194 (2013). doi:10.1038/nature11769, 1305.6062

Casassus, S., Wright, C.M., Marino, S., Maddison, S.T., Wootten, A., Roman, P., Pérez, S., Pinilla, P., Wyatt, M., Moral, V., Ménard, F., Christiaens, V., Cieza, L., van der Plas, G.: A compact concentration of large grains in the HD 142527 protoplanetary dust trap. Astrophys. J. **812**, 126 (2015). doi:10.1088/0004-637X/812/2/126, 1505.07743

Chang, P., Oishi, J.S.: On the stability of dust-laden protoplanetary vortices. Astrophys. J. **721**, 1593–1602 (2010). doi:10.1088/0004-637X/721/2/1593, 1007.2417

Chiang, E., Laughlin, G.: The minimum-mass extrasolar nebula: in situ formation of close-in super-Earths. Mon. Not. R. Astron. Soc. **431**, 3444–3455 (2013). doi:10.1093/mnras/stt424, 1211.1673

Chiang, E., Youdin, A.N.: Forming planetesimals in solar and extrasolar nebulae. Annu. Rev. Earth Planet. Sci. **38**, 493–522 (2010). doi:10.1146/annurev-earth-040809-152513, 0909.2652

Cleeves, L.I., Bergin, E.A., Harries, T.J.: Indirect detection of forming protoplanets via chemical asymmetries in disks. Astrophys. J. **807**, 2 (2015). doi:10.1088/0004-637X/807/1/2, 1505.07470

Crida, A., Morbidelli, A., Masset, F.: On the width and shape of gaps in protoplanetary disks. Icarus **181**, 587–604 (2006). doi:10.1016/j.icarus.2005.10.007, astro-ph/0511082

Currie, T., Cloutier, R., Brittain, S., Grady, C., Burrows, A., Muto, T., Kenyon, S.J., Kuchner, M.J.: Resolving the HD 100546 protoplanetary system with the gemini planet imager: evidence for multiple forming, accreting planets. Astrophys. J. Lett. **814**, L27 (2015). doi:10.1088/2041-8205/814/2/L27, 1511.02526

Cuzzi, J.N., Dobrovolskis, A.R., Champney, J.M.: Particle-gas dynamics in the midplane of a protoplanetary nebula. Icarus **106**, 102 (1993). http://adsabs.harvard.edu/cgi-bin/nph-bib_query?bibcode=1993Icar..106..102C&db_key=AST

de Boer, J., Salter, G., Benisty, M., Vigan, A., Boccaletti, A., Pinilla, P., Ginski, C., Juhasz, A., Maire, A.L., Messina, S., Desidera, S., Cheetham, A., Girard, J.H., Wahhaj, Z., Langlois, M., Bonnefoy, M., Beuzit, J.L., Buenzli E., Chauvin, G., Dominik, C., Feldt, M., Gratton, R., Hagelberg, J., Isella A., Janson, M., Keller, C.U., Lagrange, A.M., Lannier, J., Menard, F., Mesa, D., Mouillet, D., Mugrauer, M., Peretti, S., Perrot, C., Sissa, E., Snik, F., Vogt, N., Zurlo, A., SPHERE Consortium: Multiple rings in the transition disk and companion candidates around RX J1615.3-3255. High contrast imaging with VLT/SPHERE. Astron. Astrophys. **595**, A114 (2016). doi:10.1051/0004-6361/201629267, 1610.04038

de Juan Ovelar, M., Min, M., Dominik, C., Thalmann, C., Pinilla, P., Benisty M., Birnstiel, T.: Imaging diagnostics for transitional discs. Astron. Astrophys. **560**, A111 (2013). doi:10.1051/0004-6361/201322218, 1309.1039

de Juan Ovelar, M., Pinilla, P., Min, M., Dominik, C., Birnstiel, T.: Constraining turbulence mixing strength in transitional discs with planets using SPHERE and ALMA. Mon. Not. R. Astron. Soc. **459**, L85–L89 (2016). doi:10.1093/mnrasl/slw051, 1603.09357

Dipierro, G., Pinilla, P., Lodato, G., Testi, L.: Dust trapping by spiral arms in gravitationally unstable protostellar discs. Mon. Not. R. Astron. Soc. **451**, 974–986 (2015a). doi:10.1093/mnras/stv970, 1504.08099

Dipierro, G., Price, D., Laibe, G., Hirsh, K., Cerioli, A., Lodato, G.: On planet formation in HL Tau. Mon. Not. R. Astron. Soc. **453**, L73–L77 (2015b). doi:10.1093/mnrasl/slv105, 1507.06719

Dittrich, K., Klahr, H., Johansen, A.: Gravoturbulent planetesimal formation: the positive effect of long-lived zonal flows. Astrophys. J. **763**, 117 (2013). doi:10.1088/0004-637X/763/2/117, 1211.2095

Dominik, C., Blum, J., Cuzzi, J.N., Wurm, G.: Growth of dust as the initial step toward planet formation. In: Protostars and Planets V, pp. 783–800 (2007). astro-ph/0602617

Dong, R., Zhu, Z., Rafikov, R.R., Stone, J.M.: Observational signatures of planets in protoplanetary disks: spiral arms observed in scattered light imaging can be induced by planets. Astrophys. J. Lett. **809**, L5 (2015a). doi:10.1088/2041-8205/809/1/L5, 1507.03596

Dong, R., Zhu, Z., Whitney, B.: Observational signatures of planets in protoplanetary disks I. Gaps opened by single and multiple young planets in disks. Astrophys. J. **809**, 93 (2015b). doi:10.1088/0004-637X/809/1/93, 1411.6063

Dong, R., Zhu, Z., Fung, J., Rafikov, R., Chiang, E., Wagner, K.: An M dwarf companion and its induced spiral arms in the HD 100453 protoplanetary disk. Astrophys. J. Lett. **816**, L12 (2016). doi:10.3847/2041-8205/816/1/L12, 1512.04949

Draine, B.T.: On the submillimeter opacity of protoplanetary disks. Astrophys. J. **636**, 1114–1120 (2006). doi:10.1086/498130, astro-ph/0507292

Drążkowska, J., Dullemond, C.P.: Can dust coagulation trigger streaming instability? Astron. Astrophys. **572**, A78 (2014). doi:10.1051/0004-6361/201424809, 1410.3832

Drążkowska, J., Windmark, F., Dullemond, C.P.: Planetesimal formation via sweep-up growth at the inner edge of dead zones. Astron. Astrophys. **556**, A37 (2013). doi:10.1051/0004-6361/201321566, 1306.3412

Duffell, P.C., Dong, R.: Shallow cavities in multiple-planet systems. Astrophys. J. **802**, 42 (2015). doi:10.1088/0004-637X/802/1/42, 1412.3560

Dullemond, C.P., Dominik, C.: The effect of dust settling on the appearance of protoplanetary disks. Astron. Astrophys. **421**, 1075–1086 (2004). doi:10.1051/0004-6361:20040284, astro-ph/0405226

Dullemond, C.P., Monnier, J.D.: The inner regions of protoplanetary disks. Annu. Rev. Astron. Astrophys. **48**, 205–239 (2010). doi:10.1146/annurev-astro-081309-130932, 1006.3485

Dullemond, C.P., Hollenbach, D., Kamp, I., D'Alessio, P.: Models of the structure and evolution of protoplanetary disks. In: Protostars and Planets V, pp. 555–572 (2007). astro-ph/0602619

Dzyurkevich, N., Flock, M., Turner, N.J., Klahr, H., Henning, T.: Trapping solids at the inner edge of the dead zone: 3-D global MHD simulations. Astron. Astrophys. **515**, A70 (2010). doi:10.1051/0004-6361/200912834, 1002.2521

Espaillat, C., Muzerolle, J., Najita, J., Andrews, S., Zhu, Z., Calvet, N., Kraus, S., Hashimoto, J., Kraus, A., D'Alessio, P.: An observational perspective of transitional disks. In: Protostars and Planets VI, pp. 497–520 (2014). doi:10.2458/azu_uapress_9780816531240-ch022, 1402.7103

Fedele, D., van den Ancker, M.E., Henning, T., Jayawardhana, R., Oliveira, J.M.: Timescale of mass accretion in pre-main-sequence stars. Astron. Astrophys. **510**, A72 (2010). doi:10.1051/0004-6361/200912810, 0911.3320

Fedele, D., Carney, M., Hogerheijde, M.R., Walsh, C., Miotello, A., Klaassen P., Bruderer, S., Henning, T., vanDishoeck, E.F.: ALMA unveils rings and gaps in the protoplanetary system HD 169142: signatures of two giant protoplanets (2017). ArXiv e-prints 1702.02844

Finkbeiner, D.P., Davis, M., Schlegel, D.J.: Extrapolation of galactic dust emission at 100 microns to cosmic microwave background radiation frequencies using FIRAS. Astrophys. J. **524**, 867–886 (1999). doi:10.1086/307852, astro-ph/9905128

Flock, M., Ruge, J.P., Dzyurkevich, N., Henning, T., Klahr, H., Wolf, S.: Gaps, rings, and non-axisymmetric structures in protoplanetary disks. From simulations to ALMA observations. Astron. Astrophys. **574**, A68 (2015). doi:10.1051/0004-6361/201424693, 1411.2736

Fromang, S., Balbus, S.A., Terquem, C., De Villiers, J.P.: Evolution of self-gravitating magnetized disks. II. Interaction between magnetohydrodynamic turbulence and gravitational instabilities. Astrophys. J. **616**, 364–375 (2004). doi:10.1086/424829, astro-ph/0409404

Fukagawa, M., Tsukagoshi, T., Momose, M., Saigo, K., Ohashi, N., Kitamura, Y., Inutsuka, S.i., Muto, T., Nomura, H., Takeuchi, T., Kobayashi, H., Hanawa, T., Akiyama, E., Honda, M., Fujiwara, H., Kataoka, A., Takahashi, S.Z., Shibai, H.: Local enhancement of the surface density in the protoplanetary ring surrounding HD 142527. Publ. Astron. Soc. Jpn. **65**, L14 (2013). doi:10.1093/pasj/65.6.L14, 1309.7400

Garufi, A., Quanz, S.P., Avenhaus, H., Buenzli, E., Dominik, C., Meru, F., Meyer, M.R., Pinilla, P., Schmid, H.M., Wolf, S.: Small vs. large dust grains in transitional disks: do different cavity sizes indicate a planet? SAO 206462 (HD 135344B) in polarized light with VLT/NACO. Astron. Astrophys. **560**, A105 (2013). doi:10.1051/0004-6361/201322429, 1311.4195

Ginski, C., Stolker, T., Pinilla, P., Dominik, C., Boccaletti, A., de Boer, J., Benisty, M., Biller, B., Feldt, M., Garufi, A., Keller, C.U., Kenworthy, M., Maire, A.L., Ménard, F., Mesa, D., Milli, J., Min, M., Pinte, C., Quanz, S.P., van Boekel, R., Bonnefoy, M., Chauvin, G., Desidera, S., Gratton, R., Girard, J.H.V., Keppler, M., Kopytova, T., Lagrange, A.M., Langlois, M., Rouan, D., Vigan, A.: Direct detection of scattered light gaps in the transitional disk around HD 97048 with VLT/SPHERE. Astron. Astrophys. **595**, A112 (2016). doi:10.1051/0004-6361/201629265, 1609.04027

Goldreich, P., Ward, W.R.: The formation of planetesimals. Astrophys. J. **183**, 1051–1062 (1973)

Gonzalez, J.F., Pinte, C., Maddison, S.T., Ménard, F., Fouchet, L.: Planet gaps in the dust layer of 3D protoplanetary disks. II. Observability with ALMA. Astron. Astrophys. **547**, A58 (2012). doi:10.1051/0004-6361/201218806, 1208.5436

Goodman, J., Pindor, B.: Secular instability and planetesimal formation in the dust layer. Icarus **148**, 537–549 (2000)

Guilloteau, S., Dutrey, A., Piétu, V., Boehler, Y.: A dual-frequency sub-arcsecond study of protoplanetary disks at mm wavelengths: first evidence for radial variations of the dust properties. Astron. Astrophys. **529**, A105 (2011). doi:10.1051/0004-6361/201015209, 1103.1296

Gundlach, B., Blum, J.: The stickiness of micrometer-sized water-ice particles. Astrophys. J. **798**, 34 (2015). doi:10.1088/0004-637X/798/1/34, 1410.7199

Haisch, K.E. Jr., Lada, E.A., Lada, C.J.: Disk frequencies and lifetimes in young clusters. Astrophys. J. Lett. **553**, L153–L156 (2001). doi:10.1086/320685, astro-ph/0104347

Hammer, M., Kratter, K.M., Lin, M.K.: Slowly-growing gap-opening planets trigger weaker vortices. Mon. Not. R. Astron. Soc. **466**, 3533–3543 (2017). doi:10.1093/mnras/stw3000, 1610.01606

Hernández, J., Hartmann, L., Megeath, T., Gutermuth, R., Muzerolle, J., Calvet, N., Vivas, A.K., Briceño, C., Allen, L., Stauffer, J., Young, E., Fazio, G.: A spitzer space telescope study of disks in the young σ Orionis cluster. Astrophys. J. **662**, 1067–1081 (2007). doi:10.1086/513735, astro-ph/0701476

Hildebrand, R.H.: The determination of cloud masses and dust characteristics from submillimetre thermal emission. Q. J. R. Astron. Soc. **24**, 267 (1983)

Isella, A., Pérez, L.M., Carpenter, J.M., Ricci, L., Andrews, S., Rosenfeld, K.: An azimuthal asymmetry in the LkHα 330 disk. Astrophys. J. **775**, 30 (2013). doi:10.1088/0004-637X/775/1/30, 1307.5848

Isella, A., Guidi, G., Testi, L., Liu, S., Li, H., Li, S., Weaver, E., Boehler, Y., Carperter, J.M., De Gregorio-Monsalvo, I., Manara, C.F., Natta, A., Pérez, L.M., Ricci, L., Sargent, A., Tazzari, M., Turner, N.: Ringed structures of the HD 163296 protoplanetary disk revealed by alma. Phys. Rev. Lett. **117**, 251101 (2016). doi:10.1103/PhysRevLett.117.251101, http://link.aps.org/doi/10.1103/PhysRevLett.117.251101

Jin, S., Li, S., Isella, A., Li, H., Ji, J.: Modeling dust emission of HL Tau disk based on planet-disk interactions. Astrophys. J. **818**, 76 (2016). doi:10.3847/0004-637X/818/1/76, 1601.00358

Johansen, A., Youdin, A.: Protoplanetary disk turbulence driven by the streaming instability: nonlinear saturation and particle concentration. Astrophys. J. **662**, 627–641 (2007). doi:10.1086/516730, arXiv:astro-ph/0702626

Johansen, A., Oishi, J.S., Mac Low, M.M., Klahr, H., Henning, T., Youdin, A.: Rapid planetesimal formation in turbulent circumstellar disks. Nature **448**, 1022–1025 (2007). doi:10.1038/nature06086

Johansen, A., Youdin, A., Klahr, H.: Zonal flows and long-lived axisymmetric pressure bumps in magnetorotational turbulence. Astrophys. J. **697**, 1269–1289 (2009a). doi:10.1088/0004-637X/697/2/1269, 0811.3937

Johansen, A., Youdin, A., Mac Low, M.: Particle clumping and planetesimal formation depend strongly on metallicity. Astrophys. J. Lett. **704**, L75–L79 (2009b). doi:10.1088/0004-637X/704/2/L75, 0909.0259

Johansen, A., Blum, J., Tanaka, H., Ormel, C., Bizzarro, M., Rickman, H.: The multifaceted planetesimal formation process. In: Protostars and Planets VI, pp. 547–570 (2014). doi:10.2458/azu_uapress_9780816531240-ch024, 1402.1344

Johansen, A., Mac Low, M.M., Lacerda, P., Bizzarro, M.: Growth of asteroids, planetary embryos, and Kuiper belt objects by chondrule accretion. Sci. Adv. **1**, 1500109 (2015). doi:10.1126/sciadv.1500109, 1503.07347

Juhász, A., Benisty, M., Pohl, A., Dullemond, C.P., Dominik, C., Paardekooper, S.J.: Spiral arms in scattered light images of protoplanetary discs: are they the signposts of planets? Mon. Not. R. Astron. Soc. **451**, 1147–1157 (2015). doi:10.1093/mnras/stv1045, 1412.3412

Kama, M., Pinilla, P., Heays, A.N.: Spirals in protoplanetary disks from photon travel time. Astron. Astrophys. **593**, L20 (2016). doi:10.1051/0004-6361/201628924, 1608.03147

Kempf, S., Pfalzner, S., Henning, T.K.: N-particle-simulations of dust growth. I. Growth driven by Brownian motion. Icarus **141**, 388–398 (1999). doi:10.1006/icar.1999.6171

Kenyon, S.J., Hartmann, L.: Spectral energy distributions of T Tauri stars - disk flaring and limits on accretion. Astrophys. J. **323**, 714–733 (1987). doi:10.1086/165866

Klahr, H.H., Bodenheimer, P.: Turbulence in accretion disks: vorticity generation and angular momentum transport via the global baroclinic instability. Astrophys. J. **582**, 869–892 (2003). doi:10.1086/344743, astro-ph/0211629

Kley, W., Dirksen, G.: Disk eccentricity and embedded planets. Astron. Astrophys. **447**, 369–377 (2006). doi:10.1051/0004-6361:20053914, astro-ph/0510393

Kley, W., Nelson, R.P.: Planet-disk interaction and orbital evolution. Annu. Rev. Astron. Astrophys. **50**, 211–249 (2012). doi:10.1146/annurev-astro-081811-125523, 1203.1184

Kratter, K., Lodato, G.: Gravitational instabilities in circumstellar disks. Annu. Rev. Astron. Astrophys. **54**, 271–311 (2016). doi:10.1146/annurev-astro-081915-023307, 1603.01280

Kratter, K.M., Matzner, C.D., Krumholz, M.R., Klein, R.I.: On the role of disks in the formation of stellar systems: a numerical parameter study of rapid accretion. Astrophys. J. **708**, 1585–1597 (2010a). doi:10.1088/0004-637X/708/2/1585, 0907.3476

Kratter, K.M., Murray-Clay, R.A., Youdin, A.N.: The Runts of the Litter: why planets formed through gravitational instability can only be failed binary stars. Astrophys. J. **710**, 1375–1386 (2010b). doi:10.1088/0004-637X/710/2/1375, 0909.2644

Kraus, A.L., Ireland, M.J.: LkCa 15: a young exoplanet caught at formation? Astrophys. J. **745**, 5 (2012). doi:10.1088/0004-637X/745/1/5, 1110.3808

Krause, M., Blum, J.: Growth and form of planetary seedlings: results from a sounding rocket microgravity aggregation experiment. Phys. Rev. Lett. 93(2):021103 (2004). doi:10.1103/PhysRevLett.93.021103

Kretke, K.A., Lin, D.N.C.: Grain retention and formation of planetesimals near the snow line in MRI-driven turbulent protoplanetary disks. Astrophys. J. Lett. **664**, L55–L58 (2007). doi:10.1086/520718, 0706.1272

Lesur, G., Hennebelle, P., Fromang, S.: Spiral-driven accretion in protoplanetary discs. I. 2D models. Astron. Astrophys. **582**, L9. (2015). doi:10.1051/0004-6361/201526734, 1509.04859

Li, H., Finn, J.M., Lovelace, R.V.E., Colgate, S.A.: Rossby wave instability of thin accretion disks. II. Detailed linear theory. Astrophys. J. **533**, 1023–1034 (2000). doi:10.1086/308693, astro-ph/9907279

Li, H., Li, S., Koller, J., Wendroff, B.B., Liska, R., Orban, C.M., Liang, E.P.T., Lin, D.N.C.: Potential vorticity evolution of a protoplanetary disk with an embedded protoplanet. Astrophys. J. **624**, 1003–1009 (2005). doi:10.1086/429367, astro-ph/0503404

Lin, M.K.: Testing large-scale vortex formation against viscous layers in three-dimensional discs. Mon. Not. R. Astron. Soc. **437**, 575–587 (2014). doi:10.1093/mnras/stt1909, 1310.1393

Lin, D.N.C., Papaloizou, J.: Tidal torques on accretion discs in binary systems with extreme mass ratios. Mon. Not. R. Astron. Soc. **186**, 799–812 (1979). doi:10.1093/mnras/186.4.799

Lin, D.N.C., Papaloizou, J.: On the tidal interaction between protoplanets and the primordial solar nebula. II - self-consistent nonlinear interaction. Astrophys. J. **307**, 395–409 (1986a). doi:10.1086/164426

Lin, D.N.C., Papaloizou, J.: On the tidal interaction between protoplanets and the protoplanetary disk. III - orbital migration of protoplanets. Astrophys. J. **309**, 846–857 (1986b). doi:10.1086/164653

Lin, D.N.C., Pringle, J.E.: A viscosity prescription for a self-gravitating accretion disc. Mon. Not. R. Astron. Soc. **225**, 607–613 (1987). doi:10.1093/mnras/225.3.607

Lin, M.K., Youdin, A.: Cooling requirements for the vertical shear instability in protoplanetary disks (2015). arXiv:150502163 1505.02163

Lodato, G., Rice, W.K.M.: Testing the locality of transport in self-gravitating accretion discs. Mon. Not. R. Astron. Soc. **351**, 630–642 (2004). doi:10.1111/j.1365-2966.2004.07811.x, astro-ph/0403185

Lovelace, R.V.E., Li, H., Colgate, S.A., Nelson, A.F.: Rossby wave instability of Keplerian accretion disks. Astrophys. J. **513**, 805–810 (1999). doi:10.1086/306900, astro-ph/9809321

Lyra, W., Lin, M.K.: Steady state dust distributions in disk vortices: observational predictions and applications to transitional disks. Astrophys. J. **775**, 17 (2013). doi:10.1088/0004-637X/775/1/17, 1307.3770

Lyra, W., Turner, N.J., McNally, C.P.: Rossby wave instability does not require sharp resistivity gradients. Astron. Astrophys. **574**, A10 (2015). doi:10.1051/0004-6361/201424919, 1410.8092

Masset, F.: FARGO: a fast eulerian transport algorithm for differentially rotating disks. Astron. Astrophys. Suppl. Ser. **141**, 165–173 (2000). doi:10.1051/aas:2000116, astro-ph/9910390

Masset, F.S., Papaloizou, J.C.B.: Runaway migration and the formation of hot Jupiters. Astrophys. J. **588**, 494–508 (2003). doi:10.1086/373892, astro-ph/0301171

Mathis, J.S., Rumpl, W., Nordsieck, K.H.: The size distribution of interstellar grains. Astrophys. J. **217**, 425–433 (1977). doi:10.1086/155591

Mayama, S., Hashimoto, J., Muto, T., Tsukagoshi, T., Kusakabe, N., Kuzuhara M., Takahashi, Y., Kudo, T., Dong, R., Fukagawa, M., Takami, M., Momose, M., Wisniewski, J.P., Follette, K., Abe, L., Akiyama, E., Brandner, W., Brandt T., Carson, J., Egner, S., Feldt, M., Goto, M., Grady, C.A., Guyon, O., Hayano, Y., Hayashi, M., Hayashi, S., Henning, T., Hodapp, K.W., Ishii, M., Iye, M., Janson, M., Kandori, R., Kwon, J., Knapp, G.R., Matsuo, T., McElwain, M.W., Miyama, S., Morino, J.I., Moro-Martin, A., Nishimura, T., Pyo T.S., Serabyn, E., Suto, H., Suzuki, R., Takato, N., Terada, H., Thalmann, C., Tomono, D., Turner, E.L., Watanabe, M., Yamada, T., Takami, H., Usuda, T., Tamura, M.: Subaru imaging of asymmetric features in a transitional disk in Upper Scorpius. Astrophys. J. Lett. **760**, L26 (2012). doi:10.1088/2041-8205/760/2/L26, 1211.3284

Mulders, G.D., Paardekooper, S.J., Panić, O., Dominik, C., van Boekel, R., Ratzka, T.: Planet or brown dwarf? Inferring the companion mass in HD 100546 from the wall shape using mid-infrared interferometry. Astron. Astrophys. **557**, A68 (2013). doi:10.1051/0004-6361/201220930, 1306.4264

Musiolik, G., Teiser, J., Jankowski, T., Wurm, G.: Ice grain collisions in comparison: CO2, H2O, and their mixtures. Astrophys. J. **827**, 63 (2016). doi:10.3847/0004-637X/827/1/63, 1608.05017

Nakagawa, Y., Sekiya, M., Hayashi, C.: Settling and growth of dust particles in a laminar phase of a low-mass solar nebula. Icarus **67**, 375–390 (1986). doi:10.1016/0019-1035(86)90121-1

Nesvorný, D., Youdin, A.N., Richardson, D.C.: Formation of Kuiper Belt binaries by gravitational collapse. Astron. J. **140**, 785–793 (2010). doi:10.1088/0004-6256/140/3/785, 1007.1465

O'dell, C.R., Wen, Z.: Postrefurbishment mission Hubble Space Telescope images of the core of the Orion Nebula: Proplyds, Herbig-Haro objects, and measurements of a circumstellar disk. Astrophys. J. **436**, 194–202 (1994). doi:10.1086/174892

Okuzumi, S., Momose, M., Sirono, S.i., Kobayashi, H., Tanaka, H.: Sintering-induced dust ring formation in protoplanetary disks: application to the HL Tau disk. Astrophys. J. **821**, 82 (2016). doi:10.3847/0004-637X/821/2/82, 1510.03556

Ormel, C.W., Cuzzi, J.N.: Closed-form expressions for particle relative velocities induced by turbulence. Astron. Astrophys. **466**, 413–420 (2007). doi:10.1051/0004-6361:20066899, astro-ph/0702303

Owen, J.E., Clarke, C.J.: Two populations of transition discs? Mon. Not. R. Astron. Soc. **426**, L96–L100 (2012). doi:10.1111/j.1745-3933.2012.01334.x, 1205.5564

Paardekooper, S.J., Mellema, G.: Dust flow in gas disks in the presence of embedded planets. Astron. Astrophys. **453**, 1129–1140 (2006). doi:10.1051/0004-6361:20054449, astro-ph/0603132

Paszun, D., Dominik, C.: The influence of grain rotation on the structure of dust aggregates. Icarus **182**, 274–280 (2006). doi:10.1016/j.icarus.2005.12.018, astro-ph/0601262

Pérez, L.M., Carpenter, J.M., Chandler, C.J., Isella, A., Andrews, S.M., Ricci, L., Calvet, N., Corder, S.A., Deller, A.T., Dullemond, C.P., Greaves J.S., Harris, R.J., Henning, T., Kwon, W., Lazio, J., Linz, H., Mundy, L.G., Sargent, A.I., Storm, S., Testi, L., Wilner, D.J.: Constraints on the radial variation of grain growth in the AS 209 circumstellar disk. Astrophys. J. Lett. **760**, L17 (2012). doi:10.1088/2041-8205/760/1/L17, 1210.5252

Perez, S., Casassus, S., Ménard, F., Roman, P., van der Plas, G., Cieza L., Pinte, C., Christiaens, V., Hales, A.S.: CO gas inside the protoplanetary disk cavity in HD 142527: disk structure from ALMA. Astrophys. J. **798**, 85 (2015a). doi:10.1088/0004-637X/798/2/85, 1410.8168

Perez, S., Dunhill, A., Casassus, S., Roman, P., Szulágyi, J., Flores, C., Marino, S., Montesinos, M.: Planet formation signposts: observability of circumplanetary disks via gas kinematics. Astrophys. J. Lett. **811**, L5 (2015b). doi:10.1088/2041-8205/811/1/L5, 1505.06808

Perri, F., Cameron, A.G.W.: Hydrodynamic instability of the solar nebula in the presence of a planetary core. Icarus **22**, 416–425 (1974). doi:10.1016/0019-1035(74)90074-8

Pfalzner, S., Steinhausen, M., Menten, K.: Short dissipation times of proto-planetary disks: an artifact of selection effects? Astrophys. J. Lett. **793**, L34 (2014). doi:10.1088/2041-8205/793/2/L34, 1409.0978

Pinilla, P., Benisty, M., Birnstiel, T.: Ring shaped dust accumulation in transition disks. Astron. Astrophys. **545**, A81 (2012a). doi:10.1051/0004-6361/201219315, 1207.6485

Pinilla, P., Birnstiel, T., Ricci, L., Dullemond, C.P., Uribe, A.L., Testi, L., Natta, A.: Trapping dust particles in the outer regions of protoplanetary disks. Astron. Astrophys. **538**, A114 (2012b). doi:10.1051/0004-6361/201118204, 1112.2349

Pinilla, P., Birnstiel, T., Benisty, M., Ricci, L., Natta, A., Dullemond, C.P., Dominik, C., Testi, L.: Explaining millimeter-sized particles in brown dwarf disks. Astron. Astrophys. **554**, A95 (2013). doi:10.1051/0004-6361/201220875, 1304.6638

Pinilla, P., Benisty, M., Birnstiel, T., Ricci, L., Isella, A., Natta, A., Dullemond, C.P., Quiroga-Nuñez, L.H., Henning, T., Testi, L.: Millimetre spectral indices of transition disks and their relation to the cavity radius. Astron. Astrophys. **564**, A51 (2014). doi:10.1051/0004-6361/201323322, 1402.5778

Pinilla, P., Birnstiel, T., Walsh, C.: Sequential planet formation in the HD 100546 protoplanetary disk? Astron. Astrophys. **580**, A105 (2015a). doi:10.1051/0004-6361/201425539, 1506.02383

Pinilla, P., de Boer, J., Benisty, M., Juhász, A., de Juan Ovelar, M., Dominik, C., Avenhaus, H., Birnstiel, T., Girard, J.H., Huelamo, N., Isella A., Milli, J.: Variability and dust filtration in the transition disk J160421.7-213028 observed in optical scattered light. Astron. Astrophys. **584**, L4 (2015b). doi:10.1051/0004-6361/201526981, 1510.00412

Pinilla, P., de Juan Ovelar, M., Ataiee, S., Benisty, M., Birnstiel, T., van Dishoeck, E.F., Min, M.: Gas and dust structures in protoplanetary disks hosting multiple planets. Astron. Astrophys. **573**, A9 (2015c). doi:10.1051/0004-6361/201424679, 1410.5963

Pinilla, P., van der Marel, N., Pérez, L.M., van Dishoeck, E.F., Andrews, S., Birnstiel, T., Herczeg, G., Pontoppidan, K.M., van Kempen, T.: Testing particle trapping in transition disks with ALMA. Astron. Astrophys. **584**, A16 (2015d). doi:10.1051/0004-6361/201526655, 1509.03040

Pinilla, P., Flock, M., Ovelar, M.d.J., Birnstiel, T.: Can dead zones create structures like a transition disk? Astron. Astrophys. **596**, A81 (2016a). doi:10.1051/0004-6361/201628441, 1610.02044

Pinilla, P., Klarmann, L., Birnstiel, T., Benisty, M., Dominik, C., Dullemond C.P.: A tunnel and a traffic jam: How transition disks maintain a detectable warm dust component despite the presence of a large planet-carved gap. Astron. Astrophys. **585**, A35 (2016b). doi:10.1051/0004-6361/201527131, 1511.04105

Piso, A.M.A., Youdin, A.N.: On the minimum core mass for giant planet formation at wide separations (2013). arXiv13110011P 1311.0011

Pohl, A., Pinilla, P., Benisty, M., Ataiee, S., Juhász, A., Dullemond, C.P., Van Boekel, R., Henning, T.: Scattered light images of spiral arms in marginally gravitationally unstable discs with an embedded planet. Mon. Not. R. Astron. Soc. **453**, 1768–1778 (2015). doi:10.1093/mnras/stv1746, 1508.04443

Pollack, J.B., Hubickyj, O., Bodenheimer, P., Lissauer, J.J., Podolak, M., Greenzweig, Y.: Formation of the giant planets by concurrent accretion of solids and gas. Icarus **124**, 62–85 (1996). doi:10.1006/icar.1996.0190

Quanz, S.P., Amara, A., Meyer, M.R., Kenworthy, M.A., Kasper, M., Girard, J.H.: A young protoplanet candidate embedded in the circumstellar disk of HD 100546. Astrophys. J. Lett. **766**, L1 (2013a). doi:10.1088/2041-8205/766/1/L1, 1302.7122

Quanz, S.P., Avenhaus, H., Buenzli, E., Garufi, A., Schmid, H.M., Wolf, S.: Gaps in the HD 169142 protoplanetary disk revealed by polarimetric imaging: signs of ongoing planet formation? Astrophys. J. Lett. **766**, L2 (2013b). doi:10.1088/2041-8205/766/1/L2, 1302.3029

Quanz, S.P., Amara, A., Meyer, M.R., Girard, J.H., Kenworthy, M.A., Kasper, M.: Confirmation and characterization of the protoplanet HD 100546 b-direct evidence for gas giant planet formation at 50 AU. Astrophys. J. **807**, 64 (2015). doi:10.1088/0004-637X/807/1/64, 1412.5173

Rafikov, R.R.: Planet migration and gap formation by tidally induced shocks. Astrophys. J. **572**, 566–579 (2002). doi:10.1086/340228, astro-ph/0110540

Rafikov, R.R.: Properties of gravitoturbulent accretion disks. Astrophys. J. **704**, 281–291 (2009). doi:10.1088/0004-637X/704/1/281, 0901.4739

Regály, Z., Juhász, A., Sándor, Z., Dullemond, C.P.: Possible planet-forming regions on submillimetre images. Mon. Not. R. Astron. Soc. **419**, 1701–1712 (2012). doi:10.1111/j.1365-2966.2011.19834.x, 1109.6177

Reggiani, M., Quanz, S.P., Meyer, M.R., Pueyo, L., Absil, O., Amara, A., Anglada, G., Avenhaus, H., Girard, J.H., Carrasco Gonzalez, C., Graham, J., Mawet, D., Meru, F., Milli, J., Osorio, M., Wolff, S., Torrelles, J.M.: Discovery of a companion candidate in the HD 169142 transition disk and the possibility of multiple planet formation. Astrophys. J. Lett. **792**, L23 (2014). doi:10.1088/2041-8205/792/1/L23, 1408.0813

Ribas, Á., Bouy, H., Merín, B.: Protoplanetary disk lifetimes vs. stellar mass and possible implications for giant planet populations. Astron. Astrophys. **576**, A52 (2015). doi:10.1051/0004-6361/201424846, 1502.00631

Ricci, L., Testi, L., Natta, A., Brooks, K.J.: Dust grain growth in ρ-Ophiuchi protoplanetary disks. Astron. Astrophys. **521**, A66 (2010). doi:10.1051/0004-6361/201015039, 1008.1144

Ricci, L., Mann, R.K., Testi, L., Williams, J.P., Isella, A., Robberto, M., Natta, A., Brooks, K.J.: The (sub-)millimeter SED of protoplanetary disks in the outskirts of the Orion nebula cluster. Astron. Astrophys. **525**, A81 (2011). doi:10.1051/0004-6361/201015789, 1010.1677

Ricci, L., Testi, L., Natta, A., Scholz, A., de Gregorio-Monsalvo, I.: ALMA Observations of ρ-Oph 102: grain growth and molecular gas in the disk around a young brown dwarf. Astrophys. J. Lett. **761**, L20 (2012). doi:10.1088/2041-8205/761/2/L20, 1211.6743

Ricci, L., Testi, L., Natta, A., Scholz, A., de Gregorio-Monsalvo, I., Isella A.: Brown dwarf disks with ALMA. Astrophys. J. **791**, 20 (2014). doi:10.1088/0004-637X/791/1/20, 1406.0635

Rice, W.K.M., Lodato, G., Pringle, J.E., Armitage, P.J., Bonnell, I.A.: Accelerated planetesimal growth in self-gravitating protoplanetary discs. Mon. Not. R. Astron. Soc. **355**, 543–552 (2004). doi:10.1111/j.1365-2966.2004.08339.x, astro-ph/0408390

Rice, W.K.M., Lodato, G., Armitage, P.J.: Investigating fragmentation conditions in self-gravitating accretion discs. Mon. Not. R. Astron. Soc. **364**, L56–L60 (2005). doi:10.1111/j.1745-3933.2005.00105.x, astro-ph/0509413

Rice, W.K.M., Armitage, P.J., Wood, K., Lodato, G.: Dust filtration at gap edges: implications for the spectral energy distributions of discs with embedded planets. Mon. Not. R. Astron. Soc. **373**, 1619–1626 (2006). doi:10.1111/j.1365-2966.2006.11113.x, astro-ph/0609808

Rodmann, J., Henning, T., Chandler, C.J., Mundy, L.G., Wilner, D.J.: Large dust particles in disks around T Tauri stars. Astron. Astrophys. **446**, 211–221 (2006). doi:10.1051/0004-6361:20054038, astro-ph/0509555

Rosotti, G.P., Juhasz, A., Booth, R.A., Clarke, C.J.: The minimum mass of detectable planets in protoplanetary discs and the derivation of planetary masses from high-resolution observations. Mon. Not. R. Astron. Soc. **459**, 2790–2805 (2016). doi:10.1093/mnras/stw691, 1603.02141

Ruge, J.P., Flock, M., Wolf, S., Dzyurkevich, N., Fromang, S., Henning, T., Klahr, H., Meheut, H.: Gaps, rings, and non-axisymmetric structures in protoplanetary disks: emission from large grains. Astron. Astrophys. **590**, A17 (2016). doi:10.1051/0004-6361/201526616, 1603.05179

Safronov, V.S.: Evoliutsiia doplanetnogo oblaka. Nakua, Moscow (1969)

Sallum, S., Follette, K.B., Eisner, J.A., Close, L.M., Hinz, P., Kratter, K., Males, J., Skemer, A., Macintosh, B., Tuthill, P., Bailey, V., Defrère D., Morzinski, K., Rodigas, T., Spalding, E., Vaz, A., Weinberger, A.J.: Accreting protoplanets in the LkCa 15 transition disk. Nature **527**, 342–344 (2015). doi:10.1038/nature15761, 1511.07456

Schäfer, U., Yang, C.C., Johansen, A.: Initial mass function of planetesimals formed by the streaming instability (2016). ArXiv:161102285 1611.02285

Sekiya, M.: Quasi-equilibrium density distributions of small dust aggregations in the solar nebula. Icarus **133**, 298–309 (1998). http://adsabs.harvard.edu/cgi-bin/nph-bib_query?bibcode=1998Icar..133..298S&db_key=AST

Shakura, N.I., Sunyaev, R.A.: Black holes in binary systems. Observational appearance. Astron. Astrophys. **24**, 337–355 (1973)

Shariff, K., Cuzzi, J.N.: Gravitational instability of solids assisted by gas drag: slowing by turbulent mass diffusivity. Astrophys. J. **738**, 73 (2011). doi:10.1088/0004-637X/738/1/73, 1106.3145

Simon, J.B., Armitage, P.J.: Efficiency of particle trapping in the outer regions of protoplanetary disks. Astrophys. J. **784**, 15 (2014). doi:10.1088/0004-637X/784/1/15, 1402.1489

Simon, J.B., Armitage, P.J., Li, R., Youdin, A.N.: The mass and size distribution of planetesimals formed by the streaming instability. I. The role of self-gravity. Astrophys. J. **822**, 55 (2016). doi:10.3847/0004-637X/822/1/55, 1512.00009

Smoluchowski, M.V.: Drei Vortrage uber Diffusion, Brownsche Bewegung und Koagulation von Kolloidteilchen. Z. Phys. **17**, 557–585 (1916)

Stolker, T., Dominik, C., Avenhaus, H., Min, M., de Boer, J., Ginski, C., Schmid, H.M., Juhasz, A., Bazzon, A., Waters, L.B.F.M., Garufi, A., Augereau J.C., Benisty, M., Boccaletti, A., Henning, T., Langlois, M., Maire, A.L., Ménard, F., Meyer, M.R., Pinte, C., Quanz, S.P., Thalmann, C., Beuzit, J.L., Carbillet, M., Costille, A., Dohlen, K., Feldt, M., Gisler, D., Mouillet, D., Pavlov, A., Perret, D., Petit, C., Pragt, J., Rochat, S., Roelfsema, R., Salasnich, B., Soenke, C., Wildi, F.: Shadows cast on the transition disk of HD 135344B. Multiwavelength VLT/SPHERE polarimetric differential imaging. Astron. Astrophys. **595**, A113 (2016). doi:10.1051/0004-6361/201528039, 1603.00481

Strom, K.M., Strom, S.E., Edwards, S., Cabrit, S., Skrutskie, M.F.: Circumstellar material associated with solar-type pre-main-sequence stars - A possible constraint on the timescale for planet building. Astron. J. **97**, 1451–1470 (1989). doi:10.1086/115085

Takahashi, S.Z., Inutsuka, S.i.: Two-component secular gravitational instability in a protoplanetary disk: a possible mechanism for creating ring-like structures. Astrophys. J. **794**, 55 (2014). doi:10.1088/0004-637X/794/1/55, 1312.6870

Takahashi, S.Z., Inutsuka, S.i.: An origin of multiple ring structure and hidden planets in HL Tau: a unified picture by secular gravitational instability. Astron. J. **152**, 184 (2016). doi:10.3847/0004-6256/152/6/184, 1604.05450

Takeuchi, T., Lin, D.N.C.: Radial flow of dust particles in accretion disks. Astrophys. J. **581**, 1344–1355 (2002). doi:10.1086/344437, astro-ph/0208552

Tatulli, E., Benisty, M., Ménard, F., Varnière, P., Martin-Zaïdi, C., Thi, W.F., Pinte, C., Massi, F., Weigelt, G., Hofmann, K.H., Petrov, R.G.: Constraining the structure of the planet-forming region in the disk of the Herbig Be star HD 100546. Astron. Astrophys. **531**, A1 (2011). doi:10.1051/0004-6361/201016165, 1104.0905

Tazzari, M., Testi, L., Ercolano, B., Natta, A., Isella, A., Chandler, C.J., Pérez, L.M., Andrews, S., Wilner, D.J., Ricci, L., Henning, T., Linz, H., Kwon, W., Corder, S.A., Dullemond, C.P., Carpenter, J.M., Sargent, A.I., Mundy, L., Storm, S., Calvet, N., Greaves, J.A., Lazio, J., Deller, A.T.: Multiwavelength analysis for interferometric (sub-)mm observations of protoplanetary disks. Radial constraints on the dust properties and the disk structure. Astron. Astrophys. **588**, A53 (2016). doi:10.1051/0004-6361/201527423, 1512.05679

Terebey, S., Shu, F.H., Cassen, P.: The collapse of the cores of slowly rotating isothermal clouds. Astrophys. J. **286**, 529–551 (1984). doi:10.1086/162628

Testi, L., Birnstiel, T., Ricci, L., Andrews, S., Blum, J., Carpenter, J., Dominik, C., Isella, A., Natta, A., Williams, J.P., Wilner, D.J.: Dust evolution in protoplanetary disks. In: Protostars and Planets VI, pp. 339–361 (2014). doi:10.2458/azu_uapress_9780816531240-ch015, 1402.1354

Turner, N.J., Fromang, S., Gammie, C., Klahr, H., Lesur, G., Wardle, M., Bai X.N.: Transport and accretion in planet-forming disks. In: Protostars and Planets VI, pp. 411–432 (2014). doi:10.2458/azu_uapress_9780816531240-ch018, 1401.7306

Ubach, C., Maddison, S.T., Wright, C.M., Wilner, D.J., Lommen, D.J.P., Koribalski B.: Grain growth signatures in the protoplanetary discs of Chamaeleon and Lupus. Mon. Not. R. Astron. Soc. **425**, 3137–3161 (2012). doi:10.1111/j.1365-2966.2012.21603.x, 1207.0260

Uribe, A.L., Klahr, H., Flock, M., Henning, T.: Three-dimensional magnetohydrodynamic simulations of planet migration in turbulent stratified disks. Astrophys. J. **736**, 85 (2011). doi:10.1088/0004-637X/736/2/85, 1105.2235

van der Marel, N., van Dishoeck, E.F., Bruderer, S., Birnstiel, T., Pinilla, P., Dullemond, C.P., van Kempen, T.A., Schmalzl, M., Brown, J.M., Herczeg, G.J., Mathews, G.S., Geers, V.: A major asymmetric dust trap in a transition disk. Science **340**, 1199–1202 (2013). doi:10.1126/science.1236770, 1306.1768

van der Marel, N., Pinilla, P., Tobin, J., van Kempen, T., Andrews, S., Ricci L., Birnstiel, T.: A concentration of centimeter-sized grains in the Ophiuchus IRS 48 dust trap. Astrophys. J. Lett. **810**, L7 (2015). doi:10.1088/2041-8205/810/1/L7, 1508.01003

van der Marel, N., van Dishoeck, E.F., Bruderer, S., Andrews, S.M., Pontoppidan K.M., Herczeg, G.J., van Kempen, T., Miotello, A.: Resolved gas cavities in transitional disks inferred from CO isotopologs with ALMA. Astron. Astrophys. **585**, A58 (2016). doi:10.1051/0004-6361/201526988, 1511.07149

van der Plas, G., Wright, C.M., Ménard, F., Casassus, S., Canovas, H., Pinte, C., Maddison, S.T., Maaskant, K., Avenhaus, H., Cieza, L., Perez, S., Ubach, C.: Cavity and other radial substructures in the disk around HD 97048. Astron. Astrophys. **597**, A32 (2017). doi:10.1051/0004-6361/201629523, 1609.02488

van Dishoeck, E.F., van der Marel, N., Bruderer, S., Pinilla, P.: Quantifying the Gas inside dust cavities in transitional disks: implications for young planets. In: Iono, D., Tatematsu, K., Wootten, A., Testi, L. (eds.) Revolution in Astronomy with ALMA: The Third Year, Astronomical Society of the Pacific Conference Series, vol. 499, p. 281 (2015). 1505.01947

Varnière, P., Tagger, M.: Reviving Dead Zones in accretion disks by Rossby vortices at their boundaries. Astron. Astrophys. **446**, L13–L16 (2006). doi:10.1051/0004-6361:200500226, astro-ph/0511684

Vorobyov, E.I., Basu, S.: Secular evolution of viscous and self-gravitating circumstellar discs. Mon. Not. R. Astron. Soc. **393**, 822–837 (2009). doi:10.1111/j.1365-2966.2008.14376.x, 0812.1306

Wada, K., Tanaka, H., Suyama, T., Kimura, H., Yamamoto, T.: Collisional growth conditions for dust aggregates. Astrophys. J. **702**, 1490–1501 (2009). doi:10.1088/0004-637X/702/2/1490

Wada, K., Tanaka, H., Suyama, T., Kimura, H., Yamamoto, T.: The rebound condition of dust aggregates revealed by numerical simulation of their collisions. Astrophys. J. **737**, 36 (2011). doi:10.1088/0004-637X/737/1/36

Walsh, C., Juhász, A., Pinilla, P., Harsono, D., Mathews, G.S., Dent, W.R.F., Hogerheijde, M.R., Birnstiel, T., Meeus, G., Nomura, H., Aikawa, Y., Millar, T.J., Sandell, G.: ALMA hints at the presence of two companions in the disk around HD 100546. Astrophys. J. Lett. **791**, L6 (2014). doi:10.1088/2041-8205/791/1/L6, 1405.6542

Walsh, C., Juhász, A., Meeus, G., Dent, W.R.F., Maud, L.T., Aikawa, Y., Millar, T.J., Nomura, H.: ALMA reveals the anatomy of the mm-sized dust and molecular gas in the HD 97048 disk. Astrophys. J. **831**, 200 (2016). doi:10.3847/0004-637X/831/2/200, 1609.02011

Ward, W.R.: On planetesimal formation: the role of collective particle behavior. In: Origin of the Earth and Moon, pp. 75–84. University of Arizona Press, Tucson (2000)

Weidenschilling, S.J.: Aerodynamics of solid bodies in the solar nebula. Mon. Not. R. Astron. Soc. **180**, 57–70 (1977). doi:10.1093/mnras/180.1.57

Weintraub, D.A., Sandell, G., Duncan, W.D.: Submillimeter measurements of T Tauri and FU Orionis stars. Astrophys. J. Lett. **340**, L69–L72 (1989). doi:10.1086/185441

Weizsäcker, C.F.V.: Über die Entstehung des Planetensystems. Mit 2 Abbildungen. Z. Astrophys. **22**, 319 (1943)

Whipple, F.L.: On certain aerodynamic processes for asteroids and comets. In: Elvius, A. (ed.) From Plasma to Planet, p. 211. Wiley, New York (1972)

Williams, J.P., Cieza, L.A.: Protoplanetary disks and their evolution. Annu. Rev. Astron. Astrophys. **49**, 67–117 (2011). doi:10.1146/annurev-astro-081710-102548, 1103.0556

Windmark, F., Birnstiel, T., Güttler, C., Blum, J., Dullemond, C.P., Henning, T.: Planetesimal formation by sweep-up: how the bouncing barrier can be beneficial to growth. Astron. Astrophys. **540**, A73 (2012). doi:10.1051/0004-6361/201118475, 1201.4282

Wright, C.M., Maddison, S.T., Wilner, D.J., Burton, M.G., Lommen, D., van Dishoeck, E.F., Pinilla, P., Bourke, T.L., Menard, F., Walsh, C.: Resolving structure of the disc around HD100546 at 7 mm with ATCA. Mon. Not. R. Astron. Soc. **453**, 414–438 (2015). doi:10.1093/mnras/stv1619, 1506.01147

Wyatt, M.C.: Evolution of debris disks. Annu. Rev. Astron. Astrophys. **46**, 339–383 (2008). doi:10.1146/annurev.astro.45.051806.110525

Yang, C.C., Johansen, A., Carrera, D.: Concentrating small particles in protoplanetary disks through the streaming instability (2016). ArXiv e-prints 1611.07014

Youdin, A.N.: Planetesimal formation without thresholds I (2005). astro-ph/0508659

Youdin, A.N.: From grains to planetesimals. In: Montmerle, T., Ehrenreich, D., Lagrange, A.-M. (eds.) EAS Publications Series, EAS Publications Series, vol. 41, pp. 187–207 (2010). doi:10.1051/eas/1041016

Youdin, A.N.: On the formation of planetesimals via secular gravitational instabilities with turbulent stirring. Astrophys. J. **731**, 99 (2011). doi:10.1088/0004-637X/731/2/99, 1102.4620

Youdin, A.N., Chiang, E.I.: Particle pileups and planetesimal formation. Astrophys. J. **601**, 1109–1119 (2004)

Youdin, A.N., Goodman, J.: Streaming instabilities in protoplanetary disks. Astrophys. J. **620**, 459–469 (2005)

Youdin, A., Johansen, A.: Protoplanetary disk turbulence driven by the streaming instability: linear evolution and numerical methods. Astrophys. J. **662**, 613–626 (2007). doi:10.1086/516729, arXiv:astro-ph/0702625

Youdin, A.N., Kenyon, S.J.: From disks to planets. In: Planets, Stars and Stellar Systems. Volume 3: Solar and Stellar Planetary Systems, pp. 1–62. Springer, Dordrecht (2013). doi:10.1007/978-94-007-5606-9_1

Youdin, A.N., Lithwick, Y.: Particle stirring in turbulent gas disks: Including orbital oscillations. Icarus **192**, 588–604 (2007). doi:10.1016/j.icarus.2007.07.012, arXiv:0707.2975

Youdin, A.N., Shu, F.H.: Planetesimal formation by gravitational instability. Astrophys. J. **580**, 494–505 (2002)

Zhang, K., Isella, A., Carpenter, J.M., Blake, G.A.: Comparison of the dust and gas radial structure in the transition disk [PZ99] J160421.7-213028. Astrophys. J. **791**, 42 (2014). doi:10.1088/0004-637X/791/1/42, 1406.6974

Zhu, Z., Stone, J.M.: Dust trapping by vortices in transitional disks: evidence for non-ideal magnetohydrodynamic effects in protoplanetary disks. Astrophys. J. **795**, 53 (2014). doi:10.1088/0004-637X/795/1/53, 1405.2790

Zhu, Z., Nelson, R.P., Hartmann, L., Espaillat, C., Calvet, N.: Transitional and pre-transitional disks: gap opening by multiple planets? Astrophys. J. **729**, 47 (2011). doi:10.1088/0004-637X/729/1/47, 1012.4395

Zhu, Z., Nelson, R.P., Dong, R., Espaillat, C., Hartmann, L.: Dust filtration by planet-induced gap edges: implications for transitional disks. Astrophys. J. **755**, 6 (2012). doi:10.1088/0004-637X/755/1/6, 1205.5042

Chapter 5
Dust Coagulation with Porosity Evolution

Akimasa Kataoka

Abstract Theoretical studies, laboratory experiments, and numerical simulations have shown several barriers in planetesimal formation, including radial drift, fragmentation, and bouncing problems. Also, observations of protoplanetary disks have shown the radial drift problems of millimeter-sized bodies at the outer orbital radius. The key to solving these problems is understanding porosity evolution. Dust grains become fluffy by coagulation in protoplanetary disks and this alters both the dynamic and optical properties of dust aggregates. First, we revealed the overall porosity evolution from micron-sized dust grains to kilometer-sized planetesimals; dust grains form extremely porous dust aggregates, where the filling factor is $\sim 10^{-4}$, and then they are compressed by their collisions, the disk gas, and their self-gravity. In the coagulation process, they circumvent all the barriers if the monomers are 0.1-μm icy bodies. The mass and porosity of the final product are consistent with those of the comets, which are believed to be the remnants of planetesimals. We further performed coagulation simulations including porosity evolution. We found that planetesimals can form inside 10 AU, avoiding the radial drift barrier. Furthermore, we calculated the opacity evolution of porous dust aggregates and found that the observed radio emission could be explained either by compact dust grains or by fluffy dust aggregates.

5.1 Introduction

Dust coagulation is the first step in planet formation. Dust grains are solid materials in molecular cloud cores composed of silicates, water ice, carbonaceous materials and others. These solid bodies measure $\sim$0.1–1 μm (e.g., Mathis et al. 1977). During the star formation process via gravitational collapse of a molecular cloud core, the tiny solid bodies are dragged by gas and accrete to the circumstellar disk around

---

A. Kataoka (✉)
Zentrum für Astronomie der Universität Heidelberg, Institut für Theoretische Astrophysik, Albert-Ueberle-Str. 2, 69120 Heidelberg, Germany

National Astronomical Observatory of Japan, Osawa 2-21-1, Mitaka, Tokyo 181-8588, Japan
e-mail: akimasa.kataoka@nao.ac.jp

M. Pessah, O. Gressel (eds.), *Formation, Evolution, and Dynamics of Young Solar Systems*, Astrophysics and Space Science Library 445,
DOI 10.1007/978-3-319-60609-5_5

the young star. In the gas disk, dust grains stick to each other to form larger bodies, owing to their frequent collisions due to high solid density. This hit-and-stick process ultimately causes planets to form (e.g., Weidenschilling 1977; Hayashi 1981). We aim to reveal that this dust coagulation process forms planets using astronomical observations, theoretical studies, and laboratory experiments.

There are an enormous number of problems during dust coagulation depending on size. The size of the solid bodies ranges from 0.1 μm to 1000 km, which is 13 orders of magnitude in size and thus 39 orders of magnitude in mass. The dynamics of the solid bodies is determined by the friction between gas and dust, which depends of course on the size of the solid bodies. Thus, there are different problems corresponding to each grain size range.

The first problem is the charge barrier, where negatively charged particles cannot collide with other particles because the electrostatic repulsive force is high enough to avoid the collisions (Okuzumi 2009; Okuzumi et al. 2011a,b). The radial drift problem appears when dust grains range from millimeters to meters in size. These dust grains are decoupled from the gas, which results in facing the gas drag in an azimuthal direction. This reduces the angular momentum of the grains and they migrate to the central star, prohibiting further growth because the time scale of radial migration is shorter than the growth time scale (Adachi et al. 1976; Weidenschilling 1977). There are also problems due to microphysics of dust collisions, which are fragmentation and bouncing. If the collisional energy of dust grains is high enough to disrupt the grains, dust grains cannot grow further. The threshold velocity of fragmentation is a few meters per second for silicate-like aggregates, much smaller than the typical highest collision velocity of $\sim$50 m s^{-1} in disks (Blum and Münch 1993; Blum and Wurm 2008; Wada et al. 2007, 2009, 2013). Dust grains of some size ranges do not stick, but rather bounce (Güttler et al. 2010; Zsom et al. 2010). Even if these problems are solved and kilometer-size planetesimals are successfully formed, there are problems with the migration of protoplanets and others, which are beyond the scope of this chapter.

There have been many attempts to solve these problems: streaming instability (e.g., Johansen et al. 2007), direct coagulation with high dust-to-gas ratio (e.g., Brauer et al. 2008), pressure trapping (e.g., Whipple 1972; Pinilla et al. 2012), vortex trapping (e.g., Lyra et al. 2009), dead zone edge (Drążkowska et al. 2013), lucky particle approach (Windmark et al. 2012), growth by condensation close to snowlines (Ros and Johansen 2013), direct formation of planets through gravitational instability (Boss 2000), etc. We do not exclude any of these possibilities, but we discuss another topic in this chapter: porosity evolution.

Dust grains have usually been assumed to have a constant internal density during dust coagulation for simplicity. Readers can picture this by thinking about the evolution of water droplets. Two spherical water droplets stick to each other to form larger spherical water droplets with the same density. However, dust grains are not liquid but solid. The dust grains are mainly composed of water ice, which has a sublimation temperature of $\sim$150 K at the pressure in the disk, which corresponds to a location at $\sim$1 AU from the central star. Even inside the snowline, the dust grains are composed of silicates, which have a sublimation temperature of $\sim$1700 K. Therefore, when dust grains stick to each other, they naturally form pores inside.

The aggregates of dust grains have lower internal density, which equates to a lower volume filling factor. We note that the internal density $\rho_{\rm int}$ is written as $\rho_{\rm int} = \rho_0 f$, where ρ_0 is the material density and f is the filling factor, and the porosity is given as $p = 1 - f$. This has a great effect on dust coagulation because the enhanced cross-section causes the frequent collisions to reduce the growth timescale and increase the friction against the disk gas to change the dynamics. Furthermore, it also changes the optical properties, which may change the interpretations of observations.

In this chapter, we discuss porosity evolution during dust coagulation in protoplanetary disks based on the recent progress in numerical studies. As a result, we argue that treating porosity evolution naturally solves many severe problems in planetesimal formation. In addition, we also discuss how it affects planetesimal formation and observations of protoplanetary disks. Finally, we also mention future directions of the study of porosity evolution.

5.2 Porosity Evolution

First, we describe our assumptions regarding dust grains. The first dust grains are assumed to be spherical dust grains with the same radius, called monomers. The monomers are assumed to be 0.1 μm in size and composed of water ice.

Porosity evolution is divided into four stages, as shown in Fig. 5.1. The first stage is the hit-and-stick. The relative velocity between two small dust grains is dominated by their thermal Brownian motion, which results in collisions with a velocity of $\sim$mm/s. This collisional energy is not sufficient to restructure the aggregates because of the collisions. Therefore, one of the two colliding dust aggregates stays at the point where it first hits the colliding aggregate. Thus, at this stage, the porosity evolution is understood to be geometric growth. The filling factor evolution of sequential collisional growth can be understood as a function of the mass ratio of the two colliding aggregates. If the mass ratio of colliding aggregates is always unity, which corresponds to hitting the copy of the same aggregates from a random direction, this yields dust aggregates with the fractal dimension of $\sim$1.9 (e.g., Mukai et al. 1992). If the mass ratio approaches 0, at the other extreme, which corresponds to a monomer always hitting an aggregate, the fractal dimension yields $\sim$3. The former case is called ballistic cluster-cluster aggregation (BCCA) whereas the latter is called ballistic particle-cluster aggregation (BPCA). The nature of dust coagulation in planet formation is between these two types of aggregation and the question is which case describes the initial growth of dust grains well. Ormel et al. (2007) have performed a pioneering work on porosity evolution. In terms of porosity evolution, dust aggregates form porous dust aggregates, which have a filling factor of $\sim$1%, due to the simple interpolation for collisions in the intermediate aggregates between BCCA and BPCA. This was investigated later by numerical simulations of dust collisions (Okuzumi et al. 2009, 2012), which shows that similar-mass collisions lead to many more fluffy aggregates with fractal dimensions close to $\sim$2.

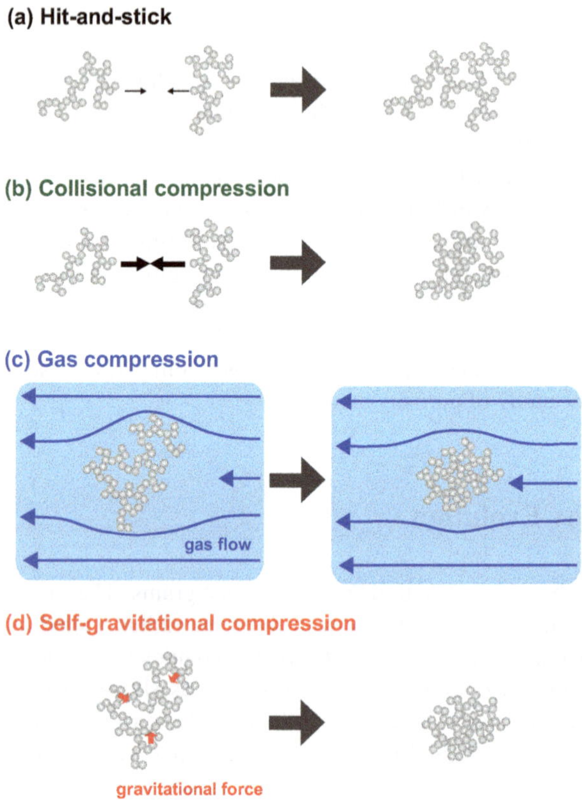

Fig. 5.1 Schematic illustration of each stage of porosity evolution. (**a**) The dust aggregate hits another aggregate to coagulate. This reduces the dust density, and occurs in the early stage of dust growth. (**b**) When the collisional velocity is high enough to disrupt the dust aggregates, they are compressed. (**c**) The dust aggregate has a relative velocity with respect to the gas, and they feel the ram pressure of the gas. The ram pressure statically compresses the dust aggregates. (**d**) When the dust aggregates becomes so massive that they do not support their structures, they are compressed by their self-gravity. This figure is published as Fig. 2 in Kataoka et al. (2013a)

As long as the dust coagulation occurs with similar-sized collisions, therefore, the dust aggregates become more and more porous because of hit-and-stick growth.

The second stage is collisional compression. Once the collisional compression energy exceeds a critical energy to move one connection of monomers, the first restructuring occurs. The critical energy is that used to roll the monomers in contact, which is defined as rolling energy $E_{\rm roll}$ (Dominik and Tielens 1997). This corresponds to the onset of collisional compressions. The mass and filling factor of the dust aggregates in typical protoplanetary disks are derived to be $m \sim 10^{-4}$ g and $f \sim 10^{-5}$. However, even when collisional compression occurs, dust aggregates are not compactified as much: the filling factor remains constant or continues to decrease (Suyama et al. 2008, 2012; Okuzumi et al. 2012). This can be understood as follows. When two dust aggregates collide, they create a new void region, which severely reduces the filling factor. Most of the collisional energy is consumed to compress this newly created void region. As a result, the pre-existing void regions are not very compactified. Even the collisional velocity increases with increasing Stokes number. This effect is not effective enough to compress the aggregates (Okuzumi et al. 2012); therefore, there must be other mechanisms to compress the porous dust aggregates.

The third and fourth stages are quasi-static compression processes. Dust aggregates always feel ram pressure from the disk gas. This ram pressure causes the aggregates to compress. For small filling factors, the compressive strength has been obtained by numerical simulations and is derived to be

$$\rho_{\mathrm{eq}} = \left(\frac{r_0^3}{E_{\mathrm{roll}}} P\right)^{1/3} \rho_0, \tag{5.1}$$

where ρ_{eq} is the internal mass density of the aggregate, r_0 the monomer radius, E_{roll} the rolling energy, P the compressive strength, and ρ_0 the material density of the dust grains (Kataoka et al. 2013b). The ram pressure can be written as

$$P_{\mathrm{gas}} = \frac{mv}{\pi r^2} \frac{1}{t_{\mathrm{s}}}, \tag{5.2}$$

where P_{gas} is the ram pressure of the disk gas, m the mass of the aggregate, v the relative velocity between the gas and the aggregate, r the radius of the aggregate, and t_{s} the stopping time. Equalizing these two equations, we can obtain the equilibrium filling factor (or internal density), which represents the lower limit of the filling factor. In the same manner, the fourth stage of the self-gravitational compression can be understood. When dust aggregates become sufficiently massive, they are compressed by their own gravity. The pressure of the self-gravity of the

$$P_{\mathrm{grav}} = \frac{Gm^2}{\pi r^4}, \tag{5.3}$$

where P_{grav} is the pressure due to self-gravity, G is the gravitational constant.

Combining these stages, we finally reveal the overall filling factor evolution in protoplanetary disks, as shown in Fig. 5.2 (Kataoka et al. 2013a). The 0.1-μm-sized dust grains become extremely fluffy dust aggregates with a filling factor of $\sim 10^{-5}$. The dust aggregates coagulate to form larger bodies maintaining the equilibrium filling factor at the ram pressure of the gas, which results in the filling factor of $\sim 10^{-4}$. Finally, when the dust aggregates become as massive as 10^{12} g, dust aggregates are compressed to form relatively compact bodies with $f = 0.1$. Note that we stop the calculation at $f = 0.1$ because the static compressive strength formula is applicable where $f \lesssim 0.1$, whereas the pathway of the filling factor is expected to approach the filling factor of unity.

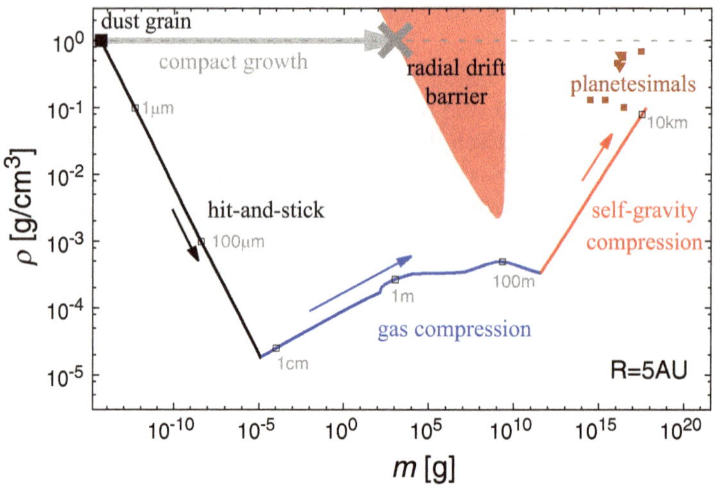

Fig. 5.2 The overall porosity evolution from 0.1 μm-sized dust grains to 10-km-sized planetesimals. The orbital radius is at 5 AU. The *black*, *blue*, and *red solid lines* correspond to hit-and-stick, gas compression, and self-gravitational compression of dust aggregates. The *red shaded region* represents where the growth time scale is shorter than the drift time scale. This figure is published as a part of Fig. 3 in Kataoka et al. (2013a)

Interestingly, the mass and filling factor of the resultant aggregates are in relatively good agreement with those of comets, which are believed to be the remnants of planetesimals (A'Hearn 2011). In this way, the overall filling factor evolution has been revealed, with the intermediate stage of extremely fluffy dust aggregates. We discuss how they avoid the radial drift barrier indicated in Fig. 5.2 in the next section.

5.3 Growth Barriers in Dust Coagulation

We have revealed the overall filling factor evolution from dust grains to planetesimals. Here, we discuss how this fluffy growth is beneficial to avoiding growth barriers in planetesimal formation.

5.3.1 Radial Drift Barrier

The radial drift problem is that the dust growth time scale is larger than the migration time scale of dust grains, especially when the Stokes number becomes unity. The growth time scale is strongly affected by the gas drag law. The gas drag law is in the Epstein regime if the aggregate radius is less than the mean free path of the gas, whereas it is in the Stokes regime if the radius is larger than the mean free path of the gas. As long as the gas drag is in the Epstein regime, both growth and drift time scales depend on the mass-to-area ratio of dust aggregates. Therefore, even if the filling factor changes, the growth time scale is always larger than the drift time scale: the radial drift barrier is not avoidable. However, if the radius of dust aggregates exceeds the mean free path of the gas, then the gas drag law changes to the Stokes regime. In this regime, the growth time scale becomes shorter. On the pathway of the filling factor evolution, when the dust aggregates have a Stokes number of unity, the radius of the aggregates is $\sim$100 m, which is much larger than the mean free path of the gas, $\sim$1 m. More details can be found in Okuzumi et al. (2012). The region where the growth time scale is shorter than the drift time scale is shown in Fig. 5.2. The revealed pathway of filling factor evolution clearly avoids the radial drift barrier. Whether the radial drift problem is avoided depends on the location of the protoplanetary disks. Roughly speaking, rapid coagulation to avoid the radial drift barrier occurs within $\sim$10–20 AU but outside of the water snow line (see Okuzumi et al. 2012; Kataoka et al. 2013a).

5.3.2 Fragmentation Barrier

When two dust aggregates collide with a sufficiently high velocity, they are disrupted and lose their mass during the collision, which causes a problem with regard to grain growth. Theoretically, the minimum amount of energy required for catastrophic disruption of dust aggregates is calculated by assuming that the collision energy is equal to the energy needed to cut all the connections between the monomers in the aggregates, which is

$$E_{\rm imp} \sim 10 n_{\rm k} E_{\rm break}, \tag{5.4}$$

where $E_{\rm imp}$ is the collision energy, $n_{\rm k}$ is the number of all connections of the monomers, and $E_{\rm break}$ is the energy required to cut one connection among monomers that are in contact (Chokshi et al. 1993; Dominik and Tielens 1997; Wada et al. 2007). As $E_{\rm imp}$ is proportional to the mass, which is proportional to a number of monomers, and $n_{\rm k}$ should be proportional to a number of monomers, the breaking-up velocity does not depend on mass. Therefore, for the catastrophic disruption, the threshold velocity has been intensively studied. Although the threshold velocity of fragmentation does not depend on the aggregate mass, it depends on materials and the size of the monomers.

Enormous efforts have been made with laboratory experiments that show that the threshold velocity of silicate grains is a few m s^{-1} (e.g., Blum and Münch 1993) and a few tens of m s^{-1} for ice (Gundlach and Blum 2015). In the series of numerical simulations (Wada et al. 2007, 2008, 2011), on the other hand, the final form of the threshold velocity for the net growth is revealed to be

$$v_{\mathrm{col,\,crit}} = \begin{cases} 80\,(r_0/0.1\,\mu\mathrm{m})^{-5/6} & [\mathrm{m\,s^{-1}}] \text{ for ice,} \\ 8\,(r_0/0.1\,\mu\mathrm{m})^{-5/6} & [\mathrm{m\,s^{-1}}] \text{ for silicate,} \end{cases} \tag{5.5}$$

where r_0 is the monomer radius (Wada et al. 2013). This is in excellent agreement with the laboratory experiments.

We turn now from microphysics to astrophysical applications. The typical maximum collision velocity between dust aggregates depends on the disk models. The collisional velocity becomes maximal when the Stokes number of dust aggregates becomes unity. The maximum collisional velocity is mainly determined by the gas turbulent velocity at the midplane of the disk. The midplane temperature is significantly lower than the surface that is irradiated by stellar radiation (e.g., Chiang and Goldreich 1997). Considering this cold midplane, the typical maximum collisional velocity is around $30\,\mathrm{m\,s^{-1}}$. This indicates that the fragmentation problem is significant for silicate dust grains because the fragmentation threshold velocity is $\sim$1–8 $\mathrm{m\,s^{-1}}$ if we assume the monomer radius to be between 0.1 and 1 μm. However, the fragmentation problem can be avoided if we assume icy particles to have a small monomer size, because the fragmentation threshold velocity ranges from $\sim$10 to 80 $\mathrm{m\,s^{-1}}$ if we assume the monomer size to be between 0.1 and 1 μm. Therefore, as long as we assume 0.1-μm-sized icy particles, the fragmentation is not a problem for planetesimal formation.

Here, we should note how this could be consistent with the observations of protoplanetary disks, which suggest that they might be full of small grains (e.g, Dullemond and Dominik 2005). If we assume that the fragmentation is completely inefficient, this is not consistent with the observations. One way of solving this problem is to introduce partial fragmentations. The fragmentation threshold velocity in the series of numerical simulations (e.g., Wada et al. 2013) is defined by whether the target aggregates gain mass from the projectile, called net growth. Therefore, even if the collision velocity is lower than the fragmentation velocity, the dust aggregates still produce small fragments, and this may explain the observations. Another possibility is to introduce high mass ratio fragmentation, which could lead to many fragments being produced (Seizinger et al. 2013; Krijt et al. 2015). However, we should also mention that Wada et al. (2013) argue that the growth efficiency via fragmentation does not depend on the mass ratio of the colliding aggregates.

5.3.3 Bouncing Barrier

Bouncing behavior has been reported in laboratory experiments: when two dust aggregates collide, they do not stick, but rebound, which causes a problem with dust growth (Güttler et al. 2010; Zsom et al. 2010). On the other hand, bouncing behavior has not been observed in numerical simulations. However, Wada et al. (2011) found that extremely highly packed compact aggregates can bounce, even in numerical simulations. The transition between sticking and bouncing is determined by an average coordination number of dust aggregates; this is equivalent to how many monomers are connected to a single monomer. The critical coordination number for bouncing is 6 (Wada et al. 2011), which is realized only when dust aggregates are compact. The corresponding filling factor is uncertain, but is around $f = 0.3$–0.6 (Wada et al. 2011; Seizinger and Kley 2013; Kothe et al. 2013). Therefore, as long as the filling factor is less than $\sim$0.3–0.6, bouncing does not occur. This is always valid if the filling factor evolution is as described in the previous section. Thus, bouncing is not a barrier to growth if we include porosity evolution in dust coagulation.

5.3.4 Short Summary and Discussion

We have discussed the notion that planetesimal formation via direct coagulation is possible if we include porosity evolution. The conditions are such that the monomer size is $\sim 0.1\,\mu\mathrm{m}$, the monomers are composed of ice to avoid the fragmentation barrier, and the location of planetesimal formation should be within 10–20 AU in protoplanetary disks to avoid the radial drift barrier. Of course, this does not fully solve the planetesimal formation problems, but this certainly gives us a theory that planetesimals form under certain conditions. Furthermore, this asks us another question: what are we looking at through radio observations? If porosity evolution is correct, it is applicable even for the outer radius, which means that the emission may be coming not from compact spheres, which has usually been assumed, but from fluffy aggregates. This also gives us an opportunity to test the fluffy growth scenario. We discuss this point in the next section.

Also, we discuss the assumptions and limitations of the numerical and theoretical modeling that we used in this work. First of all, the monomers are always assumed to have a single size in the numerical simulations. In reality, however, the monomers also have a size distribution and whether the behavior is characterized by a representative size is uncertain. Furthermore, owing to the limitation of the number of monomers that we can treat in numerical simulations, the parameter space of mass and filling factor has not been completely studied. Moreover, the mass ratio of the projectile and the target and off-center collisions has been studied, but is still under discussion. Readers should also see the papers that we did not discuss in this chapter (Paszun and Dominik 2009; Seizinger et al. 2013; Ringl et al. 2012), and should be cautious about the assumptions that we and other authors have made.

5.4 Opacity Evolution

Observations of protoplanetary disks have revealed the dust and gas structure. To measure grain size in particular, the spectral index at millimeter wavelengths is key. The opacity of dust grains at millimeter wavelengths becomes flatter with increasing grain size (Miyake and Nakagawa 1993). The opacity index β has been shown to be ~0–1 for protoplanetary disks (Beckwith and Sargent 1991) if the disk is optically thin. This indicates a significant dust growth to millimeters in size from the interstellar medium where the grain size is in microns (Pollack et al. 1994). These results are supported by later interferometric observations with high spatial resolution (e.g., Andrews and Williams 2007; Isella et al. 2009; Ricci et al. 2010a,b; Guilloteau et al. 2011). However, if the dust grains are not compact, are they really millimetric? To answer the question, in this section, we describe the dependence of the absorption opacity of dust aggregates on their filling factor.

To calculate the opacity of dust aggregates precisely, we need to use special calculations such as discrete dipole approximation (DDA; Draine 1988) and the T-matrix method (TMM; Mackowski and Mishchenko 1996). There are many applications of calculations of dust grains in the context of dust grains in disks (e.g., Mukai et al. 1992; Min et al. 2005, 2007). Here, we want to investigate at least the millimeter-sized grains, which are composed of 10^{12} of 0.1 μm-sized monomers. However, the calculations are limited by computer power and it is impossible to calculate the opacity of dust aggregates composed of such a large number of monomers. We therefore use a simpler method, which is the effective medium theory (EMT).

The EMT method is a mixing rule to compute the effective refractive index of the aggregates. The EMT approximation should be valid as long as the substructure (i.e., monomer scale) is much smaller than the wavelengths. In addition, especially for the absorption opacity, the EMT has good agreement with DDA and TMM (Kozasa et al. 1992). Here, we consider the absorption opacity of dust aggregates that are composed of micron-sized monomers at millimeter wavelengths. Therefore, the EMT approximation should be valid.

Figure 5.3 represents the absorption opacity of dust aggregates where the aggregate radius a and the filling factor f have values of $(a,f) = (1\,\text{mm}, 1.0)$, $(1\,\text{cm}, 10^{-1})$, $(10\,\text{cm}, 10^{-2})$, $(1\,\text{m}, 10^{-3})$, $(10\,\text{m}, 10^{-4})$. The absorption opacities are a good match for the five lines except for the compact grains at ~1 mm. This indicates that as long as the product of af is constant, the absorption opacity is degenerated even if the filling factor is changed. We have also derived a semi-analytical formula and confirm the fact that the absorption opacity is characterized by af except for the compact grains at the wavelengths of $\lambda = 2\pi a$ (see Kataoka et al. 2014). We also note that Cuzzi et al. (2014) obtained similar conclusions on the opacity of porous dust aggregates.

This tells us that we could possibly distinguish between compact spherical grains and porous dust aggregates with the filling factor of $\lesssim 0.1$. Figure 5.4 shows an opacity index profile as a function of $a_{\max}f$. The opacity index β is ~1.7 for

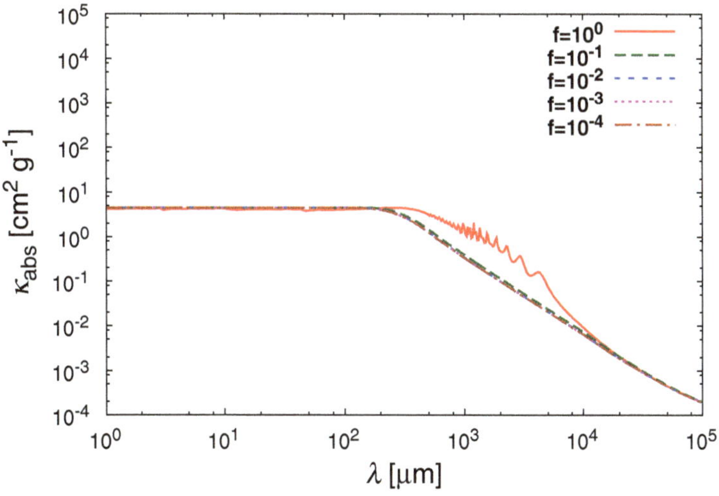

Fig. 5.3 The absorption opacity of dust aggregates is $af = 1$ mm, where the aggregate radius is a and the filling factor is f. There are five parameter sets, where $(a, f) = (1\,\text{mm}, 1.0)$, $(1\,\text{cm}, 10^{-1})$, $(10\,\text{cm}, 10^{-2})$, $(1\,\text{m}, 10^{-3})$, $(10\,\text{m}, 10^{-4})$ presented in the same figure. This figure is published as a part of Fig. 3 in Kataoka et al. (2014)

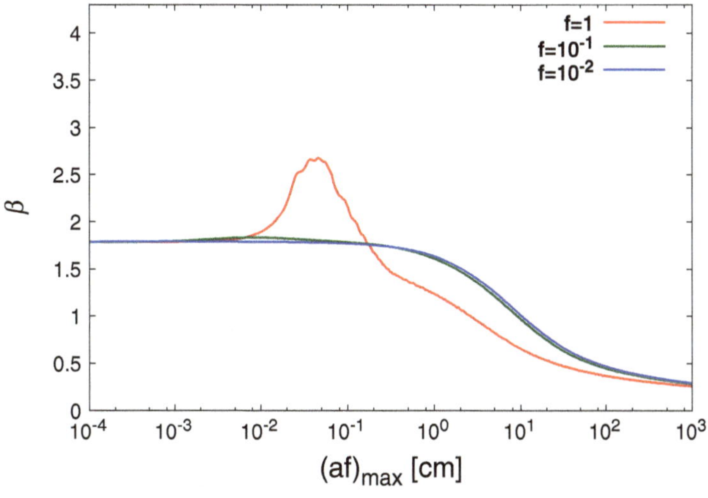

Fig. 5.4 The dust opacity index β as a function of $a_{\max} f$, where $a_{\max}$ is the maximum aggregate radius when af is constant. The aggregate size distribution is assumed to be $n \propto (af)^{-2}$. This figure is published as a part of Fig. 11 in Kataoka et al. (2014)

both compact grains and porous dust aggregates. In the case of compact grains, as dust size increases, the opacity index β is enhanced to be $\sim$2.7 at the size of $a_{\max} f \sim 0.7$ mm, and then decreases with increasing size. In the case of porous dust aggregates, however, the opacity index β monotonically decreases with increasing size. Therefore, if the opacity index of an intermediate size such as $a_{\max} f \sim 0.7$ mm is constrained, we can distinguish between spherical compact bodies and porous dust aggregates $f \lesssim 0.1$. This could be possible by measuring the radial dependence of β in protoplanetary disks. The growth timescale is shorter at inner radii, and longer for outer radii. Therefore, the radial profile of β roughly corresponds to the β as a function of different grain sizes. There have been some observational constraints on the radial profile by using interferometric observations at radio wavelengths, which shows that β is small inside the disk and large outside the disk (e.g., Isella et al. 2010; Guilloteau et al. 2011; Pérez et al. 2012, 2015; Tazzari et al. 2016). The results are still consistent with both compact and porous dust models. Future observations with higher angular resolution and greater sensitivity would reveal whether grains are compact or not.

5.5 Future Directions

Based on numerical and theoretical modeling, we argue that dust grains coagulate to form extremely fluffy aggregates in protoplanetary disks. These fluffy aggregates account for the observed radio emission, but the porosity and mass are highly degenerated in emission and it is hard to distinguish between compact grains and fluffy aggregates from continuum emission.

First, we must determine the size of the monomers. The fragmentation barrier is avoided if the monomer size is as small as 0.1 μm. However, the barrier is still severe if the monomer size is as large as 1 μm. Therefore, it is important to determine the monomer size in protoplanetary disks. There have been some studies aimed at constraining the micron-sized particles. For instance, infrared emission of dense molecular cores is well explained by intensity scattered by micron-sized dust grains (e.g., coreshine effect; Pagani et al. 2010). For another example, infrared-scattered emission of protoplanetary disks is significantly fainter than expected, which is explained well by the backward scattering of a few micron-sized particles (Mulders et al. 2013). However, this analysis is based on the assumption that dust grains are compact. If they are aggregates, the infrared scattering opacity could be enhanced by the aggregated structure (Köhler et al. 2012) and the scattering properties are determined by the monomer size (Min et al. 2016; Tazaki et al. 2016).

Second, the porosity of dust aggregates should be measured in observations of protoplanetary disks. The radial profile of the opacity index β could distinguish between spherical bodies and porous aggregates. However, once the filling factor becomes less than ∼0.1, it is impossible distinguish. To measure the porosity, we have to know information on both the absorption and the scattering opacity: the absorption opacity tells us the mass-to-area ratio (equivalent to *af*) of dust aggregates (Kataoka et al. 2014), whereas the scattering opacity tells us the size of the aggregates (Tazaki et al. 2016). Although there had been no way of constraining the scattering opacity of large grains or aggregates, we recently proposed that millimeter-wave polarization of protoplanetary disks could be due to dust scattering (Kataoka et al. 2015).

Finally, it is also important to compare the porous planetesimals as a result of porosity evolution with comets in the solar system. Comets are believed to be remnants of planetesimals. We have already compared the mass and filling factor relation, which shows a filling factor of 10–50% (see A'Hearn 2011, and references therein). Nowadays, thanks to the Rosetta mission, an enormous amount of information on the comet 67P/Churyumov-Gerasimenko has been successfully obtained, which could constrain the formation history of comets (e.g., Thomas et al. 2015).

Acknowledgements A.K. appreciates the enormous contributions by the collaborators Hidekazu Tanaka, Satoshi Okuzumi, Koji Wada, and Cornelis P. Dullemond.

References

Adachi, I., Hayashi, C., Nakazawa, K.: The gas drag effect on the elliptical motion of a solid body in the primordial solar nebula. Prog. Theor. Phys. **56**, 1756–1771 (1976). doi:10.1143/PTP.56.1756

A'Hearn, M.F.: Comets as building blocks. Annu. Rev. Astron. Astrophys. **49**, 281–299 (2011). doi:10.1146/annurev-astro-081710-102506

Andrews, S.M., Williams, J.P.: A submillimeter view of circumstellar dust disks in ρ Ophiuchi. Astrophys. J. **671**, 1800–1812 (2007). doi:10.1086/522885, 0708.4185

Beckwith, S.V.W., Sargent, A.I.: Particle emissivity in circumstellar disks. Astrophys. J. **381**, 250–258 (1991). doi:10.1086/170646

Blum, J., Münch, M.: Experimental investigations on aggregate-aggregate collisions in the early solar nebula. Icarus **106**, 151 (1993). doi:10.1006/icar.1993.1163

Blum, J., Wurm, G.: The growth mechanisms of macroscopic bodies in protoplanetary disks. Annu. Rev. Astron. Astrophys. **46**, 21–56 (2008). doi:10.1146/annurev.astro.46.060407.145152

Boss, A.P.: Possible rapid gas giant planet formation in the solar nebula and other protoplanetary disks. Astrophys. J. **536**, L101–L104 (2000). doi:10.1086/312737

Brauer, F., Dullemond, C.P., Henning, T.: Coagulation, fragmentation and radial motion of solid particles in protoplanetary disks. Astron. Astrophys. **480**, 859–877 (2008). doi:10.1051/0004-6361:20077759, 0711.2192

Chiang, E.I., Goldreich, P.: Spectral energy distributions of T Tauri stars with passive circumstellar disks. Astrophys. J. **490**, 368–376 (1997). doi:10.1086/304869, astro-ph/9706042

Chokshi, A., Tielens, A.G.G.M., Hollenbach, D.: Dust coagulation. Astrophys. J. **407**, 806–819 (1993). doi:10.1086/172562

Cuzzi, J.N., Estrada, P.R., Davis, S.S.: Utilitarian opacity model for aggregate particles in protoplanetary nebulae and exoplanet atmospheres. Astrophys. J. Suppl. **210**, 21 (2014). doi:10.1088/0067-0049/210/2/21, 1312.1798

Dominik, C., Tielens, A.G.G.M.: The physics of dust coagulation and the structure of dust aggregates in space. Astrophys. J. **480**, 647–673 (1997). doi:10.1086/303996

Draine, B.T.: The discrete-dipole approximation and its application to interstellar graphite grains. Astrophys. J. **333**, 848–872 (1988). doi:10.1086/166795

Drążkowska, J., Windmark, F., Dullemond, C.P.: Planetesimal formation via sweep-up growth at the inner edge of dead zones. Astron. Astrophys. **556**, A37 (2013). doi:10.1051/0004-6361/201321566, 1306.3412

Dullemond, C.P., Dominik, C.: Dust coagulation in protoplanetary disks: a rapid depletion of small grains. Astron. Astrophys. **434**, 971–986 (2005). doi:10.1051/0004-6361:20042080, astro-ph/0412117

Guilloteau, S., Dutrey, A., Piétu, V., Boehler, Y.: A dual-frequency sub-arcsecond study of protoplanetary disks at mm wavelengths: first evidence for radial variations of the dust properties. Astron. Astrophys. **529**, A105 (2011). doi:10.1051/0004-6361/201015209, 1103.1296

Gundlach, B., Blum, J.: The stickiness of micrometer-sized water-ice particles. Astrophys. J. **798**(1) (2015). http://iopscience.iop.org/article/10.1088/0004-637X/798/1/34/meta

Güttler, C., Blum, J., Zsom, A., Ormel, C.W., Dullemond, C.P.: The outcome of protoplanetary dust growth: pebbles, boulders, or planetesimals? I. Mapping the zoo of laboratory collision experiments. Astron. Astrophys. **513**, A56 (2010). doi:10.1051/0004-6361/200912852, 0910.4251

Hayashi, C.: Structure of the solar nebula, growth and decay of magnetic fields and effects of magnetic and turbulent viscosities on the nebula. Prog. Theor. Phys. Suppl. **70**, 35–53 (1981). doi:10.1143/PTPS.70.35

Isella, A., Carpenter, J.M., Sargent, A.I.: Structure and evolution of pre-main-sequence circumstellar disks. Astrophys. J. **701**, 260–282 (2009). doi:10.1088/0004-637X/701/1/260, 0906.2227

Isella, A., Carpenter, J.M., Sargent, A.I.: Investigating planet formation in circumstellar disks: CARMA observations of Ry Tau and Dg Tau. Astrophys. J. **714**, 1746–1761 (2010). doi:10.1088/0004-637X/714/2/1746, 1003.4318

Johansen, A., Oishi, J.S., Mac Low, M.M., Klahr, H., Henning, T., Youdin, A.: Rapid planetesimal formation in turbulent circumstellar disks. Nature **448**, 1022–1025 (2007). doi:10.1038/nature06086, 0708.3890

Kataoka, A., Tanaka, H., Okuzumi, S., Wada, K.: Fluffy dust forms icy planetesimals by static compression. Astron. Astrophys. **557**, L4 (2013a). doi:10.1051/0004-6361/201322151, 1307.7984

Kataoka, A., Tanaka, H., Okuzumi, S., Wada, K.: Static compression of porous dust aggregates. Astron. Astrophys. **554**, A4 (2013b). doi:10.1051/0004-6361/201321325, 1303.3351

Kataoka, A., Okuzumi, S., Tanaka, H., Nomura, H.: Opacity of fluffy dust aggregates. Astron. Astrophys. **568**, A42 (2014). doi:10.1051/0004-6361/201323199, 1312.1459

Kataoka, A., Muto, T., Momose, M., Tsukagoshi, T., Fukagawa, M., Shibai, H., Hanawa, T., Murakawa, K., Dullemond, C.P.: Millimeter-wave polarization of protoplanetary disks due to dust scattering. Astrophys. J. **809**, 78 (2015). doi:10.1088/0004-637X/809/1/78, 1504.04812

Köhler, M., Stepnik, B., Jones, A.P., Guillet, V., Abergel, A., Ristorcelli, I., Bernard, J.P.: Dust coagulation processes as constrained by far-infrared observations. Astron. Astrophys. **548**, A61 (2012). doi:10.1051/0004-6361/201218975

Kothe, S., Blum, J., Weidling, R., Güttler, C.: Free collisions in a microgravity many-particle experiment. III. The collision behavior of sub-millimeter-sized dust aggregates. Icarus **225**, 75–85 (2013). doi:10.1016/j.icarus.2013.02.034, 1302.5532

Kozasa, T., Blum, J., Mukai, T.: Optical properties of dust aggregates. I - Wavelength dependence. Astron. Astrophys. **263**, 423–432 (1992)

Krijt, S., Ormel, C.W., Dominik, C., Tielens, A.G.G.M.: Erosion and the limits to planetesimal growth. Astron. Astrophys. **574**, A83 (2015). doi:10.1051/0004-6361/201425222, 1412.3593

Lyra, W., Johansen, A., Zsom, A., Klahr, H., Piskunov, N.: Planet formation bursts at the borders of the dead zone in 2D numerical simulations of circumstellar disks. Astron. Astrophys. **497**, 869–888 (2009). doi:10.1051/0004-6361/200811265, 0901.1638

Mackowski, D.W., Mishchenko, M.I.: Calculation of the T matrix and the scattering matrix for ensembles of spheres. J. Opt. Soc. Am. A **13**, 2266–2278 (1996). doi:10.1364/JOSAA.13.002266

Mathis, J.S., Rumpl, W., Nordsieck, K.H.: The size distribution of interstellar grains. Astrophys. J. **217**, 425–433 (1977). doi:10.1086/155591

Min, M., Hovenier, J.W., de Koter, A.: Modeling optical properties of cosmic dust grains using a distribution of hollow spheres. Astron. Astrophys. **432**, 909–920 (2005). doi:10.1051/0004-6361:20041920, astro-ph/0503068

Min, M., Waters, L.B.F.M., de Koter, A., Hovenier, J.W., Keller, L.P., Markwick-Kemper, F.: The shape and composition of interstellar silicate grains. Astron. Astrophys. **462**, 667–676 (2007). doi:10.1051/0004-6361:20065436, astro-ph/0611329

Min, M., Rab, C., Woitke, P., Dominik, C., Ménard, F.: Multiwavelength optical properties of compact dust aggregates in protoplanetary disks. Astron. Astrophys. **585**, A13 (2016). doi:10.1051/0004-6361/201526048, 1510.05426

Miyake, K., Nakagawa, Y.: Effects of particle size distribution on opacity curves of protoplanetary disks around T Tauri stars. Icarus **106**, 20 (1993). doi:10.1006/icar.1993.1156

Mukai, T., Ishimoto, H., Kozasa, T., Blum, J., Greenberg, J.M.: Radiation pressure forces of fluffy porous grains. Astron. Astrophys. **262**, 315–320 (1992)

Mulders, G.D., Min, M., Dominik, C., Debes, J.H., Schneider, G.: Why circumstellar disks are so faint in scattered light: the case of HD 100546. Astron. Astrophys. **549**, A112 (2013). doi:10.1051/0004-6361/201219522, 1210.4132

Okuzumi, S.: Electric charging of dust aggregates and its effect on dust coagulation in protoplanetary disks. Astrophys. J. **698**, 1122–1135 (2009). doi:10.1088/0004-637X/698/2/1122, 0901.2886

Okuzumi, S., Tanaka, H., Sakagami, M.: Numerical modeling of the coagulation and porosity evolution of dust aggregates. Astrophys. J. **707**, 1247–1263 (2009). doi:10.1088/0004-637X/707/2/1247, 0911.0239

Okuzumi, S., Tanaka, H., Takeuchi, T., Sakagami, M.: Electrostatic barrier against dust growth in protoplanetary disks. I. Classifying the evolution of size distribution. Astrophys. J. **731**, 95 (2011a). doi:10.1088/0004-637X/731/2/95, 1009.3199

Okuzumi, S., Tanaka, H., Takeuchi, T., Sakagami, M.: Electrostatic barrier against dust growth in protoplanetary disks. II. Measuring the size of the "frozen" zone. Astrophys. J. **731**, 96 (2011b). doi:10.1088/0004-637X/731/2/96, 1009.3101

Okuzumi, S., Tanaka, H., Kobayashi, H., Wada, K.: Rapid coagulation of porous dust aggregates outside the snow line: a pathway to successful icy planetesimal formation. Astrophys. J. **752**, 106 (2012). doi:10.1088/0004-637X/752/2/106, 1204.5035

Ormel, C.W., Spaans, M., Tielens, A.G.G.M.: Dust coagulation in protoplanetary disks: porosity matters. Astron. Astrophys. **461**, 215–232 (2007). doi:10.1051/0004-6361:20065949, astro-ph/0610030

Pagani, L., Steinacker, J., Bacmann, A., Stutz, A., Henning, T.: The ubiquity of micrometer-sized dust grains in the dense interstellar medium. Science **329**, 1622 (2010). doi:10.1126/science.1193211, 1110.4180

Paszun, D., Dominik, C.: Collisional evolution of dust aggregates. From compaction to catastrophic destruction. Astron. Astrophys. **507**, 1023–1040 (2009). doi:10.1051/0004-6361/200810682, 0909.3168

Pérez, L.M., Carpenter, J.M., Chandler, C.J., Isella, A., Andrews, S.M., Ricci, L., Calvet, N., Corder, S.A., Deller, A.T., Dullemond, C.P., Greaves, J.S., Harris, R.J., Henning, T., Kwon, W., Lazio, J., Linz, H., Mundy, L.G., Sargent, A.I., Storm, S., Testi, L., Wilner, D.J.: Constraints on the radial variation of grain growth in the AS 209 circumstellar disk. Astrophys. J. **760**, L17 (2012). doi:10.1088/2041-8205/760/1/L17, 1210.5252

Pérez, L.M., Chandler, C.J., Isella, A., Carpenter, J.M., Andrews, S.M., Calvet, N., Corder, S.A., Deller, A.T., Dullemond, C.P., Greaves, J.S., Harris, R.J., Henning, T., Kwon, W., Lazio, J., Linz, H., Mundy, L.G., Ricci, L., Sargent, A.I., Storm, S., Tazzari, M., Testi, L., Wilner, D.J.: Grain growth in the circumstellar disks of the young stars CY Tau and DoAr 25. Astrophys. J. **813**(1) (2015)

Pinilla, P., Birnstiel, T., Ricci, L., Dullemond, C.P., Uribe, A.L., Testi, L., Natta, A.: Trapping dust particles in the outer regions of protoplanetary disks. Astron. Astrophys. **538**, A114 (2012). doi:10.1051/0004-6361/201118204, 1112.2349

Pollack, J.B., Hollenbach, D., Beckwith, S., Simonelli, D.P., Roush, T., Fong, W.: Composition and radiative properties of grains in molecular clouds and accretion disks. Astrophys. J. **421**, 615–639 (1994). doi:10.1086/173677

Ricci, L., Testi, L., Natta, A., Brooks, K.J.: Dust grain growth in ρ-Ophiuchi protoplanetary disks. Astron. Astrophys. **521**, A66 (2010a). doi:10.1051/0004-6361/201015039, 1008.1144

Ricci, L., Testi, L., Natta, A., Neri, R., Cabrit, S., Herczeg, G.J.: Dust properties of protoplanetary disks in the Taurus-Auriga star forming region from millimeter wavelengths. Astron. Astrophys. **512**, A15 (2010b). doi:10.1051/0004-6361/200913403, 0912.3356

Ringl, C., Bringa, E.M., Bertoldi, D.S., Urbassek, H.M.: Collisions of porous clusters: a granular-mechanics study of compaction and fragmentation. Astrophys. J. **752**, 151 (2012). doi:10.1088/0004-637X/752/2/151

Ros, K., Johansen, A.: Ice condensation as a planet formation mechanism. Astron. Astrophys. **552**, A137 (2013). doi:10.1051/0004-6361/201220536, 1302.3755

Seizinger, A., Kley, W.: Bouncing behavior of microscopic dust aggregates. Astron. Astrophys. **551**, A65 (2013). doi:10.1051/0004-6361/201220946, 1301.3629

Seizinger, A., Krijt, S., Kley, W.: Erosion of dust aggregates. Astron. Astrophys. **560**, A45 (2013). doi:10.1051/0004-6361/201322773, 1310.4311

Suyama, T., Wada, K., Tanaka, H.: Numerical simulation of density evolution of dust aggregates in protoplanetary disks. I. Head-on collisions. Astrophys. J. **684**, 1310–1322 (2008). doi:10.1086/590143

Suyama, T., Wada, K., Tanaka, H., Okuzumi, S.: Geometric cross sections of dust aggregates and a compression model for aggregate collisions. Astrophys. J. **753**, 115 (2012). doi:10.1088/0004-637X/753/2/115, 1205.1894

Tazaki, R., Tanaka, H., Okuzumi, S., Kataoka, A., Nomura, H.: Light scattering by fractal dust aggregates. I. Angular dependence of scattering. Astrophys. J. **823**, 70 (2016). doi:10.3847/0004-637X/823/2/70, 1603.07492

Tazzari, M., Testi, L., Ercolano, B., Natta, A., Isella, A., Chandler, C.J., Pérez, L.M., Andrews, S., Wilner, D.J., Ricci, L., Henning, T., Linz, H., Kwon, W., Corder, S.A., Dullemond, C.P., Carpenter, J.M., Sargent, A.I., Mundy, L., Storm, S., Calvet, N., Greaves, J.A., Lazio, J., Deller, A.T.: Multiwavelength analysis for interferometric (sub-)mm observations of protoplanetary disks. Radial constraints on the dust properties and the disk structure. Astron. Astrophys. **588**, A53 (2016). doi:10.1051/0004-6361/201527423, 1512.05679

Thomas, N., Sierks, H., Barbieri, C., Lamy, P.L., Rodrigo, R., Rickman, H., Koschny, D., Keller, H.U., Agarwal, J., A'Hearn, M.F., Angrilli, F., Auger, A.T., Barucci, M.A., Bertaux, J.L., Bertini, I., Besse, S., Bodewits, D., Cremonese, G., Da Deppo, V., Davidsson, B., De Cecco, M., Debei, S., El-Maarry, M.R., Ferri, F., Fornasier, S., Fulle, M., Giacomini, L., Groussin, O., Gutierrez, P.J., Güttler, C., Hviid, S.F., Ip, W.H., Jorda, L., Knollenberg, J., Kramm, J.R., Kührt, E., Küppers, M., La Forgia, F., Lara, L.M., Lazzarin, M., Moreno, J.J.L., Magrin, S., Marchi, S., Marzari, F., Massironi, M., Michalik, H., Moissl, R., Mottola, S., Naletto, G., Oklay, N., Pajola, M., Pommerol, A., Preusker, F., Sabau, L., Scholten, F., Snodgrass, C., Tubiana, C., Vincent, J.B., Wenzel, K.P.: The morphological diversity of comet 67P/Churyumov-Gerasimenko. Science **347**(1), aaa0440 (2015). doi:10.1126/science.aaa0440

Wada, K., Tanaka, H., Suyama, T., Kimura, H., Yamamoto, T.: Numerical simulation of dust aggregate collisions. I. Compression and disruption of two-dimensional aggregates. Astrophys. J. **661**, 320–333 (2007). doi:10.1086/514332

Wada, K., Tanaka, H., Suyama, T., Kimura, H., Yamamoto, T.: Numerical simulation of dust aggregate collisions. II. Compression and disruption of three-dimensional aggregates in head-on collisions. Astrophys. J. **677**, 1296–1308 (2008). doi:10.1086/529511

Wada, K., Tanaka, H., Suyama, T., Kimura, H., Yamamoto, T.: Collisional growth conditions for dust aggregates. Astrophys. J. **702**, 1490–1501 (2009). doi:10.1088/0004-637X/702/2/1490

Wada, K., Tanaka, H., Suyama, T., Kimura, H., Yamamoto, T.: The rebound condition of dust aggregates revealed by numerical simulation of their collisions. Astrophys. J. **737**(1) (2011). http://iopscience.iop.org/article/10.1088/0004-637X/737/1/36/meta

Wada, K., Tanaka, H., Okuzumi, S., Kobayashi, H., Suyama, T., Kimura, H., Yamamoto, T.: Growth efficiency of dust aggregates through collisions with high mass ratios. Astron. Astrophys. **559**, A62 (2013). https://www.aanda.org/articles/aa/abs/2013/11/aa22259-13/aa22259-13.html

Weidenschilling, S.J.: Aerodynamics of solid bodies in the solar nebula. Mon. Not. R. Astron. Soc. **180**, 57–70 (1977). doi:10.1093/mnras/180.1.57. http://adsabs.harvard.edu/abs/1977MNRAS.180...57W

Whipple, F.L.: On certain aerodynamic processes for asteroids and comets. In: Elvius, A. (ed.) From Plasma to Planet, p. 211. Wiley, New York (1972)

Windmark, F., Birnstiel, T., Güttler, C., Blum, J., Dullemond, C.P., Henning, T.: Planetesimal formation by sweep-up: how the bouncing barrier can be beneficial to growth. Astron. Astrophys. **540**, A73 (2012). doi:10.1051/0004-6361/201118475, 1201.4282

Zsom, A., Ormel, C.W., Güttler, C., Blum, J., Dullemond, C.P.: The outcome of protoplanetary dust growth: pebbles, boulders, or planetesimals? II. Introducing the bouncing barrier. Astron. Astrophys. **513**, A57 (2010). doi:10.1051/0004-6361/200912976, 1001.0488

Chapter 6
Chondrules: Ubiquitous Chondritic Solids Tracking the Evolution of the Solar Protoplanetary Disk

Martin Bizzarro, James N. Connelly, and Alexander N. Krot

Abstract Chondrite meteorites are samples of primitive asteroidal bodies that have escaped melting and differentiation. The only record of our Solar System's formative stages comes from the earliest solids preserved in chondrites, namely millimetre- to centimetre-sized calcium-aluminium-rich inclusions (CAIs) and chondrules. These solids formed by transient heating events during the lifetime of the solar protoplanetary disk. Collectively, CAIs and chondrules provide time-sequenced samples allowing us to probe the composition of the disk material that accreted to form planetesimals and planets. Here, we showcase the current state-of-the-art data with respect to the chronology and stable isotopic compositions of individual chondrules from various chondrite groups and discuss how these data can be used to provide novel insights into the thermal and chemical evolution of the solar protoplanetary disk, including mass transport processes.

6.1 Introduction

Protoplanetary disks are flattened, rotating structures consisting of cool dust and gas surrounding most young low-mass stars, and are a consequence of the requirement to conserve angular momentum during the gravitational collapse of a prestellar core. Astronomical observations suggest that these disks may only exist for a few million years (Evans et al. 2009), a timescale that corresponds to the period where observable mass accretion to the central star occurs (Williams and Cieza 2011).

M. Bizzarro (✉) • J.N. Connelly
Centre for Star and Planet Formation, Natural History Museum of Denmark, University of Copenhagen, Copenhagen, Denmark
e-mail: bizzarro@snm.ku.dk; connelly@snm.ku.dk

A.N. Krot
Hawai'i Institute of Geophysics and Planetology, School of Ocean and Earth Science and Technology, University of Hawaii at Mānoa, Honolulu, HI, USA
e-mail: sasha@higp.hawaii.edu

M. Pessah, O. Gressel (eds.), *Formation, Evolution, and Dynamics of Young Solar Systems*, Astrophysics and Space Science Library 445,
DOI 10.1007/978-3-319-60609-5_6

Some material, however, coalesces into centimetre-sized particles that accrete to form larger asteroidal bodies, which represent the building blocks of planetary systems. Thus, the study of these accreting protoplanetary disks provides direct insights into the initial conditions for planet formation.

In the Solar System, a record of the earliest evolutionary stages of the protoplanetary disk is preserved in chondritic meteorites (chondrites), which are fragments of asteroids that avoided melting and differentiation. Based on the bulk chemical and isotopic compositions, mineralogy and petrography, 15 chondrite groups and grouplets comprise three major chondrite classes, carbonaceous (CI, CM, CR, CV, CK, CO, CB, CH), ordinary (H, L, LL), enstatite (EH, EL), as well as K and R, are currently recognised. Most chondrites consist of chondrules, refractory inclusions [Ca,Al-rich inclusions (CAIs) and amoeboid olivine aggregates (AOAs)], and fine-grained matrix (Fig. 6.1a–d). The only exception are CB carbonaceous chondrites, which lack fine-grained matrix and contain anomalously high abundance of Fe,Ni-metal (up to 70 vol%; Fig. 6.1e, f).

In primitive (unmetamorphosed and unaltered) chondrites, fine-grained matrices are complex mixtures of micrometre-sized crystalline magnesian olivine and low-Ca pyroxene, amorphous ferromagnesian silicates, Fe,Ni-metal, sulfides, and organics (Greshake 1997; Abreu and Brearley 2010). Although primitive chondrite matrices are typically considered as volatile-rich thermally unprocessed materials, a significant fraction of this material was probably vaporised and recondensed during high-temperature transient heating events associated with formation of chondrules and refractory inclusions (Scott and Krot 2005).

CAIs represent the oldest Solar System dated solids and, thus, define its age at 4567.3±0.16 Myr (Connelly et al. 2012). It is commonly accepted that CAIs formed in a hot (ambient temperature above ~1300 K) disk region characterised by approximately solar oxygen isotopic composition near the proto-Sun by evaporation, condensation and aggregation processes during a brief time interval that corresponded to high stellar mass accretion rates ($\sim 10^{-5}\,M_{\odot}\,year^{-1}$) (Krot et al. 2009). Formation of CAIs near the proto-Sun is also indicated by the presence in these objects of the short-lived radioisotope ^{10}Be (half-life of ~1.4 Myr) formed by solar energetic particle irradiation (McKeegan et al. 2000). Some CAIs were subsequently melted, most in the same disk region. Following their formation, CAIs were transported to large radial distances where they accreted into chondritic and cometary parent bodies. Most chondrules formed by melting (typically incomplete) of solid precursor material during transient heating events (peak temperature of ~2000 K) of unknown nature in different, relatively cold dust-rich regions throughout the protoplanetary disk during its entire lifetime (Connelly et al. 2012). Therefore, CAIs and chondrules provide time-sequenced samples allowing us to probe the composition of the disk material that accreted to form planetesimals and planets.

Judging by their sheer abundance in chondrites, which in some cases reaches 80% by volume, chondrules must be the product of one of the most energetic processes that operated in the early Solar System. The majority of chondrules formed as melt droplets in high-density regions of the protoplanetary disk and accumulated

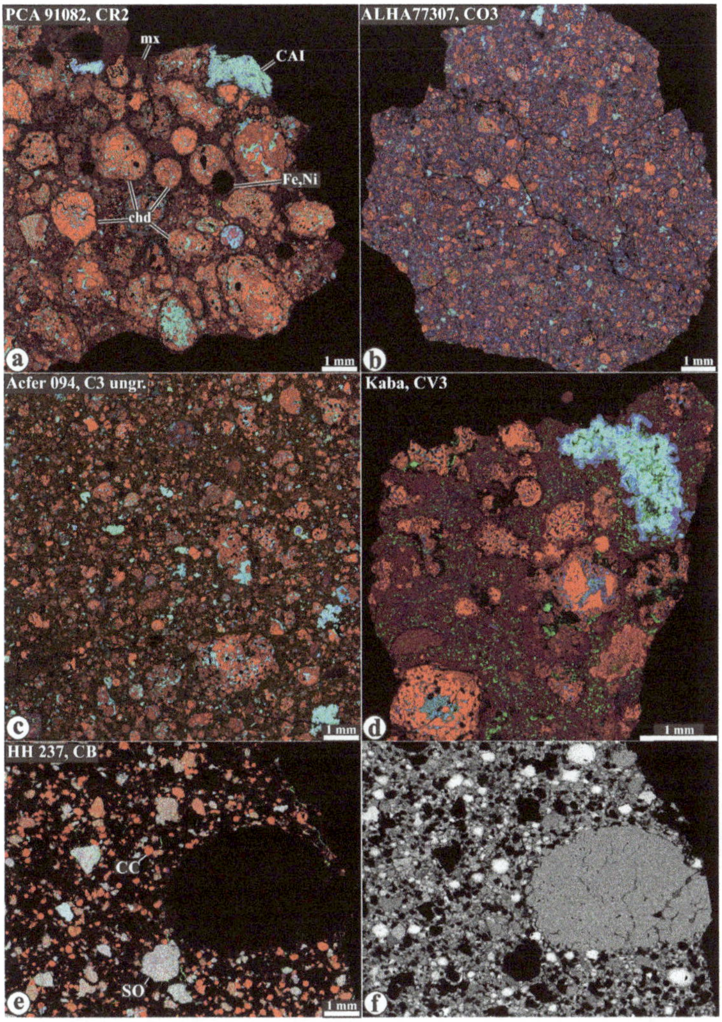

Fig. 6.1 Combined X-ray elemental maps in Mg (*red*), Ca (*green*) and Al (*blue*) of the CR (Renazzo type), CO (Ornans type), CV (Vigarano type), CB (Bencubbin type) and Acfer 094 (ungrouped) carbonaceous chondrites; (**d**) X-ray elemental map in Ni of HH 237. Typical chondrites consist of chondrules (*reddish*), Ca,Al-rich inclusions (CAIs, *bluish* and *greenish*), and Fe,Ni-metal (Fe,Ni, *black* in "(**a**)–(**e**)"), all surrounded by fine-grained matrix (mx). There are significant variations in chondrule sizes and chondrule/matrix ratio among the chondrite groups. The majority of chondrules have porphyritic textures (see Fig. 6.2a–d). The CB metal-rich carbonaceous chondrites have no matrix, contain very rare CAIs and anomalously high abundance of Fe,Ni-metal (up to 70 vol%); chondrules have exclusively non-porphyritic, cryptocrystalline (CC) and skeletal olivine (SO) textures (see Fig. 6.2e, f)

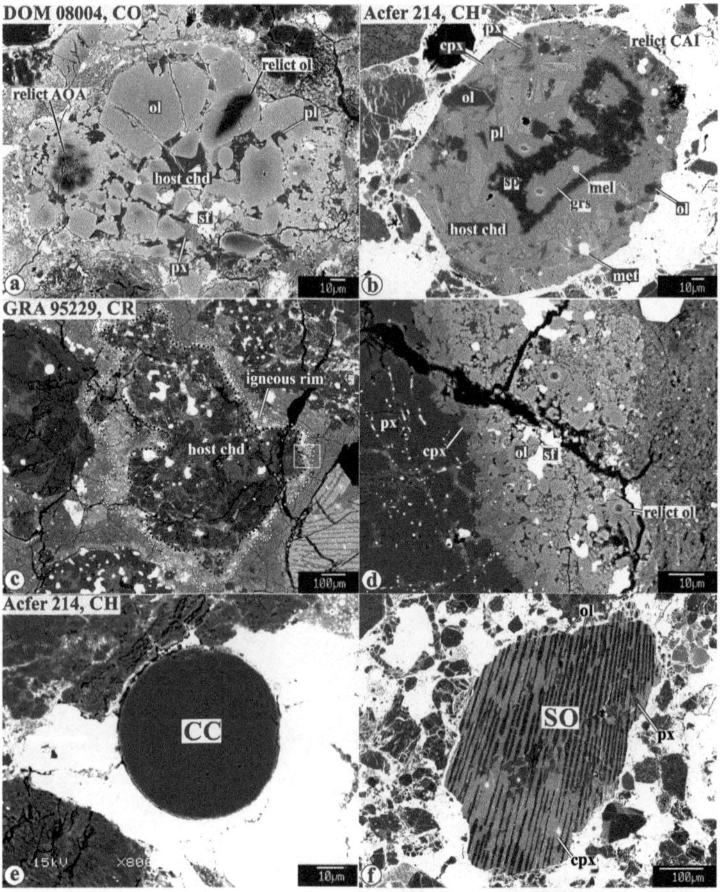

Fig. 6.2 Backscattered electron images of (**a–d**) porphyritic and (**e, f**) non-porphyritic chondrules in CO, CR, and CH carbonaceous chondrites. The porphyritic chondrules consist of ferromagnesian olivine and pyroxenes, glassy or crystalline mesostasis, Fe,Ni-metal and sulfides. Porphyritic chondrules commonly contain relict grains (**a, b**), indicative of incomplete melting of chondrule precursors. The relict grains include fragments of CAIs, AOAs, and chondrules of earlier generations. Some chondrules are surrounded by finer-grained igneous rims, indicative of repeatable melting events experienced by these chondrules (**c, d**). The non-porphyritic chondrules contain neither relict grains nor igneous rims, suggesting that they experienced crystallisation from complete melts. cpx = high-Ca pyroxene; Fe,Ni = Fe,Ni-metal; grs = grossite; mel = melilite; ol = olivine; pl = plagioclase; px = low-Ca pyroxene; sf = Fe-sulfide; sp = spinel

in the disk mid-plane together with other chondritic components. Chondrules are mainly composed of olivine ($(Fe,Mg)_2SiO_4$) and pyroxene ($(Fe,Mg)SiO_3$) minerals, which crystallised within minutes to days between ~1800 and ~1300 K (Scott 2007). Several heat sources have been proposed for the thermal processing of chondrule precursors, including shock waves (Boss and Graham 1993; Connolly and Love 1998; Hood 1998; Connolly et al. 2006), current sheets (Joung et al. 2004), x-winds (Shu et al. 1997), magnetised disk wind (Salmeron and Ireland 2012), and colliding planetesimals (Asphaug et al. 2011; Sanders and Scott 2012). A long-standing paradigm used to constrain chondrule-formation models is the so-called chemical complementarity that apparently exists between chondrules and matrix in individual chondrite groups (Hezel and Palme 2010; Palme et al. 2015; Ebel et al. 2016). In this model, it is proposed that chondrules and matrix are genetically related and formed in highly localised regions of the protoplanetary disk. The chronology of chondrule formation is typically based on the short-lived ^{26}Al to ^{26}Mg decay system [^{26}Al decays to ^{26}Mg with a half-life of 0.705 Myr (Norris et al. 1983)]. Assuming that ^{26}Al was uniformly distributed in the protoplanetary disk with the canonical ^{26}Al/^{27}Al ratio of $\sim 5\times 10^{-5}$ commonly observed in CAIs, the ^{26}Al-^{26}Mg systematics of chondrules suggest that these objects formed >1 Myr after CAIs and rapidly accreted into chondrite parent bodies together with matrix in discrete events during the lifetime of the disk. In this view, chondrule formation is restricted to the inner regions of the solar protoplanetary disk.

However, a number of recent studies investigating the absolute chronology of chondrule formation as well as the isotopic systematics of individual chondrules from various chondritic meteorites require a reassessment of current thinking with respect to the formation history of chondrules as well as the parent asteroids of chondrite meteorites. For example, the absolute isotopic dates of individual chondrules suggest that the formation of these objects started contemporaneously with the condensation and melting of CAIs and lasted $\sim$3.5 Myr (Connelly et al. 2012), which indicate the existence of multiple generations of chondrules within individual chondrites. Moreover, variability in the titanium and chromium stable isotope compositions of chondrules from individual chondrites suggests that these objects or their precursor was formed in distinct regions of the protoplanetary disk and subsequently transported to the accretion regions of their respective parent bodies (Trinquier et al. 2009; Van Kooten et al. 2016; Olsen et al. 2016). These data are at odds with the traditional view of a short formation history for chondrule population from individual chondrites, the basic concept of chondrule-matrix complementarity as well as the timescales and style of chondrite parent body accretion. In this contribution, we review the current state-of-the-art data with respect to the chronology and stable isotopic compositions of individual chondrules from various chondrite groups and discuss how these data can be used to provide novel insights into the thermal and chemical evolution of the solar protoplanetary disk, including mass transport processes.

6.2 Basic Petrological and Chemical Features of Chondrules

In typical (i.e. not metal-rich) chondrites, most chondrules have porphyritic textures (Figs. 6.1a–d and 6.2a–d) and consist of olivine and/or low-Ca pyroxene phenocrysts surrounded by glassy or microcrystalline mesostasis. Chondrules show large variations in mineral and bulk chemical compositions, allowing us to distinguish magnesian (Fa = atomic ratio Fe/(Fe+Mg)×100 or Fs = atomic ratio Fe/(Fe+Mg+Ca)×100 < 10 mol%; type I), ferroan (Fa or Fs > 10 mol%; type II), and aluminum-rich (bulk Al_2O_3 > 10 wt%) chondrules. Chondrules with non-porphyritic textures (cryptocrystalline and barred/skeletal olivine) are dominant in CB chondrites, abundant in CH chondrites (Figs. 6.1e, f and 6.2e, f), but relatively rare in other chondrite groups. Chondrules are on average depleted in siderophile (iron-loving), chalcophile (sulfide-loving) and moderately volatile (having condensation temperature < 1000 K) lithophile (silicate-loving) elements compared to whole rock; refractory lithophile elements (Ca, Al, Ti, and rare earth elements (REEs)] are approximately solar (Grossman et al. 1988).

On a typical three-isotope oxygen diagram, $\delta^{17}O$ vs. $\delta^{18}O$ [$\delta^{i}O = ({}^{i}O/{}^{16}O_{sample} / {}^{i}O/{}^{16}O_{SMOW} - 1) \times 1000$, i = 17 or 18], chondrules generally plot along a slope ~1.0 mass-independent fractionation line within ±5‰ of the terrestrial fractionation line (see Fig. 6.3). In unmetamorphosed chondrites, olivine and low-Ca pyroxene phenocrysts and mesostasis in an individual chondrule have similar oxygen-isotope compositions (Tenner et al. 2015). Porphyritic chondrules are thought to have formed by incomplete melting of solid precursors resulting in

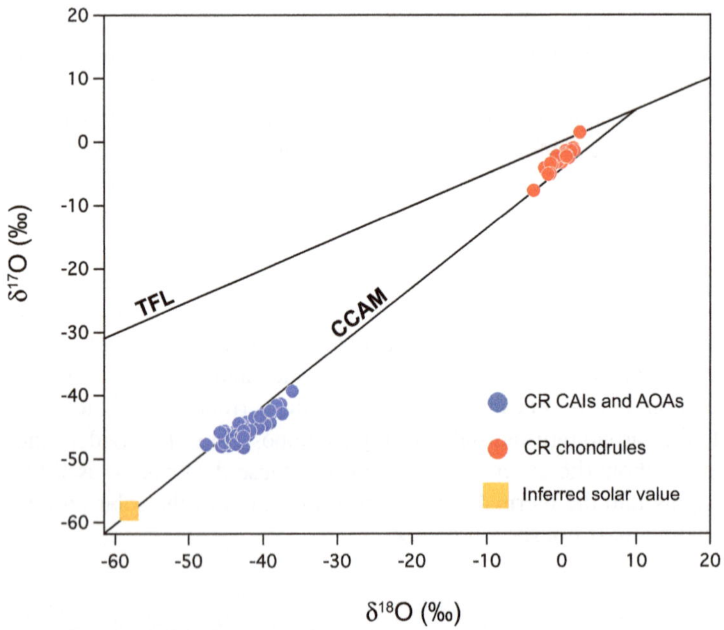

Fig. 6.3 Oxygen-isotope compositions of CAIs and amoeboid olivine aggregates (AOA) and chondrules from CR carbonaceous chondrites. The refractory inclusions and chondrules are isotopically uniform, but have different oxygen-isotope compositions, suggesting formation in isotopically distinct reservoirs. Refractory inclusions are ^{16}O-enriched relative to chondrules but ^{16}O-depleted relative to the Sun (McKeegan et al. 2011). The terrestrial fractionation line (TFL) and Carbonaceous Chondrite Anhydrous Mineral (CCAM) line are shown for reference. Data from Tenner et al. (2015), Schrader et al. (2014), Krot et al. (2006), and Makide et al. (2009)

preservation of several nuclei in a chondrule melt. In contrast, non-porphyritic chondrules crystallised from melts formed by nearly complete melting of solid precursors and/or by gas-melt condensation.

Porphyritic chondrules commonly contain relict grains that did not crystallise from a host chondrule melt, and, therefore, provide constraints on the nature of chondrule precursor materials and chondrule-forming mechanism(s). Relict grains in chondrules can be distinguished based on their textures, mineralogy, chemical and oxygen isotopic compositions. Based on these characteristics, coarse-grained relict grains (>10–100 μm) identified in porphyritic chondrules include CAIs, AOAs, fragments of chondrules of earlier generations (Fig. 6.2a, b), and, possibly fragments of thermally processed planetesimals (see Fig. 1 of Libourel and Krot 2007). Most relict ferromagnesian olivine and pyroxene grains have oxygen-isotope compositions that differ from those of the host chondrule phenocrysts and mesostasis. However, the oxygen-isotope compositions of most relict olivine and pyroxene grains are generally similar to those of typical chondrules suggesting their close genetic relationship; these relict grains are most likely fragments of chondrules of earlier generations (Tenner et al. 2015; Schrader et al. 2014; Jones et al. 2004; Krot et al. 2005b; Russell et al. 2005; Kita et al. 2010; Rudraswami et al. 2011).

Because of high dissolutions rates of olivine in chondrule melts (Soulié et al. 2012), several micron-sized relict grains could have survived only in chondrules that experienced a very small degree of melting. Chondrules in CV, CR and ordinary chondrites are typically surrounded by finer-grained ferromagnesian silicate igneous rims (Krot and Wasson 1995; Krot et al. 2004). Oxidation states of igneous rims (Fa or Fs contents in their olivines and pyroxenes) are generally similar to those of the host chondrules, suggesting formation under similar redox conditions. These rims appear to have formed by melting of relatively fine-grained solids (<10 μm) that accreted on the surface of previously solidified chondrules (Krot and Wasson 1995), indicative or repeatable transient heating events experienced by chondrules. Like host chondrules, igneous chondrule rims typically contain relict grains (Nagashima et al. 2015).

6.3 Chronology of Chondrule Formation

6.3.1 U-Corrected Pb-Pb Dating

Of the various radiometric clocks, U-corrected Pb-Pb dating is the only method that provides a high-resolution assumption-free chronology of the first 10 Myr of the Solar System. It is based on two isotopes of U that decay in a chain to stable Pb isotopes, namely ^{235}U to ^{207}Pb with a half-life of ~0.7 Gyr and ^{238}U to ^{206}Pb with a half-life of ~4 Gyr. This results in $^{207}Pb_R/^{206}Pb_R$ (where R = radiogenic) ratios that correspond to the amount of time passed since the system closed, by Eq. (6.1)

$$\frac{^{207}Pb_R}{^{206}Pb_R} = \left(\frac{^{235}U}{^{238}U}\right)\left(\frac{e^{\lambda_1 t}-1}{e^{\lambda_2 t}-1}\right), \tag{6.1}$$

where λ_1 and λ_2 are the decay constants for ^{235}U and ^{238}U, respectively; and t represents time. The $^{207}Pb_R/^{206}Pb_R$ ratio of an inclusion is calculated by extrapolating from an array of measured Pb isotopic values in $^{204}Pb/^{206}Pb$–$^{207}Pb/^{206}Pb$ space that represent varying mixtures of radiogenic Pb and its initial Pb isotopic composition, which should approximate that of the Solar System's initial Pb isotope composition (Fig. 6.4). Thus, ages are defined through the internal isochron approach. It is apparent from Eq. (6.1) that knowledge of the $^{238}U/^{235}U$ of an object is required to accurately define its Pb-Pb age. Although traditionally assumed to be 137.88 in all Solar System materials, the $^{238}U/^{235}U$ values of CAIs have been demonstrated to vary by approximately 0.35%, which corresponds to offsets in calculated Pb-Pb ages of up to ~5 Myr (Brennecka et al. 2010). The observation of U isotope variability, attributed to the decay of the short-lived ^{247}Cm nuclide (^{247}Cm decays to ^{235}U

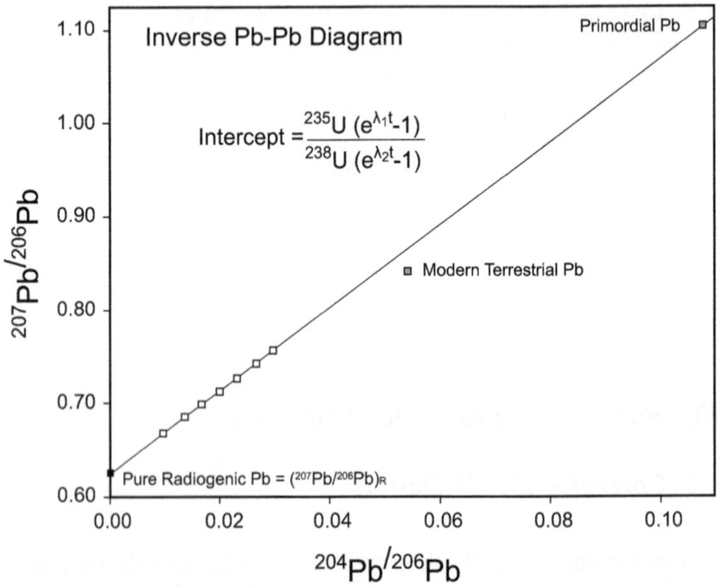

Fig. 6.4 Inverse Pb-Pb diagram. The radiogenic $^{207}Pb/^{206}Pb$ ratio [$(^{207}Pb/^{206}Pb)_R$] can be calculated by projecting a line through data points with variable mixtures of radiogenic Pb and initial Pb to the y intercept, where the initial Pb is theoretically zero.This variability is created by strategically analysing related fragments, minerals, or acid leachates with variable U/Pb ratios. (Subscript R refers to radiogenic; λ_1 and λ_2 represent the decay constants of ^{235}U and ^{238}U, respectively; and *t* represents time)

with a half-life of ~15.6 Myr), voided all published Pb-Pb ages for Solar System materials that were based on an assumed $^{238}U/^{235}U$ ratio and made clear the need to have measurements of the U isotopic compositions for all materials dated by the Pb-Pb method. However, more recent work suggests that the $^{238}U/^{235}U$ variability may be limited to the CAI reservoir, where Cm/U fractionation or, alternatively, mass-dependent $^{238}U/^{235}U$ variations can occur via evaporation and/or condensation processes (Connelly et al. 2012; Tissot et al. 2016). This view is supported by the homogeneous $^{238}U/^{235}U$ of chondrules and bulk planetary materials, which is used to define the solar $^{238}U/^{235}U$ value of 137.786 (Connelly et al. 2012; Brennecka et al. 2015).

The majority of chondrules are believed to have formed by melting of disk dust and rapidly (10–1000 $K\,h^{-1}$) cooled (Scott 2007). The so-called *nebular chondrules* are taken here as representing chondrules formed within the protoplanetary disk before dust and gas dissipation. A second group of chondrules apparently formed later by planetary collisions, namely chondrules from the metal-rich CB chondrites. Initial attempts to date nebular chondrules pooled a number of objects that resulted in ages that were 4563.66±0.63 (Amelin et al. 2002) and 4564.32±0.81 Myr (Connelly and Bizzarro 2009) (both ages adjusted for a U isotopic composition of 137.786). These ages can only reflect the average age of the chondrules pooled for these studies. Large chondrules from the Gujba CB chondrite, believed to have formed by a collision between two planetesimals, have been dated as individual inclusions with an average age of 4562.49±0.21 Myr (Bollard et al. 2015).

The only published report on the U-corrected Pb-Pb ages of individual nebular chondrules indicates a protracted formation history, with an age range for five chondrules from 4567.32±0.42 Myr to 4564.71±0.30 Myr (Connelly et al. 2012) (Fig. 6.5). The isochrons for all five chondrules project back to an initial Pb isotopic composition that is close to the presumed primordial isotopic composition of the Solar System, but with some variability interpreted to reflect a complex formation history involving more than one melting event for some chondrule precursors. In all cases, the isochrons projected well above the field for modern terrestrial Pb indicating that this was not the source of non-radiogenic Pb in the fractions used to define the isochrons. This study concluded that the ages derived for the five well-defined isochrons with sensible extrapolations to initial Pb compositions represent the age of the last crystallisation of these chondrules. They ranged in ages from the oldest chondrule that overlaps the brief formation age of CAIs to ~2.6 Myr after. Thus, the first chondrules formed contemporaneously with CAIs and the energy source responsible for the formation of chondrules lasted for at least ~2.6 Myr following condensation of CAIs. A subsequent study of 17 individual chondrules from CR and ordinary chondrites has confirmed the oldest ages of chondrules and extends the age range of nebular chondrules formation to ~3.6 Myr after CAIs (Bollard et al. 2017). The timescales for the melting of disk solids inferred from the Pb-Pb dates is comparable to the ~3 Myr median lifetime of disks around low-mass stars indicated by astronomical observations of young stellar objects within star-forming regions (Evans et al. 2009). Thus, the formation of chondrules may reflect a generic process intrinsically linked to the secular evolution of accretionary disks

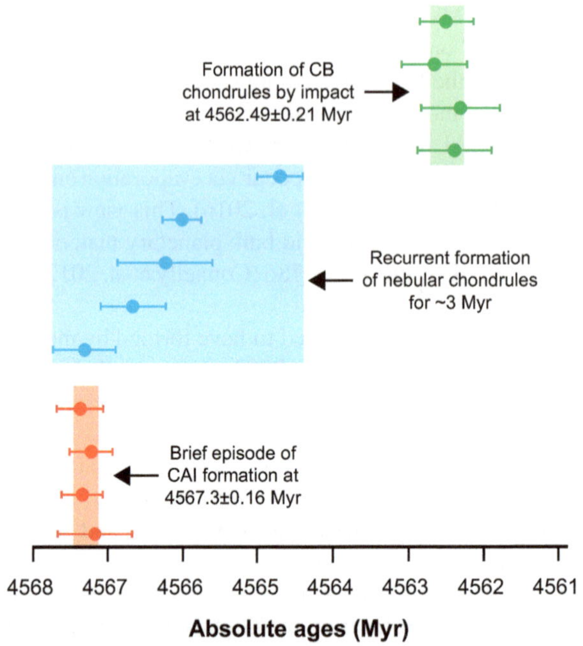

Fig. 6.5 U-corrected Pb-Pb ages of individual chondrules and CAIs from various carbonaceous and ordinary chondrites. Figure modified from Connelly et al. (2012)

not unique to our Solar System. If chondrules are indeed an important ingredient promoting the growth of asteroids and planetary embryos, the U-corrected Pb-Pb dates indicate that they existed in the disk over the timescale required to assemble Mars-sized objects.

6.3.2 The ^{26}Al-^{26}Mg Decay System

With a half-life of ~705,000 years (Norris et al. 1983), the ^{26}Al-to-^{26}Mg decay system is one of the most widely used relative chronometers to understand solid formation in the early Solar System. Indeed, the short-half life of ^{26}Al coupled to significant Al/Mg fractionation by condensation and evaporation processes as well as melting and solidification allows for the determination of highly precise relative ages. For example, it has been proposed that the epoch of primary condensation of CAIs from CV chondrites may have been as short as ~4500 years based on the ^{26}Al-^{26}Mg system (Larsen et al. 2011; Jacobsen et al. 2008). These objects define the canonical $^{26}Al/^{27}Al$ ratio of $\sim 5\times 10^{-5}$, which is widely thought to reflect the initial ^{26}Al abundance of the Solar System. Some CAIs define lower initial $^{26}Al/^{27}Al$ ratios (Fig. 6.6), which may reflect secondary disturbance of the ^{26}Al-^{26}Mg system,

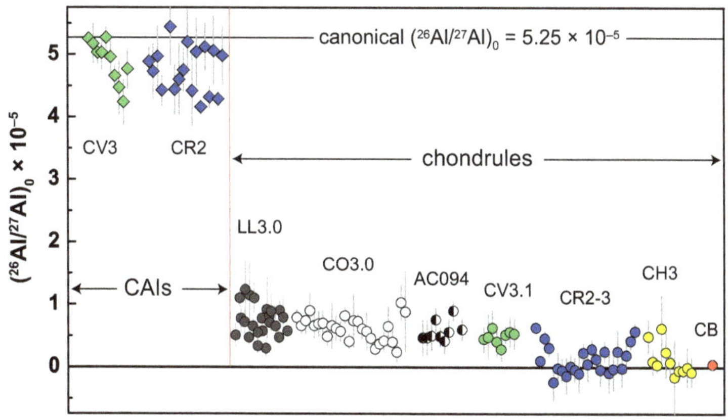

Fig. 6.6 The inferred initial $^{26}Al/^{27}Al$ ratios $[(^{26}Al/^{27}Al)_0]$ in igneous CAIs and chondrules in the least metamorphosed ordinary and carbonaceous chondrites. CAIs record $(^{26}Al/^{27}Al)_0$ of 4-5$\times10^{-5}$. In contrast, the $(^{26}Al/^{27}Al)_0$ of chondrules are systematically lower than in CAIs, $<1.5\times10^{-5}$. The lower initial is interpreted as reflecting a reduced initial abundance of ^{26}Al in the precursor material of chondrules relative to the CAI-forming region. AC094, Acfer 094. Data from Makide et al. (2009), Villeneuve et al. (2009), Larsen et al. (2011), Hutcheon and Hutchison (1989), Kita et al. (2000, 2013), Kita and Ushikubo (2012), Kunihiro et al. (2004), Kurahashi et al. (2008), MacPherson et al. (2012), Olsen et al. (2013), Ushikubo et al. (2013), and Krot and Nagashima (2017)

or alternatively, remelting events within ~300,000 years of CAI condensation (MacPherson et al. 2012). In contrast, chondrules recorded systematically lower initial $^{26}Al/^{27}Al$ ratios. Assuming that the canonical $^{26}Al/^{27}Al$ ratio represents initial abundance of ^{26}Al for the Solar System as a whole, it is possible to derive a relative chronology of chondrule formation with respect to the time of condensation of CV CAIs. Using this approach, a number of studies have suggested a time delay of ~1–2 Myr between the formation of CAIs and chondrules, a period commonly referred to as the CAI-chondrule age gap (Krot et al. 2009). Such an age gap, however, is not apparent in the U-corrected Pb-Pb ages of individual chondrules.

Luu et al. (2015) recently suggested that the bulk Al-Mg systematics of a number of chondrules from the non-pristine and aqueously altered Allende carbonaceous chondrites can be used to date the timing of condensation of chondrule precursors. This is based on the observations that a number of chondrules form an array in the Al-Mg diagram that corresponds to an initial $^{26}Al/^{27}Al$ value of 1.2$\times10^{-5}$, which the authors interpret as reflecting the timing of cessation of condensation of chondrule precursors ~1.5 Myr after formation of the Solar System first solids. This conclusion, however, is based on the assumption of ^{26}Al homogeneity, which has not been demonstrated by any study and is not constrained by their data. Moreover, the linear relationship is essentially defined by Al-rich chondrules, which are known to contain a recycled CAI component in the Allende meteorite (Krot et al. 2009). Thus, it is unlikely that the array defines a meaningful isochron. We suggest that the acceptable MSWD (mean square weighted deviations) defined by the linear array is a results of the low precision of the Mg-isotope data relative to the state-of-the-art (Bizzarro et al. 2011), which is a factor of 10 better. As such, higher precision data is required to better understand the significance of this array. Finally, we note that the initial ^{26}Al/^{27}Al value of 1.2×10^{-5} defined by the bulk Allende chondrules is comparable to the initial ^{26}Al/^{27}Al value of 1.33×10^{-5} proposed by Schiller et al. (2015a) as the initial ^{26}Al abundance for the inner protoplanetary disk. In this interpretation, the bulk Allende chondrule isochron may represent the timing of formation of chondrule precursors contemporaneously with CAI formation.

A number of recent studies have cast serious doubts on the assumption of ^{26}Al homogeneity, suggesting instead that initial heterogeneity in the ^{26}Al/^{27}Al ratio—perhaps up to 80% of the canonical value—may have existed throughout the inner Solar System (Larsen et al. 2011; Schiller et al. 2015a). In particular, Schiller et al. (2015a) recently provided a detailed comparison of the U-corrected Pb-Pb and internal ^{26}Al-^{26}Mg isochron ages for three rapidly cooled angrite meteorites. Their results demonstrate that the ^{26}Al-^{26}Mg ages obtained for angrites are systematically younger by ~1.5 Myr relative to their absolute ages, establishing that the angrite parent body accreted from precursor material typified by a reduced initial abundance of ^{26}Al relative to the canonical ratio. The three angrites concordantly define an initial ^{26}Al/^{27}Al of $1.33^{+0.21}_{-0.18}\times10^{-5}$ for the precursor of their parent body, which is identical to the estimate of $1.61\pm0.32\times10^{-5}$ inferred from the μ^{26}Mg* compositions of young angrites (Larsen et al. 2011). Preliminary reports on the comparison of the U-corrected Pb-Pb and internal ^{26}Al-^{26}Mg isochron ages of individual chondrules from various chondrites also yield systematically younger ^{26}Al-^{26}Mg ages relative to the Pb-Pb dates (Bizzarro et al. 2014). Collectively, these data support the view that the bulk of the material that accreted to form asteroidal bodies and planetary embryos was characterised by a reduced initial ^{26}Al/^{27}Al ratios relative to the canonical value. The reduced inner Solar System initial abundance of ^{26}Al relative to that of CAIs has been interpreted as reflecting the thermal processing and unmixing of presolar components with contrasting thermal properties (Trinquier et al. 2009; Paton et al. 2013; Schiller et al. 2015b). In addition, it has been proposed that bodies accreted beyond the orbits of the gas giants contain significant amount of ^{26}Al-free, thermally unprocessed molecular cloud matter (Van Kooten et al. 2016). Given the mounting evidence for initial ^{26}Al heterogeneity, it appears unlikely that the ^{26}Al-^{26}Mg system can provide an accurate chronology of the early Solar System. In this respect, the apparent ~1–2 Myr age gap inferred from the ^{26}Al-^{26}Mg system may simply reflect a reduced initial ^{26}Al abundance in chondrule precursors, which would bring the chondrule ^{26}Al-^{26}Mg age distribution in line with the U-corrected Pb-Pb dates (Bollard et al. 2017).

6.3.3 The ^{182}Hf-^{182}W Decay System

The ^{182}Hf short-lived radionuclide decays to ^{182}W with a half-life of ~9 Myr. Due to the contrasting geochemical behaviour of Hf (lithophile) and W (siderophile), these two elements are fractionated by silicate-metal segregation processes that occurred during the accretion, differentiation and early evolution of asteroidal bodies, planetary embryos and planets (Kleine et al. 2009). In contrast to ^{26}Al, which requires late-stage addition of stellar debris to the Sun's parental molecular cloud, the initial Solar System inventory of ^{182}Hf is believed to reflect long-term, steady-state galactic stellar nucleosynthesis before the formation of the protosolar molecular cloud (Wasserburg et al. 2006; Holst et al. 2013). If correct, this supports the view that the ^{182}Hf nuclide was homogeneously distributed in the protoplanetary disk at the time of formation of canonical CAIs. However, given the low abundance of W in chondritic components such as chondrules, it is not possible to date individual objects thereby necessitating the pooling of a significant amount of chondrules to obtain sufficient mounts of W. Using this approach, Budde et al. (2016) recently attempted to provide a chronology of the Allende CV chondrite chondrule formation based on the ^{182}Hf-^{182}W system. In particular, these authors inferred that the $\sim$3 Myr duration of chondrule formation documented by Connelly et al. (2012) is inconsistent with presumed chemical and isotopic complementary between chondrules and matrix that supports models of local and brief chondrule formation (Hezel and Palme 2010; Palme et al. 2015; Ebel et al. 2016). They hypothesised that the Pb-Pb ages reflect late stage parent body alteration without providing any specific mechanism to explain isochrons as old as CAIs. Instead, they used the short-lived ^{182}Hf-^{182}W decay system applied to 100s or 1000s of chondrules as well as matrix and bulk samples in an attempt to obtain the true age range of chondrules. A similar approach was used by Becker et al. (2015). However, using matrix, bulk samples and bulk chondrules to define ^{182}Hf-^{182}W isochrons requires independent evidence for a single, closed-system Hf-W fractionation event affecting the matrix and chondrules to fulfil the basic requirement of a meaningful isochron.

Budde et al. (2016) used arguments for complementarity between chondrules and matrix, including their own ^{183}W data, to infer a brief formation interval for chondrules and their immediate accretion to form chondrites with cogenetic matrix. Invoking this model to infer coeval fractionation of Hf and W between chondrules and matrix, these authors used these entities and mixtures of them to define a Hf-W array that corresponds to an age of 2.2 ± 0.8 Myr after the formation of CAIs that they accept as the time window of chondrule formation. This overlaps the result of Becker et al. (2015) who defined a chrondrule formation window of -0.9 ± 2.8 Myr relative to the formation age of CAIs. However, a recent Mg and Cr isotope study of individual CV chondrules does not support the required complementarity between matrix and chondrules (Olsen et al. 2016). Furthermore, linearity of large multi-chondrule fractions in Hf-W space is predicted if the age distributions of the various populations are similar, even if they are not coeval. In this case, their age would approximate the average age of chondrules if the Hf/W ratio of the matrix is close to the composition of the bulk Solar System. Finally, we note that two earlier studies (Amelin and Krot 2007; Connelly et al. 2008) have investigated the Pb-Pb isotopic dates of Allende chondrule populations to derive average ages for chondrule formation of 1.67 ± 0.9 Myr and 2.85 ± 0.45 Myr after CAI formation [calculated using $^{238}U/^{235}U$ of 137.786 (Connelly et al. 2012)]. These ages are consistent with that reported by Budde et al. (2016). This may indicate that both systems remained closed in each chondrule after their respective final nebular heating event and, therefore, both are capable of returning primary age information about chondrules formation, even in an aqueously altered and metamorphosed meteorite like Allende. However, only the Pb-Pb system is capable of dating individual chondrules to determine the true age range of nebular chondrule formation.

6.4 Mechanism and Style of Asteroidal Accretion

The short time interval inferred for the formation of chondrules within individual chondrite groups based on the ^{26}Al-^{26}Mg system can be used to argue for a rapid accretion of chondrite parent bodies. In this model, chondrules formed in high-density regions that were possibly self-gravitating, which resulted in the rapid collapse and accretion of these objects into chondritic parent bodies (Alexander et al. 2008). This implies that chondrule formation and asteroidal accretion are intrinsically linked processes. However, the protracted timescale for the formation of chondrules inferred from the assumption-free U-corrected Pb-Pb dating method (Connelly et al. 2012; Bollard et al. 2017) is inconsistent with this model. Indeed, the ~3 Myr formation interval for chondrules from various chondrite groups indicates prolonged accretion timescales for chondritic parent bodies.

Recent numerical simulations suggest that the formation of asteroidal bodies and planetary embryos may be a two-step process, where first generation planetesimal seeds of ~50 km diameter form rapidly via streaming instabilities (Johansen et al. 2007) followed by the protracted gas-drag-assisted accretion of chondrules during the lifetime of the protoplanetary disk (Johansen et al. 2015). The gas-drag assisted accretion of chondrules onto planetesimals is a process analogous to pebble accretion, which is the accretion of centimetre- to metre-sized particles loosely bound to the gas onto planetesimals seeds (Bitsch et al. 2015). In these simulations, the largest planetesimals of a population with a characteristic radius of ~100 km undergo run-away accretion of chondrules forming Mars-sized planetary embryos within a timescale of ~3–5 Myr. This timescale is in agreement with the timing of formation and differentiation of Mars inferred from ^{182}Hf-^{182}W chronology (Dauphas and Pourmand 2011).

A model of continuous asteroidal accretion during the lifetime of the protoplanetary disk has important implications for the thermal evolution of asteroidal bodies given that the accretion process is completed beyond the time when ^{26}Al can provide enough energy to induce heating and differentiation (Larsen et al. 2016). Thus, protracted asteroidal accretion predicts the existence of partially differentiated asteroidal bodies, namely onion-shell structured bodies with differentiated interiors consisting of silicate mantles and metallic cores surrounded by unmelted chondritic crusts (Weiss and Elkins-Tanton 2013). Although controversial, this proposal is apparently supported by the discovery of remnant magnetism in chondritic meteorites suggesting the existence of dynamo field (Carporzen et al. 2011), which can only occur through the establishment of a convecting metallic core.

6.5 Accretion Regions of Chondrite Parent Bodies

Large-scale nucleosynthetic isotopic heterogeneity exists among inner Solar System solids, planets, and asteroids, most noticeably for neutron-rich isotopes of the iron-group elements such as ^{48}Ca, ^{50}Ti, ^{54}Cr and ^{62}Ni (Birck 2004). In particular, significant ^{54}Cr variability has been documented between carbonaceous chondrite groups and meteorites originating from asteroidal bodies believed to have formed in the accretion regions of terrestrial planets such as ordinary and enstatite chondrites and the majority of differentiated meteorites (Fig. 6.7). Bulk carbonaceous chondrites are characterised by excesses in $\mu^{54}Cr$ ranging from $+57\pm11$ to $+156\pm6$ ppm relative to the terrestrial composition (Trinquier et al. 2007). In contrast, enstatite and ordinary chondrites as well as Mars, the Moon and most differentiated

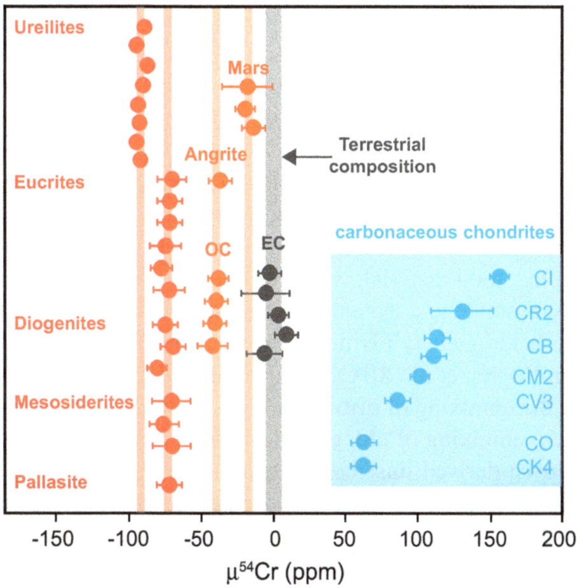

Fig. 6.7 $\mu^{54}Cr$ composition of bulk planetary reservoirs. The $\mu^{54}Cr$ notation reflects per 10^6 deviations of the $^{54}Cr/^{52}Cr$ ratio of a sample relative to the terrestrial composition. Carbonaceous chondrites formed from material that is distinct from the bulk of the material thought to have accreted in the terrestrial planet region. Thus, the $\mu^{54}C$ values can be used to track genetic relationships between early solids and bulk Solar System reservoirs. *EC*, enstatite chondrites. *OC*, ordinary chondrites. Uncertainties for ureilite data are smaller than symbols. Data from Trinquier et al. (2007) and Yamakawa et al. (2010)

meteorites record μ^{54}Cr values that span from the terrestrial composition to deficits of ~100 ppm. Thus, carbonaceous chondrites have apparently sampled, on average, material formed in a distinct reservoir in the terms of their μ^{54}Cr composition compared to the enstatite and ordinary chondrite population. The dichotomy in ^{54}Cr composition observed between carbonaceous and non-carbonaceous material is also mirrored by a number of other nuclides, including ^{43}Ca, ^{48}Ca, ^{46}Ti, ^{50}Ti, ^{62}Ni and ^{88}Sr (Paton et al. 2013; Schiller et al. 2015b; Regelous et al. 2008; Trinquier et al. 2009). This variability, which is interpreted as reflecting the selective unmixing of nucleosynthetic components during the earliest stages of Solar System formation (Trinquier et al. 2009; Paton et al. 2013; Schiller et al. 2015b; Van Kooten et al. 2016), provides a means of probing genetic relationships between early formed solids, asteroids and planetary bodies.

It has been suggested that the observed dichotomy in the abundance of the ^{54}Cr tracer between carbonaceous and non-carbonaceous chondrites essentially reflects distinct accretion regions of their parent bodies, namely that the carbonaceous material formed beyond the snow line whereas the non-carbonaceous material originated Sunward of the snow line (Larsen et al. 2016; Warren 2011). Accretion of the carbonaceous chondrites beyond the snow line is in accord with the much higher water content of these meteorites (Robert and Epstein 1982; Kerridge 1985) compared to that of enstatite and ordinary chondrites (McNaughton et al. 1981; Robert et al. 1987; Hutson and Ruzicka 2000). Moreover, recent dynamical models for the early evolution of the Solar System suggest that the parent asteroids of carbonaceous chondrites formed between and beyond the accretion regions of the giant planets and were implanted in the asteroid belt following the final outward migration of Jupiter (Walsh et al. 2011).

In the inner Solar System, bulk planetary materials with solar or near-solar ^{27}Al/^{24}Mg ratios record positively correlated variability in μ^{26}Mg* and μ^{54}Cr. This correlation is interpreted as reflecting progressive thermal processing of in-falling ^{26}Al-rich molecular cloud material, which resulted in preferential loss by sublimation of thermally unstable and isotopically anomalous presolar carriers, producing residual isotopic heterogeneity (Trinquier et al. 2009; Paton et al. 2013; Schiller et al. 2015b; Van Kooten et al. 2016). In this model, the correlated μ^{26}Mg*-μ^{54}Cr array represents the unmixing of distinct dust populations with contrasting thermal properties, namely unmixing of old, galactically inherited homogeneous dust from a young supernovae-derived dust component formed shortly prior to or during the evolution of the giant molecular cloud parental to the protosolar molecular cloud core. We illustrate this concept in Fig. 6.8, which depicts the μ^{26}Mg*-μ^{54}Cr systematics of three reservoirs, namely a bulk Solar System reservoir, the CAI-forming gas and a residual disk solid reservoir. Thermal processing of dust of solar composition occurs at T_1, which results in the establishment of two complementary reservoirs, namely the CAI forming gas enriched in ^{26}Al and ^{54}Cr and a residual disk solid reservoir depleted in ^{26}Al and ^{54}Cr relative to the bulk solar composition. The timing of the main thermal processing event (T_1) is thought to have occurred shortly prior to CAI condensation. Following their establishment, the μ^{26}Mg* isotopic

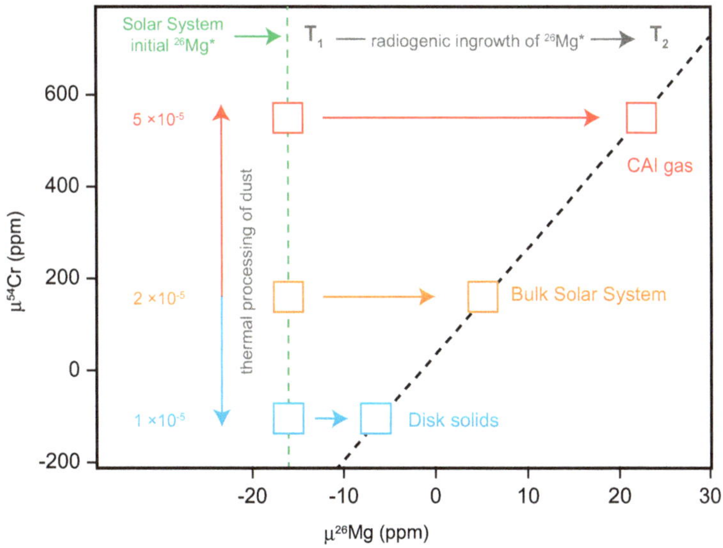

Fig. 6.8 μ^{26}Mg*–μ^{54}Cr variation diagram. Diagram depicting the effect of early thermal processing of presolar components on the μ^{26}Mg* and μ^{54}Cr isotope composition of early Solar System reservoirs with solar ^{27}Al/^{24}Mg ratios. A thermal processing event occurs at T_1, which leads to enrichments in ^{26}Al and ^{54}Cr in a hypothetical gas and depletion in ^{26}Al and ^{54}Cr in the residual disk solids. The μ^{26}Mg* compositions of the reservoirs then evolve according to their initial ^{26}Al inventory to their present-day compositions at time T_2, generating a positively correlated array

composition of the respective reservoirs evolve according to their initial ^{26}Al/^{27}Al to their present-day, measured compositions (T_2). Thermal processing at early times thus results in a positively correlated array in the measured μ^{26}Mg* and μ^{54}Cr compositions of bulk planetary materials with solar or near-solar ^{27}Al/^{24}Mg ratios.

Van Kooten et al. (2016) recently proposed that the metal-rich carbonaceous chondrites (CB, CH and CR) accreted from material predominately located beyond the orbits of the gas giant planets. This proposal is based on the coupled μ^{54}Cr and μ^{26}Mg* compositions of metal-rich carbonaceous chondrites and their components, which is distinct from bulk inner Solar System objects. In detail, metal-rich carbonaceous chondrites and their components do not plot on the Solar System's μ^{54}Cr–μ^{26}Mg* correlation line but instead have a unique isotopic signature extending from an inner Solar System composition toward a ^{26}Mg*-depleted and ^{54}Cr-enriched component (Fig. 6.9). This composition is consistent with that expected for thermally unprocessed primordial molecular cloud material before its pollution by stellar-derived ^{26}Al. Moreover, Van Kooten et al. (2016) suggest that the μ^{54}Cr and μ^{26}Mg* compositions of these objects require significant amounts (25–50%) of primordial molecular cloud matter in their precursor material. Given that such high fractions of primordial molecular cloud material are expected to survive only in the outer Solar System, the authors infer that, similarly to cometary bodies, metal-rich carbonaceous chondrites are samples of planetesimals that accreted

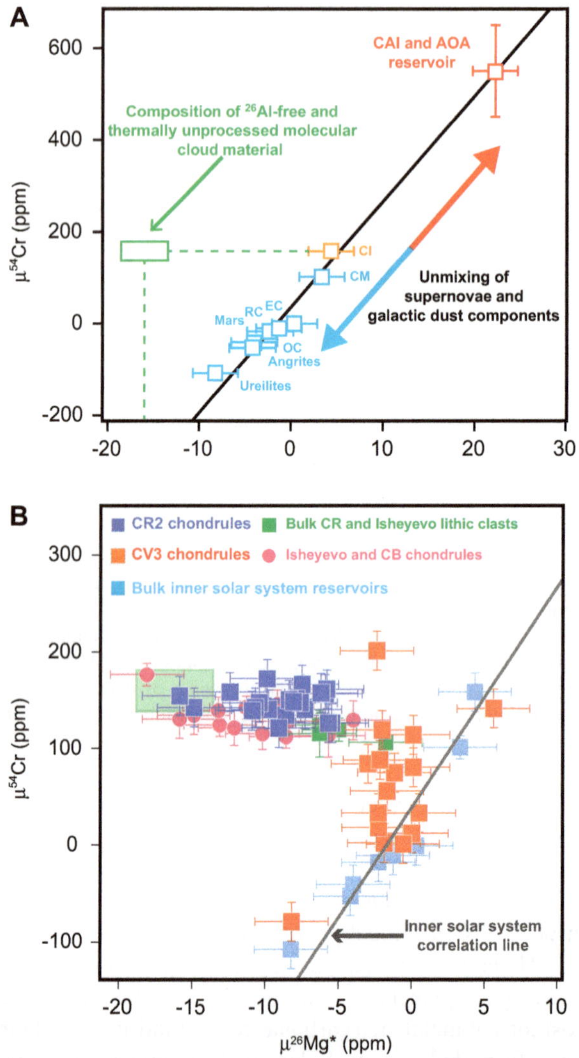

Fig. 6.9 μ^{26}Mg*–μ^{54}Cr variation diagrams. (**a**) Inner Solar System μ^{26}Mg*–μ^{54}Cr correlation. *AOA*, amoeboid olivine aggregates; *EC*, enstatite chondrites; *OC*, ordinary chondrites; *RC*, Rumuruti chondrites. The μ^{26}Mg* and μ^{54}Cr compositions of the ^{26}Al-free and thermally unprocessed molecular cloud material are defined by CI chondrites and the initial μ^{26}Mg* value of the Solar System. (**b**) The μ^{26}Mg* and μ^{54}Cr compositions of metal-rich chondrites and their components. The *green box* represents the predicted composition of the ^{26}Al-free and thermally unprocessed molecular cloud material. Modified from Van Kooten et al. (2016) and Olsen et al. (2016)

beyond the orbits of the gas giants. Collectively, the coupled μ^{54}Cr–μ^{26}Mg* systematics of various chondrite groups suggest that their parent bodies accreted in three distinct regions. The parent bodies of ordinary and enstatite chondrites are inferred to have formed Sunward of the snow line whereas the CV, CI and CM carbonaceous chondrites formed in a water-rich region located beyond the snow line, perhaps within the orbits of the gas giant planets. In contrast, metal-rich chondrites apparently formed from a distinct ^{26}Al-poor reservoir possibly located beyond the orbit of Saturn where cometary objects accreted. This proposal can be tested by investigating the μ^{54}Cr–μ^{26}Mg* systematics of objects presumed to be of cometary origin such as interplanetary dust particles or ultra-carbonaceous micrometeorites (Busemann et al. 2009; Nesvorný et al. 2010; Duprat et al. 2010).

6.6 Multiplicity of Chondrule-Forming Mechanisms

Based on petrographic and mineralogical observations as well as chemical and oxygen isotopic compositions, it is thought that porphyritic chondrules formed by melting, typically incomplete, of isotopically diverse solid precursors in dust-rich regions (dust/gas of $\sim$100–1000 $\times$ solar) of the protoplanetary disk during repeatable and localised transient heating events (Krot and Nagashima 2017; Alexander et al. 2008; Cuzzi and Alexander 2006; Alexander and Ebel 2012). Among the proposed mechanisms of chondrule formation are shock waves related to disk gravitational instability, eccentric planetesimals, and X-ray flares (Desch et al. 2005; Morris et al. 2012), magnetised turbulence in the disk (McNally et al. 2013), collisions between chondritic or differentiated planetesimals (Asphaug et al. 2011; Johnson et al. 2015), and splashing of differentiated planetesimals (Asphaug et al. 2011). Although none of the proposed mechanisms can be completely ruled out, the common presence of relict grains and Fe,Ni-metal in porphyritic chondrules as well as the large age range of chondrules from individual chondrites, including ages indistinguishable from CAIs (Connelly et al. 2012; Bollard et al. 2017), are inconsistent with formation of the majority of porphyritic chondrules by splashing of differentiated bodies.

It is well accepted that the magnesian non-porphyritic chondrules in CB chondrites are thought to have formed in a melt-gas plume generated by a hypervelocity collision between planetesimals $\sim$4.8 Myr after CV CAIs (Bollard et al. 2015; Krot et al. 2005a), with one at least one of the colliding bodies being of differentiated nature (Fedkin et al. 2015; Oulton et al. 2016). In contrast to porphyritic chondrules, the CB chondrules formed during a single-stage event and, therefore, represent single generation objects, including the magnesian cryptocrystalline CB chondrules thought to have formed by condensation as melt droplets from the impact plume (Fedkin et al. 2015). The CH metal-rich carbonaceous chondrites, thought to be genetically related to CB chondrites (Weisberg et al. 1995; Krot et al. 2002), contain multiple generations of chondrules formed by different mechanisms. In detail, the magnesian non-porphyritic chondrules formed in the CB impact plume whereas the ferromagnesian and Al-rich porphyritic chondrules formed by incomplete melting of isotopically diverse precursors from multiple heating events unrelated to the CB impact (Krot et al. 2007; Krot et al. 2017). This clearly emphasises that chondrules formed by distinct mechanisms may occur in individual chondrite groups. In this case, however, the two formation mechanisms result in easily identifiable distinct petrologic features for chondrules. Accepting that shock waves are the dominant heat source for producing chondrules in most chondrite groups, this raises the possibility that the spectrum of petrologic features observed in chondrules is the expression of the numerous potential sources of shock waves that were active during the lifetime of the protoplanetary disk.

The proposal that the combined $\mu^{26}Mg^{*}$–$\mu^{54}Cr$ systematics of Solar System objects can be used to track their formation regions predicts that the chondrule-forming process operated at various orbital distances, namely from the inner protoplanetary disk to beyond the accretion regions of the giant planets. Although the high surface densities and high energy environments typical of the inner protoplanetary disk permit chondrule formation through shock-related transient heating events (Connolly et al. 2006), thermal processing of solids in the outer Solar System may require a different source of shocks. For example, the energy required for the thermal processing of dust in the outer Solar System may result from planetary embryos bow shocks or, alternatively, impacts. This emphasises that the energy source required to melt dust resulting in the production of chondrules may be variable in both space and time during the evolution of the protoplanetary disk.

6.7 Tracking Mass Transport and Recycling

Similar to bulk asteroidal and planetary material, the stable isotopic compositions of individual chondrules allow us to determine the formation regions of their precursor material. Based on a limited dataset, Trinquier et al. (2009) identified correlated ^{46}Ti and ^{50}Ti nucleosynthetic variability in chondrules from the Allende CV3 chondrite, defining both excesses and deficits compared to the terrestrial composition. Moreover, Connelly et al. (2012) documented ^{54}Cr variability in individual Allende chondrules as well as chondrules from the NWA 5697 ordinary chondrite. At face value, these results indicate that chondrules formed from isotopically heterogeneous precursor material in different regions of the protoplanetary disk and were then transported to accretion regions of their respective parent bodies.

Accepting that the carbonaceous chondrites did indeed accrete beyond the snow line, a detailed investigation of the $\mu^{54}Cr$ compositions of individual carbonaceous chondrite chondrules can provide insights into the transport of material to the accretion region(s) or carbonaceous chondrites. Using this approach, Olsen et al. (2016) recently reported $\mu^{54}Cr$ data of 42 individual chondrules from CV and CR chondrites. Their analysis establishes that considerable $\mu^{54}Cr$ variability exists amongst individual chondrules, most noticeably for CV chondrites (Fig. 6.10). In detail, CV chondrules record a range $\mu^{54}Cr$ compositions that is comparable to that defined by our samples of chondrites, achondrites as well as meteorites from Mars and the Moon. Although it is possible that the $\mu^{54}Cr$ heterogeneity observed in CV chondrules reflects the variable incorporation of a ^{54}Cr presolar carrier, this

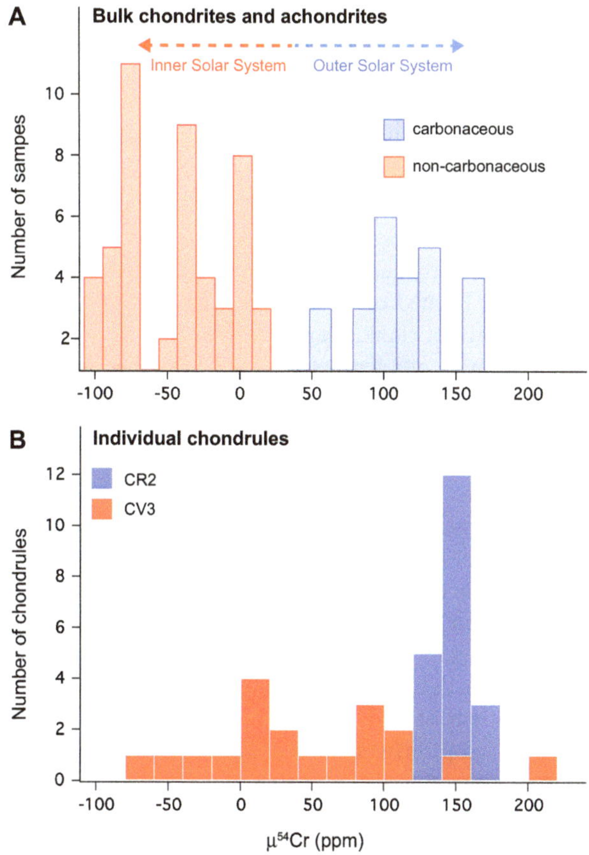

Fig. 6.10 μ^{54}Cr variability amongst Solar System solids, asteroids and planetary bodies. (**a**) μ^{54}Cr values for bulk chondrites and achondrites. The carbonaceous group comprises CI, CM, CR, CV, CK, CB chondrites as well as lithic clasts from the Isheyevo CH/CB chondrite. The non-carbonaceous group consists of ordinary and enstatite chondrites, angrites, aubrites, eucrites, diogenites, mesosiderites, acapulcoites and ureilites as well as Martian meteorites. The carbonaceous chondrites are thought to have accreted in a water-rich reservoir beyond the snow line whereas the non-carbonaceous chondrites and achondrites are believed to have formed Sunward of the snow line. (**b**) μ^{54}Cr values for individual chondrules from CV and CR carbonaceous chondrites. The scale of the *x*-axis is the same for both panels. Data from Connelly et al. (2012), Van Kooten et al. (2016), Olsen et al. (2016), Larsen et al. (2011), Trinquier et al. (2007), Yamakawa et al. (2010), and Qin et al. (2010)

interpretation is not consistent with the observed range of compositions. Indeed, given that the scale at which individual chondrules sample disk material is at least 3 orders of magnitude smaller relative to the size of the samples typically used to define the ^{54}Cr composition of chondrites and achondrites, greater variability is predicted to be recorded by individual chondrules. The comparable range of ^{54}Cr compositions observed for chondrules and bulk samples of chondrites and achondrites points to a similar process imparting the ^{54}Cr variation, namely thermal processing of their precursor material. If correct, this establishes that the ^{54}Cr values of early Solar System materials can be used to provide a high fidelity record of the formation region of their precursor material.

The range of $\mu^{54}Cr$ compositions recorded by CV chondrite chondrules, with both excesses and deficits relative to the terrestrial composition, requires that these objects formed from precursors that originated at various orbital distances, namely from the accretion regions of most differentiated asteroids and terrestrial planets to the formation regions of carbonaceous chondrites. In accord with the abundance of refractory inclusions formed close to the young Sun in CV chondrites (Krot et al. 2009), the occurrence of chondrules with inner Solar System $\mu^{54}Cr$ signatures in these chondrites requires efficient lateral outward transport of material during the earliest stages of the evolution of the solar protoplanetary disk. In contrast to CV chondrules, CR chondrules define a much more restricted range of $\mu^{54}Cr$ values characterised by compositions found in bulk carbonaceous chondrites. This observation suggests that CR chondrules formed from precursors predominantly formed in the accretion region of their parent body, with little input of material with inner Solar System $\mu^{54}Cr$ signatures. This is consistent with the low abundance of CAIs in CR chondrites thereby supporting the view of limited transport of inner Solar System solids to their accretion region. Collectively, the $\mu^{54}Cr$ systematics of individual chondrules suggest that variable recycling of thermally processed inner disk solids to the accretion of carbonaceous chondrites may be an important process regulating the compositions of the various carbonaceous chondrite groups.

The coupled $\mu^{54}Cr$ and $\mu^{26}Mg^*$ data for individual chondrules from CV and CR chondrites support the idea that the accretion region of metal-rich chondrites and their components must have been isolated from that of CV chondrites (Van Kooten et al. 2016; Olsen et al. 2016). Indeed, chondrules from CV chondrites show broadly correlated $\mu^{54}Cr$-$\mu^{26}Mg^*$ variability, similarly to bulk inner Solar System reservoirs (Fig. 6.9). The lack of evidence for admixing of appreciable amounts of thermally unprocessed primordial molecular cloud material in the precursors of CV chondrules suggest that their accretion region(s) was spatially isolated from that of metal-rich chondrites. In other words, inward transport of outer Solar System millimetre-sized solids to the accretion region of CV chondrites appears to have been limited for a significant period of the disk lifetime. A possible mechanism to limit the inward migration of outer Solar System material is the formation of gas giants opening gaps in the protoplanetary disk (Müller and Kley 2013), providing that the gas giants are able to accrete most of the solids attempting to cross the gap. Recent numerical simulations show that rapid formation of gas and ice giants can occur by accretion of centimetre-to-metre-sized particles by the mechanism of pebble accretion (Lambrechts and Johansen 2012). In these models, pebbles are concentrated by aerodynamic drag and gravitationally collapse to form objects of up to 1000 km in diameter; these planetary embryos can then efficiently accrete leftover pebbles and directly form the cores of giant planets. Therefore, the growth of giant cores by pebble accretion provides an efficient means of limiting the influx of material by generating partial disk gaps and pressure bumps outside of their orbits (Lambrechts et al. 2014).

6.8 Outward Mass Transport Mechanisms

The μ^{54}Cr data of individual chondrules from CV and CR chondrites suggest variable amounts of transport and recycling of inner Solar System material to the accretion regions of their respective parent bodies. Two classes of models have been invoked to explain how high temperature refractory material was redistributed throughout the disk to be incorporated into primitive bodies. One class of models are disk models, which explore how the inward transport of mass and angular momentum may result in outward transport in the early evolution of protoplanetary disks. For example, it has been suggested that in viscously evolving disks, turbulence can combine with the large-scale flows of the disk to carry refractory material outward against the inward motions associated with gas drag and accretion thereby providing a means to preserve and diffuse material at larger orbital distances (Cuzzi et al. 2003; Ciesla 2007, 2010). In these models, the transport of material is apparently most efficient in highly turbulent disks, which may limit the efficiency of this mechanism to the earliest stages. Recent studies suggest that the accretion of differentiated planetesimals was initiated within a few 10^5 years of CAI formation (Schiller et al. 2011; Kruijer et al. 2014; Schiller et al. 2015a; Larsen et al. 2016). Thus, a consequence of outward diffusion via the disk midplane is the incorporation of refractory material within early formed asteroidal and planetary embryos. However, the coupled ^{54}Cr and ^{50}Ti systematics of Earth, Mars, most differentiated asteroids as well as ordinary and enstatite chondrites (Trinquier et al. 2009) suggest a paucity of CAI material in their precursors, implying that outward diffusive transport through the midplane may not have been the dominant transport mechanism.

The second type of outward transport models is based on the magnetically driven outflows characteristic of young stellar objects. Both observations and simulations show that protostars exhibit powerful outflows of material accelerated to supersonic speeds along the polar axis of the star or as winds from the disk (Bontemps et al. 1996; Romanova et al. 2009; Sheikhnezami et al. 2012). These outflows provide an efficient mechanism for releasing the angular momentum inherited from the accretion process. Solids may be entrained and accelerated in jets and winds resulting in the outward transport of material in ballistic trajectories above the disk (Shu et al. 1996; Hu 2010). Although the role of stellar outflows in the formation of chondritic components such as CAIs and chondrules is uncertain (Desch et al. 2010), disk winds and jets are generic features of protostars and, thus, can provide a potentially efficient mechanism for outward transport and recycling of material during the entire lifetime of the protoplanetary disk (Hansen 2014).

The paucity of CAIs as well as the lack of chondrules with an inner solar system signature in the accretion region of the CR chondrite parent body, inferred to have been beyond the orbits of the gas giants (Van Kooten et al. 2016), suggested little input of thermally processed inner Solar System solids to the outer solar system. Thus, outward mass transport to large orbital distances such as the accretion region of cometary objects may have been only possible during the early, deeply embedded stage of the proto-Sun characterised by powerful high-velocity jets (Bontemps et al. 1996). In contrast, approximately 50% of the chondrules in CV3 chondrites have $\mu^{54}Cr$ compositions typical of inner Solar System solids indicating significant admixing of inner Solar System material to the accretion region of the CV parent body. The high abundance of CAI material in CV chondrites relative to CR supports this observation. Thus, outward transport and recycling of inner Solar System material to the outer part of the asteroid belt may have been more protracted and, hence, driven by lower-velocity stellar and disk winds, which may have been active for the entire duration of the accretion phase of the proto-Sun (Reipurth and Bally 2001).

6.9 The Chondrule-Matrix Complementarity

The bulk chemistry of chondrites is defined by the two major components, chondrules and matrix. A number of studies have investigated the apparent chemical relationship between chondrules and matrix in individual chondrite groups (Hezel and Palme 2010; Palme et al. 2015; Ebel et al. 2016). These studies have concluded that the average compositions of chondrules and matrix are typically different for a number of elements in an individual chondrite whereas the bulk composition, which reflects a mixtures of chondrules and matrix, has approximately a solar elemental abundance. This so-called chondrule-matrix complementary has been used to argue for a genetic link between these two components and, therefore, formation from a single reservoir. Given the short residence time of solids in a protoplanetary disk due to gas drag (Weidenschilling 1977), this model, in its simplest expression, predicts that all chondrules from a single chondrite should be in isotopic equilibrium and have the same age. Thus, the chondrule-matrix complementarity requires that chondrule formation and asteroidal accretion are intimately linked. However, the observed $\mu^{54}Cr$ variability suggests that chondrules from individual chondrite groups formed from isotopically diverse precursor material in different regions of the protoplanetary disk and were subsequently transported to the accretion regions of their respective parent bodies. This is consistent with the presence of age variability of ~3 Myr between chondrules from individual chondrites (Connelly et al. 2012), which requires transport and/or storage. At face value, these data appear inconsistent with the concept of chondrule-matrix complementary as originally envisaged, namely that all chondrules from an individual chondrites are all genetically related to the coexisting matrix.

Recent models of evolving viscous disks, however, suggest that a complementary relationship between chondrules and dust can be preserved for long time-scales provided that the decoupling between chondrules and gas is limited (Goldberg et al. 2015). In these models, various chondrule populations remained in complementarity such that the bulk contribution from each source is chemically solar and, thus, so is the final mixture. However, these experiments assume that the main transport mechanism of chondrules occurs through outward diffusion via the disk midplane. In disk models where outward transport of material is associated with stellar outflows (Shu et al. 1996; Hu 2010), the coarse-grained dust component (i.e. CAIs and chondrules) is not expected to be efficiently coupled to the gas and, thus, it is unclear how complementary can be preserved. A possibility is the observed chondrule-matrix complementarity is an expression of the generic process of chondrule formation and does not reflect a genetic link. In this view, the matrix comprises a complement related to the chondrule formation process (Alexander 2005) such that the bulk composition of the matrix is shifted from its starting composition and, thus, appears complementary to a chondrule composition. This does not require that the matrix is genetically linked to the chondrules in an individual chondrite but merely that some of it has experienced earlier chondrule formation events. In this view, fractions of the matrix in a particular chondrite may be complementary to chondrule populations in other chondritic meteorites.

6.10 Summary and Perspectives

The U-corrected Pb-Pb ages of individual CAIs and chondrules provide a robust framework to understand the chronology of solid formation in the early Solar System. CAIs formed during a brief time interval of less than ~0.2 Myr, possibly associated with the early stages of the proto-Sun characterised by high mass accretion rates. In contrast, the production of chondrules began contemporaneously with the formation of canonical CAIs and lasted for the entire lifetime of the solar protoplanetary disk (Fig. 6.11). All chondrite groups investigated contain chondrules of multiple generations with a similar age range of ~3 Myr. The bulk compositions of different chondrite parent bodies (CI, CM, CV, CO, OC and EC) record significant stable isotope heterogeneity of nucleosynthetic origin for various elements, which is interpreted as reflecting the selective destruction of isotopically anomalous presolar carriers during progressive thermal processing in the inner disk region. This nucleosynthetic variability indicates that these chondrite parent bodies accreted in spatially distinct disk regions. The level of this heterogeneity is comparable to the range of nucleosynthetic variability among individual chondrules from CV chondrites. Thus, CV chondrules and/or their precursors originated in different parts of the protoplanetary disk and were subsequently transported to the accretion region of the CV parent asteroid. Combined with the protracted formation timescales inferred by the absolute U-corrected Pb-Pb dating method, the observed variability in the stable isotope compositions of chondrules paints a dynamic picture of the early Solar System, where chondrule formation, transport and recycling occur continuously during the lifetime of the protoplanetary disk. Thus, accretion of individual chondrite parent bodies was a continuous process. Given the stable isotope heterogeneity recorded by individual chondrules, the

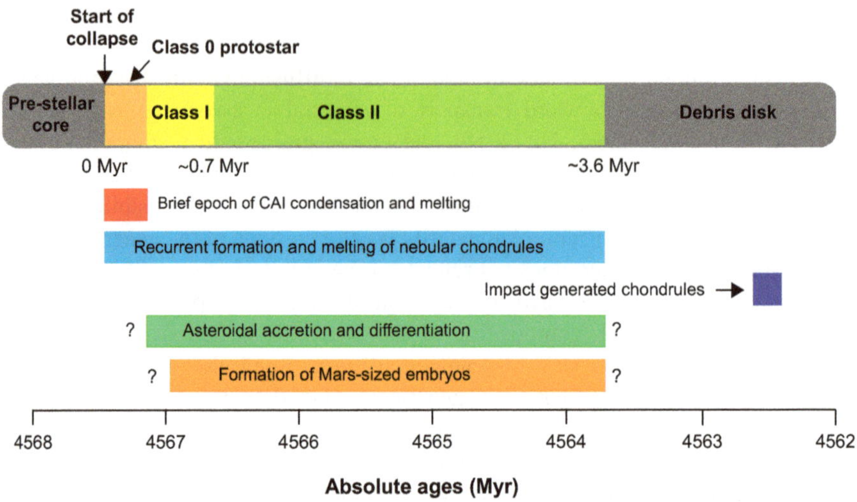

Fig. 6.11 Time scales of solid formation and disk evolution. The brief formation interval for the formation of CAIs is similar to the median lifetimes of class 0 protostars of 0.1 to 0.2 Myr inferred from astronomical observations. Therefore, the thermal regime required for CAI condensation may only have existed during the earliest stages of disk evolution typified by high mass accretion rates onto the central star. In contrast, recurrent chondrule formation occurred throughout the lifetime of the protoplanetary disk. The accretion and differentiation of asteroidal bodies, possibly leading to the growth of Mars-sized planetary embryos within ∼4–5 Myr, may have been promoted by chondrule accretion (Johansen et al. 2015). Modified from Connelly et al. (2012)

inferred continuous, layered accretion of chondrite parent bodies predicts the existence of time-dependent isotope heterogeneity within these bodies. The growth of sizeable asteroidal bodies and planetary embryos occurred during the lifetime of the protoplanetary disk, possibly promoted by the gas-drag assisted accretion of chondrules onto asteroidal seeds.

The metal-rich chondrites, namely the CR, CH and CB chondrites, formed from precursor material that has largely escaped the thermal processing recorded by the inner Solar System bodies. The metal-rich chondrites and their chondrules appear to have incorporated appreciable amounts of thermally unprocessed primordial molecular cloud material, suggesting formation in the outer part of the Solar System, beyond the orbits the gas giant planets. Thus, thermal processing of solids, including chondrule formation was not restricted to the inner disk regions but also occurred in the outer Solar System. However, the mechanism and efficiency of the thermal processing of solids at large orbital distances are poorly understood.

The stable isotope compositions of the inner and outer Solar System materials are distinct, implying limited mixing of these two reservoirs. An efficient mechanism to limit the inward transport of outer disk solids to the inner Solar System is the formation of gas giants opening gaps in the disk. Collectively, these data suggest that different chondrule-forming mechanisms may have operated at distinct times and/or regions of the protoplanetary disk. Indeed, CH chondrites contain chondrules formed by different mechanisms, including gas-melt impact plume produced chondrules as well as chondrules formed by incomplete melting during transient heating events.

Outward transport of solids could have occurred by a variety of time-dependent processes, including turbulent diffusion and stellar outflows. The outward mass transport to large orbital distances such as the accretion region of cometary objects may have been only possible during the early, deeply embedded stage of the proto-Sun characterised by powerful high-velocity jets. The paucity of CAIs in the accretion regions of inner protoplanetary disk bodies relative to the accretion regions of carbonaceous chondrites and the presence of CAI-like objects in the Jupiter family comet 81P/Wild 2 suggest that refractory inclusions were radially transported above the midplane, possibly entrained and accelerated in stellar outflows such as jets and disk winds. In contrast, outward transport and recycling of inner Solar System material to the outer part of the asteroid belt appear to be more protracted and, hence, driven by lower-velocity stellar and disk winds, which may have been active for the entire duration of the accretion phase of the proto-Sun.

The scenario presented here for the protoplantary disk evolution can be tested through key isotope measurements of various meteoritic solids. The continuous layered accretion of chondrite parent bodies can be tested by high-precision bulk Mg and Cr isotope measurements of chondrites thought to have originated at different depths of the same parent asteroid such as ordinary chondrites of petrologic types 3-6 that experienced increasingly higher peak metamorphic temperatures (300-900 °C). A better understanding of the timing and tempo of chondrule formation and, by extension, the nature of the chondrule forming mechanisms that may have operated at different times can be evaluated by acquiring a statistically significant dataset of U-corrected Pb-Pb ages of individual chondrules from various chondrite groups. The proposal of a reduced inner Solar System initial abundance of ^{26}Al relative to the canonical ratio observed in CAIs can be tested by combined Al-Mg and U-corrected Pb-Pb ages of individual chondrules from primitive chondrites. The hypothesis that metal-rich chondrites accreted in an ^{26}Al-poor region of the outer Solar System can be verified by investigating the μ^{54}Cr–μ^{26}Mg* systematics of objects presumed to be of cometary origin such as interplanetary dust particles or ultra-carbonaceous micrometeorites. Finally, the roles of stellar outflows in the outward transport of solids can be evaluated by both high-resolution astronomical observations of young disks and realistic numerical simulations of star-formation and evolving protoplanetary disks.

Acknowledgements This work was funded by grants from the Danish National Research Foundation (DNRF97) and from the European Research Council (ERC Consolidator Grant Agreement 616027-STARDUST2ASTEROIDS) to MB. We thank the referee for comments, which have improved our chapter.

References

Abreu, N.M., Brearley, A.J.: Early solar system processes recorded in the matrices of two highly pristine CR3 carbonaceous chondrites, MET 00426 and QUE 99177. Geochim. Cosmochim. Acta **74**, 1146–1171 (2010). doi:10.1016/j.gca.2009.11.009

Alexander, C.M.O.D.: Re-examining the role of chondrules in producing the elemental fractionations in chondrites. Meteorit. Planet. Sci. **40**, 943 (2005). doi:10.1111/j.1945-5100.2005.tb00166.x

Alexander, C.M.O.D., Ebel, D.S.: Questions, questions: Can the contradictions between the petrologic, isotopic, thermodynamic, and astrophysical constraints on chondrule formation be resolved? Meteorit. Planet. Sci. **47**, 1157–1175 (2012). doi:10.1111/j.1945-5100.2011.01308.x

Alexander, C.O., et al.: The formation conditions of chondrules and chondrites. Science **320**, 1617 (2008). doi:10.1126/science.1156561. http://adsabs.harvard.edu/abs/2008Sci...320.1617A

Amelin, Y., Krot, A.: Pb isotopic age of the Allende chondrules. Meteorit. Planet. Sci. **42**:1321–1335 (2007). doi:10.1111/j.1945-5100.2007.tb00577.x

Amelin, Y., Krot, A.N., Hutcheon, I.D., Ulyanov, A.A.: Lead isotopic ages of chondrules and calcium-aluminum-rich inclusions. Science **297**, 1678–1683 (2002). doi:10.1126/science.1073950

Asphaug, E., Jutzi, M., Movshovitz, N.: Chondrule formation during planetesimal accretion. Earth Planet. Sci. Lett. **308**, 369–379 (2011). doi:10.1016/j.epsl.2011.06.007

Becker, M., Hezel, D.C., Schulz, T., Elfers, B.M., Münker, C.: Formation timescales of CV chondrites from component specific Hf-W systematics. Earth Planet. Sci. Lett. **432**, 472–482 (2015). doi:10.1016/j.epsl.2015.09.049

Birck, J.L.: An overview of isotopic anomalies in extraterrestrial materials and their nucleosynthetic heritage. Rev. Mineral. Geochem. **55**(1), 25–64 (2004). doi:10.2138/gsrmg.55.1.25. http://rimg.geoscienceworld.org/content/55/1/25

Bitsch, B., Lambrechts, M., Johansen, A.: The growth of planets by pebble accretion in evolving protoplanetary discs. Astron. Astrophys. **582**, A112 (2015). doi:10.1051/0004-6361/201526463, 1507.05209

Bizzarro, M., et al.: High-precision mg-isotope measurements of terrestrial and extraterrestrial material by HR-MC-ICPMS—implications for the relative and absolute mg-isotope composition of the bulk silicate earth. J. Anal. At. Spectrom. **26**, 565 (2011). doi:10.1039/C0JA00190B. http://pubs.rsc.org/en/content/articlelanding/2011/ja/c0ja00190b

Bizzarro, M., Olsen, M., Itoh, S., Kawasaki, N., Schiller, M., Bonal, L., Yurimoto, H.: Evidence for a reduced initial abundance of ^{26}Al in chondrule forming regions and implications for the accretion timescales of protoplanets. In: 77th Annual Meeting of the Meteoritical Society, LPI Contributions, vol. 1800, p. 5125 (2014)

Bollard, J., Connelly, J.N., Bizzarro, M.: Pb-Pb dating of individual chondrules from the CB_a chondrite Gujba: assessment of the impact plume formation model. Meteorit. Planet. Sci. **50**, 1197–1216 (2015). doi:10.1111/maps.12461

Bollard, J., Connelly, J.N., Whitehouse, M.J., Pringle, E.A., Bonal, L., Jørgensen, J.K., Nordlund, Å., Moynier, F., Bizzarro, M.: Early formation of planetary building blocks inferred from Pb isotopic ages of chondrules. Sci. Adv. **3**, e1700407 (2017)

Bontemps, S., Andre, P., Terebey, S., Cabrit, S.: Evolution of outflow activity around low-mass embedded young stellar objects. Astron. Astrophys. **311**, 858–872 (1996)

Boss, A.P., Graham, J.A.: Clumpy disk accretion and chondrule formation. Icarus **106**, 168 (1993). doi:10.1006/icar.1993.1164

Brennecka, G.A., Weyer, S., Wadhwa, M., Janney, P.E., Zipfel, J., Anbar, A.D.: ^{238}U/^{235}U variations in meteorites: extant ^{247}Cm and implications for Pb-Pb dating. Science **327**, 449 (2010). doi:10.1126/science.1180871

Brennecka, G.A., Budde, G., Kleine, T.: Uranium isotopic composition and absolute ages of Allende chondrules. Meteorit. Planet. Sci. **50**, 1995 (2015). doi:10.1111/maps.12567

Budde, G., Kleine, T., Kruijer, T.S., Burkhardt, C., Metzler, K.: Tungsten isotopic constraints on the age and origin of chondrules. Proc. Natl. Acad. Sci. USA **113**(11), 2886–2891 (2016). doi:10.1073/pnas.1524980113. http://www.pnas.org/content/113/11/2886

Busemann, H., Nguyen, A.N., Cody, G.D., Hoppe, P., Kilcoyne, A.L.D., Stroud, R.M., Zega, T.J., Nittler, L.R.: Ultra-primitive interplanetary dust particles from the comet 26P/Grigg-Skjellerup dust stream collection. Earth Planet. Sci. Lett. **288**, 44–57 (2009). doi:10.1016/j.epsl.2009.09.007

Carporzen, L., Weiss, B.P., Elkins-Tanton, L.T., Shuster, D.L., Ebel, D., Gattacceca, J.: Magnetic evidence for a partially differentiated carbonaceous chondrite parent body. Proc. Natl. Acad. Sci. USA **108**(16), 6386–6389 (2011). doi:10.1073/pnas.1017165108. http://www.pnas.org/content/108/16/6386

Ciesla, F.J.: Outward transport of high-temperature materials around the midplane of the solar nebula. Science **318**, 613 (2007). doi:10.1126/science.1147273. http://adsabs.harvard.edu/abs/2007Sci...318..613C

Ciesla, F.J.: The distributions and ages of refractory objects in the solar nebula. Icarus **208**, 455–467 (2010). doi:10.1016/j.icarus.2010.02.010

Connelly, J.N., Bizzarro, M.: Pb–Pb dating of chondrules from CV chondrites by progressive dissolution. Chem. Geol. **259**(3–4), 143–151 (2009). doi:10.1016/j.chemgeo.2008.11.003. http://www.sciencedirect.com/science/article/pii/S0009254108005081

Connelly, J.N., Amelin, Y., Krot, A.N., Bizzarro, M.: Chronology of the solar system's oldest solids. Astrophys. J. Lett. **675**, L121 (2008). doi:10.1086/533586

Connelly, J.N., Bizzarro, M., Krot, A.N., Nordlund, A., Wielandt, D., Ivanova, M.A.: The absolute chronology and thermal processing of solids in the solar protoplanetary disk. Science **338**, 651 (2012). doi:10.1126/science.1226919. http://adsabs.harvard.edu/abs/2012Sci...338..651C

Connolly, H.C. Jr., Love, S.G.: The formation of chondrules: petrologic tests of the shock wave model. Science **280**, 62–67 (1998). doi:10.1126/science.280.5360.62

Connolly, H.C. Jr., Desch, S.J., Ash, R.D., Jones, R.H.: Transient heating events in the protoplanetary nebula. In: Lauretta, D.S., McSween, H.Y. (eds.) Meteorites and the Early Solar System II, pp. 383–397. University of Arizona Press, Tucson (2006)

Cuzzi, J.N., Alexander, C.M.O.D.: Chondrule formation in particle-rich nebular regions at least hundreds of kilometres across. Nature **441**, 483–485 (2006). doi:10.1038/nature04834

Cuzzi, J.N., Davis, S.S., Dobrovolskis, A.R.: Blowing in the wind. II. Creation and redistribution of refractory inclusions in a turbulent protoplanetary nebula. Icarus **166**, 385–402 (2003). doi:10.1016/j.icarus.2003.08.016

Dauphas, N., Pourmand, A.: Hf-W-Th evidence for rapid growth of Mars and its status as a planetary embryo. Nature **473**, 489–492 (2011). doi:10.1038/nature10077

Desch, S.J., Ciesla, F.J., Hood, L.L., Nakamoto, T.: Heating of chondritic materials in solar nebula shocks. In: Krot, A.N., Scott, E.R.D., Reipurth, B. (eds.) Chondrites and the Protoplanetary Disk. Astronomical Society of the Pacific Conference Series, vol. 341, p. 849. Astronomical Society of the Pacific, San Francisco (2005)

Desch, S.J., Morris, M.A., Connolly, H.C. Jr., Boss, A.P.: A critical examination of the X-wind model for chondrule and calcium-rich, aluminum-rich inclusion formation and radionuclide production. Astrophys. J. **725**, 692–711 (2010). doi:10.1088/0004-637X/725/1/692, 1011.3483

Duprat, J., Dobrica, E., Engrand, C., Aléon, J., Marrocchi, Y., Mostefaoui, S., Meibom, A., Leroux, H., Rouzaud, J.N., Gounelle, M., Robert, F.: Extreme deuterium excesses in ultracarbonaceous micrometeorites from central antarctic snow. Science **328**, 742–745 (2010). doi:10.1126/science.1184832. http://science.sciencemag.org/content/328/5979/742

Ebel, D.S., Brunner, C., Konrad, K., Leftwich, K., Erb, I., Lu, M., Rodriguez, H., Crapster-Pregont, E.J., Friedrich, J.M., Weisberg, M.K.: Abundance, major element composition and size of components and matrix in CV, CO and Acfer 094 chondrites. Geochim. Cosmochim. Acta **172**, 322–356 (2016). doi:10.1016/j.gca.2015.10.007

Evans, N.J. II, Dunham, M.M., Jørgensen, J.K., Enoch, M.L., Merín, B., van Dishoeck, E.F., Alcalá, J.M., Myers, P.C., Stapelfeldt, K.R., Huard, T.L., Allen, L.E., Harvey, P.M., van Kempen, T., Blake, G.A., Koerner, D.W., Mundy, L.G., Padgett, D.L., Sargent, A.I.: The spitzer c2d legacy results: star-formation rates and efficiencies; evolution and lifetimes. Astrophys. J. Suppl. **181**, 321 (2009). doi:10.1088/0067-0049/181/2/321. http://adsabs.harvard.edu/abs/2009ApJS..181..321E

Fedkin, A.V., Grossman, L., Humayun, M., Simon, S.B., Campbell, A.J.: Condensates from vapor made by impacts between metal-, silicate-rich bodies: comparison with metal and chondrules in CB chondrites. Geochim. Cosmochim. Acta **164**, 236–261 (2015). doi:10.1016/j.gca.2015.05.022

Goldberg, A.Z., Owen, J.E., Jacquet, E.: Chondrule transport in protoplanetary discs. Mon. Not. R. Astron. Soc. **452**, 4054–4069 (2015). doi:10.1093/mnras/stv1610

Greshake, A.: The primitive matrix components of the unique carbonaceous chondrite ACFER 094: a TEM study. Geochim. Cosmochim. Acta **61**, 437–452 (1997). doi:10.1016/S0016-7037(96)00332-8

Grossman, J.N., Rubin, A.E., Nagahara, H., King, E.A.: Properties of chondrules. In: Kerridge, J.F., Matthews, M.S. (eds.) Meteorites and the Early Solar System, pp. 619–659. University of Arizona Press, Tucson (1988)

Hansen, B.M.S.: The circulation of dust in protoplanetary discs and the initial conditions of planet formation. Mon. Not. R. Astron. Soc. **440**, 3545–3556 (2014). doi:10.1093/mnras/stu471, 1403.6552

Hezel, D.C., Palme, H.: The chemical relationship between chondrules and matrix and the chondrule matrix complementarity. Earth Planet. Sci. Lett. **294**, 85–93 (2010). doi:10.1016/j.epsl.2010.03.008

Holst, J.C., Olsen, M.B., Paton, C., Nagashima, K., Schiller, M., Wielandt, D., Larsen, K.K., Connelly, J.N., Jørgensen, J.K., Krot, A.N., Nordlund, A., Bizzarro, M.: ^{182}Hf–^{182}W age dating of a ^{26}Al-poor inclusion and implications for the origin of short-lived radioisotopes in the early Solar System. Proc. Natl. Acad. Sci. USA **110**(22), 8819–8823 (2013). doi:10.1073/pnas.1300383110. http://www.pnas.org/content/110/22/8819

Hood, L.L.: Thermal processing of chondrule and CAI precursors in planetesimal bow shocks. Meteorit. Planet. Sci. **33** (1998). doi:10.1111/j.1945-5100.1998.tb01611.x

Hu, R.: Transport of the first rocks of the solar system by X-winds. Astrophys. J. **725**, 1421–1428 (2010). doi:10.1088/0004-637X/725/2/1421, 1010.2532

Hutcheon, I.D., Hutchison, R.: Evidence from the Semarkona ordinary chondrite for Al-26 heating of small planets. Nature **337**, 238–241 (1989). doi:10.1038/337238a0

Hutson, M., Ruzicka, A.: A multi-step model for the origin of E3 (enstatite) chondrites. Meteorit. Planet. Sci. **35**, 601–608 (2000). doi:10.1111/j.1945-5100.2000.tb01440.x

Jacobsen, B., Yin, Q.Z., Moynier, F., Amelin, Y., Krot, A.N., Nagashima, K., Hutcheon, I.D., Palme, H.: ^{26}Al–^{26}Mg and ^{207}Pb–^{206}Pb systematics of Allende CAIs: canonical solar initial ^{26}Al/^{27}Al ratio reinstated. Earth Planet. Sci. Lett. **272**(1–2), 353–364 (2008). doi:10.1016/j.epsl.2008.05.003. http://www.sciencedirect.com/science/article/pii/S0012821X0800318X

Johansen, A., Oishi, J.S., Mac Low, M.M., Klahr, H., Henning, T., Youdin, A.: Rapid planetesimal formation in turbulent circumstellar disks. Nature **448**, 1022–1025 (2007). doi:10.1038/nature06086, 0708.3890

Johansen, A., Mac Low, M.M., Lacerda, P., Bizzarro, M.: Growth of asteroids, planetary embryos, and Kuiper belt objects by chondrule accretion. Sci. Adv. **1**, 1500109 (2015). doi:10.1126/sciadv.1500109, 1503.07347

Johnson, B.C., Minton, D.A., Melosh, H.J., Zuber, M.T.: Impact jetting as the origin of chondrules. Nature **517**, 339–341 (2015). doi:10.1038/nature14105

Jones, R.H., Leshin, L.A., Guan, Y., Sharp, Z.D., Durakiewicz, T., Schilk, A.J.: Oxygen isotope heterogeneity in chondrules from the Mokoia CV3 carbonaceous chondrite. Geochim. Cosmochim. Acta **68**, 3423–3438 (2004). doi:10.1016/j.gca.2004.01.013

Joung, M.K.R., Mac Low, M.M., Ebel, D.S.: Chondrule formation and protoplanetary disk heating by current sheets in nonideal magnetohydrodynamic turbulence. Astrophys. J. **606**, 532–541 (2004). doi:10.1086/381651, astro-ph/0309189

Kerridge, J.F.: Carbon, hydrogen and nitrogen in carbonaceous chondrites Abundances and isotopic compositions in bulk samples. Geochim. Cosmochim. Acta **49**, 1707–1714 (1985). doi:10.1016/0016-7037(85)90141-3

Kita, N.T., Ushikubo, T.: Evolution of protoplanetary disk inferred from ^{26}Al chronology of individual chondrules. Meteorit. Planet. Sci. **47**, 1108–1119 (2012). doi:10.1111/j.1945-5100.2011.01264.x

Kita, N.T., Nagahara, H., Togashi, S., Morishita, Y.: A short duration of chondrule formation in the solar nebula: evidence from ^{26}Al in Semarkona ferromagnesian chondrules. Geochim. Cosmochim. Acta **64**, 3913–3922 (2000). doi:10.1016/S0016-7037(00)00488-9

Kita, N.T., Nagahara, H., Tachibana, S., Tomomura, S., Spicuzza, M.J., Fournelle, J.H., Valley, J.W.: High precision SIMS oxygen three isotope study of chondrules in LL3 chondrites: role of ambient gas during chondrule formation. Geochim. Cosmochim. Acta **74**, 6610–6635 (2010). doi:10.1016/j.gca.2010.08.011

Kita, N.T., Yin, Q.Z., MacPherson, G.J., Ushikubo, T., Jacobsen, B., Nagashima, K., Kurahashi, E., Krot, A.N., Jacobsen, S.B.: ^{26}Al-^{26}Mg isotope systematics of the first solids in the early solar system. Meteorit. Planet. Sci. **48**, 1383–1400 (2013). doi:10.1111/maps.12141

Kleine, T., Touboul, M., Bourdon, B., Nimmo, F., Mezger, K., Palme, H., Jacobsen, S.B., Yin, Q.Z., Halliday, A.N.: Hf-W chronology of the accretion and early evolution of asteroids and terrestrial planets. Geochim. Cosmochim. Acta **73**, 5150–5188 (2009). doi:10.1016/j.gca.2008.11.047

Krot, A.N., Nagashima, K.: Constraints on mechanisms of chondrule formation from chondrule precursors and chronology of transient heating events in the protoplanetary disk. Geochem. J. **51**, 45–58 (2017)

Krot, A.N., Wasson, J.T.: Igneous rims on low-FeO and high-FeO chondrules in ordinary chondrites. Geochim. Cosmochim. Acta **59**, 4951–4966 (1995). doi:10.1016/0016-7037(95)00337-1

Krot, A.N., Meibom, A., Weisberg, M.K., Keil, K.: Invited review: the CR chondrite clan: implications for early solar system processes. Meteorit. Planet. Sci. **37**, 1451–1490 (2002). doi:10.1111/j.1945-5100.2002.tb00805.x

Krot, A.N., Libourel, G., Goodrich, C.A., Petaev, M.I.: Silica-rich igneous rims around magnesian chondrules in CR carbonaceous chondrites: evidence for condensation origin from fractionated nebular gas. Meteorit. Planet. Sci. **39**, 1931–1955 (2004). doi:10.1111/j.1945-5100.2004.tb00088.x

Krot, A.N., Amelin, Y., Cassen, P., Meibom, A.: Young chondrules in CB chondrites from a giant impact in the early Solar System. Nature **436**, 989–992 (2005a). doi:10.1038/nature03830

Krot, A.N., Yurimoto, H., Hutcheon, I.D., MacPherson, G.J.: Chronology of the early Solar System from chondrule-bearing calcium-aluminium-rich inclusions. Nature **434**, 998–1001 (2005b). doi:10.1038/nature03470

Krot, A.N., Libourel, G., Chaussidon, M.: Oxygen isotope compositions of chondrules in CR chondrites. Geochim. Cosmochim. Acta **70**, 767–779 (2006). doi:10.1016/j.gca.2005.08.028

Krot, A.N., Ivanova, M.A., Ulyanov, A.A.: Chondrules in the CB/CH-like carbonaceous chondrite isheyevo: evidence for various chondrule-forming mechanisms and multiple chondrule generations. Chem. Erde **67**, 283–300 (2007). doi:10.1016/j.chemer.2006.04.001. http://www.sciencedirect.com/science/article/pii/S0009281906000213

Krot, A.N., et al.: Origin and chronology of chondritic components: a review. Geochim. Cosmochim. Acta **73**, 4963 (2009)

Krot, A.N., Nagashima, K., van Kooten, E.M.M.E., Bizzarro, M.: Calcium–aluminum-rich inclusions recycled during formation of porphyritic chondrules from CH carbonaceous chondrites. Geochim. Cosmochim. Acta **201**, 185–223 (2017). doi:10.1111/j.1945-5100.2002.tb00805.x

Kruijer, T.S., Touboul, M., Fischer-Gödde, M., Bermingham, K.R., Walker, R.J., Kleine, T.: Protracted core formation and rapid accretion of protoplanets. Science **344**, 1150–1154 (2014). doi:10.1126/science.1251766

Kunihiro, T., Rubin, A.E., McKeegan, K.D., Wasson, J.T.: Initial ^{26}Al/ ^{27}Al in carbonaceous-chondrite chondrules: too little ^{26}Al to melt asteroids. Geochim. Cosmochim. Acta **68**, 2947–2957 (2004). doi:10.1016/j.gca.2004.02.006

Kurahashi, E., Kita, N.T., Nagahara, H., Morishita, Y.: ^{26}Al- ^{26}Mg systematics of chondrules in a primitive CO chondrite. Geochim. Cosmochim. Acta **72**, 3865–3882 (2008). doi:10.1016/j.gca.2008.05.038

Lambrechts, M., Johansen, A.: Rapid growth of gas-giant cores by pebble accretion. Astron. Astrophys. **544**, A32 (2012). doi:10.1051/0004-6361/201219127, 1205.3030

Lambrechts, M., Johansen, A., Morbidelli, A.: Separating gas-giant and ice-giant planets by halting pebble accretion. Astron. Astrophys. **572**, A35 (2014). doi:10.1051/0004-6361/201423814, 1408.6087

Larsen, K.K., Trinquier, A., Paton, C., Schiller, M., Wielandt, D., Ivanova, M.A., Connelly, J.N., Nordlund, Å., Krot, A.N., Bizzarro, M.: Evidence for magnesium isotope heterogeneity in the solar protoplanetary disk. Astrophys. J. Lett. **735**, L37 (2011). doi:10.1088/2041-8205/735/2/L37

Larsen, K.K., Schiller, M., Bizzarro, M.: Accretion timescales and style of asteroidal differentiation in an ^{26}Al-poor protoplanetary disk. Geochim. Cosmochim. Acta **176**, 295–315 (2016). doi:10.1016/j.gca.2015.10.036

Libourel, G., Krot, A.N.: Evidence for the presence of planetesimal material among the precursors of magnesian chondrules of nebular origin. Earth Planet. Sci. Lett. **254**, 1–8 (2007). doi:10.1016/j.epsl.2006.11.013

Luu, T.H., Young, E.D., Gounelle, M., Chaussidon, M.: Short time interval for condensation of high-temperature silicates in the solar accretion disk. Proc. Natl. Acad. Sci. **112**(5), 1298–1303 (2015). doi:10.1073/pnas.1414025112. http://www.pnas.org/content/112/5/1298

MacPherson, G.J., Kita, N.T., Ushikubo, T., Bullock, E.S., Davis, A.M.: Well-resolved variations in the formation ages for Ca–Al-rich inclusions in the early Solar System. Earth Planet. Sci. Lett. **331–332**, 43–54 (2012). doi:10.1016/j.epsl.2012.03.010. http://www.sciencedirect.com/science/article/pii/S0012821X12001240

Makide, K., Nagashima, K., Krot, A.N., Huss, G.R., Hutcheon, I.D., Bischoff, A.: Oxygen- and magnesium-isotope compositions of calcium-aluminum-rich inclusions from CR2 carbonaceous chondrites. Geochim. Cosmochim. Acta **73**, 5018–5050 (2009). doi:10.1016/j.gca.2009.01.042

McKeegan, K.D., Chaussidon, M., Robert, F.: Incorporation of short-lived ^{10}Be in a calcium-aluminum-rich inclusion from the allende meteorite. Science **289**, 1334–1337 (2000). doi:10.1126/science.289.5483.1334

McKeegan, K.D., Kallio, A.P.A., Heber, V.S., Jarzebinski, G., Mao, P.H., Coath, C.D., Kunihiro, T., Wiens, R.C., Nordholt, J.E., Moses, R.W., Reisenfeld, D.B., Jurewicz, A.J.G., Burnett, D.S.: The oxygen isotopic composition of the Sun inferred from captured solar wind. Science **332**, 1528 (2011). doi:10.1126/science.1204636

McNally, C.P., Hubbard, A., Mac Low, M.M., Ebel, D.S., D'Alessio, P.: Mineral processing by short circuits in protoplanetary disks. Astrophys. J. Lett. **767**, L2 (2013). doi:10.1088/2041-8205/767/1/L2, 1301.1698

McNaughton, N.J., Borthwick, J., Fallick, A.E., Pillinger, C.T.: Deuterium/hydrogen ratios in unequilibrated ordinary chondrites. Nature **294**, 639–641 (1981). doi:10.1038/294639a0

Morris, M.A., Boley, A.C., Desch, S.J., Athanassiadou, T.: Chondrule formation in bow shocks around eccentric planetary embryos. Astrophys. J. **752**, 27 (2012). doi:10.1088/0004-637X/752/1/27, 1204.0739

Müller, T.W.A., Kley, W.: Modelling accretion in transitional disks. Astron. Astrophys. **560**, A40 (2013). doi:10.1051/0004-6361/201322503, 1310.4398

Nagashima, K., Krot, A.N., Huss, G.R.: Oxygen-isotope compositions of chondrule phenocrysts and matrix grains in Kakangari K-grouplet chondrite: implication to a chondrule-matrix genetic relationship. Geochim. Cosmochim. Acta **151**, 49–67 (2015). doi:10.1016/j.gca.2014.12.012

Nesvorný, D., Jenniskens, P., Levison, H.F., Bottke, W.F., Vokrouhlický, D., Gounelle, M.: Cometary origin of the zodiacal cloud and carbonaceous micrometeorites. Implications for hot debris disks. Astrophys. J. **713**, 816–836 (2010). doi:10.1088/0004-637X/713/2/816, 0909.4322

Norris, T.L., Gancarz, A.J., Rokop, D.J., Thomas, K.W.: Half-life of ^{26}Al. J. Geophys. Res. **88**, B331–B333 (1983). doi:10.1029/JB088iS01p0B331

Olsen, M.B., Schiller, M., Krot, A.N., Bizzarro, M.: Magnesium isotope evidence for single stage formation of CB chondrules by colliding planetesimals. Astrophys. J. Lett. **776**, L1 (2013). doi:10.1088/2041-8205/776/1/L1

Olsen, M.B., Wielandt, D., Schiller, M., Van Kooten, E.M.M.E., Bizzarro, M.: Magnesium and ^{54}Cr isotope compositions of carbonaceous chondrite chondrules - Insights into early disk processes. Geochim. Cosmochim. Acta **191**, 118–138 (2016). doi:10.1016/j.gca.2016.07.011

Oulton, J., Humayun, M., Fedkin, A., Grossman, L.: Chemical evidence for differentiation, evaporation and recondensation from silicate clasts in Gujba. Geochim. Cosmochim. Acta **177**, 254–274 (2016). doi:10.1016/j.gca.2016.01.008

Palme, H., Hezel, D.C., Ebel, D.S.: The origin of chondrules: constraints from matrix composition and matrix-chondrule complementarity. Earth Planet. Sci. Lett. **411**, 11–19 (2015). doi:10.1016/j.epsl.2014.11.033

Paton, C., Schiller, M., Bizzarro, M.: Identification of an ^{84}Sr-depleted carrier in primitive meteorites and implications for thermal processing in the solar protoplanetary disk. Astrophys. J. Lett. **763**, L40 (2013). doi:10.1088/2041-8205/763/2/L40

Qin, L., Alexander, C.M.O., Carlson, R.W., Horan, M.F., Yokoyama, T.: Contributors to chromium isotope variation of meteorites. Geochim. Cosmochim. Acta **74**, 1122–1145 (2010). doi:10.1016/j.gca.2009.11.005

Regelous, M., Elliott, T., Coath, C.D.: Nickel isotope heterogeneity in the early Solar System. Earth Planet. Sci. Lett. **272**, 330–338 (2008). doi:10.1016/j.epsl.2008.05.001

Reipurth, B., Bally, J.: Herbig-Haro flows: probes of early stellar evolution. Annu. Rev. Astron. Astrophys. **39**, 403–455 (2001). doi:10.1146/annurev.astro.39.1.403

Robert, F., Epstein, S.: The concentration and isotopic composition of hydrogen, carbon and nitrogen in carbonaceous meteorites. Geochim. Cosmochim. Acta **46**, 81–95 (1982). doi:10.1016/0016-7037(82)90293-9

Robert, F., Javoy, M., Halbout, J., Dimon, B., Merlivat, L.: Hydrogen isotope abundances in the solar system. I - Unequilibrated chondrites. II - Meteorites with terrestrial-like D/H ratio. Geochim. Cosmochim. Acta **51**, 1787–1805 (1987). doi:10.1016/0016-7037(87)90170-0

Romanova, M.M., Ustyugova, G.V., Koldoba, A.V., Lovelace, R.V.E.: Launching of conical winds and axial jets from the disc-magnetosphere boundary: axisymmetric and 3D simulations. Mon. Not R. Astron. Soc. **399**, 1802–1828 (2009). doi:10.1111/j.1365-2966.2009.15413.x, 0907.3394

Rudraswami, N.G., Ushikubo, T., Nakashima, D., Kita, N.T.: Oxygen isotope systematics of chondrules in the Allende CV3 chondrite: high precision ion microprobe studies. Geochim. Cosmochim. Acta **75**, 7596–7611 (2011). doi:10.1016/j.gca.2011.09.035

Russell, S.S., Krot, A.N., Huss, G.R., Keil, K., Itoh, S., Yurimoto, H., MacPherson, G.J.: The genetic relationship between refractory inclusions and chondrules. In: Krot, A.N., Scott, E.R.D., Reipurth, B. (eds.) Chondrites and the Protoplanetary Disk. Astronomical Society of the Pacific Conference Series, vol. 341, p. 317. Astronomical Society of the Pacific, San Francisco (2005)

Salmeron, R., Ireland, T.R.: Formation of chondrules in magnetic winds blowing through the proto-asteroid belt. Earth Planet. Sci. Lett. **327**, 61–67 (2012). doi:10.1016/j.epsl.2012.01.033

Sanders, I.S., Scott, E.R.D.L.: The origin of chondrules and chondrites: debris from low-velocity impacts between molten planetesimals? Meteorit. Planet. Sci. **47**, 2170–2192 (2012). doi:10.1111/maps.12002

Schiller, M., et al.: Rapid timescales for magma ocean crystallization on the howardite-eucrite-diogenite parent body. Astrophys. J. Lett. **740**, L22 (2011). doi:10.1088/2041-8205/740/1/L22

Schiller, M., Connelly, J.N., Glad, A.C., Mikouchi, T., Bizzarro, M.: Early accretion of protoplanets inferred from a reduced inner solar system ^{26}Al inventory. Earth Planet. Sci. Lett. **420**, 45–54 (2015a). doi:10.1016/j.epsl.2015.03.028

Schiller, M., Paton, C., Bizzarro, M.: Evidence for nucleosynthetic enrichment of the protosolar molecular cloud core by multiple supernova events. Geochim. Cosmochim. Acta **149**, 88–102 (2015b). doi:10.1016/j.gca.2014.11.005. http://www.sciencedirect.com/science/article/pii/S001670371400667X

Schrader, D.L., Nagashima, K., Krot, A.N., Ogliore, R.C., Hellebrand, E.: Variations in the O-isotope composition of gas during the formation of chondrules from the CR chondrites. Geochim. Cosmochim. Acta **132**, 50–74 (2014). doi:10.1016/j.gca.2014.01.034

Scott, E.R.D.: Chondrites and the protoplanetary disk. Annu. Rev. Earth Plan. Sci. **35**, 577 (2007). doi:10.1146/annurev.earth.35.031306.140100. http://adsabs.harvard.edu/abs/2007AREPS..35..577S

Scott, E.R.D., Krot, A.N.: Chondrites and Their Components, p. 143. Elsevier, Amsterdam (2005)

Sheikhnezami, S., Fendt, C., Porth, O., Vaidya, B., Ghanbari, J.: Bipolar jets launched from magnetically diffusive accretion disks. I. ejection efficiency versus field strength and diffusivity. Astrophys. J. **757**, 65 (2012). doi:10.1088/0004-637X/757/1/65, 1207.6086

Shu, F.H., Shang, H., Lee, T.: Toward an astrophysical theory of chondrites. Science **271**, 1545–1552 (1996). doi:10.1126/science.271.5255.1545

Shu, F.H., Shang, H., Glassgold, A.E., Lee, T.: X-rays and fluctuating X-winds from protostars. Science **277**, 1475–1479 (1997). doi:10.1126/science.277.5331.1475

Soulié, C., Libourel, G., Tissandier, L., Hiver, J.M.: Kinetics of olivine dissolution in chondrule melts: an experimental study. In: Lunar and Planetary Science Conference, vol 43, p. 1840 (2012)

Tenner, T.J., Nakashima, D., Ushikubo, T., Kita, N.T., Weisberg, M.K.: Oxygen isotope ratios of FeO-poor chondrules in CR3 chondrites: influence of dust enrichment and H_2O during chondrule formation. Geochim. Cosmochim. Acta **148**, 228–250 (2015). doi:10.1016/j.gca.2014.09.025

Tissot, F.L.H., Dauphas, N., Grossman, L.: Origin of uranium isotope variations in early solar nebula condensates. Sci. Adv. **2**, e1501400–e1501400 (2016). doi:10.1126/sciadv.1501400, 1603.01780

Trinquier, A., Birck, J.L., Allègre, C.J.: Widespread ^{54}Cr Heterogeneity in the Inner Solar System. Astrophys. J. **655**, 1179–1185 (2007). doi:10.1086/510360

Trinquier, A., Elliott, T., Ulfbeck, D., Coath, C., Krot, A.N., Bizzarro, M.: Origin of nucleosynthetic isotope heterogeneity in the solar protoplanetary disk. Science **324**, 374 (2009). doi:10.1126/science.1168221

Ushikubo, T., Nakashima, D., Kimura, M., Tenner, T.J., Kita, N.T.: Contemporaneous formation of chondrules in distinct oxygen isotope reservoirs. Geochim. Cosmochim. Acta **109**, 280–295 (2013). doi:10.1016/j.gca.2013.01.045

Van Kooten, E.M.M.E., Wielandt, D., Schiller, M., Nagashima, K., Thomen, A., Larsen, K.K., Olsen, M.B., Nordlund, A., Krot, A.N., Bizzarro, M.: Isotopic evidence for primordial molecular cloud material in metal-rich carbonaceous chondrites. Proc. Natl. Acad. Sci. USA **113**(8), 2011–2016 (2016). doi:10.1073/pnas.1518183113. http://www.pnas.org/content/113/8/2011

Villeneuve, J., Chaussidon, M., Libourel, G.: Homogeneous distribution of ^{26}Al in the Solar System from the Mg isotopic composition of chondrules. Science **325**(5943), 985–988 (2009). doi:10.1126/science.1173907. http://www.sciencemag.org/cgi/content/abstract/325/5943/985

Walsh, K.J., Morbidelli, A., Raymond, S.N., O'Brien, D.P., Mandell, A.M.: A low mass for Mars from Jupiter's early gas-driven migration. Nature **475**, 206–209 (2011). doi:10.1038/nature10201, 1201.5177

Warren, P.H.: Stable-isotopic anomalies and the accretionary assemblage of the Earth and Mars: a subordinate role for carbonaceous chondrites. Earth Planet. Sci. Lett. **311**, 93–100 (2011). doi:10.1016/j.epsl.2011.08.047

Wasserburg, G.J., Busso, M., Gallino, R., Nollett, K.M.: Short-lived nuclei in the early solar system: possible AGB sources. Nucl. Phys. A **777**, 5–69 (2006). doi:10.1016/j.nuclphysa.2005.07.015, astro-ph/0602551

Weidenschilling, S.J.: Aerodynamics of solid bodies in the solar nebula. Mon. Not. R. Astron. Soc. **180**, 57–70 (1977). doi:10.1093/mnras/180.1.57

Weisberg, M.K., Prinz, M., Clayton, R.N., Mayeda, T.K., Grady, M.M., Pillinger, C.T.: The CR chondrite clan. Antarct. Meteorite Res. **8**, 11 (1995)

Weiss, B.P., Elkins-Tanton, L.T.: Differentiated planetesimals and the parent bodies of chondrites. Annu. Rev. Earth Plan. Sci. **41**, 529–560 (2013). doi:10.1146/annurev-earth-040610-133520

Williams, J.P., Cieza, L.A.: Protoplanetary disks and their evolution. Annu. Rev. Astron. Astrophys. **49**, 67 (2011). doi:10.1146/annurev-astro-081710-102548;

Yamakawa, A., Yamashita, K., Makishima, A., Nakamura, E.: Chromium isotope systematics of achondrites: chronology and isotopic heterogeneity of the inner solar system bodies. Astrophys. J. **720**, 150–154 (2010). doi:10.1088/0004-637X/720/1/150

Chapter 7
The Emerging Paradigm of Pebble Accretion

Chris W. Ormel

Abstract Pebble accretion is the mechanism in which small particles ("pebbles") accrete onto big bodies (planetesimals or planetary embryos) in gas-rich environments. In pebble accretion, accretion occurs by settling and depends only on the mass of the gravitating body, not its radius. I give the conditions under which pebble accretion operates and show that the collisional cross section can become much larger than in the gas-free, ballistic, limit. In particular, pebble accretion requires the pre-existence of a massive planetesimal seed. When pebbles experience strong orbital decay by drift motions or are stirred by turbulence, the accretion efficiency is low and a great number of pebbles are needed to form Earth-mass cores. Pebble accretion is in many ways a more natural and versatile process than the classical, planetesimal-driven paradigm, opening up avenues to understand planet formation in solar and exoplanetary systems.

7.1 Introduction

The goal of this chapter is to present a physically motivated understanding of pebble accretion, elucidating the role of the disk, planet, and pebble properties, and to present the conditions for which pebble accretion becomes a viable mechanism to form planets. In this work, I discuss pebble accretion from a local perspective—a planet situated at some distance from its star—and do not solve for the more formidable global problem (planet migration or the evolution of the pebble disk). However, clear conclusions can already be obtained from the local approach.

The plan of this chapter is as follows. In Sect. 7.1 I outline what is understood by pebble accretion. In Sect. 7.2 order-of-magnitude expressions for pebble accretion are derived, which are applied in Sect. 7.3 to address the question under which conditions pebble accretion is a viable mechanism. Section 7.4 highlights some applications.

C.W. Ormel (✉)
Anton Pannekoek Institute, University of Amsterdam, Amsterdam, Netherlands
e-mail: c.w.ormel@uva.nl

M. Pessah, O. Gressel (eds.), *Formation, Evolution, and Dynamics of Young Solar Systems*, Astrophysics and Space Science Library 445,
DOI 10.1007/978-3-319-60609-5_7

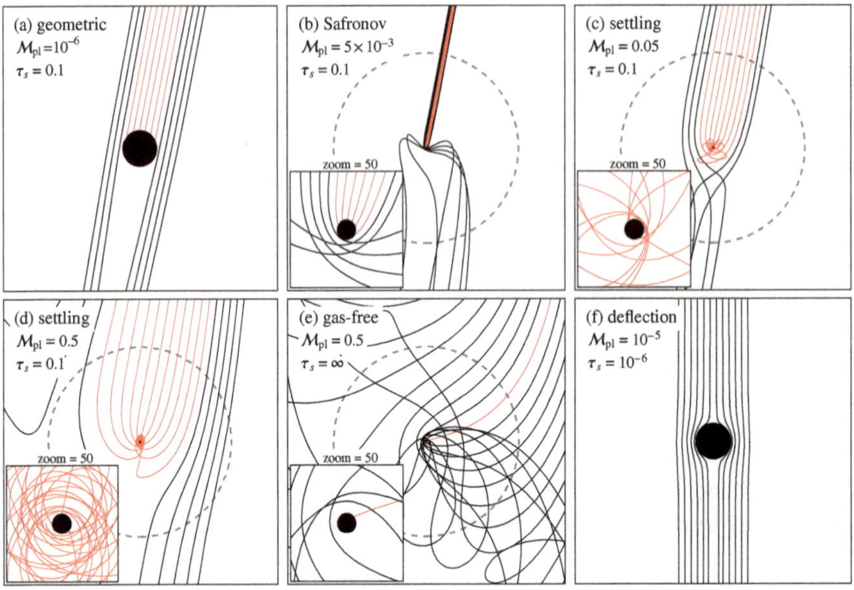

Fig. 7.1 Examples of planet-pebble interactions, viewed in the frame co-moving with the planet. In each panel the *filled circle* denotes the physical size of the planet and the *dashed circle* its Hill sphere. Pebbles, characterized by a dimensionless stopping time τ_s, enter from the *top*, because of the sub-Keplerian motion of the gas (Sect. 7.1.2). The planet mass is given in terms of M_t [Eq. (7.10)]. Trajectories in *red* accrete. Only (**c**) and (**d**) qualify as pebble accretion, while (**a**), (**b**), and (**e**) fall in the ballistic regime. In (**f**) particles are so small ($\tau_s = 10^{-6}$) that they follow gas streamlines

7.1.1 What Is Pebble Accretion (Not)?

Pebble accretion is a planet formation concept that concerns the accretion of small particles (pebbles) of negligible gravitational mass onto large, gravitating bodies: planetesimals, protoplanets, or planets.[1] In a more narrow sense, pebble accretion is an accretion process where (gas) drag and gravity play major roles. Simply put, this means that the pebble has to be *aerodynamically* small and the planet to be *gravitationally* large.

Examples of particle-planet encounters best illustrate the concept. In Fig. 7.1 several encounters are plotted for pebbles of aerodynamical size τ_s and planet mass $\mathcal{M}_{\rm pl}$, which are dimensionless quantities (their formal definition is given later in Sects. 7.1.2 and 7.2.2, respectively). In (a) the small gravitational mass hardly perturbs the pebble trajectory. Consequently, the collisional cross section is similar to the geometric cross section, $\pi R_{\rm pl}^2$, where $R_{\rm pl}$ is the radius of the gravitating body.

[1]In this work I will simply refer to the large body as "planet.'

In (b), where the planetesimal mass corresponds to a body of radius $R_{\mathrm{pl}} \approx 100\,\mathrm{km}$, gravitational effects become more significant. Trajectories focus, resulting in a cross section larger than geometrical. The hyperbolic shape of the close encounters (see inset) strongly resembles those of the well-known planet-planetesimal encounters (Safronov 1969; Wetherill 1980). Similarly, in Fig. 7.1b gas-drag close to the body is of little importance, because the encounters proceed fast. However, on longer times gas drag does re-align the pebble with the gas flow.

In Fig. 7.1c, where the mass of the gravitating body is increased by merely a factor ten, the situation differs qualitatively from (b). First, the collisional cross section has increased enormously: it already is a good fraction of the Hill sphere. Second, the way how the pebbles are accreted is very different from (b). Pebbles often revolve the planet several times, before finally accreting (see inset). Indeed, where in (a) and (b) accretion relies on the physical size of the gravitating body, this is no longer the case in (c). Even when the physical radius would shrink to zero (i.e., a true point particle) the collisional cross section would be exactly the same, because pebbles simply *settle* down the potential well. Therefore:

> Pebble accretion is characterized by settling of particles down the gravitational well of the planet. Pebble accretion only depends on the mass of the gravitating body, not its radius. It is further characterized by the absence of close, collisionless, encounters.

7.1.2 *Aerodynamically Small and Large*

Only particles tightly coupled to the gas qualify for pebble accretion. The level of coupling of a particle to the gas is customarily expressed in terms of the stopping time:

$$t_{\mathrm{stop}} = \frac{mv}{F_D}, \tag{7.1}$$

where m the mass of the particle, v its relative velocity with the gas, and F_D the gas drag law. The stopping time is simply the time needed for gas drag to align the motion of the particle to that of the gas. For example, a particle falling in a gravitational field g will attain its equilibrium (settling) velocity after a time t_{stop} at which point $g = F_D/m$ or v_{settl} (the settling velocity) $v_{\mathrm{settl}} = g t_{\mathrm{stop}}$. In general, F_D depends on velocity in a non-trivial way, but for pebble-size particles under disk conditions we typically have that F_D is linear in v. This makes t_{stop} a function of the physical properties of the pebble (its size s and internal density $\rho_\bullet$) and that of the gas, but independent of velocity. For example, in the Epstein regime we simply have $t_{\mathrm{stop}} = \rho_\bullet s / v_{\mathrm{th}} \rho_{\mathrm{gas}}$ where v_{th} is the mean thermal velocity of the gas and ρ_{gas} the gas density. A natural definition of "aerodynamical small" is that the stopping time is small in comparison to the inverse orbital frequency, Ω_K^{-1} or $\tau_s = \Omega_K t_{\rm stop} < 1$. "Heavy" bodies have $\tau_s \gg 1$ and move on Kepler orbits. For them $t_{\rm stop}$ is the time needed to damp their eccentric motions.

Formally, the steady-state solution to the equations of motions for a particle of arbitrary τ_s, accounting for gas drag and pressure forces, read (Whipple 1972; Weidenschilling 1977a; Nakagawa et al. 1986):

$$v_r = -\frac{2v_{\rm hw}\tau_s}{1+\tau_s^2} \equiv -v_{\rm drift}, \tag{7.2}$$

$$v_\phi = v_K - \frac{v_{\rm hw}}{1+\tau_s^2}, \tag{7.3}$$

where v_r is the radial motion, v_ϕ the azimuthal, v_K the Keplerian velocity and $v_{\rm hw}$, the disk headwind, the velocity offset between the gas and the Keplerian motion. Rotation is slower than Keplerian because the disk is partially pressure-supported:

$$v_{\rm hw} = -\frac{1}{2}(h_{\rm gas}/r)^2 v_K \nabla_{\log}P, \tag{7.4}$$

$$= 54\,{\rm m\,s^{-1}}\,\frac{T_1}{300\,{\rm K}}\left(\frac{\mu}{2.34}\right)^{-1}\left(\frac{M_\star}{M_\odot}\right)^{-1/2}\frac{(-\nabla_{\log}P)}{3}\left(\frac{r}{\rm AU}\right)^{(1/2-q)}, \tag{7.5}$$

where $h_{\rm gas}$ is the gas scale height, $\nabla_{\log}P = \partial \log P/\partial \log r$ the logarithmic pressure gradient, μ the mean molecular weight, T_1 the temperature at 1 au, and q the corresponding power-law index (as in $T \propto r^{-q}$). Because in many disk models $q \approx 1/2$ e.g., as obtained from a passively irradiated disk (Chiang and Goldreich 1997), we obtain that $v_{\rm hw}$ is a disk constant. For pebble accretion, the value of $v_{\rm hw}$ and its (possible) spatial and temporal variations are key parameters.

From Eq. (7.3) it follows that large bodies ($\tau_s \gg 1$) move on circular orbits ($v_r = 0$ and $v_\phi = v_K$) whereas small particles ($\tau_s < 1$), moving with the gas, have their azimuthal velocities reduced by $v_{\rm hw}$ with respect to the Keplerian motion. The large body hence overtakes these particles. From its perspective, the particles arrive from the front at velocities $\approx v_{\rm hw}$.

7.1.2.1 Pebbles

In this chapter, I consider any particle of $\tau_s < 1$ aerodynamically small. An (imprecise) lower limit may be added to the definition to distinguish drifting pebbles from "inert" dust. Usually, our definition of "pebble" then refers to particles of $10^{-3\ldots-2} \leq \tau_s \lesssim 1$. From Eq. (7.2), it is clear that these particles (indeed all particles around $\tau_s \sim 1$) have significant radial drift motions (This explains the slant seen in the incoming particle flow of the $\tau = 0.1$ pebbles of Fig. 7.1). It also implies that pebbles are constantly replenished: they are lost to the inner disk, but drift in from the outer disk. Pebble accretion, in contrast to planetesimal accretion, is therefore a global phenomenon: one has to consider the evolution of the dust population throughout the entire disk (Birnstiel et al. 2010; Okuzumi et al. 2012; Testi et al. 2014).

7.1.2.2 Planets

In our context a "planet" is any body with $\tau_s \gg 1$ moving on a circular orbit, large enough for gravity to become important.

7.1.3 The Case for Pebble Accretion

The case for pebble accretion can be made either from an observational or theoretical perspective. Observationally, pebbles are the particles inferred to be responsible for the emission seen at radio wavelengths in young disks. The argument is that the thermal emission is optically thin and is therefore proportional to the opacity of the emitting material, κ—a property of the dust grains. For a reasonable estimate of the temperature, the ratio in flux density at two wavelengths—the spectral energy index—translates into a ratio in opacity. Knowing the opacity in turn constrains the size of the emitting grains (or the maximum grain size if one considers a distribution). For example, particles much larger than the wavelength $\kappa(\lambda)$ can be expected to be wavelength independent (grey absorption), whereas for small grains emission at wavelengths much longer than their size is suppressed (Rayleigh regime). The observed spectral index translates into a size of the grains that carry most of the mass. Typically, mm- to cm-particles emerge from this spectral index analysis (e.g. Natta et al. 2007; Pérez et al. 2015). Another indication for the presence of pebble-size (drifting) particles is that disks are found to be more compact in the continuum than in the gas (Andrews et al. 2012; Panić et al. 2009; Cleeves et al. 2016).

The inferred pebble size also depends on their composition and internal structure (filling factor) of the particles; porous aggregates will result in a larger size (Ormel et al. 2007; Okuzumi et al. 2009). Even greater are the uncertainties in the *total* amount of mass in pebbles, because that depends on the absolute values of the opacity and on the assumption that the emitting dust is optically thin. A number of assumptions enter the calculation of the opacity; apart from the size, κ can also be affected by porosity (Kataoka et al. 2014), composition, and perhaps temperature (Boudet et al. 2005). (It is somewhat worrying that these systematic uncertainties are rarely highlighted in studies that quote disk masses). Nevertheless, a large amount of pebbles are inferred in this way—ranging from 10^2 to possibly up to 10^3 Earth masses (Ricci et al. 2010b,a; Andrews et al. 2013; Pérez et al. 2015). At these levels, it is hard to imagine that planetesimals (which cannot be directly observed) would yet dominate the solid mass budget. Therefore, from an observational perspective, it can well be argued that pebbles are planets' primary building blocks.

Theoretically, the case for pebble accretion arises from the drawbacks of the classical, planetesimal-driven, model. In the inner disk growth is severely restricted because of the low isolation mass $M_{\mathrm{iso,clas}}$ [see, e.g., Lissauer 1987; Kokubo and Ida 2000 and Eq. (7.18)], which limits the mass of the planetary embryos to that of Mars.[2] In the outer disk $M_{\mathrm{iso,clas}}$ is sufficiently large, but here the problem is that growth is slow. First, planetesimal-driven accretion requires extremely quiescent disk to trigger runaway growth, which is already doubtful in case of moderate turbulent excitation (Nelson and Gressel 2010; Ormel and Okuzumi 2013). The second, more fundamental, problem is that planetesimal-driven growth suffers from negative feedback: a larger embryo entails a more excited planetesimal population, increasing the relative velocities of the encounters and suppressing the gravitationally focused collisional cross sections (e.g. Kokubo and Ida 2002). Unless the disk is unusually massive, this quickly suppresses embryo growth beyond $\sim$5 AU. It has been argued that collisional fragmentation would help to suppress eccentricities (and inclinations), either by gas or collisional damping (Wetherill and Stewart 1989; Goldreich et al. 2004; Fortier et al. 2013). However, a planet will carve a gap in a disk of low-e particles (i.e., when $\tau_s > 1$ and $e \approx 0$), preventing efficient accretion of planetesimals (Levison et al. 2010). Particle gaps may be avoided for aerodynamically smaller fragments, but then one really needs to address the orbital decay of this material (Kobayashi et al. 2011). In any case, when strong collisional diminution in the presence of gas has ground down planetesimals to particles of $\tau_s < 1$, encounters enter the pebble accretion regime (Chambers 2014).

7.1.4 Misconceptions About Pebble Accretion

In closing, I list a number of misconceptions about pebble accretion:

1. *Pebble accretion involves pebbles.* Geologists define pebbles as particles of diameter between 2 and 64 mm (e.g. Williams et al. 2006). Our definition of pebble is aerodynamical: particles of stopping time below $\tau_s = 1$. Therefore, in gas-rich environments, pebble accretion can take place over a very wide spectrum of particle sizes: from meter-size boulders to micron-size dust grain. Conversely, accretion of millimeter-size particles in a gas-free medium does not qualify as pebble accretion.
2. *Pebble accretion is a planetesimal formation mechanism.* Pebble accretion describes the process of accreting small particles on a gravitating body, e.g.

[2] A related problem is that the classical theory dictates a steep gradient in embryo mass (more massive embryos at larger r), which, for the solar system, is very hard to comprehend (Morbidelli et al. 2015).

a (big) planetesimal. How planetesimals themselves form is a different topic. Recent popular models hypothesize that planetesimals could form from a population of pebble-sized particles (Youdin and Goodman 2005; Johansen et al. 2007; Cuzzi et al. 2008). Planetesimal formation and pebble accretion can therefore operate sequentially. In that case, the key question is whether, say, streaming instability produces planetesimals large enough to trigger pebble accretion.

3. *Pebble accretion is inevitable.* It is sometimes alleged that the mere presence of a large reservoir of pebble-sized particles is sufficient to trigger pebble accretion. This is not the case. For pebble accretion, a sufficiently massive seed is needed as otherwise encounters will not fall in the settling regime. Formally, pebble accretion must satisfy the settling condition (see Sect. 7.2.1).
4. *Pebble accretion is fast.* It is true that pebble accretion is characterized by large collisional cross sections (Fig. 7.1c, d). However, the radial orbital decay of pebbles potentially renders the process inefficient: most pebbles simply cross the planet's orbit, without experiencing any interaction. Therefore, pebble accretion depends on the pebble mass flux and, in particular, on how many pebbles are globally available. The low efficiency problem is especially severe when pebbles do not reside in a thin layer, i.e., for turbulent disks.

7.2 The Physics of Pebble Accretion

In this section I outline the key requirements for pebble accretion and derive order of magnitude expressions for pebble accretion rates based on timescales analysis. These expressions are, within orders of unity, consistent with recent works (Ormel and Klahr 2010; Ormel and Kobayashi 2012; Lambrechts and Johansen 2012, 2014; Guillot et al. 2014; Ida et al. 2016).

7.2.1 Requirements and Key Expressions

An intuitive understanding of pebble accretion can be obtained from the timescales involved in an encounter between a gravitating body (planetesimal, planet) and a test particle. These are (see Fig. 7.2):

- The *encounter time* $t_{\rm enc}$. The duration of the encounter or the time over which the particle experiences most of the gravitational force. It is given by $t_{\rm enc} = 2b/v_\infty$, where v_∞ is the (unperturbed) velocity at impact parameter b.
- The *settling time* $t_{\rm settl}$. The time needed for a particle to sediment to the planet. The settling time is evaluated at the minimum distance b of the unperturbed encounter. There, the settling velocity reads $v_{\rm settl} = g_{\rm pl} t_{\rm stop}$ where the planet acceleration $g_{\rm pl}$ is evaluated at b. Hence $t_{\rm settl} = b/v_{\rm settl} = b^3/GM_{\rm pl}t_{\rm stop}$.
- The *stopping time*, $t_{\rm stop}$. The aerodynamical size of the pebble.

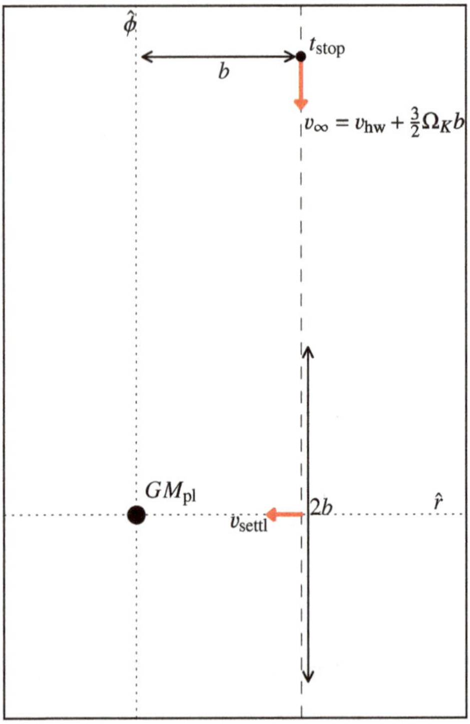

Fig. 7.2 Sketch of the pebble-planet interaction, viewed in the frame of the circularly moving planet (*center*). Pebbles, characterized by their aerodynamical size or stopping time, $t_{\rm stop}$, approach a planet of gravitational mass $GM_{\rm pl}$ at impact parameter b. The relative velocity, v_∞, is given by a combination of the disk headwind, $v_{\rm hw}$, and the Keplerian shear. The unperturbed trajectory is indicated by the *dashed line*. Key timescales are the duration of the encounter, $t_{\rm enc} = 2b/v_\infty$, the settling timescale, $t_{\rm settl} = b/v_{\rm settl}$, where $v_{\rm settl}$ is the sedimentation velocity evaluated at closest approach, and the stopping time, $t_{\rm stop}$, of the pebble

Settling Condition The interaction will operate in the settling regime when:

1. The encounter is long enough for particles to couple to the gas during the encounter $t_{\rm stop} < t_{\rm enc}$; and
2. The encounter is long enough for particles to settle to the planet, $t_{\rm settl} < t_{\rm enc}$.

These conditions can also be combined into $t_{\rm stop} + t_{\rm settl} < t_{\rm enc}$.

When either of the above does not materialize, there are no settling encounters and (according to our definition) there is no pebble accretion. The first condition expresses that gas drag matters during the interaction, which is where pebble accretion differs from planetesimal accretion. The second condition tells whether pebbles can actually sediment to the planet.

It is clear that a massive planet promotes settling, since the second condition becomes easier to fulfill and the first condition works better at larger b in any case. However, regarding the particle size the conditions work against each other. Very small particles (small $t_{\rm stop}$) are always well coupled to the gas (condition 1), yet it can take them too long to settle (condition 2), unless b is small. On the other hand, large particles (large $t_{\rm stop}$) satisfy the second condition at larger impact parameter, but they may nevertheless not qualify as "settling," because they fail to meet condition 1.

7.2.2 *Pebble Accretion Regimes*

In order to derive the pebble accretion rates, the following strategy is employed. First, by equating $t_{\rm settl}$ and $t_{\rm enc}$ the largest impact parameter ($b_{\rm col}$) is found that obeys condition 2. Then, a posteriori, it is verified whether $b_{\rm col}$ also satisfies condition 1.

The relative velocity v_∞ between pebble and planet follows from the headwind and the Keplerian shear:

$$v_\infty \simeq v_{\rm hw} + \frac{3}{2}\Omega_K b. \tag{7.6}$$

Therefore, an impact parameter of $b \sim \frac{2}{3}v_{\rm hw}/\Omega_K$ divides two velocity regimes:

- The shear regime, valid at large b;
- The wind regime, valid when impact parameters are small.

These regimes are referred to as "Hill" and "Bondi," respectively, by Lambrechts and Johansen (2012). Note that in the shear regime encounters last a dynamical timescale, $t_{\rm enc} \sim \Omega_K^{-1}$, meaning that particles $\tau_{\rm s} < 1$ satisfy condition 1.

7.2.2.1 Shear (Hill) Limit

Equating $t_{\rm enc} = \Omega_K^{-1}$ with the settling timescale, we obtain

$$b_{\rm sh} \sim \left(\frac{GM_{\rm pl}t_{\rm stop}}{\Omega_K}\right)^{1/3} \sim \tau_s^{1/3} R_{\rm Hill} \qquad (M_{\rm pl} \gtrsim M_{\rm hw/sh}), \tag{7.7}$$

where $R_{\rm Hill}$ is the Hill radius. For $\tau_s \sim 1$ particles the impact parameter is comparable to the Hill radius, greatly exceeding the gas free limit [see Fig. 7.1e and Eq. (7.15)]. For $\tau_s < 1$ impact cross sections decrease but not by much; particles down to $\tau_s = 0.01$ still accrete at impact parameter $\approx$20% of the Hill radius.

7.2.2.2 Headwind (Bondi) Limit

In the headwind limit, $t_{\rm enc} = 2b/v_{\rm hw} = t_{\rm settl}$ gives

$$b_{\rm hw} \sim \sqrt{\frac{2GM_{\rm pl}t_{\rm stop}}{v_{\rm hw}}} \qquad (M_* \lesssim M_{\rm pl} \lesssim M_{\rm hw/sh}). \tag{7.8}$$

The impact parameter increases as the square root of the stopping time and the planet mass, more steeply than in the shear limit. The transition from the headwind (valid for small $t_{\rm stop}$ or $M_{\rm pl}$) to the shear limit occurs at the mass $M_{\rm hw/sh}$ where $b_{\rm hw} = b_{\rm sh}$:

$$M_{\rm hw/sh} = \frac{v_{\rm hw}^3}{8G\Omega_K^2 t_{\rm stop}} = \frac{1}{8}\frac{M_t}{\tau_s}, \tag{7.9}$$

where

$$M_t = \frac{v_{\rm hw}^3}{G\Omega_K} = 1.6 \times 10^{-3}\, M_\oplus \left(\frac{v_{\rm hw}}{50\ {\rm m\ s^{-1}}}\right)^3 \left(\frac{M_\star}{M_\odot}\right)^{-1/2} \left(\frac{r}{\rm AU}\right)^{3/2} \tag{7.10}$$

is a fiducial mass that measures the relative importance of headwind *vs* shear. Note that M_t is larger in the outer disk.

The headwind regime applies for $M_{\rm pl} \le M_{\rm hw/sh}$. In addition, condition 1 also curtails the validity of the headwind regime. Equating $t_{\rm enc} = 2b_{\rm hw}/v_{\rm hw}$ with $t_{\rm stop}$ it follows that pebble accretion shuts off for $M_{\rm pl} < M_*$ where

$$M_* = \frac{v_{\rm hw}^3 t_{\rm stop}}{8G} = \frac{1}{8} M_t \tau_s. \tag{7.11}$$

Interactions where $M_{\rm pl} < M_*$ follow ballistic trajectories, where accretion relies on hitting the surface of the planet (Fig. 7.1b). In that case the impact parameter can be obtained from the usual Safronov-type gravitational focusing with $v_{\rm hw}$ for the relative velocity, $b_{\rm Saf} \simeq R_{\rm pl} v_{\rm esc}/v_{\rm hw}$ where $v_{\rm esc} = \sqrt{2GM_{\rm pl}/R_{\rm pl}}$ is the surface escape velocity of the planet and $R_{\rm pl}$ its radius.

7.2.2.3 Aerodynamic Deflection

When the gravitational mass of the planetesimal becomes small ($v_{\rm esc} < v_{\rm hw}$) a natural minimum impact parameter is the physical radius $R_{\rm pl}$. However, very small particles, very tightly coupled to the gas, will follow gas streamlines, avoiding accretion (Sekiya and Takeda 2003; Sellentin et al. 2013). This is referred to as aerodynamic deflection and is well known in the literature of, e.g., atmospheric sciences (Slinn 1976). It reduces the collisional cross section below the geometrical limit. The importance of aerodynamic deflection is quantified by the Stokes number[3] ${\rm Stk} = v_{\rm hw} t_{\rm stop}/R_{\rm pl}$. Particles of ${\rm Stk} \ll 1$ avoid accretion as they react to the gas flow on times ($t_{\rm stop}$) smaller than the crossing time of the planetesimal $t_{\rm enc} \sim R_{\rm pl}/v_{\rm hw}$.

[3]This is the only point where I define a Stokes number. In many works the dimensionless stopping τ_s is referred to as the Stokes number, while in the turbulent literature the Stokes number is defined as the ratio between the stopping time and an eddy turnover time.

However for gravitating bodies there is always a channel to accrete particles by settling, because the large $t_{\rm enc}$—a lower limit to $t_{\rm enc}$ is always $\sim R_{\rm pl}/v_{\rm hw}$—enables the settling condition. To zeroth order, Eq. (7.8) still applies, even in cases where $b_{\rm hw} \ll R_{\rm pl}$. A more detailed analysis should account for the modification of the flow pattern in the vicinity of the gravitating body, which depends on the Reynolds number (Johansen et al. 2015; Visser and Ormel 2016).

7.2.3 The Accretion Rate

In both the headwind and the shear regimes, the accretion rate $\dot{M} = \pi b^2 v_\infty \rho_P$, is:

$$\dot{M}_{\rm 3D} \sim 2\pi G M_{\rm pl} t_{\rm stop} \rho_P = 6\pi R_{\rm Hill}^3 \tau_s \Omega_K \rho_P \tag{7.12}$$

(as immediately follows from equating $t_{\rm settl} = t_{\rm enc}$ and solving for $b^2 v_\infty$) where ρ_P is the density in pebbles. There is no transition between the shear and headwind regimes in terms of the accretion rate. However, Eq. (7.12) did assume that the pebbles are spread out in a thick disk; the accretion is 3D. When the pebbles reside in a thin layer, the accretion becomes rather:

$$\dot{M}_{\rm 2D} \sim 2 v_\infty b_{\rm PA} \Sigma_P = \begin{cases} \sqrt{8 G M_{\rm pl} t_{\rm stop} v_{\rm hw}} \Sigma_P & \text{headwind reg.} \\ 2 R_{\rm Hill}^2 \Omega_K \tau_s^{2/3} \Sigma_P & \text{shear reg.} \end{cases} \tag{7.13}$$

where Σ_P is the surface density in pebbles. It is instructive to contrast these rates with the classical expressions for Safronov focusing

$$\dot{M}_{\rm Saf} = \pi \left(\frac{v_{\rm esc}}{v_{\rm hw}} \right)^2 R_{\rm pl}^2 v_{\rm hw} \rho_P = \frac{2\pi R_{\rm pl} G M_{\rm pl}}{v_{\rm hw}} \rho_P, \tag{7.14}$$

assuming that the surface escape velocity $v_{\rm esc} > v_{\rm hw}$ and the 3D limit; and with the gas-free, planar, zero eccentricity limit:

$$\dot{M}_{\rm gas-free} \approx 11 \sqrt{R_{\rm pl} R_{\rm Hill}^3} \Omega_K \Sigma_P \tag{7.15}$$

(Nishida 1983; Ida and Nakazawa 1989).

Although pebble accretion is fast and the rates increases with $M_{\rm pl}$, the rates are not superlinear. If $\dot{M} \propto M^\kappa$ then $\kappa = 1$, $1/2$ and $2/3$ in Eq. (7.12) and Eq. (7.13), respectively. Only Safronov focusing is a runaway growth phenomenon ($\kappa = 4/3$ for constant internal density); pebble accretion is not. However, the transition between Safronov-focusing and pebble accretion proceeds at a super-linear pace, since $\dot{M}_{\rm 3D}$ is usually much larger than $\dot{M}_{\rm Saf}$ [see also Eq. (7.19)].

7.2.4 The Pebble Flux

The pebble accretion rates given above scale with the amount of pebbles that are locally available (Σ_P or ρ_P). Because of their drift this quantity is expected to vary with time. It is useful to express the surface density in terms of the pebble mass flux through the disk $\dot{M}_{\mathrm{P,disk}}$:

$$\Sigma_P = \frac{\dot{M}_{\mathrm{P,disk}}}{2\pi r v_{\mathrm{drift}}(\tau_s)}, \tag{7.16}$$

where I have ignored diffusive transport.

A simple model for the mass flux $\dot{M}_{\mathrm{P,disk}}$ can be obtained from a timescale analysis (Birnstiel et al. 2012; Lambrechts and Johansen 2014). This entails that at any radius r dust grains coagulate into pebbles that start drifting at a size where the growth timescale t_{growth} exceeds the pebble drift timescale $t_{\mathrm{drift}} = r/v_{\mathrm{drift}}$. This results in a mass flux $\dot{M}_{\mathrm{P,disk}} = 2\pi r_0 \Sigma_0 v_{\mathrm{dr,0}}$, where the subscript "0" refers to the radius where the pebbles enter the drift-dominated regime—the pebble front (Lambrechts and Johansen 2014)—and Σ_0 is the initial density in solids. Clearly, $r_0 > r$ with $r_0(t)$ increasing with time, since coagulation proceeds slower in the outer disk.

The drift-limited solution ($t_{\mathrm{grow}} = t_{\mathrm{drift}}$; Birnstiel et al. 2012) also gives the (aerodynamic) size of the pebbles for $r < r_0$ (the region where pebbles drift). Typically, pebble sizes are then $\tau_s \sim 10^{-2}$ (Lambrechts and Johansen 2014). However, it must be emphasized that all of this depends, to considerable extent, on dust coagulation physics—sticking properties, relative velocities, fragmentation threshold, internal structure, etc.; see Johansen et al. (2014) for a review—and also on the structure of the evolving outer disk. In this review I will not adopt a global pebble model, but formulate conclusions based on the local conditions of a growing planetary embryo.

7.2.5 The Pebble Isolation Mass

When the planet mass becomes large it will start to perturb the disk, changing the radial pressure gradient ($\nabla_{\log} P$) in its vicinity. Clearly, when the perturbation becomes non-linear—planets massive enough such that their Hill radius exceeds the scale height of the disk, $R_{\mathrm{Hill}} > h_{\mathrm{gas}}$—a gap will open and a pressure maximum emerges upstream (Lin and Papaloizou 1986). Pebbles then stop their drift at the pressure maximum. The gap opening condition can be rewritten $M_{\mathrm{pl}}/M_\star > (h_{\mathrm{gas}}/r)^3$, which translates into a pebble isolation mass of

$$M_{\mathrm{P,iso}} \sim 40\, M_\oplus \left(\frac{h_{\mathrm{gas}}/r}{0.05}\right)^3. \tag{7.17}$$

for a solar-mass star. In a more detailed analysis, based on radiation hydrodynamical simulations, Lambrechts et al. (2014) and Bitsch et al. (2015a) argue for a numerical pre-factor of $20\,M_\oplus$. For comparison, the classical isolation mass (valid for bodies that do not drift) is (Kokubo and Ida 2000):

$$M_{\rm iso,clas} \sim 0.16\,M_\oplus \left(\frac{\tilde{b}}{10}\right)^{3/2} \left(\frac{\Sigma_{\rm solid}}{10\,{\rm g\,cm^{-2}}}\right)^{3/2} \left(\frac{r}{\rm AU}\right)^{3} \left(\frac{M_\star}{M_\odot}\right)^{-1/2}, \tag{7.18}$$

where $\tilde{b}$ is distance between protoplanets in mutual Hill radii.

The pebble isolation mass is of great importance for the formation of giant planets. Since pebble accretion halts for $M_{\rm pl} > M_{\rm P,iso}$, it sets an upper limit to the heavy elements mass of giant planets. Reaching the pebble isolation mass, however, does not spell an end to giant planet formation. In contrast, it may even accelerate it, because the pre-planetary envelope has lost an important source of accretional heating and opacity. Gas runaway accretion sets in once envelope and core mass are similar (Rafikov 2006) and this may well be triggered soon after the pebble isolation mass is reached.

7.2.6 Summary: Accretion Regimes

Figure 7.3 summarizes the accretion regimes as function of the pebble dimensionless stopping time τ_s (x-axis) and the planet mass (y-axis). For the planet Eq. (7.10) has been used to convert to a dimensionless mass $\mathcal{M}_{\rm pl}$. Lines of $M_{\rm pl} = M_*$ and $M_{\rm pl} = M_{\rm hw/sh}$ are invariant in terms of the dimensionless $(\tau_s, \mathcal{M}_{\rm pl})$, but not in terms of the physical mass. Now consider a small planetesimal (e.g. 1 km) that accretes pebbles of a certain stopping time (e.g. $\tau_s = 0.1$). Initially, it accretes those pebbles with the geometric cross section ($\sigma_{\rm col} \sim \pi R_{\rm pl}^2$), but at:

- $M_{\rm pl} = R_{\rm pl} v_{\rm hw}^2/2G$ ($v_{\rm esc} = v_{\rm hw}$) accretion switches to the so-called Safronov regime, where gravitational focusing enhances the collisional cross section by a factor of $(v_{\rm esc}/v_{\rm hw})^2$. Pebble accretion commences at
- $M_{\rm pl} = M_*$ [Eq. (7.11)], where ballistic accretion gives way to accretion by settling. This transition is abrupt.[4] Specifically, the boost obtained from crossing the $M_{\rm pl} = M_*$ line is:

$$\left(\frac{\dot{M}_{\rm 3D}}{\dot{M}_{\rm Saf}}\right)_{M_{\rm pl}=M_*} \sim \frac{\tau_s^{2/3}}{R_{\rm pl}/R_{\rm Hill}} \sim 100, \tag{7.19}$$

[4]According to the expressions derived above, the transition is discontinuous. In reality, it is just very steep. Numerically, one finds that the cross section for accretion by settling exponentially decreases when $M_{\rm pl} < M_*$ (e.g. Ormel and Kobayashi 2012). Correspondingly, the (numerical) transition between the settling and ballistic regime is given at the point where the rates due to settling and ballistic interactions equal.

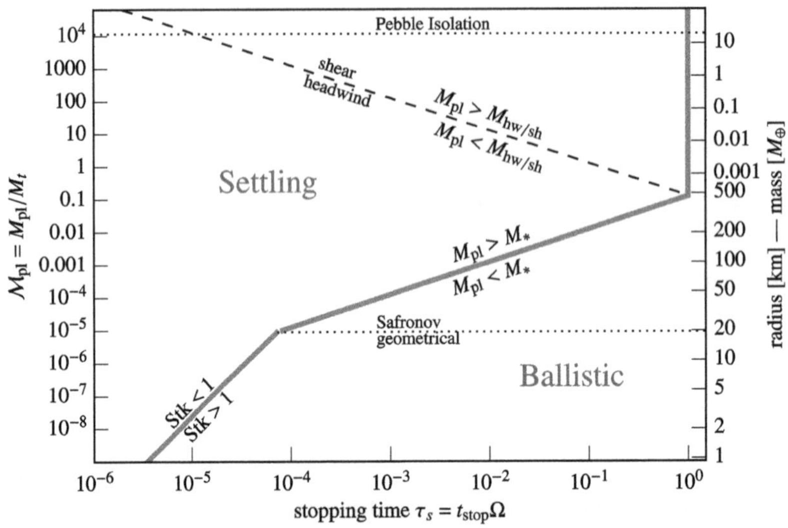

Fig. 7.3 Sketch of the accretion regimes as function of the pebble aerodynamical properties (*x*-axis) and that of the planet(esimal) (*y*-axis). The dimensionless planet mass $\mathcal{M}_{pl}$ (*left y*-axis) is defined as $\mathcal{M}_{pl} = M_{pl}/M_t$, where M_t is given by Eq. (7.10). The conversion to physical masses (*right y*-axis) holds for a radius of 1 AU and an internal density of $\rho_\bullet = 3\,\mathrm{g\,cm^{-3}}$. The primary dividing line is $M_p = M_*$, distinguishing ballistic encounters, where gas-drag effects are unimportant, from settling encounters, where particles accrete by sedimentation. The line $M_{pl} = M_{hw/sh}$ indicates where Keplerian shear becomes important and the line Stk = 1 where aerodynamical deflection matters

at 1 AU. Accretion proceeds in the headwind regime at a rate given by Eqs. (7.12) or (7.13), dependent on the thickness of the pebble disk. At

- $M_{pl} = M_{hw/sh}$ [Eq. (7.9)], pebble accretion switches to the shear limit. Pebble accretion continues until
- $M_{pl} = M_{P,iso}$ [Eq. (7.17)], where the pebble isolation mass is reached.

The ballistic:settling transition heralds the onset of pebble accretion. From Eq. (7.10) it follows that the transition occurs at a larger mass for increasing orbital radius. Visser and Ormel (2016) have calculated a more precise expression for the radius R_{PA} where pebble accretion commences:

$$R_{PA} \approx 520\,\mathrm{km} \left(\frac{v_{hw}}{50\,\mathrm{m\,s^{-1}}}\right) \left(\frac{\rho_\bullet}{\mathrm{g\,cm^{-3}}}\right)^{-0.36} \left(\frac{r}{\mathrm{AU}}\right)^{0.42} \tau_s^{0.28}. \tag{7.20}$$

Once pebble accretion commences, the jump in $\dot{M}$ is also more dramatic in the outer disk (Eq. (7.19), since R_{pl}/R_{Hill} is lower). In addition, for the outer disk, aerodynamic deflection is less of an issue since pebbles will have a larger stopping time, because of the lower gas density.

7.3 Results

I illustrate the outcome of pebble accretion with a series of contour plots, where contours of a quantity Q are plotted as function of planet mass and particle size. Instead of the order-of-magnitude expressions derived above, I employ more precise expressions that have been calibrated to numerical integrations (Ormel and Klahr 2010; Ormel and Kobayashi 2012; Visser and Ormel 2016). But many of the key features highlighted in Fig. 7.3 will resurface.

I adopt the following disk profiles for the gas surface density and midplane temperature:

$$\Sigma_{\rm gas} = 10^3\,{\rm g\,cm^{-2}} \left(\frac{r}{\rm AU}\right)^{-1}; \qquad T = 300\,{\rm K} \left(\frac{r}{\rm AU}\right)^{-1/2} \tag{7.21}$$

and take the disk headwind to be $v_{\rm hw} = 50\,{\rm m\,s^{-1}}$. I further assume that planetesimals' internal density increases according to:

$$\rho_\bullet(R_{\rm pl}) = 0.08\,{\rm g\,cm^{-3}} \left(\frac{R_{\rm pl}}{\rm km}\right)^{0.5}, \tag{7.22}$$

which crudely interpolates between "porous planetesimals" and rocky planets. The internal density of pebbles is taken to be $\rho_\bullet = 1\,{\rm g\,cm^{-3}}$. The standard value of the turbulent strength is $\alpha_T = 10^{-4}$.

7.3.1 *The Collision Cross Section*

First, in Fig. 7.4, the collision cross section, normalized to the geometrical cross section, is plotted, $f_{\rm coll} = (b/R)^2$, for $r = 1$ AU. The lower x-axis now gives the pebble radius (in cm) and the upper x-axis translates this to dimensionless stopping time. Remark that the tick marks for τ_s are much denser spaced beyond $\tau_s = 10^{-2}$, because of the transition to Stokes drag. The y-axis again gives the planet mass or radius. The thick, dashed, black line denotes the transition between the ballistic and the settling regimes.

Many of the features of Fig. 7.3 can also be identified in Fig. 7.4. Below $R_{\rm pl} \sim$ 100 km the cross section is that of the geometrical cross section (white area). Also, a steepening of the ballistic:settling transition (the dashed line) occurs below 1 mm, where aerodynamic deflection becomes important. Indeed, the collisional cross section can become very low. While micron-size dust grains may be produced by (colliding) planetesimals, they are not accreted by them! However, in a turbulent medium particles will collide more easily to the planetesimal (Homann et al. 2016). This effect is not included here.

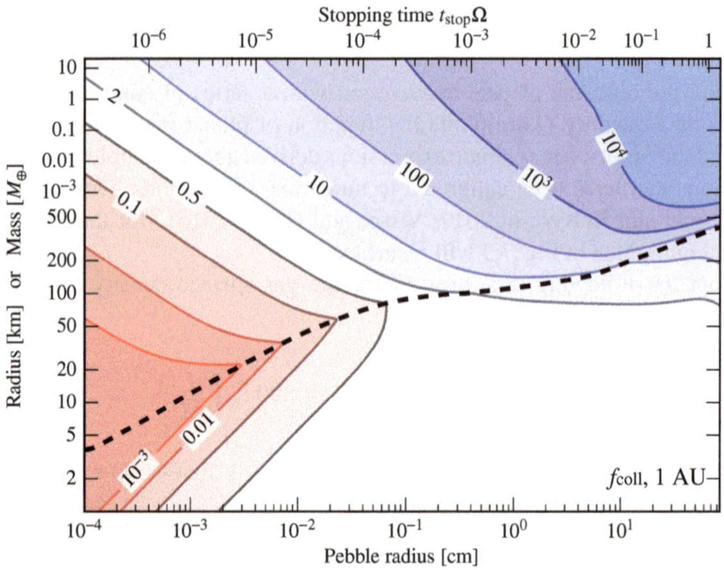

Fig. 7.4 Collision factor $f_{\rm coll}$: the collision cross section with respect to the geometrical cross section at 1 AU. The *thick dashed line* delineates the ballistic regime from the settling regime. Note the sharp increase in $f_{\rm coll}$ across this line at higher $t_{\rm stop}$

In any case, it is clear that aerodynamically very small particles ($\tau_s \sim 10^{-5}$–10^{-6}) are hard to accrete and that settling (pebble accretion) of small particles gives low $f_{\rm coll}$ (upper left corner). This simply reflects the strong coupling to the gas, which is not accreted. Rates are much larger for bodies that accrete larger pebbles ($\tau_s \sim 10^{-3}$–1) in the settling regime, which happens when the planetesimal crosses ~100 km. In particular, at this ballistic:settling transition accretion rates jump dramatically. For example, from $R_{\rm pl} = 200$ to 400 km and $\tau_s = 0.1$ $f_{\rm coll}$ increases a 100-fold [see also Eq. (7.19)]. Once $f_{\rm coll} \sim 10^4$ (top right corner) pebbles are accreted at the (maximum) Hill cross section, $\sigma_{\rm col} \sim R_{\rm Hill}^2$.

7.3.2 *Accretion Efficiencies: 2D and 3D*

Does a large accretion cross section also imply a high accretion rate? For pebble accretion this is a difficult question to answer since pebbles drift. The accretion rate [Eq. (7.12)] depends on the surface density of pebbles Σ_P that are locally available. Due to their drift, pebbles are constantly rejuvenated: old pebbles leave the accretion region while new pebbles from the outer disk drift in. The question how large accretion rates are at a certain time therefore depends on the evolution of all solids. Analytical approaches (which work for smooth disk; see Sect. 7.2.4) or numerical ones (Birnstiel et al. 2012; Drążkowska et al. 2016; Sato et al. 2016; Krijt et al. 2016) have also been developed.

However, even though the accretion rate requires a global model, a useful local quantity can still be defined: the pebble accretion efficiency or accretion probability $P_{\rm eff}$. This is simply the pebble accretion rate on the planet [Eq. (7.12)] divided by the pebble accretion rate through the disk:

$$P_{\rm eff} = \frac{\dot{M}}{\dot{M}_{\rm P,disk}} = \frac{\dot{M}}{2\pi r v_{\rm drift} \Sigma_P}, \tag{7.23}$$

where $\dot{M}$ is given by either Eq. (7.12) or Eq. (7.13). Note that Eq. (7.23) is independent of Σ_P. An efficiency of $P_{\rm eff} \geq 1$ guarantees accretion of the pebble. On the other hand, when $P_{\rm eff} \ll 1$ most pebbles avoid capture to continue their radial drift to the star.

In Fig. 7.5a contours of $P_{\rm eff}$ are plotted in the 2D-limit ($\dot{M} = \dot{M}_{\rm 2d}$). In the 2D limit it is assumed that the pebbles have settled to the disk midplane, which would be the case for a laminar disk. As is clear from Fig. 7.5a a pebble is more likely to

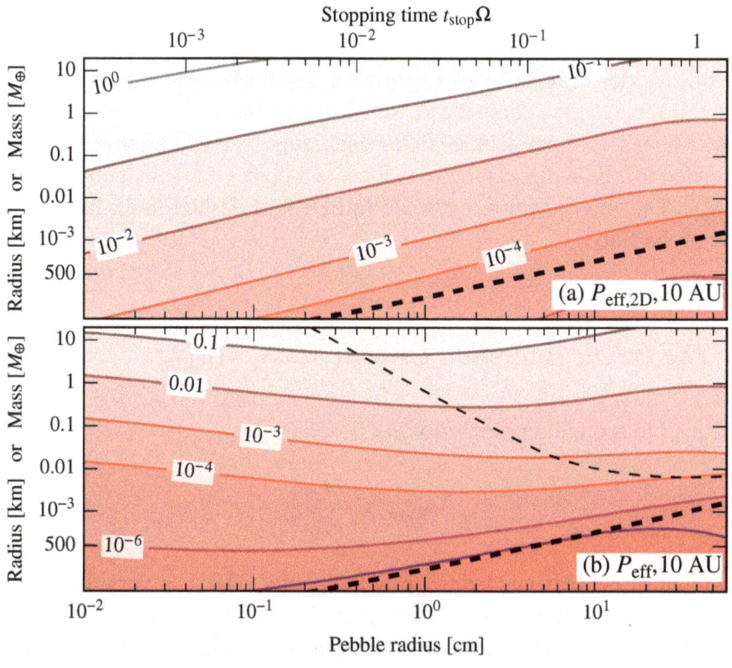

Fig. 7.5 The efficiency of pebble accretion or accretion probability $P_{\rm eff}$ for (**a**) the 2D limit (pebbles reside in the midplane) and (**b**) for the general 3D case with our standard $\alpha_T = 10^{-4}$ for 10 AU. Values around 1 indicate very efficient accretion, while $P_{\rm eff} \ll 1$ indicates most pebbles drift to the interior disk, instead of being accreted. The *thin-dashed line* in (**b**) indicates the transition between 2D and 3D accretion ($h_{\rm p} = b_{\rm col}$)

be accreted by a larger planet(esimal), which is obvious. Much less obvious is the trend in the horizontal direction. On the one hand, large pebbles result in larger cross sections, increasing $\dot{M}_{2d}$, but in the 2D limit $\dot{M}_{2D}$ is sublinear in τ_s, $\dot{M}_{2D} \propto \tau_s^{1/2}$ or $\propto \tau_s^{2/3}$ [Eq. (7.13)]. On the other hand, the drift velocity is linear in τ_s [Eq. (7.2)]. Hence, the drift dependence wins out: smaller pebbles are more likely to be accreted.

However, even for moderate turbulence accretion may well proceed in the 3D limit: the pebbles are distributed in a vertical layer of thickness larger than the impact radius $b_{\rm col}$. Correspondingly, a scale height correction is applied to get a net accretion rate of:

$$\dot{M} = \dot{M}_{2D} \frac{b_{\rm col}}{b_{\rm col} + h_{\rm peb}\sqrt{8/\pi}}, \tag{7.24}$$

where $h_{\rm peb}$ is the scale height of the pebble layer (Dubrulle et al. 1995; Cuzzi et al. 1993; Youdin and Lithwick 2007):

$$h_{\rm peb} = \sqrt{\frac{\alpha_T}{\alpha_T + \tau_s}} h_{\rm gas}. \tag{7.25}$$

The form of $\dot{M}$ adopted in Eq. (7.24) ensures the 2D and 3D expressions in the limits of $b_{\rm col} \gg h_{\rm peb}$ (2D) and $b_{\rm col} \ll h_{\rm peb}$ (3D), respectively.

The scale height-corrected accretion probability is presented in Fig. 7.5b. The 3D probability is always lower than the 2D limit and the lines are more horizontal. Even Earth-mass planets accrete pebbles rather inefficiently, meaning that a large pebble flux is needed for these planets to grow. For fixed planet mass the pebble size where $b_{\rm col} = h_{\rm peb}$, i.e., the transition from 2D to 3D (thin dashed line), has the highest accretion probability.

7.3.3 *The Pebble Accretion Growth Mass, $M_{\rm P,grw}$*

From the pebble accretion probability $P_{\rm eff}$, I define

$$M_{\rm P,grw} = \frac{M_{\rm pl}}{P_{\rm eff}}, \tag{7.26}$$

as the amount of pebbles needed to grow the planet. This is a very useful quantity since it immediately highlights where growth is slow or fast. For example, a very large $M_{\rm P,grw}$—e.g., thousands of Earth masses—indicates a bottleneck for the growth of the planet, because it is unlikely that so many pebbles are available. On the other hand, low $M_{\rm P,grw}$ likely indicate that growth is rapid. To obtain the actual growth timescale, $M_{\rm P,grw}$ should be divided by the pebble flux $\dot{M}_{\rm P,disk}$. Guillot et al. (2014) already introduced Eq. (7.26) as the *filtering mass*: $M_{\rm P,grw}$ is also the mass in planetesimals needed to ensure accretion of a single pebble.

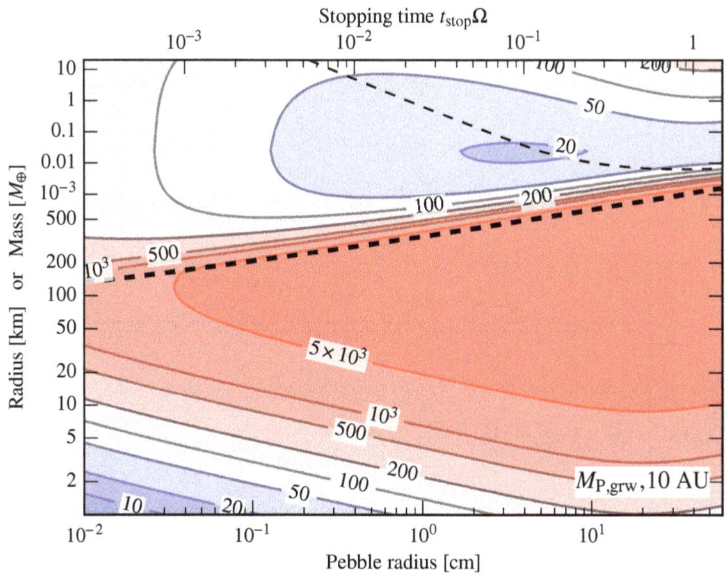

Fig. 7.6 Required mass in pebbles to grow a planet by pebble accretion as function of pebble size (x-axis) and planet mass (y-axis) at 10 AU for the standard disk model [Eq. (7.21)] and $\alpha_T = 10^{-4}$. Contours denote the total amount of pebbles in Earth masses needed to e-fold the planet's mass and include the pebbles that are drifting past without accretion. In *red regions*, growth is likely to stall. Above the ballistic:settling dividing line (*thick dashed*) growth is significantly boosted but nevertheless requires tens-to-hundreds of Earth masses in pebbles. Above the *thin dashed line* pebble accretion reaches its 2D limit

In Fig. 7.6 $M_{\rm P,grw}$ is plotted for $r = 10$ AU. What is immediately obvious is that $M_{\rm P,grw}$ is very large just below the pebble accretion initiation threshold (the black dashed line). Clearly, geometric and Safronov accretion are not effective in growing planetesimals large; and the initiation of pebble accretion relies on the presence of a massive-enough seed that is produced by a process other than sweep up of small particles. Such a seed may result from classical self-coagulation mechanisms (i.e., runaway growth of planetesimals) or, more directly, from the high-mass tail of the planetesimal formation mechanism, e.g., by streaming or gravitational instabilities (Cuzzi et al. 2010; Johansen et al. 2015; Simon et al. 2016; Schäfer et al. 2017).

Even in the settling regime, the required pebble masses are substantial. Also note that $M_{\rm P,grw}$ refers only to one e-folding growth in mass. In general, growth from an initial mass $M_{\rm ini}$ to a final mass $M_{\rm fin}$ takes a total mass of

$$M_{\rm P,tot} = \int_{M_{\rm ini}}^{M_{\rm fin}} M_{\rm P,grw}(\tau_s, M) \mathrm{d} \log M \tag{7.27}$$

in pebbles. For example, growth from 10^{-3} to 10 Earth masses involves almost 10 e-foldings. Clearly, giant planet formation by pebble accretion requires massive disks;

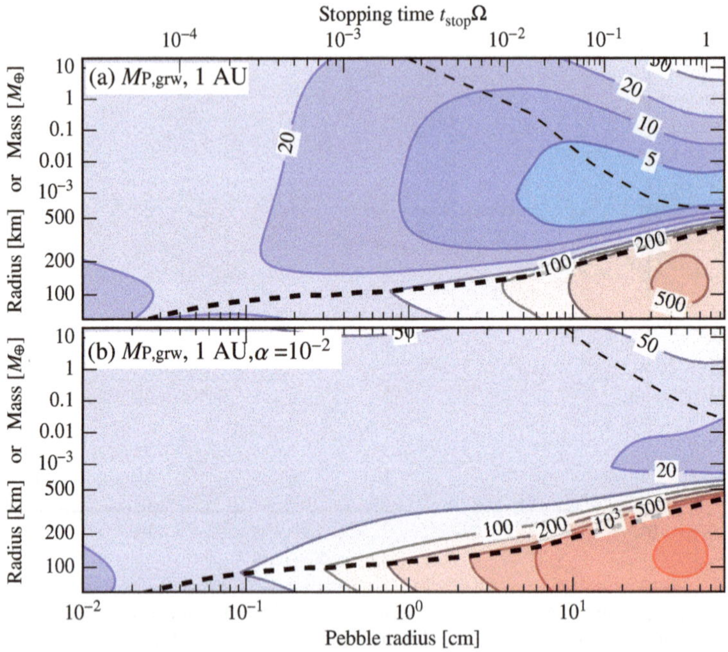

Fig. 7.7 Required pebble masses at 1 AU at standard (**a**) and high (**b**) turbulence levels. The $\alpha_T = 10^{-4}$ case is shown in Fig. 7.6

at least several hundreds of Earth masses are needed to form a $10\,M_\oplus$ core. Also, the pebbles need to be of the right (aerodynamic) size. Pebbles of $\tau_s \approx 1$ are not very suitable, as they drift too fast. Smaller pebbles are preferred.

Figure 7.7a shows contours of $M_{P,grw}$ at 1 AU. At 1 AU $M_{P,grw}$ is significantly lower than at 10 AU, because the pebbles are more efficiently accreted. The larger P_{eff} is caused by (1) a higher probability of encountering the planet because of the smaller circumference ($2\pi r$); and (2) a reduced scale height of the pebble layer. However, it is especially below the ballistic:settling line where the change is the largest. Windmark et al. (2012) and Garaud et al. (2013) have hypothesized that some particles could cross the fragmentation/bouncing barriers that operate around $\tau_s = 1$, because of fortuitously colliding with particles at low collision energies. To grow into planetesimals, these "lucky" particles, however, still need to grow fast. Ignoring the radial drift problem, and with some tuning of the parameters this could work at 1 AU, but for $r \gg 1$ AU sweep up growth in the geometric and settling regimes becomes simply too slow, as is seen in Fig. 7.6.

A major determinant for the efficacy of pebble accretion, and a key unknown, is the turbulence strength parameter, α_T. Since disks are observed to accrete onto their host stars, it has been hypothesized that disks are turbulent, with the turbulent viscosity providing the angular momentum transport. For example, the magneto-rotational instability (Balbus and Hawley 1991), which operates in sufficiently ionized disks, provides $\alpha_T \sim 10^{-2}$. However, the turbulence could also be hydrodynamically driven, such as the recently postulated vertical shear instability (Nelson et al. 2013; Stoll and Kley 2014), the spiral wave instability (Bae et al. 2016), or the baroclinic instability (Klahr and Bodenheimer 2006), which provide perhaps $\alpha_T \sim 10^{-4}$. Or one could imagine layered accretion (Gammie 1996), of which the disk wind model has recently become popular (Bai 2014, 2016). In that case the midplane—relevant here—stays laminar. Given these uncertainties—indeed α_T is very hard to constrain observationally (Teague et al. 2016)—it is best to consider α_T as a free parameter and test how it affects the pebble accretion rates.

In Fig. 7.7b $M_{\rm P,grw}$ is plotted for $\alpha_T = 10^{-2}$, which significantly reduces the spatial density in pebbles in the midplane and increases $M_{\rm P,grw}$ compared to the nominal $\alpha = 10^{-4}$ value. For $\tau_s \lesssim \alpha_T = 10^{-2}$ $M_{\rm P,grw}$ is very flat. The reason is that accretion is now in the 3D limit, where $\dot{M}$ is linear in both τ_s and $M_{\rm pl}$—dependencies that cancel upon conversion to $M_{\rm P,grw}$. Still, at 1 AU pebble accretion is relatively efficient. However, at 10 AU the dependence on α_T becomes more extreme. In Fig. 7.8 contours of $M_{\rm P,grw}$ are plotted for $\alpha_T = 10^{-2}$ and $\alpha_T = 10^{-6}$, (the nominal $\alpha_T = 10^{-4}$ case is plotted in Fig. 7.6). From Fig. 7.8 it follows that the mass requirements are very high for turbulent disks ($\alpha_T = 10^{-2}$), but more comfortable for laminar disks ($\alpha_T = 10^{-6}$). For the outer disk in particular, the ability of pebble accretion to spawn planets strongly depends on the level of turbulence.

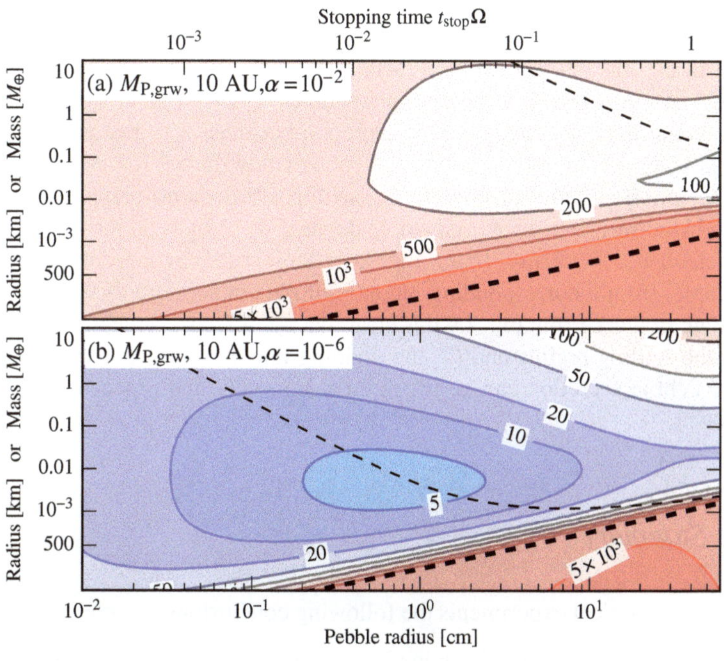

Fig. 7.8 Required pebble masses at 10 AU for (**a**) strong turbulence and weak turbulence (**b**)

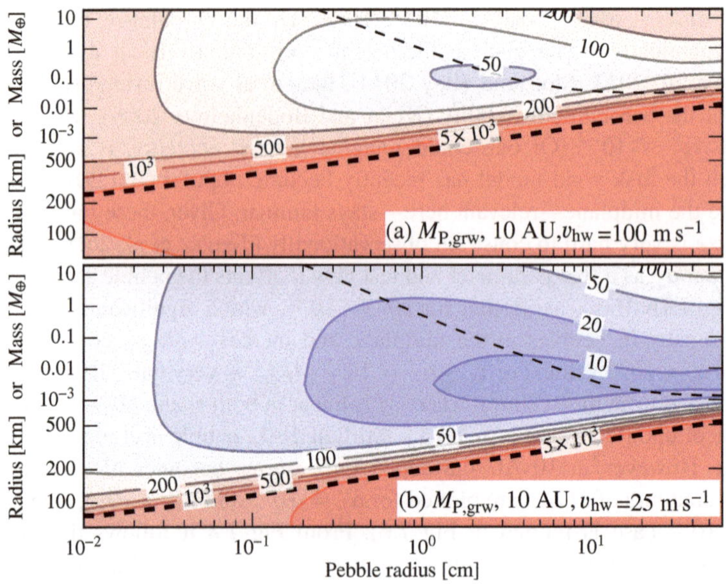

Fig. 7.9 This figure illustrates the dependence of the required pebble mass on the headwind parameter, v_{hw}—compare with Fig. 7.6, above

In addition to α_T, the pebble accretion efficiency is sensitive to the radial pressure gradient in the gas [$\nabla_{\log}P$; see Eq. (7.4)]. In the case where the pressure gradient reverses—i.e., at a pressure maximum—pebbles will no longer drift inwards. It is clear that at these locations pebble accretion becomes very fast and the planet' growth is given by the rate at which the pebbles flow in, $\dot{M}_{P,disk}$. More generally, pebble accretion is quite sensitive to v_{hw} as Fig. 7.9 demonstrates. In Fig. 7.9 the disk headwind has been increased (a) or decreased (b) by a factor 2 with respect to the default $v_{hw} = 50\,\mathrm{m\,s^{-1}}$ (see Fig. 7.6). An increase by a factor 2 may arise, for example, from a corresponding increase in the temperature of the disk (hotter disks rotate slower). From Fig. 7.9 it is clear that v_{hw} affects (1) the pebble accretion rates in the settling regime and (2) the dividing line between ballistic and settling regimes. Pebble accretion can be triggered more easily and proceeds faster when v_{hw} is lower.

7.3.4 Summary

From these numerical experiments the following conclusions emerge:

1. Pebble accretion is generally an inefficient process. Not all pebbles are accreted.
2. The efficiency of pebble accretion and its capability to produce large planets strongly depends on the vertical thickness of the pebble layer, which is deter-

mined by the disk turbulence. In particular in the outer disk, pebbles simply drift past the planet without experiencing an interaction. Efficiencies are also boosted in regions where the radial gas pressure gradient is small.
3. Pebble accretion is more efficient for particles of $\tau_s \lesssim 0.1$. Specifically, for a given mass, pebble accretion is most efficient for particle sizes where the growth modes switches from 2D to 3D, i.e., where $h_{\rm peb} \approx b_{\rm col}$.
4. Pebble accretion requires an initial seed that must lie above the $M_{\rm pl} = M_*$ line [Eq. (7.20)] distinguishing ballistic from settling encounters. Such a seed must be produced from the planetesimal formation process or (failing that) by classical coagulation among planetesimals. The pebble accretion initiation mass is much larger in the outer disk.

7.4 Applications

I close this chapter with a brief recent overview of applications of pebble accretion.

7.4.1 *Solar System*

Lambrechts et al. (2014) argue that that the pebble isolation mass [Eq. (7.17)]—the upper limit at which planetary cores can accrete pebbles—naturally explains the heavy element contents in the solar system's outer planets. After this mass is reached, pebble accretion shuts off; and the lack of accretional heating (and arguably opacity) will trigger runaway accretion of H/He gas. In their model, Uranus and Neptune never reached $M_{\rm P,iso}$ and accreted pebbles during the lifetime of the disk. Their arguments rely on an efficient accretion of pebbles (indeed they do consider the 2D limit), which means that turbulence levels had to be low or that the solar nebula contained massive amounts of pebbles.

A key parameter in the pebble accretion scenario is the size and number of the initial seeds. Here, the classic planetesimal-formation model has the advantage that growth proceeds through a runaway growth phase, where the biggest bodies outcompete their smaller siblings, since $\dot{M}_{\rm Saf}$ is superlinear [Eq. (7.14)]. However, pebble accretion is a more "democratic" process; embryos will tend to stay similar in terms of mass (Sect. 7.2.3). This was demonstrated by the N-body simulations of Kretke and Levison (2014), in which the pebbles were shared more-or-less evenly among the growing embryos, resulting in a bunch of Mars-to-Earth size planets, but not the $\sim 10\,M_\oplus$ needed to form gas giants.

A way to remedy this problem is to invoke classical planet formation concepts. In Levison et al. (2015a) pebbles were fed into the simulation on much longer timescales ($\sim$Myr), resulting in the dynamical excitation of especially the smallest embryos. Pebbles then were preferentially accreted by the largest embryos, recovering the "winner-takes-it-all" feature that giant planet formation requires. This, so-called "viscously-stirred pebble accretion" (essentially a blend between classical planetesimal accretion and pebble accretion) was also applied to the inner solar system (Levison et al. 2015b). Here, the authors claim to have found a possible solution to the persistent "small Mars" problem (e.g. Raymond et al. 2009) by reversing the mass-order of planetary embryos: more massive embryos can form further in because accretion is more efficient. Although encouraging, it must be emphasized that these N-body models contain a great number of free parameters—ranging from the initial size-distribution of planetesimals, their location in the disk, to the properties of the pebbles and the gas—which means a proper investigation would imply a (prohibitively?) large scan of the parameter space.

Adopting a more basic approach, Morbidelli et al. (2015) claimed that solar system's "great dichotomy"—small Mars next to big Jupiter—naturally follows from pebble accretion. The key feature is the iceline. Pebbles (τ_s) as well as the pebble flux ($\dot{M}_{\mathrm{P,disk}}$) are large exterior to it; the first because of the fragmentation velocity threshold of silicate *vs* icy grains and the second because the pebbles lose their ice after crossing the iceline. Morbidelli et al. (2015) find that the $\tau_s = 10^{-1.5}$ pebbles beyond the iceline accrete more efficiently than the smaller pebbles ($\tau_s \approx 10^{-2.5}$) interior to it (Fig. 7.6, albeit for different disk parameters, gives the gist of their result). Hence, Jupiter's core grows faster than Mars' and eventually, when it opens a gap, will starve the inner disk of pebbles. A similar result, using a more elaborate model, was found by Chambers (2016).

7.4.2 *Exoplanetary Systems*

7.4.2.1 Super Earths

One of the key surprises of the last decade has been the discovery of the super-Earth planet population (Fressin et al. 2013; Petigura et al. 2013): close-in planets ($r \sim 0.1$ AU) of mass between Earth and Neptune, which are rock-dominated but often have a gaseous envelope (Lopez et al. 2012). Because super-Earths orbit their host stars at close distance, there is no longer a timescale problem for their assembly—even without gravitational focusing coagulation proceeds fast. In the classical *in situ* model there is however a mass budget problem and to form super-Earths *in situ* the canonical Minimum-Mass Solar Nebula surface density profile (Weidenschilling 1977b; Hayashi 1981) has to be cranked up (Chiang and Laughlin 2013). Another challenge is that the isolation masses at ~ 0.1 AU are very small and that embryos need to merge at a later stage through giant impacts, while preserving their hydrogen/helium atmospheres (Inamdar and Schlichting 2015; Lee and Chiang 2016).

In the context of the pebble accretion model super-Earths formation at/near the inner disk edge seems very natural. The idea is that pebbles drift inwards until they meet a pressure maximum. This could simply be the disk edge (provided pebbles do not evaporate!) or an MRI active/dead transition region (Kretke and Lin

2007; Chatterjee and Tan 2014; Hu et al. 2016). At the pressure maximum, mass piles up until gravitational instability is triggered. Pebble accretion then proceeds until the pebble isolation mass [Eq. (7.17)], which indeed evaluates to super-Earth masses. An advantage of pebble migration over planet migration is that pebbles, due to their coupling to the gas, avoid trapping in mean motion resonance, which is needed to explain those very compact systems. However, to reproduce the exoplanet architecture in a framework that includes Type I migration remains challenging (Ogihara et al. 2015; Liu et al. 2017).

7.4.2.2 Distant Planets

Another hallmark of the exoplanet field has been the discovery of distant planets, of which the HR-8799 system is the poster boy (Marois et al. 2008, 2010). HR-8799 harbors four super-Jupiter planets at, respectively, 15, 24, 38, and 68 AU—too far to form with the classical core-accretion model. It has therefore been proposed that these planets formed from a gravitational unstable disk (Dodson-Robinson et al. 2009; Boss 2011). Pebble accretion of the resulting clumps can accelerate their collapse, because the accretion effectively cools the clump (Nayakshin 2016). Pebble accretion may also re-invigorate the core-accretion model, provided the disk is laminar. In Fig. 7.10 the required pebble masses for moderate ($\alpha_T = 10^{-4}$) and

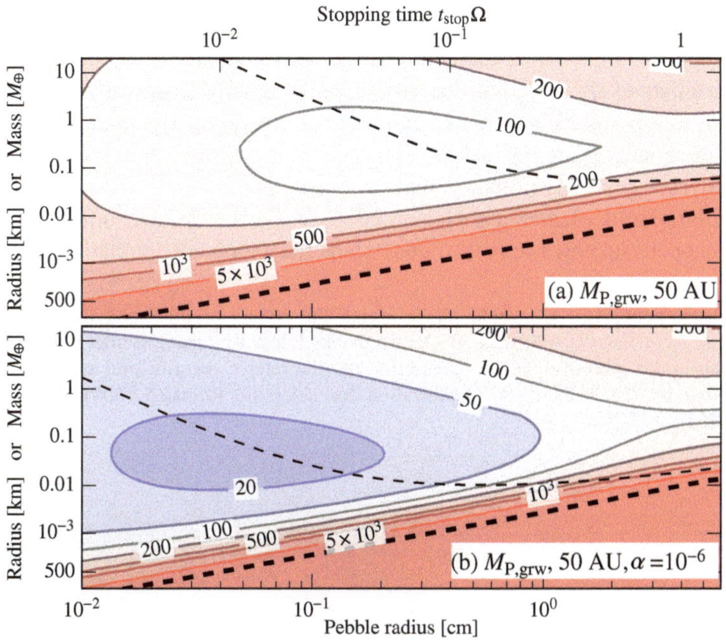

Fig. 7.10 Pebble mass reservoir to grow planets in the outer disk

very low turbulence levels ($\alpha_T = 10^{-6}$) are given. For moderate turbulence, pebble accretion requires a pebble reservoir of hundreds of Earth masses, while in the laminar case accretion becomes quite efficient when the particles are of sub-mm size. Although challenging, forming these distant planets by pebble accretion cannot be dismissed at first hand.

7.4.2.3 Population Synthesis Models

The assembly of a solid core is a critical (albeit not the only!) part of planet population synthesis models (see, e.g., Ida and Lin 2004; Mordasini et al. 2009 and their sequels). Because they rely on the classical (planetesimal-driven) scenario, sufficiently large planetary embryos can only form around $\sim$ 5 AU (because of the isolation mass constraint [Eq. (7.18)]. On the other hand, as also discussed in Sect. 7.1.3, accretion is getting progressively slower in the outer disk.[5] Consequently, there is a limited range in the disk where giant planets can form; and success relies on fortuitous placement of initial seeds in massive disks (Bitsch et al. 2015b).

Because of pebble drift, growing cores have in principle access to the entire solid mass of the disk. This, according to Bitsch et al. (2015b), makes pebble accretion attractive. From the results described in this review, it is clear that success also depends on the turbulent state of the (outer) disk and on the existence of a massive pebble reservoir; but it is true that once conducive conditions materialize there is no longer a timescale problem. In addition Bitsch et al. (2015b) find a prolific formation of super-Earths and ice giants. Recently, pebble accretion-driven models have also been adopted to understand the composition and chemistry of Jupiter-size planets (Madhusudhan et al. 2016; Ali-Dib 2017)—a potentially observable characteristic. However, any population synthesis model is as strong as its weakest link. In this regard, key concerns are the lack of a physical model for the formation of the seeds and the hyper-sensitivity of the synthesized planet population to the prescription for planet migration (Coleman and Nelson 2014). Nevertheless, accounting for pebble accretion opens up new avenues to understand the exoplanet population as a whole.

Acknowledgements C.W.O. would like to thank the editors of this book for providing the opportunity to review the subject. C.W.O. also thanks Lucia Klarmann, Beibei Liu, and Djoeke Schoonenberg for proofreading the manuscript and the referee for a helpful report. This work is supported by the Netherlands Organization for Scientific Research (NWO; VIDI project 639.042.422).

[5]Also, population synthesis models do not account for planetesimal fragmentation, which would further suppress core formation (see Sect. 7.1.3).

References

Ali-Dib, M.: A pebbles accretion model with chemistry and implications for the Solar system. Mon. Not. R. Astron. Soc. **464**, 4282–4298 (2017). doi:10.1093/mnras/stw2651, 1609.03227

Andrews, S.M., Wilner, D.J., Hughes, A.M., Qi, C., Rosenfeld, K.A., Öberg, K.I., Birnstiel, T., Espaillat, C., Cieza, L.A., Williams, J.P., Lin, S.Y., Ho, P.T.P.: The TW Hya Disk at 870 μm: comparison of CO and dust radial structures. Astrophys. J. **744**, 162 (2012). doi:10.1088/0004-637X/744/2/162, 1111.5037

Andrews, S.M., Rosenfeld, K.A., Kraus, A.L., Wilner, D.J.: The mass dependence between protoplanetary disks and their stellar hosts. Astrophys. J. **771**, 129 (2013). doi:10.1088/0004-637X/771/2/129, 1305.5262

Bae, J., Nelson, R.P., Hartmann, L.: The spiral wave instability induced by a giant planet. I. Particle stirring in the inner regions of protoplanetary disks. Astrophys. J. **833**, 126 (2016). doi:10.3847/1538-4357/833/2/126, 1610.08502

Bai, X.N.: Hall-effect-controlled gas dynamics in protoplanetary disks. I. Wind solutions at the inner disk. Astrophys. J. **791**, 137 (2014). doi:10.1088/0004-637X/791/2/137, 1402.7102

Bai, X.N.: Towards a global evolutionary model of protoplanetary disks. Astrophys. J. **821**, 80 (2016). doi:10.3847/0004-637X/821/2/80, 1603.00484

Balbus, S.A., Hawley, J.F.: A powerful local shear instability in weakly magnetized disks. I - Linear analysis. II - Nonlinear evolution. Astrophys. J. **376**, 214–233 (1991). doi:10.1086/170270

Birnstiel, T., Ricci, L., Trotta, F., Dullemond, C.P., Natta, A., Testi, L., Dominik, C., Henning, T., Ormel, C.W., Zsom, A.: Testing the theory of grain growth and fragmentation by millimeter observations of protoplanetary disks. Astron. Astrophys. **516**, L14 (2010). doi:10.1051/0004-6361/201014893, 1006.0940

Birnstiel, T., Andrews, S.M., Ercolano, B.: Can grain growth explain transition disks? Astron. Astrophys. **544**, A79 (2012). doi:10.1051/0004-6361/201219262, 1206.5802

Bitsch, B., Johansen, A., Lambrechts, M., Morbidelli, A.: The structure of protoplanetary discs around evolving young stars. Astron. Astrophys. **575**, A28 (2015a). doi:10.1051/0004-6361/201424964, 1411.3255

Bitsch, B., Lambrechts, M., Johansen, A.: The growth of planets by pebble accretion in evolving protoplanetary discs. Astron. Astrophys. **582**, A112 (2015b). doi:10.1051/0004-6361/201526463, 1507.05209

Boss, A.P.: Formation of giant planets by disk instability on wide orbits around protostars with varied masses. Astrophys. J. **731**, 74 (2011). doi:10.1088/0004-637X/731/1/74, 1102.4555

Boudet, N., Mutschke, H., Nayral, C., Jäger, C., Bernard, J.P., Henning, T., Meny, C.: Temperature dependence of the submillimeter absorption coefficient of amorphous silicate grains. Astrophys. J. **633**, 272–281 (2005). doi:10.1086/432966

Chambers, J.E.: Giant planet formation with pebble accretion. Icarus **233**, 83–100 (2014). doi:10.1016/j.icarus.2014.01.036

Chambers, J.E.: Pebble accretion and the diversity of planetary systems. Astrophys. J. **825**, 63 (2016). doi:10.3847/0004-637X/825/1/63, 1604.06362

Chatterjee, S., Tan, J.C.: Inside-out planet formation. Astrophys. J. **780**, 53 (2014). doi:10.1088/0004-637X/780/1/53, 1306.0576

Chiang, E., Laughlin, G.: The minimum-mass extrasolar nebula: in situ formation of close-in super-Earths. Mon. Not. R. Astron. Soc. **431**, 3444–3455 (2013). doi:10.1093/mnras/stt424, 1211.1673

Chiang, E.I., Goldreich, P.: Spectral energy distributions of T Tauri stars with passive circumstellar disks. Astrophys. J. **490**, 368–376 (1997). doi:10.1086/304869, astro-ph/9706042

Cleeves, L.I., Öberg, K.I., Wilner, D.J., Huang, J., Loomis, R.A., Andrews, S.M., Czekala, I.: The coupled physical structure of gas and dust in the IM lup protoplanetary disk. Astrophys. J. **832**, 110 (2016). doi:10.3847/0004-637X/832/2/110, 1610.00715

Coleman, G.A.L., Nelson, R.P.: On the formation of planetary systems via oligarchic growth in thermally evolving viscous discs. Mon. Not. R. Astron. Soc. **445**, 479–499 (2014). doi:10.1093/mnras/stu1715, 1408.6993

Cuzzi, J.N., Dobrovolskis, A.R., Champney, J.M.: Particle-gas dynamics in the midplane of a protoplanetary nebula. Icarus **106**, 102 (1993). doi:10.1006/icar.1993.1161

Cuzzi, J.N., Hogan, R.C., Shariff, K.: Toward planetesimals: dense chondrule clumps in the protoplanetary nebula. Astrophys. J. **687**, 1432–1447 (2008). doi:10.1086/591239, 0804.3526

Cuzzi, J.N., Hogan, R.C., Bottke, W.F.: Towards initial mass functions for asteroids and Kuiper Belt Objects. Icarus **208**, 518–538 (2010). doi:10.1016/j.icarus.2010.03.005, 1004.0270

Dodson-Robinson, S.E., Veras, D., Ford, E.B., Beichman, C.A.: The formation mechanism of gas giants on wide orbits. Astrophys. J. **707**, 79–88 (2009). doi:10.1088/0004-637X/707/1/79, 0909.2662

Drążkowska, J., Alibert, Y., Moore, B.: Close-in planetesimal formation by pile-up of drifting pebbles. Astron. Astrophys. **594**, A105 (2016). doi:10.1051/0004-6361/201628983, 1607.05734

Dubrulle, B., Morfill, G., Sterzik, M.: The dust subdisk in the protoplanetary nebula. Icarus **114**, 237–246 (1995). doi:10.1006/icar.1995.1058

Fortier, A., Alibert, Y., Carron, F., Benz, W., Dittkrist, K.M.: Planet formation models: the interplay with the planetesimal disc. Astron. Astrophys. **549**, A44 (2013). doi:10.1051/0004-6361/201220241, 1210.4009

Fressin, F., Torres, G., Charbonneau, D., Bryson, S.T., Christiansen, J., Dressing, C.D., Jenkins, J.M., Walkowicz, L.M., Batalha, N.M.: The false positive rate of Kepler and the occurrence of planets. Astrophys. J. **766**, 81 (2013). doi:10.1088/0004-637X/766/2/81, 1301.0842

Gammie, C.F.: Layered accretion in T Tauri disks. Astrophys. J. **457**, 355 (1996). doi:10.1086/176735

Garaud, P., Meru, F., Galvagni, M., Olczak, C.: From dust to planetesimals: an improved model for collisional growth in protoplanetary disks. Astrophys. J. **764**, 146 (2013). doi:10.1088/0004-637X/764/2/146, 1209.0013

Goldreich, P., Lithwick, Y., Sari, R.: Planet formation by coagulation: a focus on Uranus and Neptune. Annu. Rev. Astron. Astrophys. **42**, 549–601 (2004). doi:10.1146/annurev.astro.42.053102.134004, arXiv:astro-ph/0405215

Guillot, T., Ida, S., Ormel, C.W.: On the filtering and processing of dust by planetesimals. I. Derivation of collision probabilities for non-drifting planetesimals. Astron. Astrophys. **572**, A72 (2014). doi:10.1051/0004-6361/201323021, 1409.7328

Hayashi, C.: Structure of the solar nebula, growth and decay of magnetic fields and effects of magnetic and turbulent viscosities on the nebula. Prog. Theor. Phys. Suppl. **70**, 35–53 (1981). doi:10.1143/PTPS.70.35

Homann, H., Guillot, T., Bec, J., Ormel, C.W., Ida, S., Tanga, P.: Effect of turbulence on collisions of dust particles with planetesimals in protoplanetary disks. Astron. Astrophys. **589**, A129 (2016). doi:10.1051/0004-6361/201527344, 1602.03037

Hu, X., Zhu, Z., Tan, J.C., Chatterjee, S.: Inside-out planet formation. III. Planet-disk interaction at the dead zone inner boundary. Astrophys. J. **816**, 19 (2016). doi:10.3847/0004-637X/816/1/19, 1508.02791

Ida, S., Lin, D.N.C.: Toward a deterministic model of planetary formation. I. A desert in the mass and semimajor axis distributions of extrasolar planets. Astrophys. J. **604**, 388–413 (2004). doi:10.1086/381724, arXiv:astro-ph/0312144

Ida, S., Nakazawa, K.: Collisional probability of planetesimals revolving in the solar gravitational field. III. Astron. Astrophys. **224**, 303–315 (1989)

Ida, S., Guillot, T., Morbidelli, A.: The radial dependence of pebble accretion rates: a source of diversity in planetary systems. I. Analytical formulation. Astron. Astrophys. **591**, A72 (2016). doi:10.1051/0004-6361/201628099, 1604.01291

Inamdar, N.K., Schlichting, H.E.: The formation of super-Earths and mini-Neptunes with giant impacts. Mon. Not. R. Astron. Soc. **448**, 1751–1760 (2015). doi:10.1093/mnras/stv030, 1412.4440

Johansen, A., Oishi, J.S., Low, M., Klahr, H., Henning, T., Youdin, A.: Rapid planetesimal formation in turbulent circumstellar disks. Nature **448**, 1022–1025 (2007). doi:10.1038/nature06086

Johansen, A., Blum, J., Tanaka, H., Ormel, C., Bizzarro, M., Rickman, H.: The multifaceted planetesimal formation process. In: Protostars and Planets VI pp. 547–570. University of Arizona Press, Tucson (2014). doi:10.2458/azu_uapress_9780816531240-ch024, 1402.1344

Johansen, A., Mac Low, M.M., Lacerda, P., Bizzarro, M.: Growth of asteroids, planetary embryos, and Kuiper belt objects by chondrule accretion. Sci. Adv. **1**, 1500109 (2015). doi:10.1126/sciadv.1500109, 1503.07347

Kataoka, A., Okuzumi, S., Tanaka, H., Nomura, H.: Opacity of fluffy dust aggregates. Astron. Astrophys. **568**, A42 (2014). doi:10.1051/0004-6361/201323199, 1312.1459

Klahr, H., Bodenheimer, P.: Formation of giant planets by concurrent accretion of solids and gas inside an anticyclonic vortex. Astrophys. J. **639**, 432–440 (2006). doi:10.1086/498928, arXiv: astro-ph/0510479

Kobayashi, H., Tanaka, H., Krivov, A.V.: Planetary core formation with collisional fragmentation and atmosphere to form gas giant planets. Astrophys. J. **738**, 35 (2011). doi:10.1088/0004-637X/738/1/35, 1106.2047

Kokubo, E., Ida, S.: Formation of protoplanets from planetesimals in the solar nebula. Icarus **143**, 15–27 (2000). doi:10.1006/icar.1999.6237

Kokubo, E., Ida, S.: Formation of protoplanet systems and diversity of planetary systems. Astrophys. J. **581**, 666–680 (2002). doi:10.1086/344105

Kretke, K.A., Levison, H.F.: Challenges in forming the solar system's giant planet cores via pebble accretion. Astron. J. **148**, 109 (2014). doi:10.1088/0004-6256/148/6/109, 1409.4430

Kretke, K.A., Lin, D.N.C.: Grain retention and formation of planetesimals near the snow line in MRI-driven turbulent protoplanetary disks. Astrophys. J. **664**, L55–L58 (2007). doi:10.1086/520718, 0706.1272

Krijt, S., Ormel, C.W., Dominik, C., Tielens, A.G.G.M.: A panoptic model for planetesimal formation and pebble delivery. Astron. Astrophys. **586**, A20 (2016). doi:10.1051/0004-6361/201527533, 1511.07762

Lambrechts, M., Johansen, A.: Rapid growth of gas-giant cores by pebble accretion. Astron. Astrophys. **544**, A32 (2012). doi:10.1051/0004-6361/201219127, 1205.3030

Lambrechts, M., Johansen, A.: Forming the cores of giant planets from the radial pebble flux in protoplanetary discs. Astron. Astrophys. **572**, A107 (2014). doi:10.1051/0004-6361/201424343, 1408.6094

Lambrechts, M., Johansen, A., Morbidelli, A.: Separating gas-giant and ice-giant planets by halting pebble accretion. Astron. Astrophys. **572**, A35 (2014). doi:10.1051/0004-6361/201423814, 1408.6087

Lee, E.J., Chiang, E.: Breeding super-Earths and birthing super-puffs in transitional disks. Astrophys. J. **817**, 90 (2016). doi:10.3847/0004-637X/817/2/90, 1510.08855

Levison, H.F., Thommes, E., Duncan, M.J.: Modeling the formation of giant planet cores. I. Evaluating key processes. Astron. J. **139**, 1297–1314 (2010). doi:10.1088/0004-6256/139/4/1297

Levison, H.F., Kretke, K.A., Duncan, M.J.: Growing the gas-giant planets by the gradual accumulation of pebbles. Nature **524**, 322–324 (2015a). doi:10.1038/nature14675

Levison, H.F., Kretke, K.A., Walsh, K.J., Bottke, W.F.: Growing the terrestrial planets from the gradual accumulation of sub-meter sized objects. Proc. Natl. Acad. Sci. **112**, 14180–14185 (2015b). doi:10.1073/pnas.1513364112, 1510.02095

Lin, D.N.C., Papaloizou, J.: On the tidal interaction between protoplanets and the protoplanetary disk. III - Orbital migration of protoplanets. Astrophys. J. **309**, 846–857 (1986). doi:10.1086/164653

Lissauer, J.J.: Timescales for planetary accretion and the structure of the protoplanetary disk. Icarus **69**, 249–265 (1987). doi:10.1016/0019-1035(87)90104-7

Liu, B., Ormel, C.W., Lin, D.N.C.: Dynamical rearrangement of super-Earths during disk dispersal I. Outline of the magnetospheric rebound model. ArXiv e-prints:170202059 (2017). 1702.02059

Lopez, E.D., Fortney, J.J., Miller, N.: How thermal evolution and mass-loss sculpt populations of super-Earths and sub-Neptunes: application to the Kepler-11 system and beyond. Astrophys. J. **761**, 59 (2012). doi:10.1088/0004-637X/761/1/59, 1205.0010

Madhusudhan, N., Bitsch, B., Johansen, A., Eriksson, L.: Atmospheric signatures of giant exoplanet formation by pebble accretion. ArXiv e-prints:161103083 (2016). 1611.03083

Marois, C., Macintosh, B., Barman, T., Zuckerman, B., Song, I., Patience, J., Lafrenière, D., Doyon, R.: Direct imaging of multiple planets orbiting the star HR 8799. Science **322**, 1348 (2008). doi:10.1126/science.1166585, 0811.2606

Marois, C., Zuckerman, B., Konopacky, Q.M., Macintosh, B., Barman, T.: Images of a fourth planet orbiting HR 8799. Nature **468**, 1080–1083 (2010). doi:10.1038/nature09684, 1011.4918

Morbidelli, A., Lambrechts, M., Jacobson, S., Bitsch, B.: The great dichotomy of the Solar System: small terrestrial embryos and massive giant planet cores. Icarus **258**, 418–429 (2015). doi:10.1016/j.icarus.2015.06.003, 1506.01666

Mordasini, C., Alibert, Y., Benz, W.: Extrasolar planet population synthesis. I. Method, formation tracks, and mass-distance distribution. Astron. Astrophys. **501**, 1139–1160 (2009). doi:10.1051/0004-6361/200810301, 0904.2524

Nakagawa, Y., Sekiya, M., Hayashi, C.: Settling and growth of dust particles in a laminar phase of a low-mass solar nebula. Icarus **67**, 375–390 (1986). doi:10.1016/0019-1035(86)90121-1

Natta, A., Testi, L., Calvet, N., Henning, T., Waters, R., Wilner, D.: Dust in protoplanetary disks: properties and evolution. Protostars and Planets V, pp. 767–781. University of Arizona Press, Tucson (2007). arXiv:astro-ph/0602041

Nayakshin, S.: Tidal Downsizing model - IV. Destructive feedback in planets. Mon. Not. R. Astron. Soc. **461**, 3194–3211 (2016). doi:10.1093/mnras/stw1404, 1510.01630

Nelson, R.P., Gressel, O.: On the dynamics of planetesimals embedded in turbulent protoplanetary discs. Mon. Not. R. Astron. Soc. **409**, 639–661 (2010). doi:10.1111/j.1365-2966.2010.17327.x, 1007.1144

Nelson, R.P., Gressel, O., Umurhan, O.M.: Linear and non-linear evolution of the vertical shear instability in accretion discs. Mon. Not. R. Astron. Soc. **435**, 2610–2632 (2013). doi:10.1093/mnras/stt1475, 1209.2753

Nishida, S.: Collisional processes of planetesimals with a protoplanet under the gravity of the proto-sun. Prog. Theor. Phys. **70**, 93–105 (1983). doi:10.1143/PTP.70.93

Ogihara, M., Morbidelli, A., Guillot, T.: A reassessment of the in situ formation of close-in super-Earths. Astron. Astrophys. **578**, A36 (2015). doi:10.1051/0004-6361/201525884, 1504.03237

Okuzumi, S., Tanaka, H., Sakagami, M.: Numerical modeling of the coagulation and porosity evolution of dust aggregates. Astrophys. J. **707**, 1247–1263 (2009). doi:10.1088/0004-637X/707/2/1247, 0911.0239

Okuzumi, S., Tanaka, H., Kobayashi, H., Wada, K.: Rapid coagulation of porous dust aggregates outside the snow line: a pathway to successful icy planetesimal formation. Astrophys. J. **752**, 106 (2012). doi:10.1088/0004-637X/752/2/106, 1204.5035

Ormel, C.W., Klahr, H.H.: The effect of gas drag on the growth of protoplanets. Analytical expressions for the accretion of small bodies in laminar disks. Astron. Astrophys. **520**, A43 (2010). doi:10.1051/0004-6361/201014903, 1007.0916

Ormel, C.W., Kobayashi, H.: Understanding how planets become massive. I. Description and validation of a new toy model. Astrophys. J. **747**, 115 (2012). doi:10.1088/0004-637X/747/2/115, 1112.0274

Ormel, C.W., Okuzumi, S.: The fate of planetesimals in turbulent disks with dead zones. II. Limits on the viability of runaway accretion. Astrophys. J. **771**, 44 (2013). doi:10.1088/0004-637X/771/1/44, 1305.1890

Ormel, C.W., Spaans, M., Tielens, A.G.G.M.: Dust coagulation in protoplanetary disks: porosity matters. Astron. Astrophys. **461**, 215–232 (2007). doi:10.1051/0004-6361:20065949, arXiv: astro-ph/0610030

Panić, O., Hogerheijde, M.R., Wilner, D., Qi, C.: A break in the gas and dust surface density of the disc around the T Tauri star IM Lupi. Astron. Astrophys. **501**, 269–278 (2009). doi:10.1051/0004-6361/200911883, 0904.1127

Pérez, L.M., Chandler, C.J., Isella, A., Carpenter, J.M., Andrews, S.M., Calvet, N., Corder, S.A., Deller, A.T., Dullemond, C.P., Greaves, J.S., Harris, R.J., Henning, T., Kwon, W., Lazio, J., Linz, H., Mundy, L.G., Ricci, L., Sargent, A.I., Storm, S., Tazzari, M., Testi, L., Wilner, D.J.: Grain growth in the circumstellar disks of the young stars CY Tau and DoAr 25. Astrophys. J. **813**, 41 (2015). doi:10.1088/0004-637X/813/1/41, 1509.07520

Petigura, E.A., Marcy, G.W., Howard, A.W.: A plateau in the planet population below twice the size of Earth. Astrophys. J. **770**, 69 (2013). doi:10.1088/0004-637X/770/1/69, 1304.0460

Rafikov, R.R.: Atmospheres of protoplanetary cores: critical mass for nucleated instability. Astrophys. J. **648**, 666–682 (2006). doi:10.1086/505695, arXiv:astro-ph/0405507

Raymond, S.N., O'Brien, D.P., Morbidelli, A., Kaib, N.A.: Building the terrestrial planets: constrained accretion in the inner Solar System. Icarus **203**, 644–662 (2009). doi:10.1016/j.icarus.2009.05.016, 0905.3750

Ricci, L., Testi, L., Natta, A., Brooks, K.J.: Dust grain growth in ρ-Ophiuchi protoplanetary disks. Astron. Astrophys. **521**, A66 (2010a). doi:10.1051/0004-6361/201015039, 1008.1144

Ricci, L., Testi, L., Natta, A., Neri, R., Cabrit, S., Herczeg, G.J.: Dust properties of protoplanetary disks in the Taurus-Auriga star forming region from millimeter wavelengths. Astron. Astrophys. **512**, A15 (2010b). doi:10.1051/0004-6361/200913403, 0912.3356

Safronov, V.S.: Evolution of the protoplanetary cloud and formation of Earth and the planets. Moscow: Nauka. Transl. 1972 NASA Tech. F-677 (1969)

Sato, T., Okuzumi, S., Ida, S.: On the water delivery to terrestrial embryos by ice pebble accretion. Astron. Astrophys. **589**, A15 (2016). doi:10.1051/0004-6361/201527069, 1512.02414

Schäfer, U., Yang, C.C., Johansen, A.: Initial mass function of planetesimals formed by the streaming instability. Astron. Astrophys. **597**, A69 (2017). doi:10.1051/0004-6361/201629561, 1611.02285

Sekiya, M., Takeda, H.: Were planetesimals formed by dust accretion in the solar nebula? Earth Planets Space **55**, 263–269 (2003)

Sellentin, E., Ramsey, J.P., Windmark, F., Dullemond, C.P.: A quantification of hydrodynamical effects on protoplanetary dust growth. Astron. Astrophys. **560**, A96 (2013). doi:10.1051/0004-6361/201321587, 1311.3498

Simon, J.B., Armitage, P.J., Li, R., Youdin, A.N.: The mass and size distribution of planetesimals formed by the streaming instability. I. The role of self-gravity. Astrophys. J. **822**, 55 (2016). doi:10.3847/0004-637X/822/1/55, 1512.00009

Slinn, W.G.N.: Precipitation scavenging of aerosol particles. Geophys. Res. Lett. **3**, 21–22 (1976). doi:10.1029/GL003i001p00021

Stoll, M.H.R., Kley, W.: Vertical shear instability in accretion disc models with radiation transport. Astron. Astrophys. **572**, A77 (2014). doi:10.1051/0004-6361/201424114, 1409.8429

Teague, R., Guilloteau, S., Semenov, D., Henning, T., Dutrey, A., Piétu, V., Birnstiel, T., Chapillon, E., Hollenbach, D., Gorti, U.: Measuring turbulence in TW Hydrae with ALMA: methods and limitations. Astron. Astrophys. **592**, A49 (2016). doi:10.1051/0004-6361/201628550, 1606.00005

Testi, L., Birnstiel, T., Ricci, L., Andrews, S., Blum, J., Carpenter, J., Dominik, C., Isella, A., Natta, A., Williams, J.P., Wilner, D.J.: Dust Evolution in Protoplanetary Disks. Protostars and Planets VI pp. 339–361. University of Arizona Press, Tucson (2014). doi:10.2458/azu_uapress_9780816531240-ch015, 1402.1354

Visser, R.G., Ormel, C.W.: On the growth of pebble-accreting planetesimals. Astron. Astrophys. **586**, A66 (2016). doi:10.1051/0004-6361/201527361, 1511.03903

Weidenschilling, S.J.: Aerodynamics of solid bodies in the solar nebula. Mon. Not. R. Astron. Soc. **180**, 57–70 (1977a)

Weidenschilling, S.J.: The distribution of mass in the planetary system and solar nebula. Astrophys. Space Sci. **51**, 153–158 (1977b). doi:10.1007/BF00642464

Wetherill, G.W.: Formation of the terrestrial planets. Annu. Rev. Astron. Astrophys. **18**, 77–113 (1980). doi:10.1146/annurev.aa.18.090180.000453

Wetherill, G.W., Stewart, G.R.: Accumulation of a swarm of small planetesimals. Icarus **77**, 330–357 (1989). doi:10.1016/0019-1035(89)90093-6

Whipple, F.L.: On certain aerodynamic processes for asteroids and comets. In: Elvius, A. (ed.) From Plasma to Planet, p. 211. Wiley, New York (1972)

Williams, S., Arsenault, M., Buczkowski, B., Reid, J., Flocks, J., Kulp, M., Penland, S., Jenkins, C.: Surficial sediment character of the Louisiana offshore continental shelf region: a GIS compilation. US Geological Survey Open-File Report 2006-1195 (2006). http://pubs.usgs.gov/of/2006/1195/index.htm

Windmark, F., Birnstiel, T., Ormel, C.W., Dullemond, C.P.: Breaking through: the effects of a velocity distribution on barriers to dust growth. Astron. Astrophys. **544**, L16 (2012). doi:10.1051/0004-6361/201220004, 1208.0304

Youdin, A.N., Goodman, J.: Streaming instabilities in protoplanetary disks. Astrophys. J. **620**, 459–469 (2005). doi:10.1086/426895, arXiv:astro-ph/0409263

Youdin, A.N., Lithwick, Y.: Particle stirring in turbulent gas disks: including orbital oscillations. Icarus **192**, 588–604 (2007). doi:10.1016/j.icarus.2007.07.012, 0707.2975

Chapter 8
White Dwarf Planetary Systems: Insights Regarding the Fate of Planetary Systems

Amy Bonsor and Siyi Xu

Abstract Planetary material accreted by white dwarfs provides critical insights regarding the composition of planetary material. The analysis of polluted white dwarfs suggests that rocky planetary material similar in composition to bulk Earth is common, and that differentiation and collisions play a key role in planetary systems. Infrared observations track the accretion of dusty planetary material in disks that share many similarities with Saturn's rings, as well as proto-planetary disks. Planetary material arrives close to the star, scattered inwards from an outer planetary systems that has survived the star's evolution. White dwarf planetary systems provide key observational constraints that enable us to study the evolution of planetary systems under extreme conditions.

8.1 Introduction

Almost all stars where exo-planets have been detected to date will one day evolve to end their lives as white dwarfs. Theoretically even the fate of our own Solar System is uncertain (Duncan and Lissauer 1998). Whilst we are unable to observe the evolution of our Solar System in real time, we can study nearby white dwarfs to try and infer what may happen to our Solar System and other planetary systems, once their host stars leave the main-sequence.

White dwarfs are the faint remnants of stars like our Sun. They have a typical mass of $0.6\,M_{\odot}$, but a radius of about Earth, and cool from a temperature of over one hundred thousand kelvin to a few thousand kelvin, in the age of the Universe. Stars like our Sun, and up to a mass of about $8\,M_{\odot}$, swell to become giant stars. They loose

A. Bonsor (✉)
Institute of Astronomy, University of Cambridge, Madingley Road, Cambridge CB3 0HA, UK
e-mail: amy.bonsor@gmail.com

S. Xu (✉)
European Southern Observatory, Karl-Schwarzschild-Strasse 2, 85748 Garching, Germany
e-mail: sxu@eso.org

M. Pessah, O. Gressel (eds.), *Formation, Evolution, and Dynamics of Young Solar Systems*, Astrophysics and Space Science Library 445,
DOI 10.1007/978-3-319-60609-5_8

a substantial fraction of their mass (between half and two thirds), mainly during the tip of the asymptotic giant branch, in a planetary nebulae. The compact remnant left behind becomes a white dwarf. Whilst the inner planetary system may be lost, swallowed in the giant star's expanding envelope, Mustill and Villaver (2012) show that rocky planets should survive exterior to ~3 AU and gas giants initially exterior to ~5 AU. Any surviving planetary system expands, due to adiabatic mass loss, and Villaver and Livio (2007) show that the planetary system interior to 15 AU should be devoid of giant planets, unless they have arrived in this region after the giant branch. Outer debris belts, outside of a few AU, will survive the star's evolution, albeit stripped of small planetesimals (up to tens of metre in size on the AGB), which are rapidly replenished by collisions during the early white dwarf phase (Bonsor and Wyatt 2010).

Observationally it is very difficult to study white dwarf planetary systems directly. Detection of outer planets, or outer debris belts is difficult due to their low luminosity. The best way to study the planetary system orbiting a white dwarf is from observations of planetary material that has been accreted by *polluted* white dwarfs. A white dwarf should have an atmosphere that is pure hydrogen or helium. The presence of heavier elements indicates the recent accretion of material from exterior to the white dwarf. Over 30% of white dwarfs exhibit heavy elements in their spectra (Zuckerman et al. 2003, 2010; Koester et al. 2014), the composition of which are consistent with rocky planetary material, in general similar to bulk Earth (Jura and Young 2014). Alongside observations of gas and dust, polluted white dwarfs provide the best way to study the fate of planetary systems.

8.2 Planetary Systems Post-main Sequence

A white dwarf may be orbited by planets and (or) debris belts, exterior to a few AU, that have survived the star's evolution. The most likely explanation for white dwarf pollution is that planetary material has been transported from this outer planetary system close enough to the star to be tidally disrupted (about a solar radius). Fragments from the disrupted planetary bodies accrete onto the star. There are few transport processes that lead to planetary material arriving this close to the white dwarf. For example, Bonsor and Wyatt (2010) show that Poynting-Robertson drag can only transport dust grains smaller than 20 mm to the star, if they start exterior to 1 AU. Scattering remains the most effective mechanism at transporting planetary material from the outer planetary system down to the star. The accreted material could originate from a range of sources, asteroids, comets or planets. We use the term planetesimal to refer to any planetary body that is in a debris belt.

8.2.1 *Dynamical Instabilities Following Stellar Mass Loss*

When a star loses mass, if the mass loss is adiabatic, the orbits of any planets or planetesimals spiral outwards. The planet's semi-major axis increases as

$$a_f = a_i \frac{M_i}{M_f}. \tag{8.1}$$

With a reduced central mass, the influence of any planetary bodies on one another increases. This can lead to dynamical instabilities. Debes and Sigurdsson (2002) illustrate how an initially stable configuration can become unstable following mass loss. Such instabilities can lead to planetary bodies scattered onto star-grazing orbits, which in turn can lead to the pollution of white dwarfs. White dwarf pollution could occur when planets (minor planets or exo-moons) themselves are scattered onto star-grazing orbits, or alternatively when planets scatter smaller bodies, either exo-moons, or planetesimals (asteroids or comets) onto star-grazing orbits. There is as yet no consensus in the literature regarding which process dominates, and probably all processes are occurring. However, important insights can be derived by considering the frequency, level and composition of the pollution. The scattering of larger bodies will be less frequent and lead to higher levels of pollution than the (continuous) scattering of smaller bodies, as shown by Wyatt et al. (2014).

8.2.1.1 Planetesimals

The observational evidence for the presence of both planetesimals (debris belts) and planets in white dwarf systems is growing (Wyatt 2008). About 30% of A stars, the progenitors of most white dwarfs, exhibit an infrared excess from a debris belt (Su et al. 2006). When the star loses mass, the influence of any planets on the planetesimals in that system increases. This can lead to the scattering of planetesimals by planets. We can consider two scenarios.

In the first, the inner edge of a planetesimal belt is carved out by a planet that has cleared its chaotic zone of material. The width of the chaotic zone, for a planet on a circular orbit, is given by Wisdom (1980):

$$\frac{\delta a_{\rm chaos}}{a_{\rm pl}} = C \left(\frac{m_{\rm pl}}{M_\odot}\right)^{2/7}, \tag{8.2}$$

where C is a constant whose value is about 1.2, $a_{\rm pl}$ is the planet semi-major axis, $m_{\rm pl}$ is the planet mass and M_* is the stellar mass. As the stellar mass decreases, the size of the chaotic zone increases. This leads to increased scattering, as shown by Bonsor et al. (2011) and Frewen and Hansen (2014) for eccentric planets. Bonsor et al. (2011) show that a single planet at 30 AU (or 60 AU post-stellar mass loss), interior to a planetesimal belt, scatters sufficient material inwards in order to explain the observed white dwarf pollution, if there exists an interior planetary system that is similarly efficient at scattering planetesimals inwards as our Solar System. Bonsor et al. (2011) assume a distribution of planetesimal belt masses that fits the observations of debris disks around A stars (Wyatt et al. 2007), evolved to the start of the white dwarf phase (Bonsor and Wyatt 2010). The accretion rates from scattering fall off steeply with time after the start of the white dwarf phase.

In the second scenario, planetesimals are scattered due to the increase in width of unstable resonances. Debes et al. (2012) show that the width of the interior mean-motion resonances with Jupiter in the asteroid belt increase following stellar mass loss. In the restricted three-body case, expanding for low eccentricities (Murray and Dermott 1999), the maximum libration width for interior first-order resonances is given by

$$\delta a_{\max} = \pm \left(\frac{16C_r e}{n}\right)^{1/2} \left(1 + \frac{C_r}{27j_2^2 e^3 n}\right)^{1/2} - \frac{2C_r a}{9j_2 e n}, \tag{8.3}$$

where n is the mean motion of the planetesimal is proportional to the stellar mass, $n = \sqrt{\frac{GM_*}{a^3}}$, a, e are the semi-major axis and eccentricity of the planetesimal, C_r and j_2 are constants from the distributing function (Murray and Dermott 1999). This width is inversely proportional to the stellar mass. As the star loses mass, many previously stable particles become unstable, as shown in Fig. 8.1. Debes et al. (2012) use N-body simulations to determine the rate at which asteroids are scattered onto star-grazing orbits in the post-main sequence Solar System. They show that this mechanism can produce the observed accretion rates from polluted white dwarfs in a planetary system with an asteroid belt than has a mass ~820 times that of the Solar System's asteroid belt.

Figure 8.1 shows that the difference in scattering rates between an exterior and interior planet is minimal, and the scattering rates are dominated by the planet and belt mass. Whilst more material survives in outer planetesimal belts, it is easier to transport material inwards from an inner planetesimal belt, and it is, therefore, currently not clear from the dynamics whether the accretion would be dominated by planetesimals that originate from planetesimal belts that are close to the star (exo-asteroid belts) or planetesimal belts that are further from the star (exo-Kuiper belts).

8.2.1.2 Planets

Dynamical instabilities and planet–planet scattering are thought to be a key feature of the early evolution of planetary systems, e.g. Raymond et al. (2009). Most instabilities occur early (<10 Myr), after which planetary systems generally settle down to stable configurations (Raymond et al. 2010). Late-time instabilities are rare (Raymond et al. 2010; Bonsor et al. 2013). Such stable configurations, however, may not remain stable following stellar mass loss on the giant branches. Analytically it can be shown that by reducing the stellar mass, the separation of two planets must

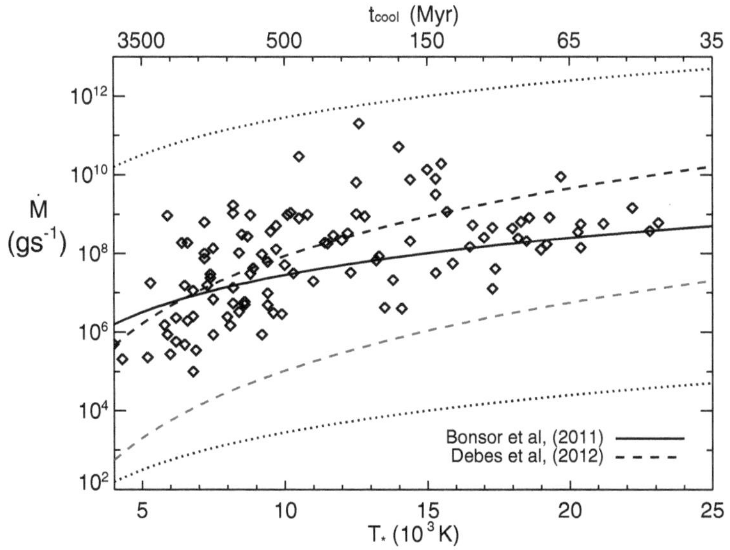

Fig. 8.1 The accretion rates derived from calcium for known white dwarfs, assuming a composition of bulk Earth (Farihi, personal communication). These are compared to the distribution of accretion rates fitted to the scattering simulations of Bonsor et al. (2011), tracking scattering following the increase in width of the chaotic zone post-stellar mass loss (*black solid line*), and Debes et al. (2012), following the increase in width of internal resonances with a planet (*black dashed line*). The *blue dashed line* shows an asteroid with the mass of the current day asteroid belt, whereas the *black dashed line* shows a mass of 820 times the current day. The *dotted lines* show the minimum and maximum of the distribution (Bonsor et al. 2011), depending on planet mass and belt mass. Any comparison with the observed distribution of accretion rates must take into account selection effects that make it harder to detect lower accretion rates for hot stars

increase in order for them to continue to be Hill stable (Debes and Sigurdsson 2002; Veras et al. 2013).

Planetary systems are chaotic and the exact dynamics depend on the exact configuration of individual systems. A wide range of numerical simulations have shown (Debes and Sigurdsson 2002; Veras et al. 2013; Mustill et al. 2014; Veras et al. 2016a) that instabilities can occur both during the giant branch evolution and in the white dwarf evolution that follows. Whilst the rate of instabilities decreases significantly with time after the end of mass loss, instabilities can still occur long into the white dwarf phase (Veras et al. 2013). Some instabilities lead to planets that approach close to the star, whilst others may scatter planets into regions where these planets then scatter planetesimals onto star-grazing orbits. The majority of planets are ejected following stellar mass loss (Veras et al. 2016a), in addition to which non-adiabatic mass loss for objects on long period orbits can lead to the *escape* of planets or comets (Veras et al. 2011; Veras and Wyatt 2012).

8.2.1.3 Exo-Moons

In our Solar System, many planets are orbited by several moons or satellites. Payne et al. (2016) show that gravitational scattering amongst planets following stellar mass loss leads to the liberation of exo-moons. These exo-moons could themselves be scattered onto star-grazing orbits, or instead be scattered into a region where they readily scatter planetesimals onto star-grazing orbits, thus, leading to white dwarf pollution.

8.2.1.4 The Influence of Binary Companions

The rate of dynamical instabilities induced by stellar mass loss necessarily decreases very steeply as a function of time after the end of stellar mass loss (or white dwarf cooling age). The presence of a binary companion can induce dynamical instabilities in a planetary system following stellar mass loss, even after Gyrs of stable evolution. This can lead to white dwarf pollution even in old white dwarfs, in a manner that is relatively independent of age.

It has been shown for main-sequence planetary systems, that whilst in general wide ($a > 1000\,$AU) binary companions do not influence the dynamics of the planetary system, the orbit of the binary varies due to Galactic tides. During periods of 'close' pericentre passage, the binary may excite the eccentricities of planets, even ejecting them (Kaib et al. 2013). The increased eccentricity of exoplanets in systems with wide binary companions has been confirmed by observations (Kaib et al. 2013). The same scenario could be applied to white dwarf planetary systems. If the white dwarf is orbited by a wide binary companion, the planetary system may remain unperturbed for billions of years, before Galactic tides alter the orbit of the companion such that it induces dynamical instabilities in the planetary system. The increase in separation of the binary following stellar mass loss increases the perturbations on the binary's orbit due to the Galactic tides, making it more likely for the first close approach of a binary to occur during the white dwarf phase. Figure 8.2 shows the evolution of a possible binary orbit and its first close approach during the white dwarf phase.

In Bonsor and Veras (2015) we show that the first close approach of a wide binary companion can lead to pollution even in Gyr old white dwarfs. The presence of a planet is not a necessity, although it can aid the scattering of planetesimals onto star-grazing orbits. The polluted white dwarf, WD 1009-184, with its companion at 6870 AU and an effective temperature of 9940 K (equivalent to a cooling age of $\sim$700 Myr) (Sion et al. 2009; Zuckerman 2014) stands out as an example system where this mechanism could be acting. Based on a wide binary fraction of $\sim$10%, we estimate that this mechanism could contribute to pollution in up to a few percent of any sample of white dwarfs. Zuckerman (2014) presents a sample of 17 white dwarfs with companions separated by more than 1000 AU, searched for Ca II with Keck/VLT and finds that 5 are polluted, a similar rate to pollution in apparently single white dwarfs ($\sim$ 30% Zuckerman et al. 2003).

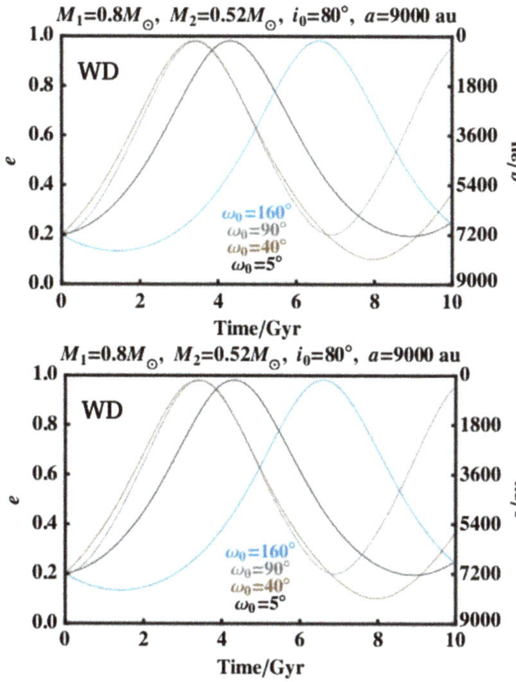

Fig. 8.2 A binary can have a first close approach (high eccentricity) during the white dwarf phase. Such a close approach could trigger instabilities that lead to white dwarf pollution. The figures show the evolution of an example binary's orbit due to Galactic tides; the *top panel* on the main-sequence and the *bottom panel*, during the white dwarf phase, following stellar mass loss (Bonsor and Veras 2015)

8.3 Circumstellar Material

Young stellar systems are characterised by the accretion of gas and dust onto the central star. Proto-planetary disks are the sites of complex gas and dust processes that potentially lead to grain growth and planet formation, as well as the clearing of the inner cavities during transition disk phases. The same stars, at the very end of their lives, when all that is left behind is a compact remnant, can again accrete material. The observations of gas and dust very close to some polluted white dwarfs tell us about these accretion processes in action. Whilst some similar physical processes are shared by the accretion of material onto white dwarfs and in proto-planetary disks, the accretion rates and scales are very different. In many respects the circumstellar disks around polluted white dwarfs share more similarities with Saturn's rings.

8.3.1 Dust

Observations in the near-infrared of a sub-set of polluted white dwarfs find excess emission, over and above that predicted for the stellar photosphere. This emission is from dusty material at temperatures of around a thousand Kelvin. If such material is heated by the white dwarf, it will reside within a solar radius of the white dwarf, i.e. within the Roche limit. Asteroids that are scattered into this zone will be disrupted by strong tidal forces. The observed dusty material is thought to originate in the disrupted bodies and accretes onto the white dwarf, leading to the observed pollution.

The first dusty white dwarf was observed in 1987, G29-38 (Zuckerman and Becklin 1987), shortly after the discovery of the first debris disk around a main-sequence star (Aumann et al. 1984), and before the first planets were discovered in 1991 (Wolszczan and Frail 1992). Initially the debate in the literature regarded whether the emission was from a brown dwarf companion or dust (Wickramasinghe et al. 1988; Tokunaga et al. 1988; Haas and Leinert 1990; Graham et al. 1990a,b; Telesco et al. 1990). The infrared emission was linked to the presence of heavy elements discovered in the spectra by Koester et al. (1997). G29-38 remains one of the best studied dusty white dwarfs, with the largest infrared excesses, observed over wavelengths from K-band (2.2 μm) to 24 μm (Reach et al. 2005).

There are now over 40 dusty white dwarfs, all of which are polluted (Farihi 2016). The frequency of infrared excess for white dwarfs has been characterised by several surveys, which reach a consensus of 1–5% (Girven et al. 2012; Farihi et al. 2009; Barber et al. 2012; Debes et al. 2011; Hoard et al. 2013; Mullally et al. 2007). The majority of the white dwarfs where an infrared excess has been detected are warm stars, with $T_* > 10{,}000\,$K, with a few notable exceptions (Xu and Jura 2012). The rates of infrared excess are strikingly lower than the fraction of stars where pollution is detected of around 30%, e.g. Zuckerman et al. (2003, 2010) and Koester et al. (2014).

Infrared spectroscopy of six dusty white dwarfs has revealed the presence of molecular emission features, most notably from silicates (Jura et al. 2009). The dusty material accreting onto these polluted white dwarfs appears to have a composition similar to rocky material in our Solar System, and similar to the compositions derived from white dwarf pollution (Xu et al. 2013). Future observations, for example using *MIRI* on *JWST*, will be well placed to further characterise the composition of the dust.

8.3.1.1 Structure of Emitting Dust

The observations constrain the temperature of the emitting dusty material; it clearly lies very close to the white dwarfs, generally within the tidal limit. The standard model, widely used in the literature to explain the observed emission is an opaque, flat dust disk, akin to Saturn's rings (Jura 2003). This dust disk must have a scale height that is significantly less than the white dwarf radius, in order that sufficient energy can be reprocessed by the disk to explain the observed infrared luminosities of the brightest objects (Reach et al. 2009). The observed silicate emission features (Jura et al. 2009) must result from an optically thin region of the disk, such as a surface layer or warped disk (Jura et al. 2007).

8.3.1.2 Accretion and Dust

The dust observed close to white dwarfs supplies the accretion that is observed as pollution. Dust spirals inwards under Poynting-Robertson drag (PR-drag). Rafikov (2011a) and Bochkarev and Rafikov (2011) show that PR-drag alone is sufficient to supply the observed accretion in most polluted white dwarfs in an opaque, flat dust disk. Rafikov (2011b) and Metzger et al. (2012) suggest the coupling between gas and dust as a potential mechanism to enhance the accretion rates for those polluted white dwarfs with accretion rates higher than possible from PR-drag alone.

In this section we focus on what infrared observations of white dwarfs can tell us about the accretion. A conundrum exists. The fraction of white dwarfs with an infrared excess (a few percent) is significantly lower than the fraction that are polluted (about 30%). Bonsor et al. (2016) show that this discrepancy cannot be explained by opaque, flat dust disks that escape detection. At 4.5 μm, observations with *Spitzer* or *WISE* should detect 95% of these disks, if they extend from $T_{in} = 1400$ K to the Roche limit.

In order to assess the frequency of dust around white dwarfs, Bonsor et al. (2016) consider an unbiased sample of 528 white dwarfs with *Spitzer* and *WISE* observations collated from the literature. Figure 8.3 plots the cumulative distribution of infrared excesses at 4.5 μm (or W2) in this sample, this is the fraction of those white dwarfs where the observations are sensitive to a given excess, that have an infrared excess above the given level. The observations are consistent with up to 3.3% of the stars in the sample having an opaque, flat dust disk that spans from

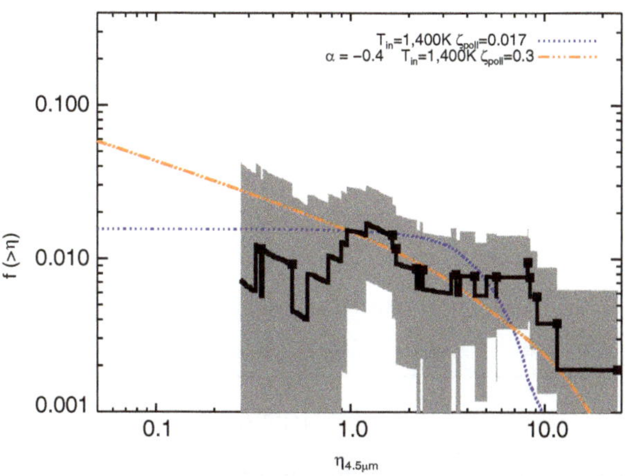

Fig. 8.3 Infrared observations of an unbiased sample of white dwarfs are consistent with the presence of a wide, opaque, flat dust disk around up to 3.3%, whilst pollution occurs for around 30% of stars. Plotted is the cumulative distribution of infrared excesses (*black*); the number of stars in the sample with an infrared excess above a given level, divided by the number of stars for which such an excess is detectable, compared to two models in which all stars have a disk that extends from $T_{in} = 1400$ K to the Roche limit, or a distribution of disk widths, where the number of disks of width $\delta r/r$ varies as $(\delta r/r)^{-0.4}$ (Bonsor et al. 2016)

$T_{in} = 1400\,K$ to the Roche limit, and has an inclination randomly selected from an isotropic distribution, as shown by the blue dashed line in Fig. 8.3. This is significantly lower than the typical pollution rate of 30%. The only way to reconcile the observations with 30% of stars having an opaque, flat dust disks, is for most of those dust disks to be narrow. If the number of disks with a width $\delta r/r$ decreases as $(\delta r/r)^{-0.4}$, 30% of the stars in the sample can have a dust disk, yet 18% must have a dust disk narrower than $10^{-3}r$, as shown by the orange dot-dashed line. Whether it is physically reasonable for so many stars to have such narrow dust disks remains unclear.

Figure 8.4 and Bonsor et al. (2016) present four alternative explanations for the absence of an infrared excess, depending on the white dwarf's temperature and accretion rate. For the oldest white dwarfs, with the longest sinking timescales, the absence of an infrared excess can be explained by an opaque, flat dust disk that has been fully accreted, yet sustains detectable pollution (green region). The absence of an infrared excess for white dwarfs with low accretion rates ($<10^7\,gs^{-1}$) can be sustained by the accretion, via Poynting-Robertson drag, of optically thin dust, whose infrared emission is below our detection limits, however, this dust must be replenished, if no variability in the infrared excess or pollution is expected (blue region). For the hottest white dwarfs, dust that is released from a disrupting body and directly heated by the stellar radiation will sublimate. Those hot polluted white dwarfs without an infrared excess could have pure gas accretion (brown

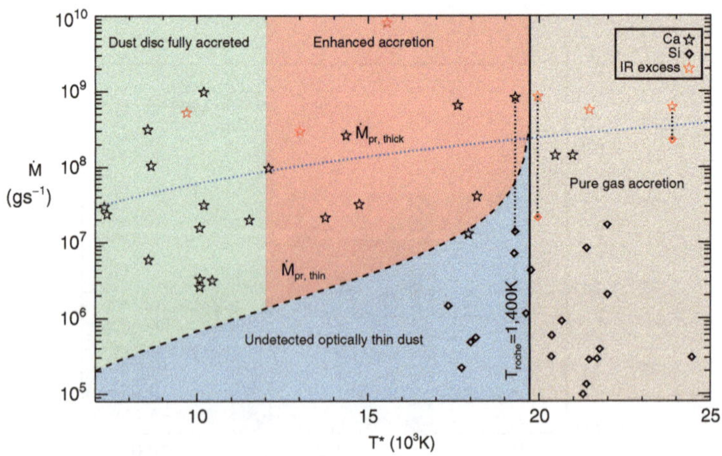

Fig. 8.4 The absence of an infrared excess for a polluted white dwarf can be explained by (1) undetected optically thin dust (*blue*) (2) pure gas accretion, as dust sublimates directly (*brown*) (3) enhanced accretion of optically thin dust (*red*) or (4) short dust disk lifetimes compared to sinking timescales (*green*). The *dashed and dotted lines* show the accretion rate due to Poynting-Robertson of undetected ($R_{4.5\,\mu m} = 0.3$) optically thin and optically thick dust disk (Rafikov 2011a), with an inner temperature of $T_{in} = 1400$ K and $R_{out} = R_{roche}$. The accretion rates over plotted are from either Ca (*stars*) (Farihi, personal communication) or Si (*diamonds*) (Koester et al. 2014), assuming a composition of bulk Earth. *Red symbols* indicate the detection of an infrared excess

region). The greatest uncertainty remaining with this scenario is whether the fast timescales for accretion of gas (Rafikov 2011b; Metzger et al. 2012) can be reconciled with the continued presence of accretion at a steady-state, potentially via the replenishment of material. Warm white dwarfs with high accretion rates are the hardest to explain. It could be that the accreted material in these objects sublimates at lower temperatures ($\sim$900 K), such that the brown region takes over the red region. Alternatively, the accretion of optically thin dust could be enhanced by the presence of gas. Future unbiased surveys that examine the occurrence of pollution, dust and gas in white dwarfs are key to understanding the full picture.

8.3.1.3 Variability

Recently, short term variability has been discovered around a dusty white dwarf WD J0959-0200 (Xu and Jura 2014). Within 300 days in 2010, its fluxes in 3.6 and 4.5 μm have decreased by $\sim 30\%$ and the flux levels have remained unchanged ever since then. WD J0959-0200 also displays calcium triplet emission from an orbiting gas disk[1] (Farihi et al. 2012). There are at least two scenarios that can explain the partial destruction of the dust disk. An impact of a small object onto a pre-existing dust disk or instability within the dust disk, possibly due to the coupling of the dust and gas (Rafikov and Garmilla 2012; Metzger et al. 2012). Further investigation is needed to understand the origin of the change of the infrared fluxes.

8.4 Compositions of the Accreting Planetesimals

Spectroscopic observations of polluted white dwarfs allow us to infer the chemical compositions of the accreting planetary material. There are about a thousand polluted white dwarfs know to-date and the majority were discovered in the Sloan Digital Sky Survey (Dufour et al. 2007; Koester et al. 2011). Calcium is the most easily detected element in the optical wavelength range and it is the only element detected in most polluted white dwarfs. In order to measure the full composition, high resolution spectroscopic observations are required. There are only a handful of such objects that have been thoroughly studied (Jura and Young 2014). In the optical range, we are mostly sensitive to the rock-forming elements, such as Ca, Mg, Si, Fe and O. The ultraviolet wavelength range is most ideal for detecting volatile elements, including C, S and N. So far, a total of 18 elements have been detected in a white dwarf's atmosphere, including C, O, Na, Mg, Al, Si, P, S, Ca, Sc, Ti, V, Cr,

[1]Out of all the dusty white dwarfs, eight of them display double peaked emission lines from a circumstellar gas disk. Some of the gas lines also change on yearly timescale (e.g. Gänsicke et al. 2006). However, the detailed discussion about these systems is beyond the scope of this review.

Mn, Fe, Ni, Zn and Sr. Here, we summarise the highlights of the compositions of extrasolar planetesimals and discuss the implications of these measurements.

8.4.1 Bulk Compositions

Our solar system is the best studied planetary system and can be used as a benchmark to understand other planetary systems. In terms of bulk composition, the best measurement comes from the analysis of meteorites, which are the building blocks of rocky planets. The standard picture is that planets formed by accreting from material in the surrounding area, which is strongly dependent on the temperature of the protoplanetary disk at a given location (Larimer 1967; Wetherill 1990). The overall compositions of the solar system objects are a result of the temperature gradient of the protoplanetary disk: dry, rocky, terrestrial planets in the inner solar system and ice and gas giants in the outer solar system. CI chondrites are considered to be the most primitive object in the solar system and have a composition that closely resembles the solar composition (Lodders 2003). Do extrasolar planetary systems have similar abundance patterns?

Figure 8.5 shows the abundances of materials that have been accreted onto polluted white dwarfs (Xu et al. 2014). The composition of bulk Earth and comet Halley are also shown for comparison. So far, the accreting material mostly resembles the composition of bulk Earth, which is dominated by four elements, O, Mg, Si and Fe. The only possible exception is #12 Ton 345, which has accreted a

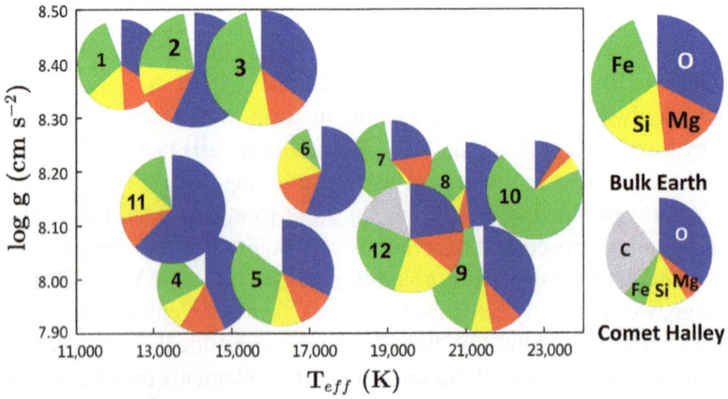

Fig. 8.5 A compilation of 12 white dwarfs with detections of at least O, Mg, Si and Fe (updated from Xu et al. 2014). The abscissa represents the effective temperature of the white dwarf, which is related its cooling age. The ordinate is surface gravity, which is connected to the main sequence mass. The mass fractions of the major elements are labelled. #12 Ton 345 is the only white dwarf that has accreted a planetesimal with a substantial amount of carbon. Here, the result from Jura et al. (2015) is shown. See Sect. 8.4.1 for more details. Most white dwarfs have accreted extrasolar planetesimals with compositions similar to bulk Earth

significant amount of carbon (Jura et al. 2015). The relatively low oxygen abundance suggests the accreting material has little H_2O. The parent body is probably an analogue to dry Kuiper Belt Objects. However, the high carbon result is a bit controversial because a different group found a much lower carbon abundance by analysing the ultraviolet spectrum of Ton 345 (Wilson et al. 2015).

In Fig. 8.5, #6 GD 61 and #11 WD J1242+5226 have accreted a significant amount of oxygen. Detailed abundance analysis shows that the oxygen is over abundant compared to the quantities required to combine with other observed elements to form typical rocky minerals, e.g. MgO, SiO_2, FeO or Fe_2O_3, CaO, Al_2O_3, etc. Therefore, the natural explanation for the oxygen over-abundance is that H_2O must be an important constituent of the accreting material (Farihi et al. 2013; Raddi et al. 2015).

Note all white dwarfs shown in Fig. 8.5 sit in a "sweet spot" of 11,000–23,000 K. At a higher temperature, pollution could come from radiative levitation, which complicates the analysis (Koester et al. 2014); at a lower temperature, dredge-up from the core could also bring material to the white dwarf's atmosphere (Brassard et al. 2007). No general trend is present in the abundance of the accreting material with respect to the white dwarf's effective temperature or the main sequence mass. It has been proposed that as the white dwarf cools/ages, the accreting material could change from asteroid-analog to comet-analog (Bonsor et al. 2011; Xu and Jura 2012). High resolution spectroscopy for a large sample of white dwarfs at different temperatures is required to study the full range of compositions in exo-planetary asteroids.

8.4.2 Differentiation and Collisions

In the solar system, there are objects with very different compositions from the chondritic abundances. For example, iron meteorites are largely made from iron and probably come from the core of a differentiated parent body. The formation of Mercury is likely to involve the collision of two objects (Benz et al. 1988). Differentiation and collisions can significantly change the compositions of extra-solar planetesimals; the formation of planets with non-chondritic compositions is inevitable (O'Neill and Palme 2008).

One idea for exploring these effects is shown in Fig. 8.6, which looks for evidence for core and crust separation. Fe and Al are a representative core and crust element, respectively. The dispersion between Fe-to-Al ratios measured in polluted white dwarfs is much larger than the ratios measured in CI chondrites and planet-hosting stars. It shows that differentiation and collisions can be common in extrasolar planetary materials (Jura et al. 2013). In addition, 16 heavy elements are detected in the parent body that is accreting onto the white dwarf GD 362. The best match solar system object with a similar composition is mesosiderite, a blend of core and crust material (Xu et al. 2013). Evidence for differentiation and collision can be found in individual objects given enough elements are detected.

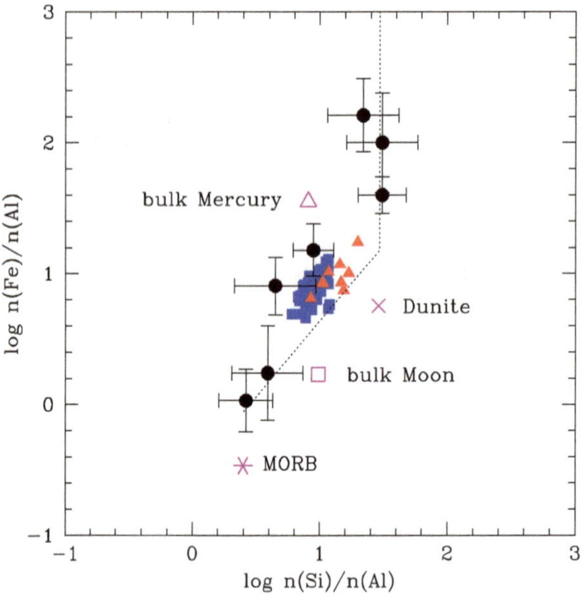

Fig. 8.6 Abundance ratios between Fe and Al versus Si and Al from materials accreted onto polluted white dwarfs (*black dots*), chondrites (*red triangles*) and planet-hosting stars (*blue squares*). A few solar system rocks are also included as magenta symbols (bulk Mercury, Dunite, bulk Moon, MORB [Mid Ocean Ridge Basalt] Jura et al. 2013). Si and Al are typical crust material while Fe is mostly contained in the planetary core. The *vertical black dotted line* represents a model with a blend of core and mantle material while the *sloped black dotted line* represents a blend of core and crust material with 10% core by mass. The toy model can fit all the observed abundances in polluted white dwarfs. Chondrites and planet-hosting stars reside in a very small phase space. In addition, the dispersion between the Fe-to-Al ratios is much larger than that of Si-to-Al ratios, likely due to differentiation and collision

Plate tectonics can be closely related to a planet's habitability and its effect has been under intense investigation (Valencia et al. 2007; Foley et al. 2012). Earth is the only known object that has undergone plate tectonics and an important outcome is the enrichment of the so-called "incompatible" elements, e.g. Ba, Sr, in Earth's continental crust compared to other crust materials in the solar system. Polluted white dwarfs can be used to search for evidence of plate tectonics on extrasolar rocky objects. A pilot study on two objects was performed and upper limits of Ba-to-Ca ratios were obtained (Jura et al. 2014). No evidence for plate tectonics was found but this search should be extended to additional targets.

8.4.3 Implications

Planetesimals are the building blocks of planets and measuring their compositions can place strong constraint on planet formation models. For example, recent studies hypothesised that carbon planets could form around carbon-enhanced stars (Bond et al. 2010; Moriarty et al. 2014). So far, all the abundances observed in polluted white dwarfs have found low carbon abundances. Whatever the planet formation pathway it is, it must be universal in both the solar system and extrasolar planetary systems.

Tracing the volatile elements and locating the snow line/soot line is a key goal in the study of protoplanetary disks (Lecar et al. 2006; Qi et al. 2013). One end result of volatiles are accreted onto the planetary objects in the system. Viewed as an ensemble, most white dwarfs are accreting from water-depleted extrasolar objects (Jura and Xu 2012). We are sampling drying rocky objects and this provides indirect evidence that snow lines can be present in other systems.

8.5 WD 1145+017: A White Dwarf with an Actively Disintegrating Asteroid

The tidal disruption of an asteroid is the standard model to explain white dwarf pollution (Jura 2003). However, the disintegrating asteroid has never been directly observed, until very recently.

8.5.1 Intriguing Light Curve

WD 1145+017 was observed in the first campaign of the extended *Kepler* mission. Multiple transits with periods around 4.5 h were detected during the 80-day observation and significant evolutions were detected (Vanderburg et al. 2015). Ground-based follow-ups came quickly and the transit properties are very unique (Croll et al. 2015; Rappaport et al. 2016; Gänsicke et al. 2016; Gary et al. 2017). (1) Transits are deep—transit depths as deep as 70% have been detected. (2) Transit profiles are asymmetric—the egress lasts much longer than ingress. Similar transit profiles have been observed for disintegrating planets around main sequence stars. (3) The transit duration is much longer than expected for a solid object. (4) Light curve changes on a daily basis and it can be completely different after several weeks. A light curve is shown in Fig. 8.7.

The general consensus is that there is an actively disintegrating object around the white dwarf WD 1145+017. The transits are produced by dust coming off from the disintegrating objects. There are still many mysteries, particularly the long term evolution of the system (Veras et al. 2016b). How many fragments are there in the system? What is the dust production mechanism? What is the typical size and composition of the dust?

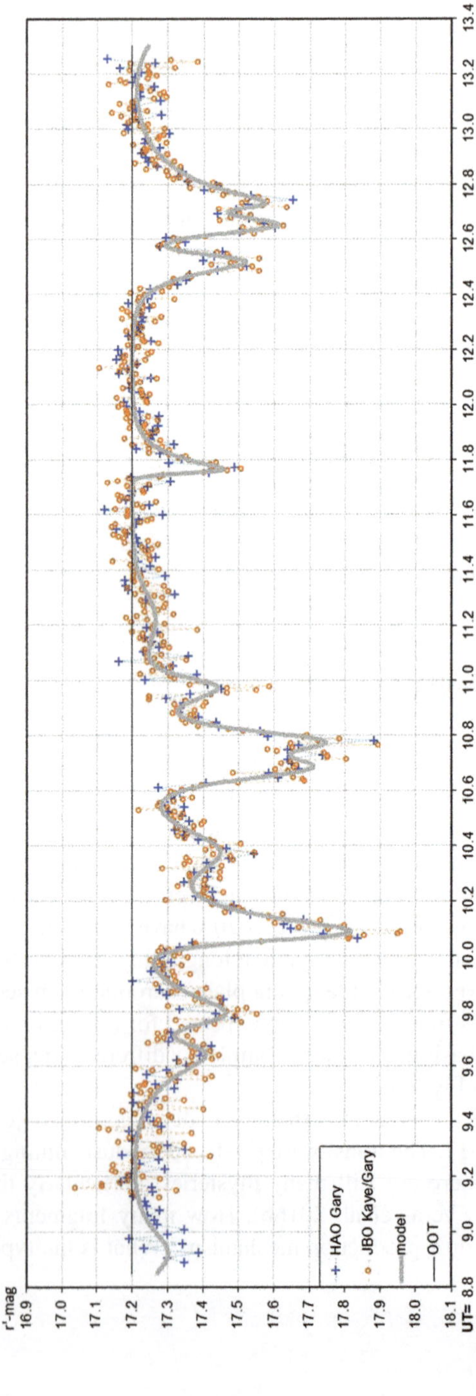

Fig. 8.7 A representative light curve of WD 1145+017 from the night of 2015 December 16, covering two full periods (updated from Rappaport et al. 2016; Gary et al. 2017). The *blue crosses* and *yellow circles* represent observations from the Gary 35 cm and Kaye 80 cm telescope, respectively. The *grey line* is the model fit to the light curve and the *black line* is the out-of-transit level

8.5.2 *Infrared Excess*

WD 1145+017 displays strong excess infrared radiation starting from the K band (Vanderburg et al. 2015). The SED can be fit by a warm disk within the tidal radius of the white dwarf, similar to the disk model discussed in Sect. 8.3.1. However, it is puzzling that the dust disk is not aligned with the transiting objects (Vanderburg et al. 2015; Xu et al. 2016). The infrared excess can also be fitted with a blackbody companion. Future observations, particularly near infrared spectroscopy, would reveal the nature of the infrared excess.

8.5.3 *Polluted Atmosphere*

High-resolution spectroscopic observations of WD 1145+017 with the *Keck Telescope* reveal 11 heavy elements, including O, Mg, Al, Si, Ca, Ti, V:, Cr, Mn, Fe and Ni, making it one of the most heavily polluted white dwarfs known (Xu et al. 2016). The overall abundance pattern is very similar to those of CI chondrites, as shown in Fig. 8.8. The relatively high oxygen abundance indicates the possibility of presence of water in the accreting material. However, further observations are required to confirm this due to the large uncertainty associated in the oxygen measurement

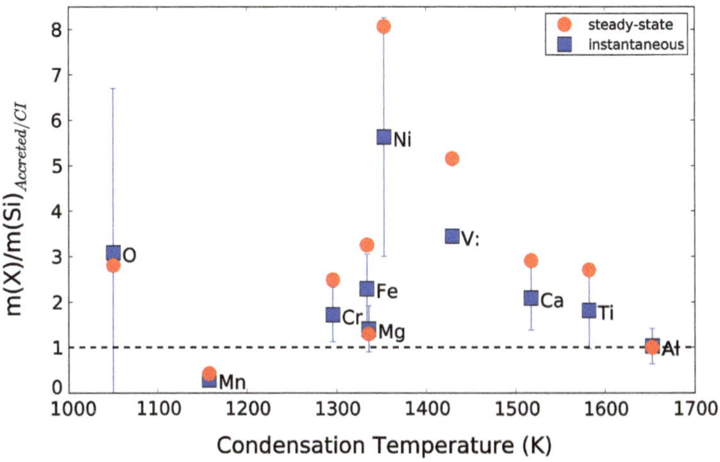

Fig. 8.8 Mass fraction of an element relative to Si in the material accreted onto WD 1145+017, normalised to the ratios in CI chondrites (Xu et al. 2016). Two models to derive the intrinsic abundances of the accreting material are presented, including the steady-state approximation and instantaneous model. Within the uncertainties, the overall abundance ratios are very similar to those observed in CI chondrites, consistent with abundances observed in other polluted white dwarfs (see Fig. 8.5)

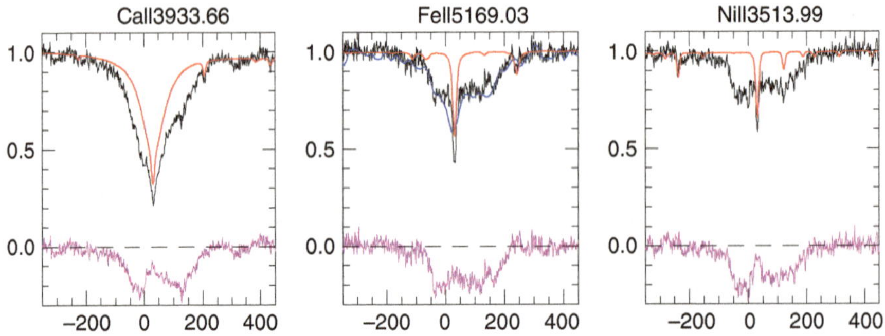

Fig. 8.9 Example of circumstellar feature detected around WD 1145+017 from Xu et al. (2016). The figure is shown in velocity space in the white dwarf's reference frame. The *black and blue lines* represents data taken with Keck/HIRES and a slightly lower resolution spectrograph ESI a few weeks later. The *red line* represents the best model fit to the photospheric lines. The *lower solid magenta line* is the HIRES data minus the model—residual from circumstellar gas absorption

(Xu et al. 2016). High resolution ultraviolet spectroscopy would allow us to measure the abundances of volatile elements, such as C, S and possibly N so we could assess the volatile content of the accreting material.

8.5.4 Circumstellar Gas

High-resolution spectroscopic observations from *Keck* also discovered numerous absorption features from Mg, Ca, Ti, Cr, Mn, Fe and Ni around WD 1145+017. These features are broad, with a line width of $\sim$300 km s^{-1}, asymmetric and deep, with an average depth about 15% below the continuum, as shown in Fig. 8.9. They all come from transitions with a lower energy level between 0–3 eV (Xu et al. 2016). These absorption lines are formed from circumstellar gas around WD 1145+017, likely associated with the disintegrating objects.

Previously, circumstellar gas has been detected around other white dwarfs, but none of them are morphologically similar to WD 1145+017. By monitoring the circumstellar features, Redfield et al. (2016) found that they change on a very short timescale. Further investigation is needed to understand the changes of the circumstellar feature.

8.6 Conclusions

In this chapter, we provide evidence that planetary systems around white dwarfs are common. Studying these systems can provide us with vital information regarding the formation and evolution of planetary systems, in particular the bulk compositions of extrasolar rocky objects. There is no other technique in the foreseeable future that will provide such good bulk compositions for rocky planetary material outside our Solar system. The recent discovery of an actively disintegrating asteroid around WD 1145+017 provides an exciting opportunity to study the real time tidal disruption processes and further our understanding of white dwarf planetary systems.

Looking into the future, *Gaia* will significantly increase the white dwarf population. Ongoing and future spectroscopic surveys, e.g. the *Sloan Digital Sky Survey*, WEAVE and 4MOST, will return many more polluted white dwarfs. The *James Webb Space Telescope* will allow us to measure the dust composition and directly link that to the atmospheric pollution. In addition, upcoming transit surveys and time-domain studies, such as *TESS*, *LSST* and *Pan-STARRS*, will likely return many more objets similar to WD 1145+017.

Acknowledgements We thank Bruce Gary and Saul Rappaport for updating Fig. 8.7 for this chapter.

References

Aumann, H.H., Beichman, C.A., Gillett, F.C., de Jong, T., Houck, J.R., Low, F.J., Neugebauer, G., Walker, R.G., Wesselius, P.R.: Discovery of a shell around Alpha Lyrae. Astrophys. J. Lett. **278**, L23–L27 (1984), doi:10.1086/184214

Barber, S.D., Patterson, A.J., Kilic, M., Leggett, S.K., Dufour, P., Bloom, J.S., Starr, D.L.: The frequency of debris disks at white dwarfs. Astrophys. J. 760(1), Article ID. 26, 11 pp. (2012). doi:10.1088/0004-637X/760/1/26

Benz, W., Slattery, W.L., Cameron, A.G.W.: Collisional stripping of Mercury's mantle. Icarus **74**, 516–528 (1988). doi:10.1016/0019-1035(88)90118-2

Bochkarev, K.V., Rafikov, R.R.: Global modeling of radiatively driven accretion of metals from compact debris disks onto white dwarfs. Astrophys. J. 741(1), Article ID. 36, 9 pp. (2011). doi:10.1088/0004-637X/741/1/36

Bond, J.C., O'Brien, D.P., Lauretta, D.S.: The compositional diversity of extrasolar terrestrial planets. I. In situ simulations. Astrophys. J. **715**, 1050–1070 (2010). doi:10.1088/0004-637X/715/2/1050, 1004.0971

Bonsor, A., Veras, D.: A wide binary trigger for white dwarf pollution. Mon. Not. R. Astron. Soc. **454**, 53–63 (2015). doi:10.1093/mnras/stv1913, 1508.05715

Bonsor, A., Wyatt, M.: Post-main-sequence evolution of A star debris discs. Mon. Not. R. Astron. Soc. **409**, 1631–1646 (2010). doi:10.1111/j.1365-2966.2010.17412.x, 1007.4517

Bonsor, A., Mustill, A.J., Wyatt, M.C.: Dynamical effects of stellar mass-loss on a Kuiper-like belt. Mon. Not. R. Astron. Soc. **414**, 930–939 (2011), doi:10.1111/j.1365-2966.2011.18524.x, 1102.3185

Bonsor, A., Raymond, S.N., Augereau, J.C.: The short-lived production of exozodiacal dust in the aftermath of a dynamical instability in planetary systems. Mon. Not. R. Astron. Soc. **433**, 2938–2945 (2013). doi:10.1093/mnras/stt933, 1306.0592

Bonsor, A., Farihi, J., Wyatt, M.C., Van Lieshout, R.: Mon. Not. R. Astron. Soc. **468**, 154 (2017)

Brassard, P., Fontaine, G., Dufour, P., Bergeron, P.: The origin and evolution of DQ white dwarfs: the carbon pollution problem revisited. In: Napiwotzki, R., Burleigh, M.R. (eds.) 15th European Workshop on White Dwarfs, Astronomical Society of the Pacific Conference Series, vol. 372, p. 19 (2007)

Croll, B., Dalba, P.A., Vanderburg, A., Eastman, J., Rappaport, S., DeVore, J., Bieryla, A., Muirhead, P.S., Han, E., Latham, D.W., Beatty, T.G., Wittenmyer, R.A., Wright, J.T., Johnson, J.A., McCrady, N.: Multiwavelength transit observations of the candidate disintegrating planetesimals orbiting WD 1145+017. (2015). ArXiv e-prints 1510.06434

Debes, J.H., Sigurdsson, S.: Are there unstable planetary systems around white dwarfs? Astrophys. J. **572**, 556–565 (2002). doi:10.1086/340291, astro-ph/0202273

Debes, J.H., Hoard, D.W., Kilic, M., Wachter, S., Leisawitz, D.T., Cohen, M., Kirkpatrick, J.D., Griffith, R.L.: The WIRED survey. I. A bright IR excess due to dust around the heavily polluted white Dwarf Galex J193156.8+011745. Astrophys. J. **729**(1), Article ID. 4, 6 pp. (2011). doi:10.1088/0004-637X/729/1/4

Debes, J.H., Walsh, K.J., Stark, C.: The link between planetary systems, dusty white dwarfs, and metal-polluted white dwarfs. Astrophys. J. **747**, 148 (2012). doi:10.1088/0004-637X/747/2/148, 1201.0756

Dufour, P., Bergeron, P., Liebert, J., Harris, H.C., Knapp, G.R., Anderson, S.F., Hall, P.B., Strauss, M.A., Collinge, M.J., Edwards, M.C.: On the spectral evolution of cool, helium-atmosphere white dwarfs: detailed spectroscopic and photometric analysis of DZ stars. Astrophys. J. **663**, 1291–1308 (2007). doi:10.1086/518468, astro-ph/0703758

Duncan, M.J., Lissauer, J.J.: The effects of post-main-sequence solar mass loss on the stability of our planetary system. Icarus **134**, 303–310 (1998). doi:10.1006/icar.1998.5962

Farihi, J.: Circumstellar debris and pollution at white dwarf stars. New Astron. Rev. **71**, 9–34 (2016) doi:10.1016/j.newar.2016.03.001

Farihi, J., Jura, M., Zuckerman, B.: Infrared signatures of disrupted minor planets at white dwarfs. Astrophys. J. **694**, 805–819 (2009). doi:10.1088/0004-637X/694/2/805, 0901.0973

Farihi, J., Gänsicke, B.T., Steele, P.R., Girven, J., Burleigh, M.R., Breedt, E., Koester, D.: A trio of metal-rich dust and gas discs found orbiting candidate white dwarfs with K-band excess. Mon. Not. R. Astron. Soc. **421**, 1635–1643 (2012). doi:10.1111/j.1365-2966.2012.20421.x, 1112.5163

Farihi, J., Gänsicke, B.T., Koester, D.: Evidence for water in the rocky debris of a disrupted extrasolar minor planet. Science **342**, 218–220 (2013), doi:10.1126/science.1239447, 1310.3269

Foley, B.J., Bercovici, D., Landuyt, W.: The conditions for plate tectonics on super-Earths: Inferences from convection models with damage. Earth Planet. Sci. Lett. **331**, 281–290 (2012). doi:10.1016/j.epsl.2012.03.028

Frewen, S.F.N., Hansen, B.M.S.: Eccentric planets and stellar evolution as a cause of polluted white dwarfs. Mon. Not. R. Astron. Soc. **439**, 2442–2458 (2014). doi:10.1093/mnras/stu097, 1401.5470

Gänsicke, B.T., Marsh, T.R., Southworth J, Rebassa-Mansergas A (2006) A gaseous metal disk around a white dwarf. Science 314:1908, doi:10.1126/science.1135033, astro-ph/0612697

Gänsicke, B.T., Aungwerojwit, A., Marsh, T.R., Dhillon, V.S., Sahman, D.I., Veras, D., Farihi, J., Chote, P., Ashley, R., Arjyotha, S., Rattanasoon, S., Littlefair, S.P., Pollacco, D., Burleigh, M.R.: High-speed photometry of the disintegrating planetesimals at WD1145+017: evidence for rapid dynamical evolution. Astrophys. J. Lett. **818**, L7 (2016). doi:10.3847/2041-8205/818/1/L7, 1512.09150

Gary, B.L., Rappaport, S., Kaye, T.G., Alonso, R., Hambschs, F.J.: WD 1145+017 photometric observations during eight months of high activity. Mon. Not. R. Astron. Soc. **465**, 3267–3280 (2017). doi:10.1093/mnras/stw2921, 1608.00026

Girven, J., Brinkworth, C.S., Farihi, J., Gänsicke, B.T., Hoard, D.W., Marsh, T.R., Koester, D.: Constraints on the lifetimes of disks resulting from tidally destroyed rocky planetary bodies. Astrophys. J. **749**, 154 (2012). doi:10.1088/0004-637X/749/2/154, 1202.3784

Graham, J.R., Matthews, K., Neugebauer, G., Soifer, B.T.: The infrared excess of G29-38 - a brown dwarf or dust? Astrophys. J. **357**, 216–223 (1990a). doi:10.1086/168907

Graham, J.R., Reid, I.N., McCarthy, J.K., Rich, R.M.: Does G29-38 have a massive companion? Astrophys. J. Lett. **357**, L21–L24 (1990b). doi:10.1086/185757

Haas, M., Leinert, C.: Search for the suspected brown dwarf companion to Giclas 29-38 using IR-slit-scans. Astron. Astrophys. **230**, 87–90 (1990)

Hoard, D.W., Debes, J.H., Wachter, S., Leisawitz, D.T., Cohen, M.: The WIRED survey. IV. New dust disks from the McCook & Sion White Dwarf catalog. Astrophys. J. **770**(1), Article ID. 21, 14 pp. (2013). doi:10.1088/0004-637X/770/1/21

Jura, M.: A tidally disrupted asteroid around the white dwarf G29-38. Astrophys. J. Lett. **584**, L91–L94 (2003). doi:10.1086/374036, astro-ph/0301411

Jura, M., Xu, S.: Water fractions in extrasolar planetesimals. Astron. J. **143**, 6 (2012). doi:10.1088/0004-6256/143/1/6, 1110.1774

Jura, M., Young, E.D.: Extrasolar cosmochemistry. Annu. Rev. Earth Planet. Sci. **42**, 45–67 (2014). doi:10.1146/annurev-earth-060313-054740

Jura, M., Farihi, J., Zuckerman, B., Becklin, E.E.: Infrared emission from the dusty disk orbiting GD 362, an externally polluted white dwarf. Astron. J **133**, 1927–1933 (2007). doi:10.1086/512734, astro-ph/0701469

Jura, M., Farihi, J., Zuckerman, B.: Six white dwarfs with circumstellar silicates. Astron. J. **137**, 3191–3197 (2009). doi:10.1088/0004-6256/137/2/3191, 0811.1740

Jura, M., Xu, S., Young, E.D.: ^{26}Al in the early solar system: not so unusual after all. Astrophys. J. Lett. **775**, L41 (2013). doi:10.1088/2041-8205/775/2/L41, 1308.6325

Jura, M., Klein, B., Xu, S., Young, E.D.: A pilot search for evidence of extrasolar earth-analog plate tectonics. Astrophys. J. Lett. **791**, L29 (2014). doi:10.1088/2041-8205/791/2/L29, 1407.5923

Jura, M., Dufour, P., Xu, S., Zuckerman, B., Klein, B., Young, E.D., Melis, C.: Evidence for an anhydrous carbonaceous extrasolar minor planet. Astrophys. J. **799**, 109 (2015). doi:10.1088/0004-637X/799/1/109, 1411.5036

Kaib, N.A., Raymond, S.N., Duncan, M.: Planetary system disruption by Galactic perturbations to wide binary stars. Nature **493**, 381–384 (2013). doi:10.1038/nature11780, 1301.3145

Koester, D., Provencal, J., Shipman, H.L.: Metals in the variable DA G29-38. Astron. Astrophys. **320**, L57–L59 (1997)

Koester, D., Girven, J., Gänsicke, B.T., Dufour, P.: Cool DZ white dwarfs in the SDSS. Astron. Astrophys. **530**, A114 (2011). doi:10.1051/0004-6361/201116816, 1105.0268

Koester, D., Gänsicke, B.T., Farihi, J.: The frequency of planetary debris around young white dwarfs. Astron. Astrophys. **566**, A34 (2014). doi:10.1051/0004-6361/201423691, 1404.2617

Larimer, J.W.: Chemical fractionations in meteorites – I. Condensation of the elements. Geochim. Cosmochim. Acta **31**, 1215–1238 (1967). doi:10.1016/S0016-7037(67)80013-9

Lecar, M., Podolak, M., Sasselov, D., Chiang, E.: On the location of the snow line in a protoplanetary disk. Astrophys. J. **640**, 1115–1118 (2006). doi:10.1086/500287, astro-ph/0602217

Lodders, K.: Solar system abundances and condensation temperatures of the elements. Astrophys. J. **591**, 1220–1247 (2006). doi:10.1086/375492

Metzger, B.D., Rafikov, R.R., Bochkarev, K.V.: Global models of runaway accretion in white dwarf debris discs. Mon. Not. R. Astron. Soc. **423**, 505–528 (2012). doi:10.1111/j.1365-2966.2012.20895.x, 1202.0557

Moriarty, J., Madhusudhan, N., Fischer, D.: Chemistry in an evolving protoplanetary disk: effects on terrestrial planet composition. Astrophys. J. **787**, 81 (2014). doi:10.1088/0004-637X/787/1/81, 1405.3253

Mullally, F., Kilic, M., Reach, W.T., Kuchner, M.J., von Hippel, T., Burrows, A., Winget, D.E.: A Spitzer white dwarf infrared survey. Astrophys. J. Suppl. Ser. **171**, 206–218 (2007). doi:10.1086/511858, astro-ph/0611588

Murray, C.D., Dermott, S.F.: Solar System Dynamics. Cambridge University Press, Cambridge (1999)

Mustill, A.J., Villaver, E.: Foretellings of Ragnarök: world-engulfing asymptotic giants and the inheritance of white dwarfs. Astrophys. J. **761**, 121 (2012). doi:10.1088/0004-637X/761/2/121, 1210.0328

Mustill, A.J., Veras, D., Villaver, E.: Long-term evolution of three-planet systems to the post-main sequence and beyond. Mon. Not. R. Astron. Soc. **437**, 1404–1419 (2014). doi:10.1093/mnras/stt1973, 1310.3168

O'Neill, H.S.C., Palme, H.: Collisional erosion and the non-chondritic composition of the terrestrial planets. Philos. Trans. R. Soc. Lond. A **366**, 4205–4238 (2008). doi:10.1098/rsta.2008.0111

Payne, M.J., Veras, D., Holman, M.J., Gänsicke, B.T.: Liberating exomoons in white dwarf planetary systems. Mon. Not. R. Astron. Soc. **457**, 217–231 (2016). doi:10.1093/mnras/stv2966, 1603.09344

Qi, C., Öberg, K.I., Wilner, D.J., D'Alessio, P., Bergin, E., Andrews, S.M., Blake, G.A., Hogerheijde, M.R., van Dishoeck, E.F.: Imaging of the CO snow line in a solar nebula analog. Science **341**, 630–632 (2013). doi:10.1126/science.1239560, 1307.7439

Raddi, R., Gänsicke, B.T., Koester, D., Farihi, J., Hermes, J.J., Scaringi, S., Breedt, E., Girven, J.: Likely detection of water-rich asteroid debris in a metal-polluted white dwarf. Mon. Not. R. Astron. Soc. **450**, 2083–2093 (2015). doi:10.1093/mnras/stv701, 1503.07864

Rafikov, R.R.: Metal accretion onto white dwarfs caused by Poynting-Robertson drag on their debris disks. Astrophys. J. Lett. **732**, L3 (2011a). doi:10.1088/2041-8205/732/1/L3, 1102.3153

Rafikov, R.R.: Runaway accretion of metals from compact discs of debris on to white dwarfs. Mon. Not. R. Astron. Soc. L287+ (2011b) doi:10.1111/j.1745-3933.2011.01096.x, 1102.4343

Rafikov, R.R., Garmilla, J.A.: Inner edges of compact debris disks around metal-rich white dwarfs. Astrophys. J. **760**, 123 (2012). doi:10.1088/0004-637X/760/2/123, 1207.7082

Rappaport, S., Gary, B.L., Kaye, T., Vanderburg, A., Croll, B., Benni, P., Foote, J.: Drifting asteroid fragments around WD 1145+017. Mon. Not. R. Astron. Soc. **458**, 3904–3917 (2012). doi:10.1093/mnras/stw612, 1602.00740

Raymond, S.N., Barnes, R., Veras, D., Armitage, P.J., Gorelick, N., Greenberg, R.: Planet-planet scattering leads to tightly packed planetary systems. Astrophys. J. Lett. **696**, L98–L101 (2009). doi:10.1088/0004-637X/696/1/L98, 0903.4700

Raymond, S.N., Armitage, P.J., Gorelick, N.: Planet-planet scattering in planetesimal disks. II. Predictions for outer extrasolar planetary systems. Astrophys. J. **711**, 772–795 (2010). doi:10.1088/0004-637X/711/2/772, 1001.3409

Reach, W.T., Kuchner, M.J., von Hippel, T., Burrows, A., Mullally, F., Kilic, M., Winget, D.E.: The dust cloud around the white dwarf G29-38. Astrophys. J. Lett. **635**, L161–L164 (2005). doi:10.1086/499561, astro-ph/0511358

Reach, W.T., Lisse, C., von Hippel, T., Mullally, F.: The dust cloud around the white dwarf G 29-38. II. Spectrum from 5 to 40 μm and mid-infrared photometric variability. Astrophys. J. **693**, 697–712 (2009). doi:10.1088/0004-637X/693/1/697, 0810.3276

Redfield, S., Farihi, J., Cauley, P.W., Parsons, S.G., Gaensicke, B.T., Duvvuri, G.: Spectroscopic evolution of disintegrating planetesimals: minutes to months variability in the circumstellar gas associated with WD 1145+017 (2016). ArXiv e-prints 1608.00549

Sion, E.M., Holberg, J.B., Oswalt, T.D., McCook, G.P., Wasatonic, R.: The white dwarfs within 20 parsecs of the sun: kinematics and statistics. Astron. J. **138**, 1681–1689 (2009). doi:10.1088/0004-6256/138/6/1681, 0910.1288

Su, K.Y.L., Rieke, G.H., Stansberry, J.A., Bryden, G., Stapelfeldt, K.R., Trilling, D.E., Muzerolle, J., Beichman, C.A., Moro-Martin, A., Hines, D.C., Werner, M.W.: Debris disk evolution around A stars. Astrophys. J. **653**, 675–689 (2006). doi:10.1086/508649, astro-ph/0608563

Telesco, C.M., Joy, M., Sisk, C.: Observations of G29 - 38 at 10 microns. Astrophys. J. Lett. **358**, L17–L19 (1990). doi:10.1086/185769

Tokunaga, A.T., Hodapp, K.W., Becklin, E.E., Cruikshank, D.P., Rigler, M., Toomey, D., Brown, R.H., Zuckerman, B.: Infrared spectroscopy, imaging, and 10 micron photometry of Giclas 29-38. Astrophys. J. Lett. **332**, L71–L74 (1988). doi:10.1086/185269

Valencia, D., O'Connell, R.J., Sasselov, D.D.: Inevitability of plate tectonics on super-earths. Astrophys. J. Lett. **670**, L45–L48 (2007). doi:10.1086/524012, 0710.0699

Vanderburg, A., Johnson, J.A., Rappaport, S., Bieryla, A., Irwin, J., Lewis, J.A., Kipping, D., Brown, W.R., Dufour, P., Ciardi, D.R., Angus, R., Schaefer, L., Latham, D.W., Charbonneau,

D., Beichman, C., Eastman, J., McCrady, N., Wittenmyer, R.A., Wright, J.T.: A disintegrating minor planet transiting a white dwarf. Nature **526**, 546–549 (2015). doi:10.1038/nature15527, 1510.06387
Veras, D., Wyatt, M.C.: The solar system's post-main-sequence escape boundary. Mon. Not. R. Astron. Soc. **421**, 2969–2981 (2012). doi:10.1111/j.1365-2966.2012.20522.x, 1201.2412
Veras, D., Wyatt, M.C., Mustill, A.J., Bonsor, A., Eldridge, J.J.: The great escape: how exoplanets and smaller bodies desert dying stars. Mon. Not. R. Astron. Soc. **417**, 2104–2123 (2011). doi:10.1111/j.1365-2966.2011.19393.x, 1107.1239
Veras, D., Mustill, A.J., Bonsor, A., Wyatt, M.C.: Simulations of two-planet systems through all phases of stellar evolution: implications for the instability boundary and white dwarf pollution. Mon. Not. R. Astron. Soc. **431**, 1686–1708 (2013). doi:10.1093/mnras/stt289, 1302.3615
Veras, D., Mustill, A.J., Gänsicke, B.T., Redfield, S., Georgakarakos. N., Bowler, A.B., Lloyd, M.J.S.: Full-lifetime simulations of multiple unequal-mass planets across all phases of stellar evolution. Mon. Not. R. Astron. Soc. **458**, 3942–3967 (2016a). doi:10.1093/mnras/stw476, 1603.00025
Veras, D., Marsh, T.R., Gänsicke, B.T.: Dynamical mass and multiplicity constraints on co-orbital bodies around stars. Mon. Not. R. Astron. Soc. **461**(2), 1413–1420 (2016b). doi:10.1093/mnras/stw1324
Villaver, E., Livio, M.: Can planets survive stellar evolution? Astrophys. J. **661**, 1192–1201 (2007). doi:10.1086/516746, astro-ph/0702724
Wetherill, G.W.: Formation of the earth. Annu. Rev. Earth Planet. Sci. **18**, 205–256 (1990). doi:10.1146/annurev.ea.18.050190.001225
Wickramasinghe, N.C., Hoyle, F., Al-Mufti, S.: The infrared excess from the white dwarf star g29-38 - a brown dwarf or dust? Astrophys. Space Sci. **143**, 193–197 (1988). doi:10.1007/BF00636767
Wilson, D.J., Gänsicke, B.T., Koester, D., Toloza, O., Pala, A.F., Breedt, E., Parsons, S.G.: The composition of a disrupted extrasolar planetesimal at SDSS J0845+2257 (Ton 345). Mon. Not. R. Astron. Soc. **451**, 3237–3248 (2015). doi:10.1093/mnras/stv1201, 1505.07466
Wisdom, J.: The resonance overlap criterion and the onset of stochastic behavior in the restricted three-body problem. Astron. J. **85**, 1122–1133 (1980). doi:10.1086/112778
Wolszczan, A., Frail, D.A.: A planetary system around the millisecond pulsar PSR1257 + 12. Nature **355**, 145–147 (1992). doi:10.1038/355145a0
Wyatt, M.C.: Evolution of debris disks. Annu. Rev. Astron. Astrophys. **46**, 339–383 (2008). doi:10.1146/annurev.astro.45.051806.110525
Wyatt, M.C., Smith, R., Su, K.Y.L., Rieke, G.H., Greaves, J.S., Beichman, C.A., Bryden, G.: Steady state evolution of debris disks around A stars. Astrophys. J. **663**, 365–382 (2007). doi:10.1086/518404, astro-ph/0703608
Wyatt, M.C., Farihi, J., Pringle, J.E., Bonsor, A.: Stochastic accretion of planetesimals on to white dwarfs: constraints on the mass distribution of accreted material from atmospheric pollution. Mon. Not. R. Astron. Soc. **439**, 3371–3391 (2014). doi:10.1093/mnras/stu183, 1401.6173
Xu, S., Jura, M.: Spitzer observations of white dwarfs: the missing planetary debris around DZ stars. Astrophys. J. **745**, 88 (2012). doi:10.1088/0004-637X/745/1/88, 1109.4207
Xu, S., Jura, M.: The drop during less than 300 days of a dusty white dwarf's infrared luminosity. Astrophys. J. Lett. **792**, L39 (2014). doi:10.1088/2041-8205/792/2/L39, 1408.1618
Xu, S., Jura, M., Klein, B., Koester, D., Zuckerman, B.: Two beyond-primitive extrasolar planetesimals. Astrophys. J. **766**, 132 (2013). doi:10.1088/0004-637X/766/2/132, 1302.4799
Xu, S., Jura, M., Koester, D., Klein, B., Zuckerman, B.: Elemental compositions of two extrasolar rocky planetesimals. Astrophys. J. **783**, 79 (2014). doi:10.1088/0004-637X/783/2/79, 1401.4252
Xu, S., Jura, M., Dufour, P., Zuckerman, B.: Evidence for gas from a disintegrating extrasolar asteroid. Astrophys. J. Lett. **816**, L22 (2016). doi:10.3847/2041-8205/816/2/L22, 1511.05973
Zuckerman, B.: The occurrence of wide-orbit planets in binary star systems. Astrophys. J. Lett. **791**, L27 (2014). doi:10.1088/2041-8205/791/2/L27, 1407.2666

Zuckerman, B., Becklin, E.E.: Excess infrared radiation from a white dwarf - an orbiting brown dwarf? Nature **330**, 138–140 (1987). doi:10.1038/330138a0

Zuckerman, B., Koester, D., Reid, I.N., Hünsch, M.: Metal lines in DA white dwarfs. Astrophys. J. **596**, 477–495 (2003). doi:10.1086/377492

Zuckerman, B., Melis, C., Klein, B., Koester, D., Jura, M.: Ancient planetary systems are orbiting a large fraction of white dwarf stars. Astrophys. J. **722**, 725–736 (2010). doi:10.1088/0004-637X/722/1/725, 1007.2252

Chapter 9
Observational Signatures of Planet Formation in Recent Resolved Observations of Protoplanetary Disks

Ruobing Dong, Zhaohuan Zhu, and Jeffrey Fung

Abstract Planets form in gaseous protoplanetary disks surrounding newborn stars. As such, the best way to learn how planets form from observations is to directly watch them forming in disks. By doing so, we can directly address the three most fundamental questions in planet formation: *when, where, and how planets form*. In the past, due to the difficulties in directly detecting planets in disks, planet formation has been a subject of theoretical astrophysics. Now, thanks to a fleet of new instruments with unprecedented resolving power that have come online in the past decade, we have just started to be able to directly resolve structures in protoplanetary disks that are unambiguously associated with planet formation activities. By comparing these observations with theoretical models of disk–planet interactions, we will be able to constrain the locations and properties of the (unseen) planets. This is the onset of a new field—*observational planet formation*. In this chapter, we highlight selected recent progresses in this field, with emphases on numerical modeling of observational signatures of planet-induced structures in disks, as well as the comparison between models and observations.

9.1 Introduction

The study of how planets form is among the most active research areas in contemporary astronomy. Although it is widely accepted that they form in gaseous protoplanetary disks surrounding newly born stars (typically one to a few million years old), the specific processes that transform the building blocks—sub-micron-sized interstellar medium grains and various spices of gas molecule in the star

R. Dong (✉) • J. Fung
Department of Astronomy, UC Berkeley, Berkeley, CA 94705, USA
e-mail: rdong@gmail.com; jeffrey.fung@berkeley.edu

Z. Zhu
Department of Astrophysical Sciences, Princeton University, Princeton, NJ 08544, USA
e-mail: zhzhu@astro.princeton.edu

M. Pessah, O. Gressel (eds.), *Formation, Evolution, and Dynamics of Young Solar Systems*, Astrophysics and Space Science Library 445,
DOI 10.1007/978-3-319-60609-5_9

forming environment—to the final products are not well understood. Because protoplanetary disks are the birthplaces of planets, the best way to study how they form from observation is to directly watch them forming in disks. This is the only direct way to address the three most fundamental questions in this field: *when, where, and how planets form*. Unfortunately, identifying young planets in disks is difficult, as none of the major planet detection techniques invented to date is particularly suitable for this purpose.[1] Because of this, planet formation has historically been a subject of theoretical astrophysics.

A promising alternative method to look for forming planets in disks is to infer their existences and properties through gravitationally induced structures caused by disk–planet interactions. In theory, we know planets can drive spiral density waves (e.g., Goldreich and Tremaine 1980; Lin and Papaloizou 1993), clear material around their orbits to form gaps (e.g., Bryden et al. 1999), and induce vortices at the gap edge (Koller et al. 2003; Li et al. 2005; for a recent review on disk–planet interactions see Kley and Nelson 2012). These structures are on much larger physical scales and can be much more prominent than the planets themselves. If we can directly detect such structures in resolved observations of disks, and assert they are induced by (embedded and unseen) planets, these structures will be the signposts of planets.

Now, thanks to a fleet of new instruments with unprecedented resolving power that have come online in the past decade, a new field—*observational planet formation*—is emerging. These new instruments, along with new observational techniques, have enabled us to spatially resolve fine structures in the inner tens of AU of protoplanetary disks, and many of them are potentially planet-induced. High spatial resolution is the key in this novel field. Planets are generally believed to form at a few to a few tens of AU in disks; at these distances, planet-induced structures, such as spiral arms and gaps, are on the order of 1–10 AU in size. At 140 pc, the distance to nearby star formation regions (e.g., Taurus), these sizes transform to about 0.01–0.1$''$ in angular scale. For reference, the angular resolution of *HST* is about 0.1$''$ at *J*-band.

There are two spectral windows in which we can routinely achieve the needed angular resolution to resolve planet-induced structures in nearby star forming regions. The first window is optical to near-infrared (NIR). A few 8 m class ground based telescopes equipped with adaptive optics (AO) and various stellar halo suppression techniques have enabled us to take scattered light images with sufficient angular resolution to probe the inner regions of disks. The most important instruments at these wavelengths are NAos-COnica (NACO; Lenzen et al. 2003) and Spectro-Polarimetric High-contrast Exoplanet REsearch (SPHERE; Beuzit et al.

[1]The two most prolific techniques, radial velocity and transit surveys, demand the target star to be quiet and stable, so that the tiny signal introduced by the motion of a planet can be disentangled from the light curve of the star. Newborn stars tend to be too active, which largely washes out the planetary signal (although see David et al. 2016). Meanwhile, protoplanetary disks are often bright at near-to-mid-infrared wavelengths, which renders identifying planets through direct imaging difficult.

2008) onboard the Very Large Telescope (VLT), the High-Contrast Coronographic Imager for Adaptive Optics (HiCIAO; Tamura et al. 2006) onboard Subaru, and the Gemini Planet Imager (GPI; Macintosh et al. 2008) onboard Gemini. These instruments can achieve diffraction limited resolution at R-band ($\sim$0.6 μm) to K-band ($\sim$2 μm) ($\sim$0.04″ at H-band, or 6 AU at Taurus). To suppress the stellar contributions in their images, they are equipped with coronagraphs to block the central star, and more importantly, dual-beam polarimeter to take polarized intensity images (Perrin et al. 2004; Hinkley et al. 2009). They can routinely achieve an inner working angle (IWA) as small as 0.1″ from the star (14 AU at Taurus). At these wavelengths, imaging observations probe the distribution of small dust grains, μm-sized or smaller, due to their dominance in scattering starlight.

The other spectral window is mm to cm wavelengths, at which two radio interferometers, the Atacama Large Millimeter Array (ALMA) and the Expanded Very Large Array (EVLA), detect dust thermal continuum and molecular line emissions from disks. The improvements of the two arrays over the previous generation of interferometers, such as the Submillimeter Array (SMA), the Combined Array for Research in Millimeter-wave Astronomy (CARMA), and the Plateau de Bure interferometer (PdBI), are mainly threefold: they have longer baselines for finer spatial resolution (for reference, ALMA can achieve 0.02″ with a 16 km baseline at Band-6, or 1.3 mm); larger collecting area and better site condition for increased sensitivity; and more baselines for better *uv* coverage (in Fourier space). At mm to cm wavelengths, observations are mainly sensitive to the spatial and temperature structures of $\sim$mm-sized dust grains, as well as various molecule species.

As of 2016, about 30 protoplanetary disks have been spatially resolved at various wavelengths. In nearly all of them, features have been detected; they include spiral arms, gaps, cavities, and clumps. While the origins of these features are currently in debate, evidence is mounting that many of them may be the products of planet formation in disks. There are at least two conditions to unambiguously establish a linkage between an observed feature and an embedded (unseen) planet:

- theoretically predicted observational signatures of disk–planet interaction, taking into account the specific observational conditions, can quantitatively match the observation; and
- in the cases when the predicted planet cannot be directly verified using other methods, alternative explanations for the same feature that do not involve planets must be ruled out.

Currently, the best way to accomplish the first task is to produce synthetic images of planet-induced structures under realistic conditions, and compare them with real data. This process requires a combined effort of hydrodynamics and radiative transfer simulations: the former calculates the response of a disk to the presence of planets, producing the spatial distribution of gas and dust particles of various sizes; and the latter generates synthetic images of these disks given the angular resolution and sensitivity. In principle, hydro and radiative transfer equations should be coupled, as material distributions affect radiative transfer processes, and vice versa; in reality such calculations are still technically challenging at the moment, and current efforts have been largely limited to separate steps, in which density structures produced by hydro simulations are post-processed by radiative transfer simulations to produce images.

In this chapter, we will discuss three types of planet-induced structures: gaps and cavities (Sect. 9.3), spiral arms (Sect. 9.4), and vortices (Sect. 9.5). In Sect. 9.2 we will give an overview of the general scheme and numerical methods that we employ to study synthetic observational signals of planet-induced structures. In each of the later three sections, we will present numerical disk modeling of observational signatures, and compare them with actual data from observations. This chapter does not aim at an overview of this general topic; rather, it serves to highlight some of the latest progresses in this direction, with an emphasis on the works done by the authors.

9.2 General Modeling Scheme and Numerical Methods

We carry out hydrodynamic and radiative transfer simulations to produce NIR scattered light and mm dust continuum images for disk-planet models. We adopt the following general workflow:

1. We run global 2D vertically integrated hydro simulations or direct 3D hydro simulations to calculate the surface density (in 2D simulations) or volume density (in 3D simulations) structures of a disk with one or multiple planets. For scattered light, we run gas simulations only, and assume the sub-μm-sized small grains are well mixed with the gas, due to their short stopping time; for mm continuum emissions, we carry out multi-fluid simulations, with dust particles at various sizes approximated as separate fluids to calculate their spatial distribution as they interact with the gas. For the latter, grain growth/fragmentation are not taken into account.
2. The resulting density structures are fed into a radiative transfer tool to produce synthetic images. For 2D simulations, we puff up the disk in the vertical dimension by a Gaussian profile (i.e., hydrostatic equilibrium solution for vertically isothermal, thin disks) that follows a parametrized scale height profile in the radial direction $h/r \propto r^{0.25}$. For 3D simulations, the exact 3D volume density structures are fed into the radiative transfer tool.
3. The full resolution radiative transfer images are convolved with Gaussian point-spread functions (PSF) to simulate the appropriate angular resolutions for comparisons with observations.

For hydro simulations we employ three different grid-based codes, depending on the purposes:

- FARGO Masset (2000) (Zhu et al. 2011, 2012; Dong et al. 2015c; Zhu and Baruteau 2016; Baruteau and Zhu 2016),

- Athena/Athena++ Stone et al. (2008) (Zhu and Stone 2014; Zhu et al. 2014, 2015; Dong et al. 2015b, 2016c),
- PEnGUIn Fung (2015) (Fung and Dong 2015; Dong et al. 2016a,c),

The first two codes are CPU-based while the latter is GPU-based. For spiral arms, we also carry out simulations of massive disks with no planets to study the morphology of spiral arms induced by gravitational instability (GI) (Dong et al. 2015a, 2016b).

Initially, the system is axisymmetric in all simulations, and we adopt power law radial profiles as initial conditions for both the surface density and the temperature ($\Sigma \propto r^a$, $T \propto r^b$). Also, in 3D simulations the disk is isothermal in the vertical direction (except in Dong et al. (2015b), Zhu et al. (2015), in which an adiabatic equation of state with simple radiative cooling is adopted in certain cases). The planets are assumed to be on circular orbits in all cases, and planetary migration is not taken into account. At the beginning of the simulations, we gradually inject the planets into the disk, and turn on their masses in about ten orbits. Different types of planet-induced features take different amount of time to reach (semi-)steady state. Planet-induced spiral arms typically reach their final states in a few to a few tens of orbits, while gaps approach their final width and depth on viscous timescale, which can be thousands of orbits or more in a low viscosity environment. In most cases, our simulations are run to either (semi-)steady state, or typical disk lifetime.

We perform both 2D and 3D hydro simulations of disk–planet interactions. Different types of simulations have different pros and cons, and are suitable for different purposes. 2D simulations calculate the surface density of the disk (gas and/or dust) as perturbed by the planets. The resulting disk structures need to be puffed up in the vertical direction for radiative transfer postprocessing. Such simulations are numerically cheap, thus good for large parameter space explorations. However, they may not be used to model observations that are sensitive to disk surface structures, such as certain NIR scattered light observations. 3D simulations directly calculate volume density of gas and/or dust, and can be used to produce all kind of observations. However they are numerically orders of magnitude more expensive than 2D simulations, and are more suitable for modeling individual objects. The outcomes of 2D and 3D simulations in planet-induced spiral arms and gaps are compared in Zhu et al. (2015), Fung and Chiang (2016), Dong and Fung (2017a,b). 2D simulations generally produce weaker inner spiral arms in scattered light than 3D simulations, while 2D and 3D simulations produce very similar gaps in scattered light.

For the radiative transfer simulations, we mostly use the Monte Carlo radiative transfer (MCRT) tool (Whitney et al. 2013). The 3D gas and dust structures are directly read into the code (2D simulations are vertically puffed up to 3D first). As discussed in Sect. 9.1, scattered light is determined by the spatial distribution of small/sub-μm-sized grains, which we assume are always well coupled to the gas; while ALMA dust continuum images probe the distribution of the big/∼mm-sized grains, which are followed separately from the gas in our simulations. For the small grains, we use the standard interstellar medium (ISM) grains models (Kim et al. 1994), which are made of silicate, graphite, and amorphous carbon, with a size (s) distribution $n(s) \propto s^{-3.5}$ in between 0.02 and 0.025 μm. For the big grains, we assume a similar composition, and a size distribution $n(s) \propto s^{-3}$ all the way to $s = 1$ mm. Mie scattering is adopted when the particle size is smaller or comparable to the wavelength. We assume 100:1 for the gas-to-total-dust mass ratio, as the gas-to-solid mass ratio in the interstellar medium, and 9:1 for the big-to-small-dust mass ratio. We note that both are working assumptions only, and real systems may have different ratios (e.g., Ansdell et al. 2016).

The full-resolution model images produced by the MCRT simulations need to be convolved with PSFs before comparing with actual data. We use Gaussian kernels as PSFs. At H-band, we convolve the images to an angular resolution of 0.04″, or 6 AU at 140 pc (unless otherwise indicated), as a good approximation to the resolution achieved by Subaru, VLT, and Gemini with their high contrast imaging systems. We normally assume the source is at a distance of 140 pc from Earth, as the distance to nearby star forming regions such as Taurus. Within ∼200 pc from Earth, there are at least hundreds of protoplanetary disks that can be resolved by the aforementioned instruments.

Finally, the resulting model images are compared with real observations. Quantitative comparisons can be done for a variety of feature properties, such as contrast (how much brighter or fainter features are relative to the surrounding background), shape, and physical size.

9.3 Gaps

In this section, we discuss observational signatures of planet-opened gaps, and compare numerical models with gap observations. This section mainly highlights the work in Zhu et al. (2011, 2012) and Dong et al. (2015c).

9.3.1 Observations

Gap-like features have been discovered in many disks. A particularly interesting group is the so-called transitional disks (see the recent review Espaillat et al. 2014). Discovered through their unique infrared deficit from NIR to ∼10 μm, which signals a lack of warm dust in the inner few to few tens of AU (Calvet et al. 2005; Espaillat et al. 2007, 2010), these disks have been subsequently shown to harbor gaps or cavities in spatially resolved observations at NIR (Thalmann et al. 2010; Hashimoto et al. 2011, 2012; Mayama et al. 2012; Canovas et al. 2013; Garufi et al. 2013; Quanz et al. 2013b; Grady et al. 2013; Avenhaus et al. 2014b; Tsukagoshi et al. 2014; Momose et al. 2015; Pinilla et al. 2015; Akiyama et al. 2015) and mm wavelengths (Andrews et al. 2011; Casassus et al. 2013; Fukagawa et al. 2013; van der Marel et al. 2013; Pérez et al. 2014; Perez et al. 2015; Zhang et al. 2014). These gaps are wide and deep. In resolved cases, their sizes are typically tens of AU, while the surface density depletion can be up to several orders of magnitude in both the dust and the gas (e.g., Dong et al. 2012a; van der Marel et al. 2015b, 2016). Figures 9.1 and 9.2 show examples of such systems at both NIR and mm wavelengths.

The properties of the cavity, such as its size and depletion, can be measured from observations. These properties are dependent on wavelengths. In particular, the cavity sizes in dust continuum are systematically larger than they are in scattered light and in the gas (e.g., Zhang et al. 2014; van der Marel et al. 2015b, 2016). In

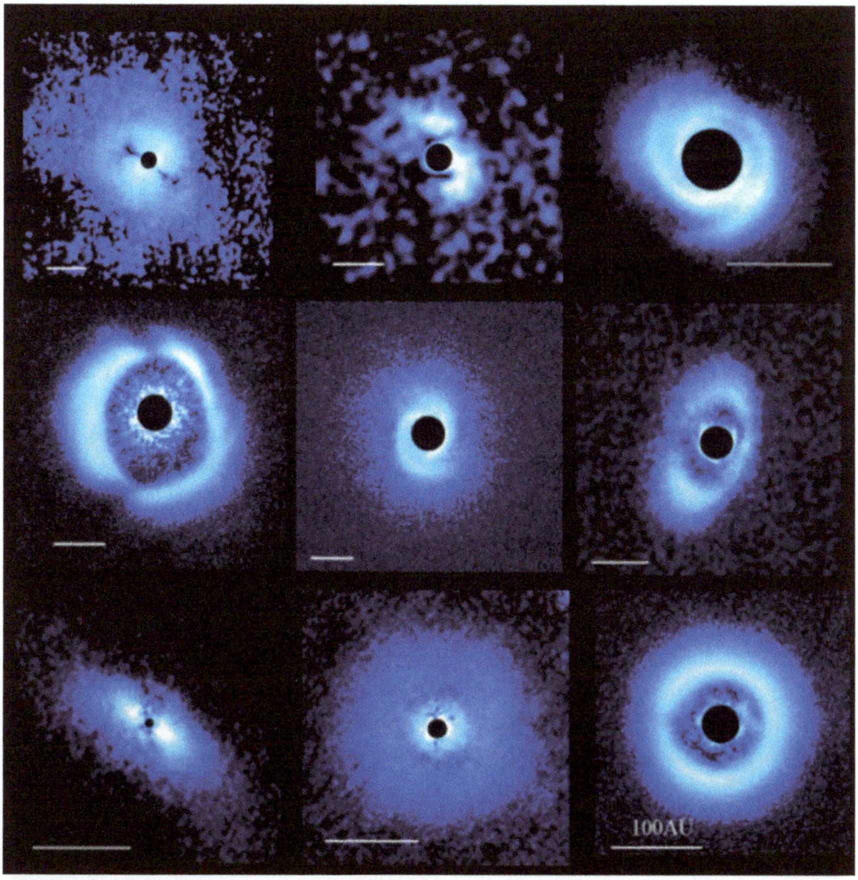

Fig. 9.1 Examples of transitional disks in NIR scattered light from the Subaru-based SEEDS survey (Tamura 2009)

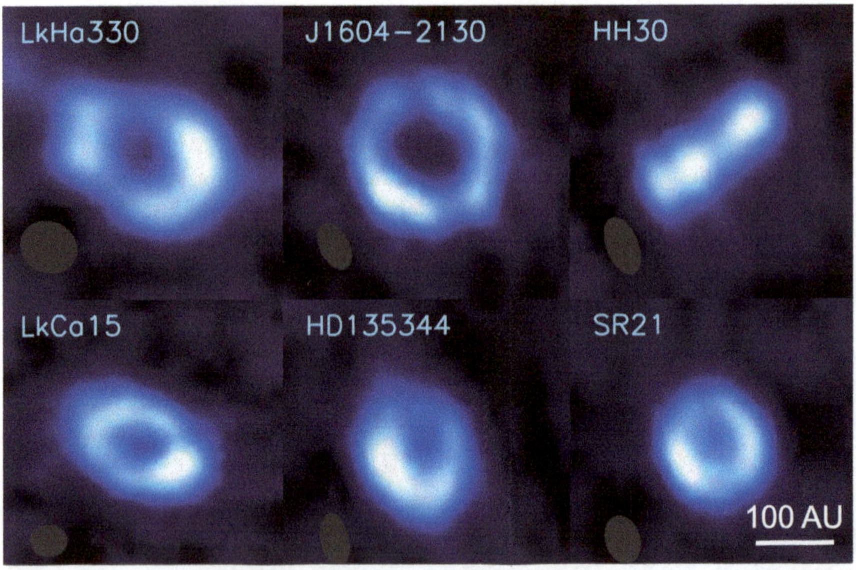

Fig. 9.2 Examples of transitional disks in mm dust continuum emissions (Fig. 11 in Williams and Cieza 2011)

extreme cases, NIR images do not show the cavities seen at mm, down to their inner working angles (e.g., Dong et al. 2012b; Follette et al. 2013).[2]

Narrow gaps have also been observed in disks. The two best examples are the multiple gaps discovered in HL Tau and TW Hya; in the latter gaps are seen in both scattered light and dust continuum emission (Akiyama et al. 2015; Rapson et al. 2015; Andrews et al. 2016), while in the former the gaps are seen in dust continuum and gas emission (ALMA Partnership et al. 2015; Yen et al. 2016; Carrasco-González et al. 2016). Different from transitional disks, these narrow gaps tend to have a lower depletion. In both HL Tau and TW Hya, the gap regions are only a few to a few tens of percents fainter than their surroundings. Generally speaking, a surface density depletion of a factor of a few is needed in order for the gap to be visible in current NIR scattered light observations, while a depletion on the order of $\sim$10% in mm-sized dust may be enough for the gap to be visible in ALMA continuum observations.

[2]The inner working angle in scattered light imaging observations of disks indicates the minimum angular separation from the star, inside which disk structures cannot be accessed.

9.3.2 Modeling

Both wide and narrow gaps can be opened by planets—a narrow gap may be opened by a single planet, and wide gaps such as those in transitional disks may be opened by multiple planets, since a single planet can hardly account for the gap width in the gas (Zhu et al. 2011; Dodson-Robinson and Salyk 2011).[3] The conditions for gap opening by a single planet, i.e., the viscous criteria and thermal criteria, are summarized in Bryden et al. (1999) (however note that Goodman and Rafikov (2001) pointed out in extremely low viscosity environment, planets with masses much lower than these conditions may still open gaps due to the nonlinear wave damping effects). The criteria for opening a common gap instead of several individual gaps in the multiple planet scenario is that the individual gap width exceeds the separation between adjacent planets (Duffell and Dong 2015). Figure 9.3 shows synthetic images of 4 planet-disk models from Dong et al. (2015c). Two of the models have a single planet (models $1 \times 0.2M_J$ and $1 \times 1M_J$), and the other two have four planets (models $4 \times 1M_J$ and $4 \times 2M_J$; the first number in the names indicates the total number of planets, and the second number indicates the mass of each planet in the unit of M_J). The parameters of the models are listed in Table 9.1.

In scattered light, the planets in $1 \times 0.2M_J$ and $1 \times 1M_J$ open narrow gaps around their orbits (∼25–35 AU), while the planets in $4 \times 1M_J$ and $4 \times 2M_J$ open a large common gap extending inward to a few AU. In contrast, the dust continuum emission in all models with $M_p \geq 1M_J$ has a large gap extending from a few AU to ∼40 AU. This is caused by the "dust filtration" effect (e.g., Rice et al. 2006; Paardekooper and Mellema 2006), as illustrated in detail in the context of transition disks in Zhu et al. (2012) (see also Pinilla et al. 2012 and de Juan Ovelar et al. 2013). As the (outermost) planet opens a gap, the gas pressure peaks at the outer gap edge. This pressure bump acts like a dam that stops and accumulates particles with a dimensionless stopping time T_s around unity. For a dust particle with a size s at the midplane of a disk, its dimensionless stopping time is

$$T_s = t_s \Omega = \frac{\pi s \rho_p}{2\Sigma_g} = 1.57 \times 10^{-2} \frac{\rho_p}{1\,\mathrm{g\,cm^{-3}}} \frac{s}{1\,\mathrm{mm}} \frac{10\,\mathrm{g\,cm^{-2}}}{\Sigma_g} . \tag{9.1}$$

where ρ_p is the grain's density and Σ_g is the gaseous disk's surface density. For the mm-sized grains, the pressure bump outside the outermost planet's orbit in all four models effectively traps the grains and piles them up. Increasing the mass of the planets in the four planet models from $1M_J$ to $2M_J$ causes the gas gap to become wider and the edge to become sharper, so the "ring" in the big grains becomes

[3]Really massive companions, such as stellar mass companions, may be able to open gaps wide enough, though their presences have generally been ruled out in direct imaging observations.

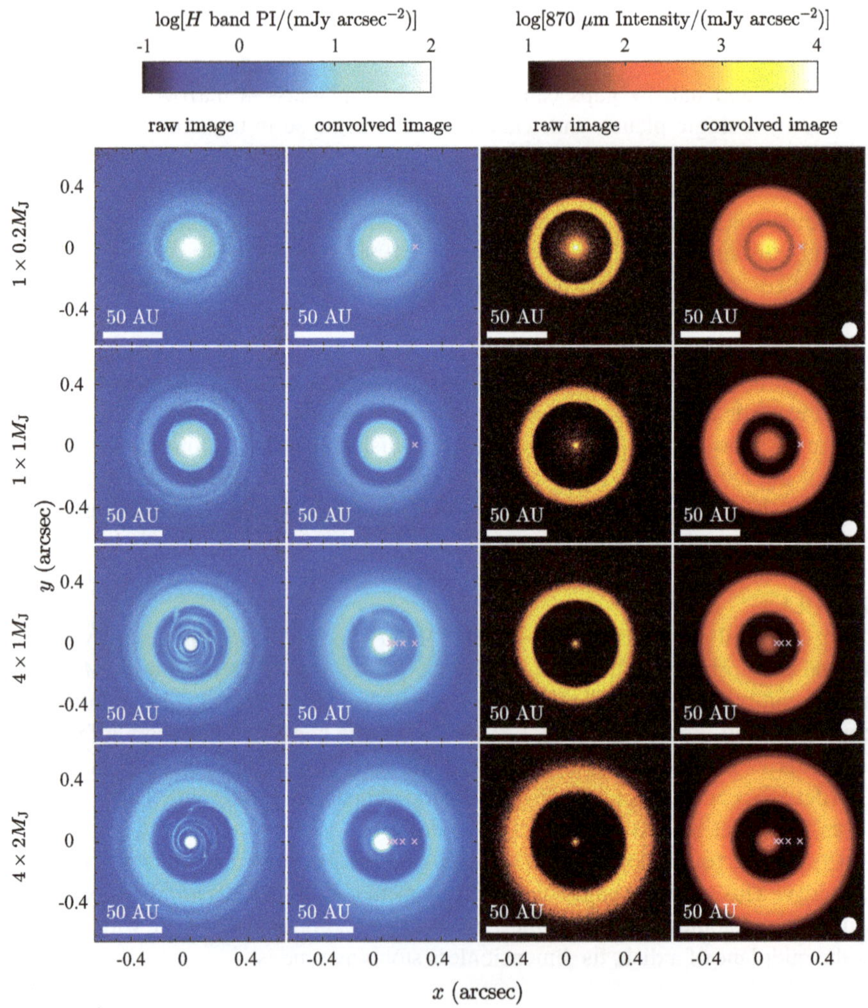

Fig. 9.3 Synthetic raw (full resolution) and convolved H-band polarized intensity images (*left two columns*) and ALMA band 7 (870 μm continuum) intensity images (*right two columns*) for our models (Table 9.1) at face on angle. The names of the models are labeled on the left. Systems are assumed to be at 140 pc. The convolved H-band images have a resolution of $0.04''$, and the convolved ALMA images have a resolution of $0.1''$ (beam size indicated at the lower right corner in the convolved mm images) to mimic realistic angular resolutions. The *grey crosses* in the convolved images mark the orbits of the planets. The figure is from Dong et al. (2015c)

wider and moves outward. This dust filtration mechanism naturally provides an explanation for the observed wavelength dependence of gap sizes in transitional disks.

Table 9.1 Model properties

Model	Planet mass M_J	Planet position AU	Evolution time yr	M_{disk}[a] $M_\odot$	Particle size[b] mm
$1 \times 0.2M_J$	0.2	30.0	4×10^5	0.04	1
$1 \times 1M_J$	1	30.0	4×10^5	0.04	1
$4 \times 1M_J$	1/1/1/1	30.0/18.9/11.9/7.5	4×10^5	0.04	1
$4 \times 2M_J$	2/2/2/2	30.0/18.9/11.9/7.5	4×10^5	0.04	1
$3 \times 0.2M_J$	0.2/0.2/0.2	12/30/65	2×10^5	0.2	1

Properties of the models
[a]The initial gas disk mass
[b]The size of the particles in the two fluid hydro simulations. Models are from Dong et al. (2015c)

Table 9.2 Comparison of NIR gap contrast between observations and models

System name	Gap contrast at H band	References
RX J1604.3-2130	3.6 ± 0.5	Mayama et al. (2012)
HD 142527	4.2 ± 1.2	Canovas et al. (2013)
SAO 206462	3.3 ± 1.9	Garufi et al. (2013)
PDS 70	3.5 ± 0.7 (along major axes)	Hashimoto et al. (2012)
$4 \times 1M_J$	3.5	Dong et al. (2015c)
$4 \times 2M_J$	6.0	Dong et al. (2015c)

The gap contrast is the ratio of the peak intensity at the outer gap edge to the minimum intensity at the trough of the gap; the quoted value is the azimuthal mean while the quoted error in observations is the standard deviation over different azimuthal angles. The first three observed systems are relatively face-on, while PDS 70 has a non-trivial inclination of 50°, so only major axes are taken into account. The floor intensity inside the gap is detected only in HD 142527. In the other three systems the possible inner edge of the gap (if exists) is blocked by the inner working angle in observations, so the measured gap contrast may be a lower limit. Observational noise is not added in our model images, so measured values may be upper limits

Common gaps opened by multiple planets can well match the observations of transitional disks. Table 9.2 shows the comparisons of the gap contrasts and widths at NIR wavelengths between the models and four transitional disk systems. The models reproduce the observed gap contrasts. For dust continuum observations, generally only upper limits on the gap contrasts exist at the moment due to limited sensitivities and angular resolutions in existing mm observations; these upper limits are consistent with model predictions.

Figure 9.4 shows another multiple-planet model ($3 \times 0.2M_J$). Different from models $4 \times 1M_J$ and $4 \times 2M_J$, the three planets are placed far apart from each other. As a result, they each open a narrow gap around their orbits. These narrow gaps qualitatively match the characteristics of HL Tau. We note that the depths of these mm gaps are significantly enhanced compared to the depths in gas, due to the dust-gas coupling effects.

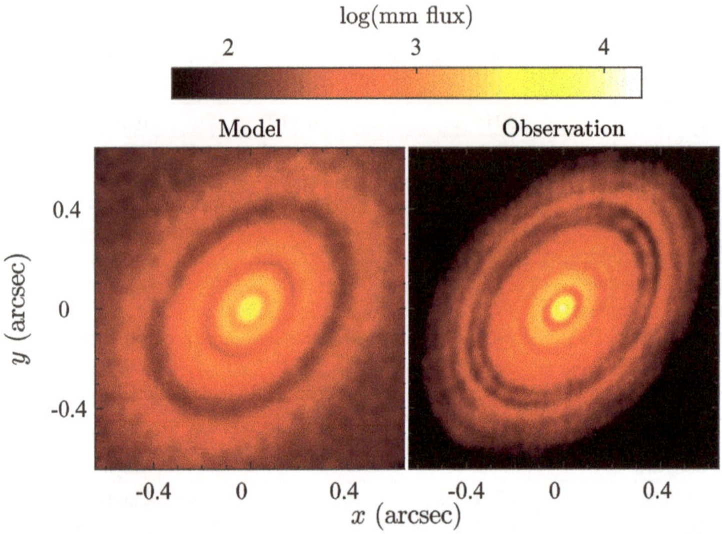

Fig. 9.4 Comparison between Model $3 \times 0.2M_{\rm J}$ with the ALMA observation of HL Tau (1.0 mm, Band 6+7, ALMA Partnership et al. 2015). Model image has an inclination of 45° and a position angle of 138°. The color scale is logarithmic and the unit is arbitrary

9.3.3 Discussion

9.3.3.1 Alternative Mechanisms for Gap-Like Features in Disks

Non-dynamical mechanisms to open big gaps as in transitional disks have been proposed, such as photoevaporation (e.g., Clarke et al. 2001; Alexander et al. 2006; Owen et al. 2012; Rosotti et al. 2013) and grain growth (e.g., Dullemond and Dominik 2005; Birnstiel et al. 2012). However, all these mechanisms have major drawbacks and cannot explain at least a subset of systems. If transitional disks are produced by photoevaporation, their accretion rates are expected to be low, because photoevaporation only effectively removes material from the disk and open gaps at later stages when the disk accretion rate is lower than the photoevaporation rate. However, the average accretion rate of transitional disks as a whole is only slightly lower than that of ordinary T Tauri or Herbig stars (Andrews 2015), which disqualifies photoevaporation as the mechanism responsible for the majority of transitional disks. Grain growth can explain the infrared dip on the SED, but cannot produce the sharp gap edges as observed (Birnstiel et al. 2012). For narrow gaps, secular gravitational instability has been proposed as an alternative gap-forming mechanism (Takahashi and Inutsuka 2014), and the locations of the gaps have been noticed to coincide with the snow lines of certain volatile species (e.g. gaps in HL Tau Zhang et al. 2015). However, detailed quantitative models are needed to compare with observations in order to constrain these theoretical models.

For ring-like or arc-like features in scatted light, disk self-shadowing has been proposed as an explanation (e.g., Garufi et al. 2014; Chen et al. 2012; Quanz et al. 2013b; Akiyama et al. 2015; Momose et al. 2015). In this scenario, the dark region in the disk is shadowed by puffed up inner disk structures, with the bright ring/arc-like structures outside the dark regions being the edges of the shadows. Through detailed radiative transfer modeling, Dong (2015) has shown that shadowing by inner disk structures, such as a puffed inner rim at the dust sublimation radius, cannot produce ring/arc-like features in scattered light; rather, they can only produce broken power laws in the radial profile of the scattered light.

9.3.3.2 Dynamics of Multi-Planet Systems in Transitional Disks

While common gaps opened by multiple planets is an attractive mechanism for explaining the gaps in transitional disks, one needs to consider if such planetary systems can be stable over the disk lifetime for their gaps to be seen, as gravitational planet–planet interactions may disrupt such systems and scatter the planets. Through N-body simulations, Dong and Dawson (2016) studied the stability of multi-planet systems that are required to open gaps consistent with transitional disks observations. In general, without the aid from gas damping, systems with a smaller number of more massive planets tend to be dynamically stable for the typical disk lifetime, while systems with a larger number of less massive planets may be unstable. Eccentricity damping from the residual gas inside the gaps can help stabilize systems by locking planets into mean motion resonances, establishing a chain of pairs resembling the older HR 8799 system (Fabrycky and Murray-Clay 2010).

9.3.3.3 The Occurrence Rate of Transitional Disks *vs* the Occurrence Rate of Giant Planets at Large Distance

If the gaps in transitional disks are common gaps opened by multiple planets, the giant planet occurrence rate at wide separations must equal or exceed the occurrence rate of transitional disks, which is determined to be $\sim$10% from various surveys (e.g., Muzerolle et al. 2010; Luhman et al. 2010). Dong and Dawson (2016) found that this can only be satisfied under some of the most favorable conditions for gap opening, namely, $h/r < 0.1$ and $\alpha \lesssim 0.001$ in the gap regions (see the dependence of gap depth on planet mass; Fung et al. 2014); more importantly, the occurrence rate of giant planets at $\sim$1–50 AU has to largely follow the radial velocity statistics (Cumming et al. 2008), not the direct imaging statistics (Brandt et al. 2014). This situation may be significantly mitigated if the disk viscosity is lower than 10^{-3}, enabling lower mass planets to open gaps through non-linear wave damping processes.

9.4 Spiral Arms

In this section, we discuss the observational signatures of spiral arms induced by giant planets (and more massive companions), and compare numerical models with observations of spirals in disks. This section mainly highlights the work in Zhu et al. (2015), Dong et al. (2015b, 2016a,c), and Fung and Dong (2015).

9.4.1 Observations

Spiral arms have been found in a few protoplanetary disks, such as AB Aur (Hashimoto et al. 2011), MWC 758 (Grady et al. 2013; Benisty et al. 2015), SAO 206462 (Muto et al. 2012; Garufi et al. 2013; Stolker et al. 2016), HD 100453 (Wagner et al. 2015), HD 100546 (Boccaletti et al. 2013; Avenhaus et al. 2014a; Currie et al. 2015; Garufi et al. 2016), AK Sco (Janson et al. 2016), V1247 Orionis (Ohta et al. 2016), LkHα 330 (Akiyama et al. 2016), and TW Hya (Rapson et al. 2015). At this moment almost all spiral arm observations are in scattered light; resolved mm observations currently suffer from insufficient spatial resolution, and cannot pinpoint the mm counterparts of these arms (Chapillon et al. 2008; Isella et al. 2010, 2013; Andrews et al. 2011; Pérez et al. 2014; Marino et al. 2015a) (but see Christiaens et al. 2014; van der Marel et al. 2016).

Among these disks, MWC 758, SAO 206462, and HD 100453 form a subgroup—they all have two roughly symmetric arms (Fig. 9.5). Their arms are global, spanning from 10s to 100 AU from the primary, and extending over at least 180° azimuthally. These arms are rather open, with pitch angles (the angle between the azimuthal direction and the tangent of the arms) of 10° or more. Most strikingly, in all cases the two arms are in approximate $m = 2$ rotational symmetry. The parent disks of all

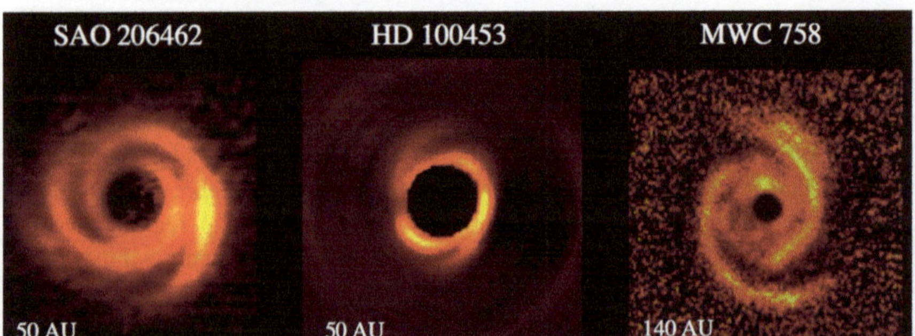

Fig. 9.5 The three near-symmetric double arm systems in scattered light: MWC 758 (Benisty et al. 2015), SAO 206462 (Garufi et al. 2013), and HD 100453 (Wagner et al. 2015)

three systems are recognized as transitional disks, with a giant cavity at the center. In all of the systems, only HD 100453 has a confirmed companion—a $0.3M_{\odot}$ M dwarf at 120 AU (the primary star is $1.7M_{\odot}$).

9.4.2 Modeling

Recent modeling work has shown that a massive companion (giant planet, brown dwarf, or stellar mass companion) can excite two prominent spiral arms inside its orbit that match the morphologies of the observed arms. Figures 9.6 and 9.7 present the surface density and synthetic H-band images of 6 disk-planet models (Dong et al. 2015b). The two inner spirals excited by a planet with $q = M_{\rm p}/M_{\star} \geq 10^{-3}$ can be visible in scattered light images under current NIR direct imaging capability, and they have a morphology resembling the observed double arm systems. Quantitatively, they match observations in (1) pitch angle, (2) contrast, (3) radial and azimuthal extends. The two outer arms can also be visible, however they are too tightly wound given typical disk scale heights. Figure 9.8 shows a comparison between the inner two arms excited by a $q = 0.006$ planet with MWC 758.

Figure 9.9 shows a comparison between a model (left) and the observation of HD 100453 (right). The companion–disk interaction model reproduces all main features in the system, in particular, the morphology of the arms, and the size of the truncated disk around the primary star. In addition, the model predicts the disk around the primary star to be rotating in a counterclockwise direction, which is the same as the companion.

The morphologies of companion-induced spiral arms can be dramatically modified by the viewing angle of the system (Dong et al. 2016a) (Fig. 9.10). A giant planet opens a gap and excites two prominent spiral arms in the inner disk; the inner disk arms, the outer disk ring, and the gap are readily detected when viewed face-on. Depending on the viewing inclination and position angle, the trailing-double-arm pattern for face-on views may appear as one trailing arm; two trailing arms on one side of the star; or two arms on the same side but winding in opposite directions.

At high inclinations, the disk ring outside the gap opened by a giant planet may manifest itself as two pseudo-arms placed symmetrically about the minor axis winding in opposite directions. This may be the origin of the two arms in systems like AK Sco. Figure 9.11 shows that a model image of a disk perturbed by a $3M_{\rm J}$ planet at 25 AU, when viewed at 70° inclination, resembles the AK Sco system as observed by SPHERE (Janson et al. 2016).

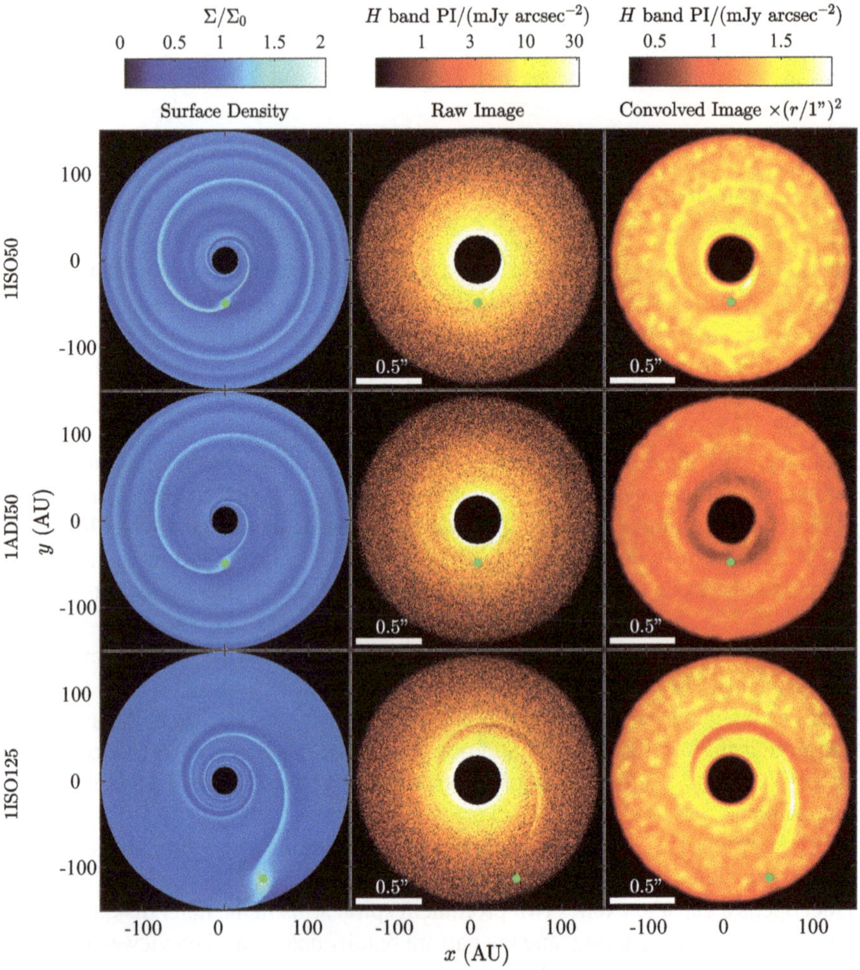

Fig. 9.6 Model results of disk–planet interaction models with planet mass $q = M_p/M_\star = 10^{-3}$, showing the relative surface density perturbation Σ/Σ_0 (*left column*, linear scale); *H*-band full resolution polarized intensity images (*middle*, logarithmic scale, the central $0.2''$ has been masked out to highlight the structures in the outer disk); and $1/r^2$-scaled convolved images (*right*, linear scale, convolved by a Gaussian PSF with FWHM$= 0.06''$). The location of the planet is marked by the *green dot*, which is located at 50 AU in the top and middle models, and 125 AU in the bottom model. The top and bottom models adopt an isothermal equation of state, while the middle model adopts an adiabatic equation of state. This figure is from Dong et al. (2015b)

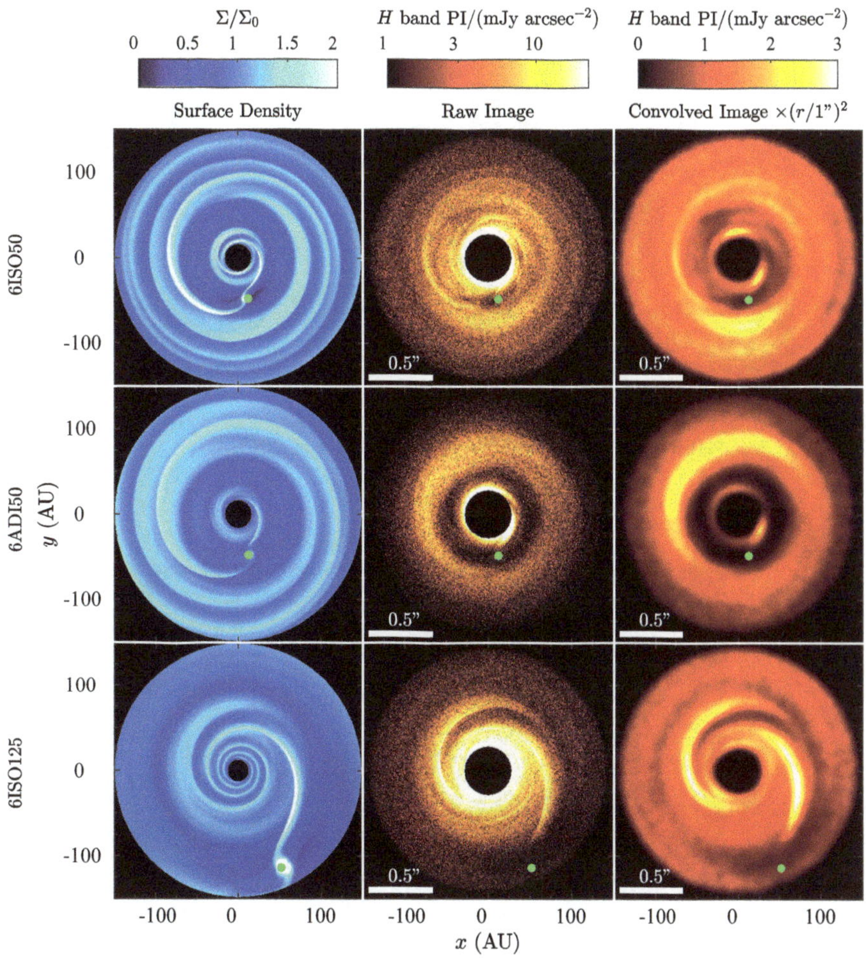

Fig. 9.7 Same as Fig. 9.6, but for the $q = 6\times10^{-3}$ models. This figure is from Dong et al. (2015b)

9.4.3 *Discussion*

9.4.3.1 Alternative Spiral Arm Theories

Although numerical modeling has shown that the observed symmetric double-arm patterns can be well explained by a giant companion outside the arms, ultimately we need to directly identify the companion to definitively confirm this scenario. So far, only the companion in the HD 100453 system has been confirmed, while a campaign to look for the predicted planet in the other two systems is currently underway.

Some companion-less alternative mechanisms have been proposed. In particular, massive disks may develop gravitational instability (GI), which can produce spiral

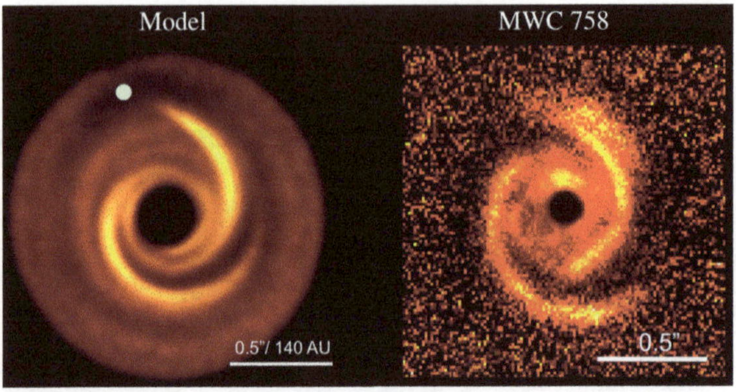

Fig. 9.8 *Left*: Synthetic image of a disk model with a $q = 0.006$ planet (*green dot*) (Dong et al. 2015b). *Right*: SPHERE observations of MWC 758 (Benisty et al. 2015)

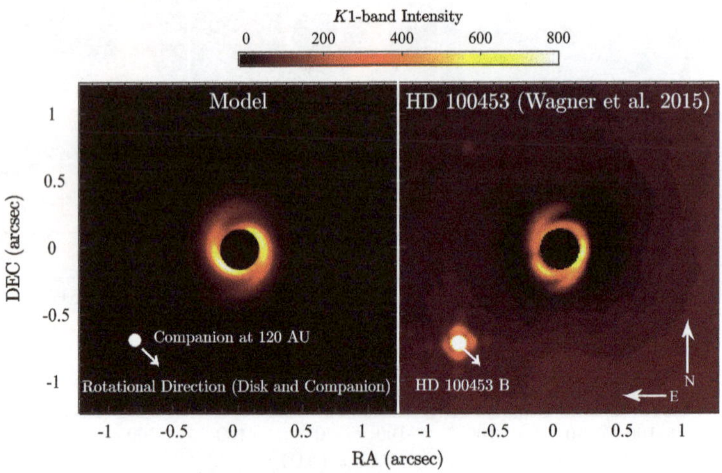

Fig. 9.9 Model image alongside the observation of the HD 100453 system at $K1$-band [the observation is taken by the infrared dual-band imager and spectrograph on VLT/SPHERE [(Wagner et al. 2015)]. HD 100453 B is at a projected separation of 120 AU from the center. The model on the left adopts all the parameters about the disk and the companion as constrained by observations. The inner $0.15''$ (IWA, 17 AU) is masked out. The units are mJy arcsec^{-2} for the model image, and arbitrary for the VLT/SPHERE observation. This figure is from Dong et al. (2016c)

arms as well. Quantitatively, GI happens when the disk satisfies the Q criterion (Goldreich and Lynden-Bell 1965):

$$Q = \frac{c_s \Omega}{\pi G \Sigma} \lesssim \text{unity}, \tag{9.2}$$

where c_s, Ω, and Σ are the disk's sound speed, orbital frequency, and surface density. Figure 9.12 shows two models of gravitationally unstable disks (Dong et al.

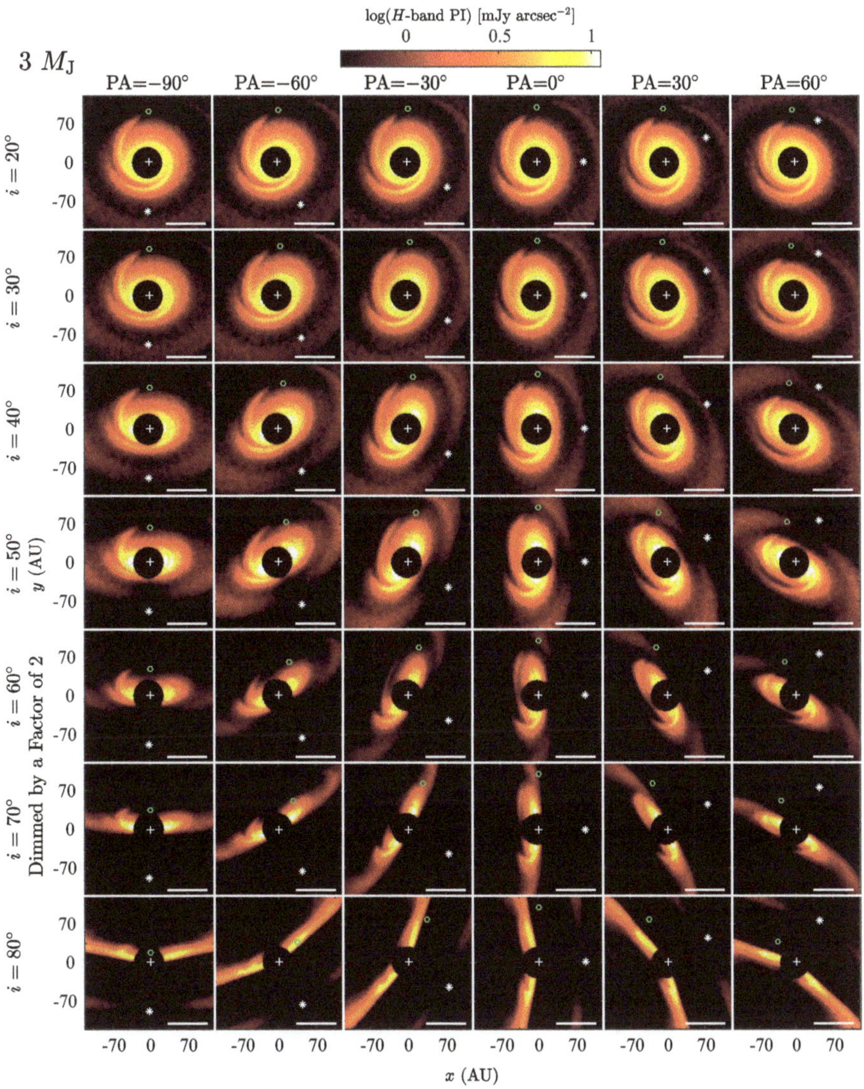

Fig. 9.10 The inner arms excited by a $3M_J$ planet in a disk in H-band polarized intensity at various viewing angles (inclinations i, vertical, and position angles PA, horizontal). The disk's nearside is marked by the *white asterisk* on the minor axis. The *white bar* at the lower left corner is 0.5″ long. Each panel is 240 AU on a side. The *open green circle* indicates the projected location of the planet. The panels at 60° and 70° inclinations are dimmed by a factor of 2 to fit within the color scaling. The figure is from Dong et al. (2016a)

2015a). Disks with a total mass 25–50% of the central star can develop prominent spiral arms that can match the observations well, while if the disk mass is below 10% of the stellar mass, GI-induced spirals are unlikely to be visible with current

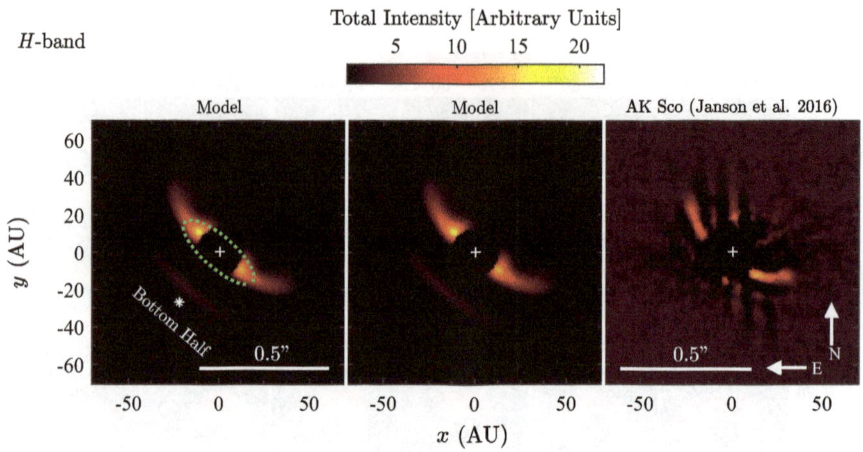

Fig. 9.11 Synthetic model image with a $3M_J$ planet (*left and middle*; identical except labels) from Dong et al. (2016a), and the VLT/SPHERE observation of AK Sco (Janson et al. 2016) (*right*) in H-band total intensity. The planet is at 25 AU. The inner circle of radius $r = 0.0925''$ is blocked off to simulate a coronagraph. The angular resolution is $0''.04$ in both images. The model image is viewed at a position angle of $-130°$ and an inclination of $70°$; these same parameters characterize AK Sco as constrained by CO observations (Czekala et al. 2015). The bottom (obscured) half of the model disk is labeled, and the planet's orbit (circular and coplanar with the disk) projected on the sky is marked by the *green dotted circle*. The apparent spiral-arm-like features ("pseudo-arms") in the model actually arise from the outer disk ring viewed at high inclination (note their morphologies are insensitive to the specific location of the planet in the azimuthal direction). The pseudo-arms in the observations appear to be sharper, possibly because of the angular differential imaging (ADI) data reduction process. This figure is from Dong et al. (2015b)

NIR capabilities. In addition, models predict these GI disks to have high accretion rates (on the order of $10^{-6} M_\odot\ \mathrm{yr}^{-1}$), driven by the strong turbulence induced by the instability. In the case of MWC 758 and SAO 206462, their low disk masses ($M_{\mathrm{disk}} \sim 0.01 \times M_\star$; Andrews et al. 2011) and low accretion rates ($10^{-8} M_\odot\ \mathrm{yr}^{-1}$) strongly disfavor this possibility.[4]

9.4.3.2 Current Theories of Giant Companion-Induced Multi-Arms

Additional information can be derived from these spiral arms based on our understanding of their propagation. Since they are essentially sound waves in disks, we can use their openness (or pitch angles) to measure the disk temperature and the companion position. However, observed spiral arms are very wide open, implying an unrealistically high disk temperature based on linear theory (Muto et al. 2012;

[4]We note that in early stage protostellar disks, GI-induced spiral arms may have been found (Liu et al. 2016; Dong et al. 2016b); however, these arms are on the scale of hundreds of AU and are different from the arms in disks at late stages discussed in this chapter.

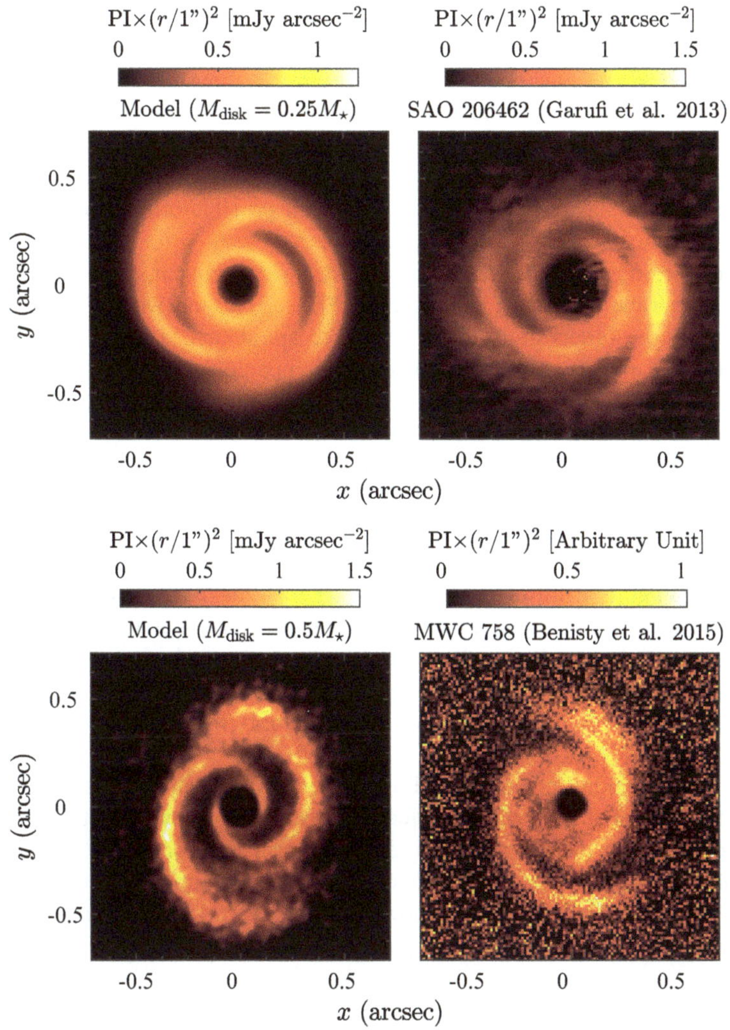

Fig. 9.12 Comparisons of GI disk model with MWC 758 (*top*; Benisty et al. 2015) and SAO 206462 (*bottom*; Garufi et al. 2013) in NIR scattered light. $M_{\rm disk} = 0.5 \times M_\star$ for the upper left model, and $M_{\rm disk} = 0.25 \times M_\star$ for the lower right model. This figure is from Dong et al. (2015a)

Benisty et al. 2015). Recently, detailed numerical simulations (Zhu et al. 2015) suggest that the openness of the spiral arms not only depends on the disk temperature but also depends on the planet mass, possibly due to the non-linear property of the spiral shocks (Goodman and Rafikov 2001; Rafikov 2002). This not only alleviates the discrepancy between observations and theory, but also implies that we could use the openness of the spiral arm to estimate the planet mass if the disk temperature is known.

It is known, based on linear calculations, that a planet excites two one-armed spirals, one in the inner disk and one outer, consequent of the propagation of waves excited at Lindblad resonances (Ogilvie and Lubow 2002). The appearance of multi-arms on one side of the disk is however only found in simulations (Juhász et al. 2015; Zhu et al. 2015). Despite continuing work on this phenomenon, a fully consistent theory that explains the generation of these spiral arms remains lacking. Some suggestions have been made regarding their possible origin. Here we briefly summarize some of the more likely possibilities.

A common suggestion is that the secondary and subsequent spirals are products of nonlinear wave coupling (Juhász et al. 2015; Fung and Dong 2015). This is motivated by the observation that multi-armed spirals appear to be only present when the companion is massive, i.e. when the companion's mass is near or exceeds the disk's thermal mass. This possibility is recently explored (Lee 2016) by applying the theory of ultraharmonic waves (Artymowicz and Lubow 1992), which are the products of the nonlinear coupling between linear waves, to planet-induced spirals. It shows that ultraharmonic waves are indeed excited by a massive planet, but it is unclear whether they match the multi-armed spiral patterns found in numerical simulations.

Another suggestion draws from the similarity between planet-induced spirals and spirals around binary systems of near-unity mass ratios (Zhu et al. 2015). For these binary systems, the gravitational potential has an enhanced component in low m Fourier modes, and therefore commonly excites multi-armed spirals. It is possible that what we observe in the planetary case is simply in transition to this binary regime.

Despite our lacking of understanding in their origin, some properties of these multi-armed spirals have been empirically derived. It is found that the azimuthal separation between the primary and secondary arm is larger when the planet is more massive, which suggests a potential method to estimate planet mass (Zhu et al. 2015). Numerical simulations show that the two arms have a roughly constant azimuthal separation that follows a simple power-law scaling (Fung and Dong 2015):

$$\phi_{\text{sep}} = 102^{\circ} \left(\frac{q}{10^{-3}} \right)^{0.2}, \tag{9.3}$$

where ϕ_{sep} is the azimuthal separation between the primary and secondary arm, and q is the planet-to-star mass ratio. This result is shown to be independent of the disk temperature.

Equation (9.3) can be used to infer planet mass from observed two-armed spirals. When applied to the SAO 206462 system, this relation suggests that the spiral-arm-producing planet at $\sim$85 AU is about $6M_{\text{J}}$ (Fung and Dong 2015). A careful application of this method will require calibration with other planet mass measuring methods. It also sheds light on the possible origin of these spirals; its independence from disk thermal properties shows that at least some aspects of these spirals are unlike those from linear or even weakly nonlinear theories.

9.5 Large Scale Disk Asymmetry

In this section, we discuss the observational signatures of vortices at the edges of planet-opened gaps, and compare numerical models with observations of lopsided disks. This section mainly summarizes the work in Zhu and Stone (2014), Zhu and Baruteau (2016), Bae et al. (2016), and Baruteau and Zhu (2016).

9.5.1 Observations

Lopsided disks, in which mm dust continuum emissions strongly concentrate on one side the of disk, have been observed in a few systems (see the recent review by Andrews 2015). The best examples are HD 142527 (Casassus et al. 2013; Fukagawa et al. 2013), IRS 48 (van der Marel et al. 2013), SR 21 (Pérez et al. 2014), SAO 206462 (Pérez et al. 2014), and LkHα330 (Isella et al. 2013) (Fig. 9.13). These large scale lopsided structures are extended and resolved in the azimuthal direction, but narrow and often unresolved in the radial direction.

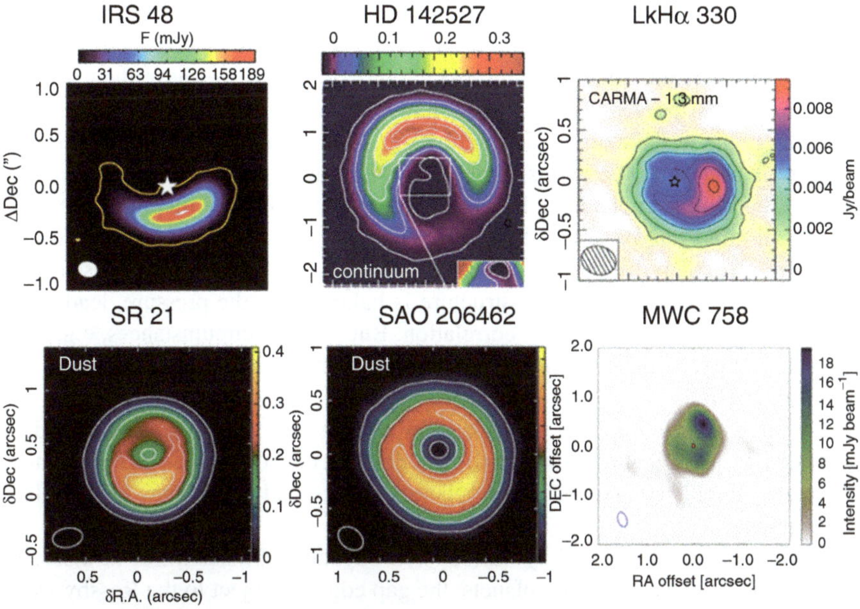

Fig. 9.13 Large scale disk asymmetry revealed at sub-mm wavelengths for several transitional disks: IRS 48—van der Marel et al. (2015a), HD 142527—Casassus et al. (2013); Fukagawa et al. (2013), LkHα330— Isella et al. (2013), SR 21—Pérez et al. (2014), SAO 206462—Pérez et al. (2014), MWC 758—Marino et al. (2015b)

Before ALMA, earlier observations by SMA and CARMA have already implied that some disks may exhibit large scale disk asymmetry (e.g. Isella et al. 2013). With ALMA's unprecedented resolution and sensitivity, this large scale disk asymmetry has been unambiguously detected. The level of asymmetry in these systems varies, from a factor of ~2–100, and the mm emission "bump" on one side of the disk can be responsible for a significant fraction of the total mm luminosity of the disk (a few ×10 to 100%). In the extreme case of Oph IRS 48, a highly asymmetric crescent-shaped dust structure exists between 45 and 80 AU from the star, and the peak emission from this structure is at least 130 times stronger than the upper limit of the opposite side of the disk (van der Marel et al. 2013). On the other hand, the contrast of disk asymmetry in SR 21 and SAO 206462 is less than a factor of 2 (Pérez et al. 2014).

9.5.2 Modeling

To explain these asymmetric dust structures, azimuthal particle trapping by vortices has been proposed (van der Marel et al. 2013). It is known that anticyclonic vortices are long lived in 2D simulations and can efficiently trap dust particles (Barge and Sommeria 1995; Tanga et al. 1996). One simple way (maybe too simple) to understand the azimuthal particle trapping is by making an analogy with particle trapping in the radial direction. In the radial direction, particles have the tendency to drift towards the gas pressure maximum. The same tendency of particle drift normally holds in the azimuthal direction too. For an anticyclonic vortex, the gas pressure is highest at the vortex center, and so particles drift to the vortex center. Despite this simple explanation has been widely adopted in the literature, we need to keep in mind that particle drift in disks only depends on the drag force which is determined solely by the velocity structure of the disk, not by the pressure. In most circumstances, the velocity structure is balanced by the pressure, leading to the "drift to pressure maximum" correlation. But in some circumstances, e.g. with disk self gravity, particles are not always trapped at the pressure maximum (Zhu and Baruteau 2016).

Although particle trapping vortices are promising for explaining large scale disk asymmetries, several questions need to be addressed: (1) how vortices are generated, (2) are they long lived in realistic protoplanetary disks, and (3) can they fully explain observations?

Vortices can be generated by various hydrodynamical instabilities in disks. In the scenario of gap opening by planets, the gap edge is subject to the Rossby wave instability (RWI) which can lead to vortex formation (Papaloizou and Pringle 1984; Lovelace et al. 1999). Spiral shocks excited by a planet push the disk material away from the planet, leading to gap opening. This process also piles up material at the gap edge, leading to a density bump which has a vortensity minimum (i.e., a location in the disk where vortensity is lower than it is in the surrounding regions) and is subject to the RWI (Koller et al. 2003; Li et al. 2005; Lin and Papaloizou

2010; Meheut et al. 2010; Lobo Gomes et al. 2015). However, if the disk can transfer mass efficiently (e.g. turbulence is strong), such a density bump will be erased during the gap opening process and vortices cannot form. Earlier 2D viscous hydrodynamical simulations (de Val-Borro et al. 2007; Fu et al. 2014) have shown that the viscous α coefficient needs to be smaller than 10^{-3} for vortex generation at the gap edge. Such α threshold has also been found in recent direct 3D MHD turbulent simulations (Zhu and Stone 2014). In ideal-MHD simulations that have strong turbulence induced by the magnetoroational instability (the equivalent α parameter is 10^{-2}), vortices cannot form at the gap edge. But when non-ideal MHD effects have been considered, e.g. ambipolar diffusion, which is expected to operate at the outer disks, disk turbulence is reduced to $\alpha \lesssim 10^{-3}$ and vortices are generated (Fig. 9.14).

After vortices are generated in a smooth disk, they will excite density waves and migrate in disks (Paardekooper et al. 2010). Thus, they may be quickly lost in the disk after being generated. On the other hand, if they are produced at the gap edge, the strong density gradient at the gap edge will trap vortices there.

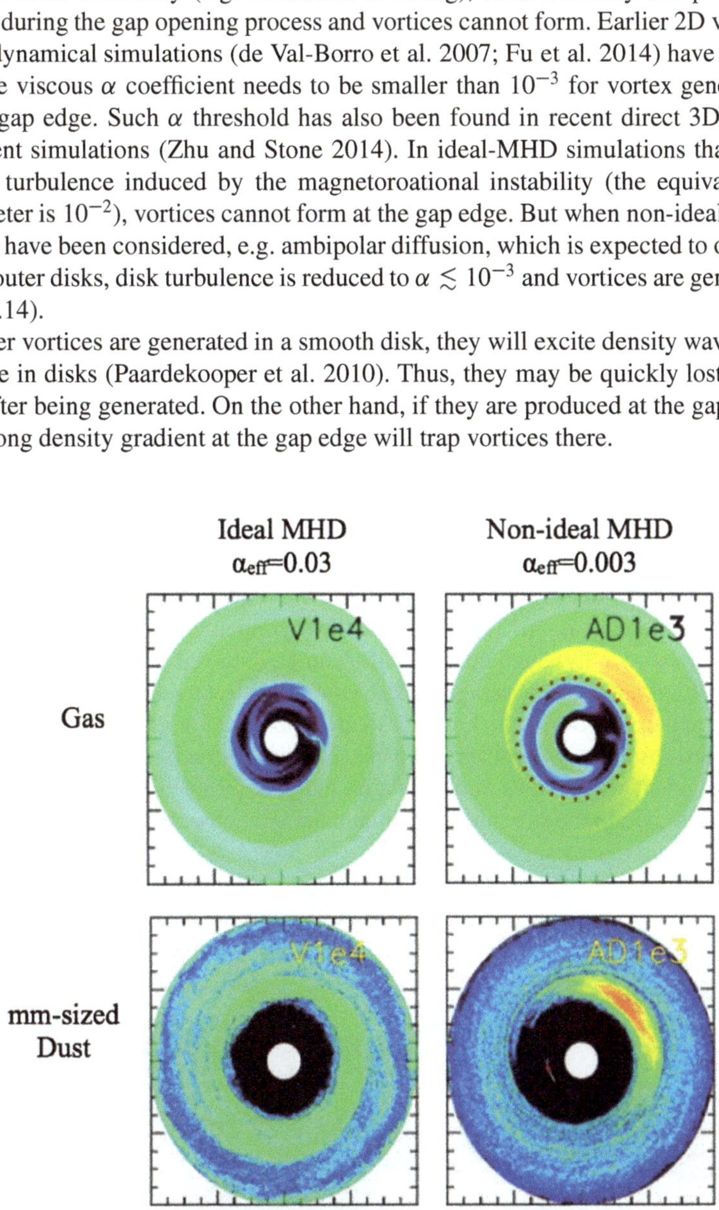

Fig. 9.14 The gas (*left panels*) and dust (*right panels*) surface density for turbulent ideal-MHD simulations ($\alpha \sim 0.03$ *upper panels*) and MHD simulations with ambipolar diffusion (*lower panels*). This figure is from Zhu and Stone (2014)

Even if a vortex does not migrate, it is subject to various instabilities. 3D vortices are subject to the elliptical instability (Lesur and Papaloizou 2009). When dust is concentrated in vortices, vortices are subject to the heavy-core instability (Chang and Oishi 2010) or parametric instability (Railton and Papaloizou 2014). Such instabilities may or may not necessarily destroy vortices. 3D simulations have shown that vortices can survive over hundreds of orbits under certain conditions (Meheut et al. 2012; Zhu et al. 2014; Zhu and Stone 2014). During such time, particles can be efficiently trapped, and the maximum dust surface density can be many times larger than the gas surface density (Raettig et al. 2015). Sometimes this trapping can be too efficient. When the total dust to gas mass ratio (not the peak density ratio) within the vortex is around unity, vortices can be destroyed (Fu et al. 2014; Crnkovic-Rubsamen et al. 2015).

Both the analytical model (Lyra and Lin 2013) and numerical simulations (Zhu and Stone 2014) have shown that dust trapping vortices could explain observed asymmetric disk features (Fig. 9.15). However, there is a modest tension on the radial width of the asymmetric disk structure. Theoretically, a steady vortex has to be narrower than two disk scale heights to be within the subsonic region of the vortex center. In simulations, the wing of the vortex can extend slightly further and the total vortex width can be four disk scale heights (Zhu et al. 2014; Fu et al. 2014). But even so, they are narrower than the observational beam of recent asymmetric disk observations (van der Marel et al. 2013). On the other hand, observations of IRS 48 (van der Marel et al. 2013) suggest that the asymmetric structure has marginally been resolved radially. Such conflict will soon be resolved with higher resolution ALMA observations. If observed disk asymmetric structures are due to vortices, they will appear narrower with higher resolution observations (Zhu and Stone 2014).

If future observations confirm that these large scale structures are due to vortices at gap edges, it will give us constraints on the disk turbulence level ($\alpha \lesssim 10^{-3}$), the gas mass within the structure (larger than the total observed dust mass), and has implications on planetesimal formation.

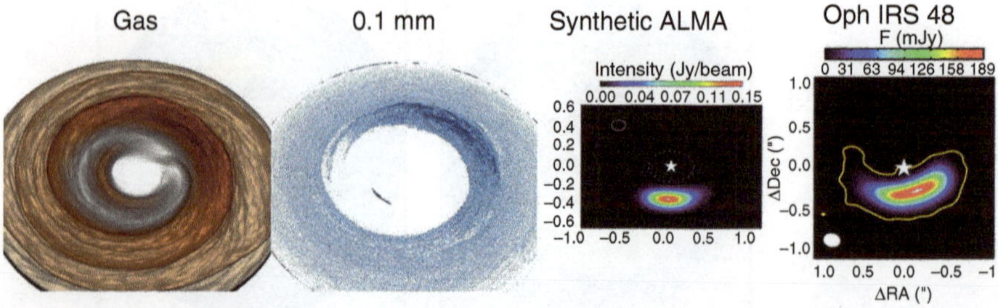

Fig. 9.15 *Left two*: Distributions of gas and dust, respectively, from global 3D MHD simulations with a planet. *Second from the right*: the synthetic ALMA image from this simulation. *Rightmost*: the ALMA observations of Oph IRS 48 from van der Marel et al. (2015a). Figures are from Zhu and Stone (2014)

9.5.3 Spiral–Vortex Interaction

In Sect. 9.4, we have discussed spiral arms as the signature of young planets in disks, while in Sect. 9.5 we have shown that young planets can also induce large scale disk asymmetry. Thus, spiral arms and large scale disk asymmetry should not be considered as mutually exclusive.

Both spiral arms and large scale disk asymmetry have been simultaneously observed in several sources. For example, NWC 758 and SAO 206462, which have two spiral arms in their near-IR scattered light images (Benisty et al. 2015; Garufi et al. 2013), show large scale disk asymmetry in their submm images (Marino et al. 2015b; Pérez et al. 2014). To see if disk–planet interaction can explain both of these features in SAO 206462, 2D, two-fluid hydrodynamic calculations with a young planet embedded in the disk was carried out (Bae et al. 2016) (Fig. 9.16). The results

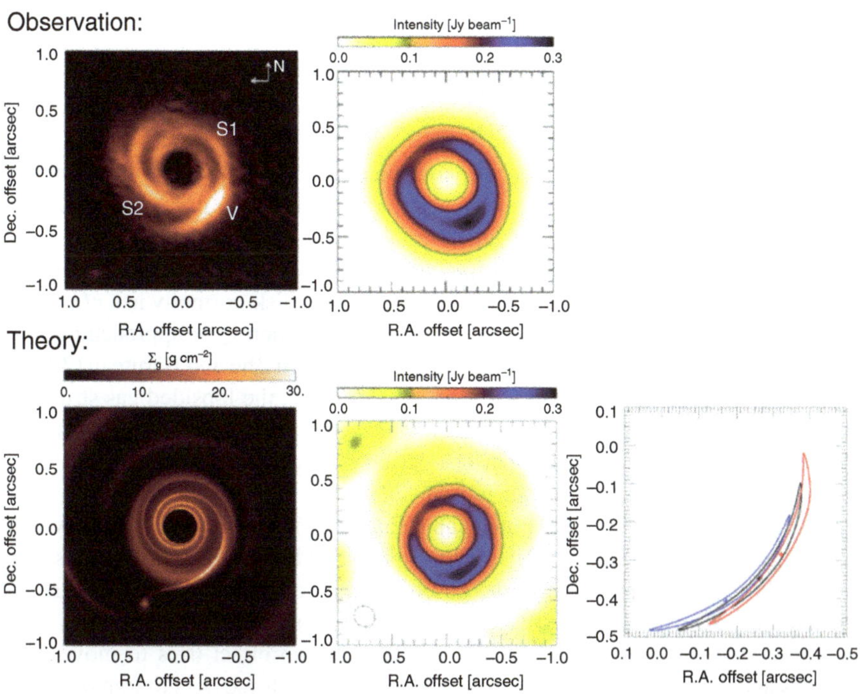

Fig. 9.16 *Upper panels*: Polarized scattered light observed in the *Ks*-band by Garufi et al. (2013) (the *left panel*) and the dust continuum emission obtained with ALMA at 690 GHz by Pérez et al. (2014) (the *right panel*). *Bottom panels*: Gas density distribution in the inner 140 AU disk from simulations (the *left panel*), and ALMA simulated image based on simulations with the cycle 0 extended configuration. The synthesized beam is displayed in the lower-left corner (the *middle panel*). The *lower right panel* shows the time evolution of the bright southwestern feature in the simulation. The *blue* and *red* contours show positions at −10 and +10 years from the black contours, respectively.The *cross symbols* indicate the positions where the density peaks in the structures. Figures are from Bae et al. (2016)

show that a planetary-mass companion located at the outer disk could be responsible for these observed structures. In this model, the planet excites the primary and secondary arms interior to its orbit and also carves a gap generating a vortex at the inner gap edge through Rossby wave instability. The vortex traps radially drifting dust particles, forming a dust-depleted cavity in the inner disk. The vortex could be responsible for the brightest southwestern peak seen in both infrared scattered light and sub-millimeter dust continuum images. Based on the model, there could be a companion with a mass of 10–15 M_J orbiting SAO 26462 at 100–120 AU.

The origin of these disk features is not only constrained by detailed model fitting, but also by their variation over time. The planet-induced spiral arms will rotate around the central star at a frequency the same as the planet orbital frequency, while the vortex will orbit around the central source at the local disk Keplerian frequency. Thus, the spiral patterns and the disk asymmetry will shift azimuthally at different speed over years, which can be potentially probed in next several years. If the vortex center in SAO 206462 is at 0.4″ = 56 AU, the peak will rotate about 1.1° per year.

9.5.4 *Disk Gravity*

Some transitional disks are quite massive, making disk self-gravity potentially important for both vortex generation and particle trapping in vortices. Linear stability analysis has shown that the m mode of the Rossby Wave Instability is suppressed when the disk's Toomre Q parameter is smaller than $r/(mh)$ (Lovelace and Hohlfeld 2013). Considering that a protoplanetary disk normally has $r/h > 10$, disk self-gravity starts to suppress vortex generation when Q is approaching ten. This has been confirmed through numerical simulations in Zhu and Baruteau (2016). Furthermore, when the disk is massive enough ($Q < 3$), the lopsided gas structure orbits around the star at a speed significantly slower than the local Keplerian speed (Zhu and Baruteau 2016), resulting in dust particles concentrating at a radius slightly beyond the vortex center (Baruteau and Zhu 2016).

One surprising result from Baruteau and Zhu (2016) is that even in a disk which is not massive and the disk's self-gravity does not play any role in the gaseous vortex structure, the gravity from the gaseous vortex could still affect the particle distribution within the vortex. Dust particles undergo horseshoe U-turns relative to the vortex due to the gravity from the vortex. Such an effect tries to move dust particles away from the vortex center, while the gas drag force tries to trap particles within the vortex. For larger particles, the horseshoe effect becomes more important and these particles are trapped further ahead of the vortex in the azimuthal direction. Shift angles up to 90° are reached for 10 cm-sized grains, and such large offsets can produce a double-peaked continuum emission observable at mm/cm wavelengths (Fig. 9.17). In a less massive disk, the horseshoe effect becomes weaker since the gravity from the vortex is weaker. On the other hand, the drag force for the same sized (e.g. cm) particle also becomes smaller because the drag force is proportional to the disk surface density for a given size dust particle in the Epstein regime. These

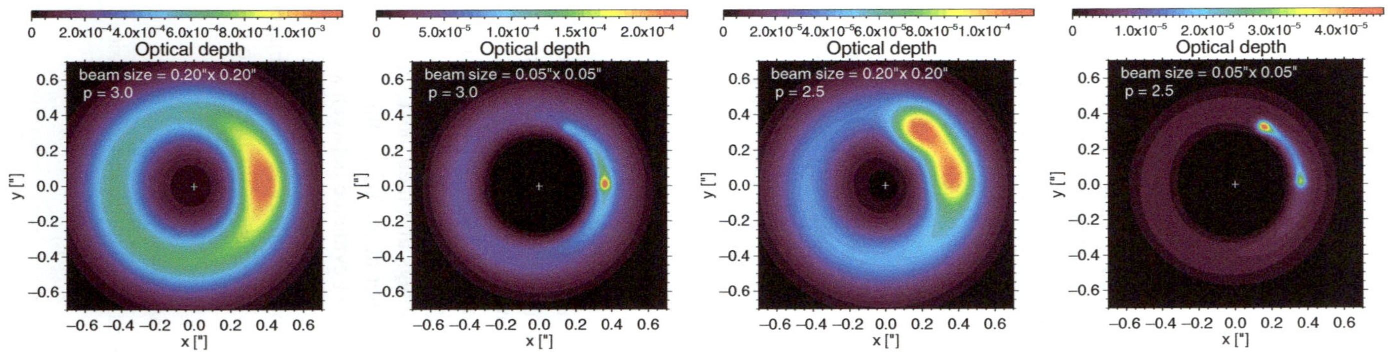

Fig. 9.17 Synthetic images at 37.5 GHz (8 mm) of the disk's optical depth obtained from the simulations where the disk Q is 160. The value of p (minus the power-law index of the grains size distribution) equals 3 in the two columns on the *left-hand side*, and 2.5 on the *right-hand side*. Convolution is done with a Gaussian beam size of FWHM set to 0.2″ or 0.05″. The *plus sign* at the origin marks the star's location. Figures are from Baruteau and Zhu (2016)

two effects compensate each other so that cm-sized grains always shift azimuthally off the vortex.

9.6 Summary and Discussion

The past few years have witnessed the emergence of this new field "observational planet formation," thanks to several key instruments that have come online recently, including Subaru/HiCIAO, Gemini/GPI, VLT/SPHERE, ALMA, and EVLA, and the advances in observational techniques. As of today, a few tens of nearby gaseous disks have been resolved in NIR scattered light and/or mm/cm dust and gas emissions, with angular resolutions as small as $0.03''$. In most cases, asymmetric structures that are possibly planet-induced have been identified. In this chapter, we highlight some recent numerical modeling results on three types of planet-induced structures: gaps, spiral arms, and vortices, and compare them with observations (if available).

Many of these planet-induced structures may have already been found, while it is still too early yet to firmly assert their planetary origin. Table 9.3 summarizes the current status, in which we use "Y" to indicate that a type of planet-induced feature may have been found in a corresponding type of observations (i.e., data is consistent with numerical models), and "N" to indicate the feature has not been found yet.

Finally, we stress that observational signatures of disk–planet interaction are signposts of planets. With only disk features, the existence of planets is after all speculative. To advance our understanding of planet formation, planet detection is crucial.

Several campaigns are underway to look for giant planets in young gaseous disk systems using direct imaging. Currently only a handful of companions (candidates) have been discovered (e.g., Huélamo et al. 2011; Biller et al. 2012; Kraus and Ireland 2012; Quanz et al. 2013a; Brittain et al. 2014; Close et al. 2014; Reggiani et al. 2014; Sallum et al. 2015). Directly imaging planets is difficult due to the high contrast ratio between the brightness of planets and their host stars, and the proximity of the planets to the stars (i.e., a planet at 30 AU is only $\sim 0.2''$ away from its star at 140 pc). So far the strongest direct evidence between structures in disks and the planet sculpting scenario is the detection of three companions inside the wide gap in LkCa 15 (Sallum et al. 2015) (two of them are accreting). Future missions, such as JWST and the planned 30 m class telescopes, are needed to discover more planets

Table 9.3 Current status of observational planet formation

	Scattered light	Dust continuum emissions	Gas emissions
Gaps and cavities	Y	Y	Y
Spiral arms	Y	?	?
Vortices	N	Y	N

in young disk systems, which is crucial to establishing planet-induced features as a new, reliable way for planet detections.

Acknowledgements We thank Jonathan Williams for sharing with us the mm transitional disk mozaic in Fig. 9.2. This project is partially supported by NASA through Hubble Fellowship grants HST-HF-51333.01-A (Z.Z.) and HST-HF-51320.01-A (R.D.) awarded by the Space Telescope Science Institute, which is operated by the Association of Universities for Research in Astronomy, Inc., for NASA, under contract NAS 5-26555. J.F. gratefully acknowledges support from the Natural Sciences and Engineering Research Council of Canada, the Center for Integrative Planetary Science at the University of California, Berkeley, and the Sagan Fellowship Program under contract with the Jet Propulsion Laboratory (JPL) funded by NASA and executed by the NASA Exoplanet Science Institution.

References

Akiyama, E., Muto, T., Kusakabe, N., Kataoka, A., Hashimoto, J., Tsukagoshi, T., Kwon, J., Kudo, T., Kandori, R., Grady, C.A., Takami, M., Janson, M., Kuzuhara, M., Henning, T., Sitko, M.L., Carson, J.C., Mayama, S., Currie, T., Thalmann, C., Wisniewski, J., Momose, M., Ohashi, N., Abe, L., Brandner, W., Brandt, T.D., Egner, S., Feldt, M., Goto, M., Guyon, O., Hayano, Y., Hayashi, M., Hayashi, S., Hodapp, K.W., Ishi, M., Iye, M., Knapp, G.R., Matsuo, T., Mcelwain, M.W., Miyama, S., Morino, J.I., Moro-Martin, A., Nishimura, T., Pyo, T.S., Serabyn, G., Suenaga, T., Suto, H., Suzuki, R., Takahashi, Y.H., Takato, N., Terada, H., Tomono, D., Turner, E.L., Watanabe, M., Yamada, T., Takami, H., Usuda, T., Tamura, M.: Discovery of a disk gap candidate at 20 AU in TW hydrae. Astrophys. J. Lett. **802**, L17 (2015). doi:10.1088/2041-8205/802/2/L17, 1503.01856

Akiyama, E., Hashimoto, J., Liu, H.B., Li, J.I.-H., Bonnefoy, M., Dong, R., Hasegawa, Y., Henning, T., Sitko, M.L., Janson, M., Feldt, M., Wisniewski, J., Kudo, T., Kusakabe, N., Tsukagoshi, T., Momose, M., Muto, T., Taki, T., Kuzuhara, M., Mayama, S., Takami, M., Ohashi, N., Grady, C.A., Kwon, J., Thalmann, C., Abe, L., Brandner, W., Brandt, T.D., Carson, J.C., Egner, S., Goto, M., Guyon, O., Hayano, Y., Hayashi, M., Hayashi, S.S., Hodapp, K.W., Ishii, M., Iye, M., Knapp, G.R., Kandori, R., Matsuo, T., McElwain, M.W., Miyama, S., Morino, J.I., Moro-Martin, A., Nishimura, T., Pyo, T.S., Serabyn, E., Suenaga, T., Suto, H., Suzuki, R., Takahashi, Y.H., Takato, N., Terada, H., Tomono, D., Turner, E.L., Watanabe, M., Yamada, T., Takami, H., Usuda, T., Tamura, M.: Spiral structure and differential dust size distribution in the LkHa 330 disk (2016). ArXiv e-prints 1607.04708

Alexander, R.D., Clarke, C.J., Pringle, J.E.: Photoevaporation of protoplanetary discs - II. Evolutionary models and observable properties. Mon. Not. R. Astron. Soc. **369**, 229–239 (2006). doi:10.1111/j.1365-2966.2006.10294.x, astro-ph/0603254

ALMA Partnership, Brogan, C.L., Pérez, L.M., Hunter, T.R., Dent, W.R.F., Hales, A.S., Hills, R.E., Corder, S., Fomalont, E.B., Vlahakis, C., Asaki, Y., Barkats, D., Hirota, A., Hodge, J.A., Impellizzeri, C.M.V., Kneissl, R., Liuzzo, E., Lucas, R., Marcelino, N., Matsushita, S., Nakanishi, K., Phillips, N., Richards, A.M.S., Toledo, I., Aladro, R., Broguiere, D., Cortes, J.R., Cortes, P.C., Espada, D., Galarza, F., Garcia-Appadoo, D., Guzman-Ramirez, L., Humphreys, E.M., Jung, T., Kameno, S., Laing, R.A., Leon, S., Marconi, G., Mignano, A., Nikolic, B., Nyman, L.A., Radiszcz, M., Remijan, A., Rodón, J.A., Sawada, T., Takahashi, S., Tilanus, R.P.J., Vila Vilaro, B., Watson, L.C., Wiklind, T., Akiyama, E., Chapillon, E., de Gregorio-Monsalvo, I., Di Francesco, J., Gueth, F., Kawamura, A., Lee, C.F., Nguyen Luong, Q., Mangum, J., Pietu, V., Sanhueza, P., Saigo, K., Takakuwa, S., Ubach, C., van Kempen, T., Wootten, A., Castro-Carrizo, A., Francke, H., Gallardo, J., Garcia, J., Gonzalez, S., Hill, T., Kaminski, T., Kurono, Y., Liu, H.Y., Lopez, C., Morales, F., Plarre, K., Schieven, G., Testi,

L., Videla, L., Villard, E., Andreani, P., Hibbard, J.E., Tatematsu, K.: The 2014 ALMA long baseline campaign: first results from high angular resolution observations toward the HL Tau region. Astrophys. J. Lett. **808**, L3 (2015). doi:10.1088/2041-8205/808/1/L3, 1503.02649

Andrews, S.M.: Observations of solids in protoplanetary disks. Publ. Astron. Soc. Pac. **127**, 961–993 (2015). doi:10.1086/683178, 1507.04758

Andrews, S.M., Wilner, D.J., Espaillat, C., Hughes, A.M., Dullemond, C.P., McClure, M.K., Qi, C., Brown, J.M.: Resolved images of large cavities in protoplanetary transition disks. Astrophys. J. **732**, 42 (2011). doi:10.1088/0004-637X/732/1/42, 1103.0284

Andrews, S.M., Wilner, D.J., Zhu, Z., Birnstiel, T., Carpenter, J.M., Pérez, L.M., Bai, X.N., Öberg, K.I., Hughes, A.M., Isella, A., Ricci, L.: Ringed substructure and a gap at 1 AU in the nearest protoplanetary disk. Astrophys. J. Lett. **820**, L40 (2016). doi:10.3847/2041-8205/820/2/L40, 1603.09352

Ansdell, M., Williams, J.P., van der Marel, N., Carpenter, J.M., Guidi, G., Hogerheijde, M., Mathews, G.S., Manara, C.F., Miotello, A., Natta, A., Oliveira, I., Tazzari, M., Testi, L., van Dishoeck, E.F., van Terwisga, S.E.: ALMA survey of lupus protoplanetary disks. I. Dust and gas masses. Astrophys. J. **828**, 46 (2016). doi:10.3847/0004-637X/828/1/46, 1604.05719

Artymowicz, P., Lubow, S.H.: Dynamics of ultraharmonic resonances in spiral galaxies. Astrophys. J. **389**, 129–140 (1992). doi:10.1086/171192

Avenhaus, H., Quanz, S.P., Meyer, M.R., Brittain, S.D., Carr, J.S., Najita, J.R.: HD100546 multi-epoch scattered light observations. Astrophys. J. **790**, 56 (2014a). doi:10.1088/0004-637X/790/1/56, 1405.6120

Avenhaus, H., Quanz, S.P., Schmid, H.M., Meyer, M.R., Garufi, A., Wolf, S., Dominik, C.: Structures in the protoplanetary disk of HD142527 seen in polarized scattered light. Astrophys. J. **781**, 87 (2014b). doi:10.1088/0004-637X/781/2/87, 1311.7088

Bae, J., Zhu, Z., Hartmann, L.: Planetary signatures in the SAO 206462 (HD 135344B) disk: a spiral arm passing through vortex? Astrophys. J. **819**, 134 (2016). doi:10.3847/0004-637X/819/2/134, 1601.04976

Barge, P., Sommeria, J.: Did planet formation begin inside persistent gaseous vortices? Astron. Astrophys. **295**, L1–L4 (1995). astro-ph/9501050

Baruteau, C., Zhu, Z.: Gas and dust hydrodynamical simulations of massive lopsided transition discs - II. Dust concentration. Mon. Not. R. Astron. Soc. **458**, 3927–3941 (2016). doi:10.1093/mnras/stv2527, 1511.03498

Benisty, M., Juhasz, A., Boccaletti, A., Avenhaus, H., Milli, J., Thalmann, C., Dominik, C., Pinilla, P., Buenzli, E., Pohl, A., Beuzit, J.L., Birnstiel, T., de Boer, J., Bonnefoy, M., Chauvin, G., Christiaens, V., Garufi, A., Grady, C., Henning, T., Huelamo, N., Isella, A., Langlois, M., Ménard, F., Mouillet, D., Olofsson, J., Pantin, E., Pinte, C., Pueyo, L.: Asymmetric features in the protoplanetary disk MWC 758. Astron. Astrophys. **578**, L6 (2015). doi:10.1051/0004-6361/201526011, 1505.05325

Beuzit, J.L., Feldt, M., Dohlen, K., Mouillet, D., Puget, P., Wildi, F., Abe, L., Antichi, J., Baruffolo, A., Baudoz, P., Boccaletti, A., Carbillet, M., Charton, J., Claudi, R., Downing, M., Fabron, C., Feautrier, P., Fedrigo, E., Fusco, T., Gach, J.L., Gratton, R., Henning, T., Hubin, N., Joos, F., Kasper, M., Langlois, M., Lenzen, R., Moutou, C., Pavlov, A., Petit, C., Pragt, J., Rabou, P., Rigal, F., Roelfsema, R., Rousset, G., Saisse, M., Schmid, H.M., Stadler, E., Thalmann, C., Turatto, M., Udry, S., Vakili, F., Waters, R.: SPHERE: a 'Planet Finder' instrument for the VLT. In: Society of Photo-Optical Instrumentation Engineers (SPIE) Conference Series, Society of Photo-Optical Instrumentation Engineers (SPIE) Conference Series, vol. 7014 (2008). doi:10.1117/12.790120

Biller, B., Lacour, S., Juhász, A., Benisty, M., Chauvin, G., Olofsson, J., Pott, J.U., Müller, A., Sicilia-Aguilar, A., Bonnefoy, M., Tuthill, P., Thebault, P., Henning, T., Crida, A.: A likely close-in low-mass stellar companion to the transitional disk star HD 142527. Astrophys. J. Lett. **753**, L38 (2012). doi:10.1088/2041-8205/753/2/L38, 1206.2654

Birnstiel, T., Andrews, S.M., Ercolano, B.: Can grain growth explain transition disks? Astron. Astrophys. **544**, A79 (2012). doi:10.1051/0004-6361/201219262, 1206.5802

Boccaletti, A., Pantin, E., Lagrange, A.M., Augereau, J.C., Meheut, H., Quanz, S.P.: Multiple spiral patterns in the transitional disk of HD 100546. Astron. Astrophys. **560**, A20 (2013). doi:10.1051/0004-6361/201322365, 1310.7092

Brandt, T.D., McElwain, M.W., Turner, E.L., Mede, K., Spiegel, D.S., Kuzuhara, M., Schlieder, J.E., Wisniewski, J.P., Abe, L., Biller, B., Brandner, W., Carson, J., Currie, T., Egner, S., Feldt, M., Golota, T., Goto, M., Grady, C.A., Guyon, O., Hashimoto, J., Hayano, Y., Hayashi, M., Hayashi, S., Henning, T., Hodapp, K.W.,Inutsuka, S., Ishii, M., Iye, M., Janson, M., Kandori, R., Knapp, G.R., Kudo, T., Kusakabe, N., Kwon, J., Matsuo, T., Miyama, S., Morino, J.I., Moro-Martín, A., Nishimura, T., Pyo, T.S., Serabyn, E., Suto, H., Suzuki, R., Takami, M., Takato, N., Terada, H., Thalmann, C., Tomono, D., Watanabe, M., Yamada, T., Takami, H., Usuda, T., Tamura, M.: A Statistical analysis of SEEDS and other high-contrast exoplanet surveys: massive planets or low-mass brown dwarfs? Astrophys. J. **794**, 159 (2014). doi:10.1088/0004-637X/794/2/159, 1404.5335

Brittain, S.D., Carr, J.S., Najita, J.R., Quanz, S.P., Meyer, M.R.: NIR spectroscopy of the HAeBe star HD 100546. III. Further evidence of an orbiting companion? Astrophys. J. **791**, 136 (2014). doi:10.1088/0004-637X/791/2/136, 1409.0804

Bryden, G., Chen, X., Lin, D.N.C., Nelson, R.P., Papaloizou, J.C.B.: Tidally induced gap formation in protostellar disks: gap clearing and suppression of protoplanetary growth. Astrophys. J. **514**, 344–367 (1999). doi:10.1086/306917

Calvet, N., D'Alessio, P., Watson, D.M., Franco-Hernández, R., Furlan, E., Green, J., Sutter, P.M., Forrest, W.J., Hartmann, L., Uchida K.I., Keller, L.D., Sargent, B., Najita, J., Herter, T.L., Barry, D.J., Hall, P.: Disks in transition in the taurus population: spitzer IRS spectra of GM aurigae and DM tauri. Astrophys. J. **630**, L185–L188 (2005). doi:10.1086/491652

Canovas, H., Ménard, F., Hales, A., Jordán, A., Schreiber, M.R., Casassus, S., Gledhill, T.M., Pinte, C.: Near-infrared imaging polarimetry of HD 142527. Astron. Astrophys. **556**, A123 (2013). doi:10.1051/0004-6361/201321924, 1306.6379

Carrasco-González, C., Henning, T., Chandler, C.J., Linz, H., Pérez, L., Rodríguez, L.F., Galván-Madrid, R., Anglada, G., Birnstiel, T., van Boekel, R., Flock, M., Klahr, H., Macias, E., Menten, K., Osorio, M., Testi, L., Torrelles, J.M., Zhu, Z.: The VLA View of the HL Tau disk: disk mass, grain evolution, and early planet formation. Astrophys. J. Lett. **821**, L16 (2016). doi:10.3847/2041-8205/821/1/L16, 1603.03731

Casassus, S., van der Plas, G., Sebastian Perez, M., Dent, W.R.F., Fomalont, E., Hagelberg, J., Hales, A., Jordán, A., Mawet, D., Ménard, F., Wootten, A., Wilner, D., Hughes, A.M., Schreiber, M.R., Girard, J.H., Ercolano, B., Canovas, H., Román, P.E., Salinas, V.: Flows of gas through a protoplanetary gap. Nature **493**, 191–194 (2013). doi:10.1038/nature11769, 1305.6062

Chang, P., Oishi, J.S.: On the stability of dust-laden protoplanetary vortices. Astrophys. J. **721**, 1593–1602 (2010). doi:10.1088/0004-637X/721/2/1593, 1007.2417

Chapillon, E., Guilloteau, S., Dutrey, A., Piétu, V.: Disks around CQ Tauri and MWC 758: dense PDR or gas dispersal? Astron. Astrophys. **488**, 565–578 (2008). doi:10.1051/0004-6361:200809523, 0805.3473

Chen, L., Kreplin, A., Wang, Y., Weigelt, G., Hofmann, K.H., Kraus, S., Schertl, D., Lagarde, S., Natta, A., Petrov, R., Robbe-Dubois, S., Tatulli, E.: Near-infrared interferometric observation of the Herbig Ae star HD 144432 with VLTI/AMBER. Astron. Astrophys. **541**, A104 (2012). doi:10.1051/0004-6361/201218818, 1203.6240

Christiaens, V., Casassus, S., Perez, S., van der Plas, G., Ménard, F.: Spiral arms in the disk of HD 142527 from CO emission lines with ALMA. Astrophys. J. Lett. **785**, L12 (2014). doi:10.1088/2041-8205/785/1/L12, 1403.1463

Clarke, C.J., Gendrin, A., Sotomayor, M.: The dispersal of circumstellar discs: the role of the ultraviolet switch. Mon. Not. R. Astron. Soc. **328**, 485–491 (2001). doi:10.1046/j.1365-8711.2001.04891.x

Close, L.M., Follette, K.B., Males, J.R., Puglisi, A., Xompero, M., Apai, D., Najita, J., Weinberger, A.J., Morzinski, K., Rodigas, T.J., Hinz, P., Bailey, V., Briguglio, R.: Discovery of Hα emission from the close companion inside the gap of transitional disk HD 142527. Astrophys. J. Lett. **781**, L30 (2014). doi:10.1088/2041-8205/781/2/L30, 1401.1273

Crnkovic-Rubsamen, I., Zhu, Z., Stone, J.M.: Survival and structure of dusty vortices in protoplanetary discs. Mon. Not. R. Astron. Soc. **450**, 4285–4291 (2015). doi:10.1093/mnras/stv828

Cumming, A., Butler, R.P., Marcy, G.W., Vogt, S.S., Wright, J.T., Fischer, D.A.: The keck planet search: detectability and the minimum mass and orbital period distribution of extrasolar planets. Publ. Astron. Soc. Pac. **120**, 531–554 (2008). doi:10.1086/588487, 0803.3357

Currie, T., Cloutier, R., Brittain, S., Grady, C., Burrows, A., Muto, T., Kenyon, S.J., Kuchner, M.J.: Resolving the HD 100546 protoplanetary system with the gemini planet imager: evidence for multiple forming, accreting planets (2015). ArXiv e-prints 1511.02526

Czekala, I., Andrews, S.M., Jensen, E.L.N., Stassun, K.G., Torres, G., Wilner, D.J.: A disk-based dynamical mass estimate for the young binary AK sco. Astrophys. J. **806**, 154 (2015). doi:10.1088/0004-637X/806/2/154, 1505.01850

David, T.J., Hillenbrand, L.A., Petigura, E.A., Carpenter, J.M., Crossfield, I.J.M., Hinkley, S., Ciardi, D.R., Howard, A.W., Isaacson, H.T., Cody, A.M., Schlieder, J.E., Beichman, C.A., Barenfeld, S.A.: A Neptune-sized transiting planet closely orbiting a 5–10-million-year-old star. Nature **534**, 658–661 (2016). doi:10.1038/nature18293, 1606.06729

de Juan Ovelar, M., Min, M., Dominik, C., Thalmann, C., Pinilla, P., Benisty, M., Birnstiel, T.: Imaging diagnostics for transitional discs. Astron. Astrophys. **560**, A111 (2013). doi:10.1051/0004-6361/201322218, 1309.1039

de Val-Borro, M., Artymowicz, P., D'Angelo, G., Peplinski, A.: Vortex generation in protoplanetary disks with an embedded giant planet. Astron. Astrophys. **471**, 1043–1055 (2007). doi:10.1051/0004-6361:20077169, 0706.3200

Dodson-Robinson, S.E., Salyk, C.: Transitional disks as signposts of young, multiplanet systems. Astrophys. J. **738**, 131 (2011). doi:10.1088/0004-637X/738/2/131, 1106.4824

Dong, R.: The effects of self-shadowing by a puffed-up inner rim in scattered light images of protoplanetary disks. Astrophys. J. **810**, 6 (2015). doi:10.1088/0004-637X/810/1/6, 1507.07925

Dong, R., Dawson, R.: Stability and occurrence rate constraints on the planetary sculpting hypothesis for "Transitional Disks". Astrophys. J. **825**, 77 (2016). doi:10.3847/0004-637X/825/1/77, 1605.02074

Dong, R., Fung, J.: How bright are planet-induced spiral arms in scattered light? Astrophys. J. **35**, 38 (2017a). ArXiv e-prints 1612.00446

Dong, R., Fung, J.: What is the mass of a gap-opening planet? Astrophys. J. **835**, 146 (2017b). ArXiv e-prints 1612.04821

Dong, R., Hashimoto, J., Rafikov, R., Zhu, Z., Whitney, B., Kudo, T., Muto, T., Brandt, T., McClure, M.K., Wisniewski, J., Abe, L., Brandner, W., Carson, J., Egner, S., Feldt, M., Goto, M., Grady, C., Guyon, O., Hayano, Y., Hayashi, M., Hayashi, S., Henning, T., Hodapp, K.W., Ishii, M., Iye, M., Janson, M., Kandori, R., Knapp, G.R., Kusakabe, N., Kuzuhara, M., Kwon, J., Matsuo, T., McElwain, M., Miyama, S., Morino, J.I., Moro-Martin, A., Nishimura, T., Pyo, T.S., Serabyn, E., Suto, H., Suzuki, R., Takami, M., Takato, N., Terada, H., Thalmann, C., Tomono, D., Turner, E., Watanabe, M., Yamada, T., Takami, H., Usuda, T., Tamura, M.: The structure of pre-transitional protoplanetary disks. I. Radiative transfer modeling of the disk+cavity in the PDS 70 system. Astrophys. J. **760**, 111 (2012a). doi:10.1088/0004-637X/760/2/111, 1209.3772

Dong, R., Rafikov, R., Zhu, Z., Hartmann, L., Whitney, B., Brandt, T., Muto, T., Hashimoto, J., Grady, C., Follette, K., Kuzuhara, M., Tanii, R., Itoh, Y., Thalmann, C., Wisniewski, J., Mayama, S., Janson, M., Abe, L., Brandner, W., Carson, J., Egner, S., Feldt, M., Goto, M., Guyon, O., Hayano, Y., Hayashi, M., Hayashi, S., Henning, T., Hodapp, K.W., Honda, M., Inutsuka, S., Ishii, M., Iye, M., Kandori, R., Knapp, G.R., Kudo, T., Kusakabe, N., Matsuo, T., McElwain, M.W., Miyama, S., Morino, J.I., Moro-Martin, A., Nishimura, T., Pyo, T.S., Suto, H., Suzuki, R., Takami, M., Takato, N., Terada, H., Tomono, D., Turner, E.L., Watanabe, M., Yamada, T., Takami, H., Usuda, T., Tamura, M.: The missing cavities in the SEEDS polarized scattered light images of transitional protoplanetary disks: a generic disk model. Astrophys. J. **750**, 161 (2012b). doi:10.1088/0004-637X/750/2/161, 1203.1612

Dong, R., Hall, C., Rice, K., Chiang, E.: Spiral arms in gravitationally unstable protoplanetary disks as imaged in scattered light. Astrophys. J. Lett.**812**, L32 (2015a). doi:10.1088/2041-8205/812/2/L32, 1510.00396

Dong, R., Zhu, Z., Rafikov, R.R., Stone, J.M.: Observational signatures of planets in protoplanetary disks II: spiral arms observed in scattered light imaging can be induced by planets. Astrophys. J. Lett. **809**, L5 (2015b). doi:10.1088/2041-8205/809/1/L5, 1507.03596

Dong, R., Zhu, Z., Whitney, B.: Observational signatures of planets in protoplanetary disks I. gaps opened by single and multiple young planets in disks. Astrophys. J. **809**, 93 (2015c). doi:10.1088/0004-637X/809/1/93, 1411.6063

Dong, R., Fung, J., Chiang, E.: How spirals and gaps driven by companions in protoplanetary disks appear in scattered light at arbitrary viewing angles. Astrophys. J. **826**, 75 (2016a). doi:10.3847/0004-637X/826/1/75, 1602.04814

Dong, R., Vorobyov, E., Pavlyuchenkov, Y., Chiang, E., Liu, H.B.: Signatures of gravitational instability in resolved images of protostellar disks. Astrophys. J. **823**, 141 (2016b). doi:10.3847/0004-637X/823/2/141, 1603.01618

Dong, R., Zhu, Z., Fung, J., Rafikov, R., Chiang, E., Wagner, K.: An M dwarf companion and its induced spiral arms in the HD 100453 protoplanetary disk. Astrophys. J. Lett. **816**, L12 (2016c). doi:10.3847/2041-8205/816/1/L12, 1512.04949

Duffell, P.C., Dong, R.: Shallow cavities in multiple-planet systems. Astrophys. J. **802**, 42 (2015). doi:10.1088/0004-637X/802/1/42, 1412.3560

Dullemond, C.P., Dominik, C.: Dust coagulation in protoplanetary disks: a rapid depletion of small grains. Astron. Astrophys. **434**, 971–986 (2005). doi:10.1051/0004-6361:20042080, astro-ph/0412117

Espaillat, C., Calvet, N., D'Alessio, P., Hernández, J., Qi, C., Hartmann, L., Furlan, E., Watson, D.M.: On the diversity of the taurus transitional disks: UX tauri A and LkCa 15. Astrophys. J. **670**, L135–L138 (2007). doi:10.1086/524360, 0710.2892

Espaillat, C., D'Alessio, P., Hernández, J., Nagel, E., Luhman, K.L., Watson, D.M., Calvet, N., Muzerolle, J., McClure, M.: Unveiling the structure of pre-transitional disks. Astrophys. J. **717**, 441–457 (2010). doi:10.1088/0004-637X/717/1/441, 1005.2365

Espaillat, C., Muzerolle, J., Najita, J., Andrews, S., Zhu, Z., Calvet, N., Kraus, S., Hashimoto, J., Kraus, A., D'Alessio, P.: An observational perspective of transitional disks (2014). ArXiv e-prints 1402.7103

Fabrycky, D.C., Murray-Clay, R.A.: Stability of the directly imaged multiplanet system HR 8799: resonance and masses. Astrophys. J. **710**, 1408–1421 (2010). doi:10.1088/0004-637X/710/2/1408. 0812.0011

Follette, K.B., Tamura, M., Hashimoto, J., Whitney, B., Grady, C., Close, L., Andrews, S.M., Kwon, J., Wisniewski, J., Brandt, T.D., Mayama, S., Kandori, R., Dong, R., Abe, L., Brandner, W., Carson, J., Currie, T., Egner, S.E., Feldt, M., Goto, M., Guyon, O., Hayano, Y., Hayashi, M., Hayashi, S., Henning, T., Hodapp, K., Ishii, M., Iye, M., Janson, M., Knapp, G.R., Kudo, T., Kusakabe, N., Kuzuhara, M., McElwain, M.W., Matsuo, T., Miyama, S., Morino, J.I., Moro-Martin, A., Nishimura, T., Pyo, T.S., Serabyn, E., Suto, H., Suzuki, R., Takami, M., Takato, N., Terada, H., Thalmann, C., Tomono, D., Turner, E.L., Watanabe, M., Yamada, T., Takami, H., Usuda, T.: Mapping H-band scattered light emission in the mysterious SR21 transitional disk. Astrophys. J. **767**, 10 (2013). doi:10.1088/0004-637X/767/1/10, 1302.5705

Fu, W., Li, H., Lubow, S., Li, S., Liang, E.: Effects of dust feedback on vortices in protoplanetary disks. Astrophys. J. Lett. **795**, L39 (2014). doi:10.1088/2041-8205/795/2/L39, 1410.4196

Fukagawa, M., Tsukagoshi, T., Momose, M., Saigo, K., Ohashi, N., Kitamura, Y., Inutsuka, S.I., Muto, T., Nomura, H., Takeuchi, T., Kobayashi, H., Hanawa, T., Akiyama, E., Honda, M., Fujiwara, H., Kataoka, A., Takahashi, S.Z., Shibai, H.: Local enhancement of the surface density in the protoplanetary ring surrounding HD 142527. Publ. Astron. Soc. Jpn. **65**, L14 (2013), doi:10.1093/pasj/65.6.L14, 1309.7400

Fung, J.: A study of protoplanetary disk dynamics using accelerated hydrodynamics simulations on graphics processing units. PhD thesis, University of Toronto, Canada (2015)

Fung, J., Chiang, E.: Gap opening in 3D: single-planet gaps. Astrophys. J. **832**, 105 (2016). doi:10.3847/0004-637X/832/2/105, 1606.02299

Fung, J., Dong, R.: Inferring planet mass from spiral structures in protoplanetary disks. Astrophys. J. Lett. **815**, L21 (2015). doi:10.1088/2041-8205/815/2/L21, 1511.01178

Fung, J., Shi, J.M., Chiang, E.: How empty are disk gaps opened by giant planets? Astrophys. J. **782**, 88 (2014). doi:10.1088/0004-637X/782/2/88, 1310.0156

Garufi, A., Quanz, S.P., Avenhaus, H., Buenzli, E., Dominik, C., Meru, F., Meyer, M.R., Pinilla, P., Schmid, H.M., Wolf, S.: Small vs. large dust grains in transitional disks: do different cavity sizes indicate a planet? SAO 206462 (HD 135344B) in polarized light with VLT/NACO. Astron. Astrophys. **560**, A105 (2013). doi:10.1051/0004-6361/201322429, 1311.4195

Garufi, A., Quanz, S.P., Schmid, H.M., Avenhaus, H., Buenzli, E., Wolf, S.: Shadows and cavities in protoplanetary disks: HD 163296, HD 141569A, and HD 150193A in polarized light. Astron. Astrophys. **568**, A40 (2014). doi:10.1051/0004-6361/201424262, 1406.7387

Garufi, A., Quanz, S.P., Schmid, H.M., Mulders, G.D., Avenhaus, H., Boccaletti, A., Ginski, C., Langlois, M., Stolker, T., Augereau, J.C., Benisty, M., Lopez, B., Dominik, C., Gratton, R., Henning, T., Janson, M., Menard, F., Meyer, M.R., Pinte, C., Sissa, E., Vigan, A., Zurlo, A., Bazzon, A., Buenzli, E., Bonnefoy, M., Brandner, W., Chauvin, G., Cheetham, A., Cudel, M., Desidera, S., Feldt, M., Galicher, R., Kasper, M., Lagrange, A.M., Lannier, J., Maire, A.L., Mesa, D., Mouillet, D., Peretti, S., Perrot, C., Salter, G., Wildi, F.: The SPHERE view of the planet-forming disk around HD100546 (2016). ArXiv e-prints 1601.04983

Goldreich, P., Lynden-Bell, D.: I. Gravitational stability of uniformly rotating disks. Mon. Not. R. Astron. Soc. **130**, 97 (1965). doi:10.1093/mnras/130.2.97

Goldreich, P., Tremaine, S.: Disk-satellite interactions. Astrophys. J. **241**, 425–441 (1980). doi:10.1086/158356

Goodman, J., Rafikov, R.R.: Planetary torques as the viscosity of protoplanetary disks. Astrophys. J. **552**, 793–802 (2001). doi:10.1086/320572, astro-ph/0010576

Grady, C.A., Muto, T., Hashimoto, J., Fukagawa, M., Currie, T., Biller, B., Thalmann, C., Sitko, M.L., Russell, R., Wisniewski, J., Dong, R., Kwon, J., Sai, S., Hornbeck, J., Schneider, G., Hines, D., Moro Martín, A., Feldt, M., Henning, T., Pott, J.U., Bonnefoy, M., Bouwman, J., Lacour, S., Mueller, A., Juhász, A., Crida, A., Chauvin, G., Andrews, S., Wilner, D., Kraus, A., Dahm, S., Robitaille, T., Jang-Condell, H., Abe, L., Akiyama, E., Brandner, W., Brandt, T., Carson, J., Egner, S., Follette, K.B., Goto, M., Guyon, O., Hayano, Y., Hayashi, M., Hayashi, S., Hodapp, K., Ishii, M., Iye, M., Janson, M., Kandori, R., Knapp, G., Kudo, T., Kusakabe, N., Kuzuhara, M., Mayama, S., McElwain, M., Matsuo, T., Miyama, S., Morino, J.I., Nishimura, T., Pyo, T.S., Serabyn, G., Suto, H., Suzuki, R., Takami, M., Takato, N., Terada, H., Tomono, D., Turner, E., Watanabe, M., Yamada, T., Takami, H., Usuda, T., Tamura, M.: Spiral arms in the asymmetrically illuminated disk of MWC 758 and constraints on giant planets. Astrophys. J. **762**, 48 (2013). doi:10.1088/0004-637X/762/1/48, 1212.1466

Hashimoto, J., Tamura, M., Muto, T., Kudo, T., Fukagawa, M., Fukue, T., Goto, M., Grady, C.A., Henning, T., Hodapp, K., Honda, M., Inutsuka, S., Kokubo, E., Knapp, G., McElwain, M.W., Momose, M., Ohashi, N., Okamoto, Y.K., Takami, M., Turner, E.L., Wisniewski, J., Janson, M., Abe, L., Brandner, W., Carson, J., Egner, S., Feldt, M., Golota, T., Guyon, O., Hayano, Y., Hayashi, M., Hayashi, S., Ishii, M., Kandori, R., Kusakabe, N., Matsuo, T., Mayama, S., Miyama, S., Morino, J.I., Moro-Martin, A., Nishimura, T., Pyo, T.S., Suto, H., Suzuki, R., Takato, N., Terada, H., Thalmann, C., Tomono, D., Watanabe, M., Yamada, T., Takami, H., Usuda, T.: Direct imaging of fine structures in giant planet-forming regions of the protoplanetary disk around AB aurigae. Astrophys. J. Lett. **729**, L17 (2011). doi:10.1088/2041-8205/729/2/L17. 1102.4408

Hashimoto, J., Dong, R., Kudo, T., Honda, M., McClure, M.K., Zhu, Z., Muto, T., Wisniewski, J., Abe, L., Brandner, W., Brandt, T., Carson, J., Egner, S., Feldt, M., Fukagawa, M., Goto, M., Grady, C.A., Guyon, O., Hayano, Y., Hayashi, M., Hayashi, S., Henning, T., Hodapp, K., Ishii, M., Iye, M., Janson, M., Kandori, R., Knapp, G., Kusakabe, N., Kuzuhara, M., Kwon, J., Matsuo, T., Mayama, S., McElwain, M.W., Miyama, S., Morino, J.I., Moro-Martin, A., Nishimura, T., Pyo, T.S., Serabyn, G., Suenaga, T., Suto, H., Suzuki, R., Takahashi, Y., Takami,

M., Takato, N., Terada, H., Thalmann, C., Tomono, D., Turner, E.L., Watanabe, M., Yamada, T., Takami, H., Usuda, T., Tamura, M.: Polarimetric imaging of large cavity structures in the pre-transitional protoplanetary disk around PDS 70: observations of the disk. Astrophys. J. Lett. **758**, L19 (2012). doi:10.1088/2041-8205/758/1/L19, 1208.2075

Hinkley, S., Oppenheimer, B.R., Soummer, R., Brenner, D., Graham, J.R., Perrin, M.D., Sivaramakrishnan, A., Lloyd, J.P., Roberts, L.C., Jr, Kuhn, J.: Speckle suppression through dual imaging polarimetry, and a ground-based image of the HR 4796A circumstellar disk. Astrophys. J. **701**, 804 (2009). doi:10.1088/0004-637X/701/1/804, 0906.3010

Huélamo, N., Lacour, S., Tuthill, P., Ireland, M., Kraus, A., Chauvin, G.: A companion candidate in the gap of the T Chamaeleontis transitional disk. Astron. Astrophys. **528**, L7 (2011). doi:10.1051/0004-6361/201016395, 1102.4982

Isella, A., Carpenter, J.M., Sargent, A.I.: Investigating planet formation in circumstellar disks: CARMA observations of Ry Tau and Dg Tau. Astrophys. J. **714**, 1746–1761 (2010). doi:10.1088/0004-637X/714/2/1746, 1003.4318

Isella, A., Pérez, L.M., Carpenter, J.M., Ricci, L., Andrews, S., Rosenfeld, K.: An azimuthal asymmetry in the LkHα 330 disk. Astrophys. J. **775**, 30 (2013). doi:10.1088/0004-637X/775/1/30, 1307.5848

Janson, M., Thalmann, C., Boccaletti, A., Maire, A.L., Zurlo, A., Marzari, F., Meyer, M.R., Carson, J.C., Augereau, J.C., Garufi, A., Henning, T., Desidera, S., Asensio-Torres, R., Pohl, A.: Detection of sharp symmetric features in the circumbinary disk around AK Sco. Astrophys. J. Lett. **816**, L1 (2016). doi:10.3847/2041-8205/816/1/L1, 1512.04552

Juhász, A., Benisty, M., Pohl, A., Dullemond, C.P., Dominik, C., Paardekooper, S.J.: Spiral arms in scattered light images of protoplanetary discs: are they the signposts of planets? Mon. Not. R. Astron. Soc. **451**, 1147–1157 (2015). doi:10.1093/mnras/stv1045, 1412.3412

Kim, S.H., Martin, P.G., Hendry, P.D.: The size distribution of interstellar dust particles as determined from extinction. Astrophys. J. **422**, 164–175 (1994). doi:10.1086/173714

Kley, W., Nelson, R.P.: Planet-disk interaction and orbital evolution. Annu. Rev. Astron. Astrophys. **50**, 211–249 (2012). doi:10.1146/annurev-astro-081811-125523, 1203.1184

Koller, J., Li, H., Lin, D.N.C.: Vortices in the co-orbital region of an embedded protoplanet. Astrophys. J. Lett. **596**, L91–L94 (2003). doi:10.1086/379032, astro-ph/0308358

Kraus, A.L., Ireland, M.J.: LkCa 15: a young exoplanet caught at formation? Astrophys. J. **745**, 5 (2012). doi:10.1088/0004-637X/745/1/5, 1110.3808

Lee, W.K.: Ultraharmonics and secondary spiral wakes induced by a planet (2016). ArXiv e-prints 1604.08941

Lenzen, R., Hartung, M., Brandner, W., Finger, G., Hubin, N.N., Lacombe, F., Lagrange, A.M., Lehnert, M.D., Moorwood, A.F.M., Mouillet, D.: NAOS-CONICA first on sky results in a variety of observing modes. In: Iye, M., Moorwood, A.F.M. (eds.) Instrument Design and Performance for Optical/Infrared Ground-based Telescopes, Proceedings of SPIE, vol. 4841, pp. 944–952 (2003). doi:10.1117/12.460044

Lesur, G., Papaloizou, J.C.B.: On the stability of elliptical vortices in accretion discs. Astron. Astrophys. **498**, 1–12 (2009). doi:10.1051/0004-6361/200811577, 0903.1720

Li, H., Li, S., Koller, J., Wendroff, B.B., Liska, R., Orban, C.M., Liang, E.P.T., Lin, D.N.C.: Potential vorticity evolution of a protoplanetary disk with an embedded protoplanet. Astrophys. J. **624**, 1003–1009 (2005). doi:10.1086/429367, astro-ph/0503404

Lin, D.N.C., Papaloizou, J.C.B.: On the tidal interaction between protostellar disks and companions. In: Levy, E.H., Lunine, J.I. (eds.) Protostars and Planets III, pp. 749–835. University of Arizona Press, Tucson (1993)

Lin, M.K., Papaloizou, J.C.B.: Type III migration in a low-viscosity disc. Mon. Not. R. Astron. Soc. **405**, 1473–1490 (2010). doi:10.1111/j.1365-2966.2010.16560.x, 1002.3742

Liu, H.B., Takami, M., Kudo, T., Hashimoto, J., Dong, R., Vorobyov, E.I., Pyo, T.S., Fukagawa, M., Tamura, M., Henning, T., Dunham, M.M., Karr, J.L., Kusakabe, N., Tsuribe, T.: Circumstellar disks of the most vigorously accreting young stars. Sci. Adv. **2**, e1500875 (2016). doi:10.1126/sciadv.1500875, 1602.04068

Lobo Gomes, A., Klahr, H., Uribe, A.L., Pinilla, P., Surville, C.: Vortex formation and evolution in planet harboring disks under thermal relaxation. Astrophys. J. **810**, 94 (2015). doi:10.1088/0004-637X/810/2/94, 1508.00903

Lovelace, R.V.E., Hohlfeld, R.G.: Rossby wave instability with self-gravity. Mon. Not. R. Astron. Soc. **429**, 529–533 (2013). doi:10.1093/mnras/sts361, 1212.0443

Lovelace, R.V.E., Li, H., Colgate, S.A., Nelson, A.F.: Rossby wave instability of Keplerian accretion disks. Astrophys. J. **513**, 805–810 (1999). doi:10.1086/306900, astro-ph/9809321

Luhman, K.L., Allen, P.R., Espaillat, C., Hartmann, L., Calvet, N.: The disk population of the Taurus star-forming region. Astrophys. J. Suppl. Ser. **186**, 111–174 (2010). doi:10.1088/0067-0049/186/1/111, 0911.5457

Lyra, W., Lin, M.K.: Steady state dust distributions in disk vortices: observational predictions and applications to transitional disks. Astrophys. J. **775**, 17 (2013). doi:10.1088/0004-637X/775/1/17, 1307.3770

Macintosh, B.A., Graham, J.R., Palmer, D.W., Doyon, R., Dunn, J., Gavel, D.T., Larkin, J., Oppenheimer, B., Saddlemyer, L., Sivaramakrishnan, A., Wallace, J.K., Bauman, B., Erickson, D.A., Marois, C., Poyneer, L.A., Soummer, R.: The Gemini Planet Imager: From science to design to construction. In: Society of Photo-Optical Instrumentation Engineers (SPIE) Conference Series, Society of Photo-Optical Instrumentation Engineers (SPIE) Conference Series, vol. 7015 (2008). doi:10.1117/12.788083

Marino, S., Casassus, S., Perez, S., Lyra, W., Roman, P.E., Avenhaus, H., Wright, C.M., Maddison, S.T.: Compact dust concentration in the MWC 758 protoplanetary disk. Astrophys. J. **813**, 76 (2015a). doi:10.1088/0004-637X/813/1/76, 1505.06732

Marino, S., Perez, S., Casassus, S.: Shadows cast by a warp in the HD 142527 protoplanetary disk. Astrophys. J. Lett. **798**, L44 (2015b). doi:10.1088/2041-8205/798/2/L44, 1412.4632

Masset, F.: FARGO: a fast eulerian transport algorithm for differentially rotating disks. Astron. Astrophys. Supp. Series **141**, 165–173 (2000). doi:10.1051/aas:2000116, astro-ph/9910390

Mayama, S., Hashimoto, J., Muto, T., Tsukagoshi, T., Kusakabe, N., Kuzuhara, M., Takahashi, Y., Kudo, T., Dong, R., Fukagawa, M., Takami, M., Momose, M., Wisniewski, J.P., Follette, K., Abe, L., Akiyama, E., Brandner, W., Brandt, T., Carson, J., Egner, S., Feldt, M., Goto, M., Grady, C.A., Guyon, O., Hayano, Y., Hayashi, M., Hayashi, S., Henning, T., Hodapp, K.W., Ishii, M., Iye, M., Janson, M., Kandori, R., Kwon, J., Knapp, G.R., Matsuo, T., McElwain, M.W., Miyama, S., Morino, J.I., Moro-Martin, A., Nishimura, T., Pyo, T.S., Serabyn, E., Suto, H., Suzuki, R., Takato, N., Terada, H., Thalmann, C., Tomono, D., Turner, E.L., Watanabe, M., Yamada, T., Takami, H., Usuda, T., Tamura, M.: Subaru imaging of asymmetric features in a transitional disk in upper scorpius. Astrophys. J. Lett. **760**, L26 (2012). doi:10.1088/2041-8205/760/2/L26, 1211.3284

Meheut, H., Casse, F., Varniere, P., Tagger, M.: Rossby wave instability and three-dimensional vortices in accretion disks. Astron. Astrophys. **516**, A31 (2010). doi:10.1051/0004-6361/201014000, 1004.0302

Meheut, H., Meliani, Z., Varniere, P., Benz, W.: Dust-trapping Rossby vortices in protoplanetary disks. Astron. Astrophys. **545**, A134 (2012). doi:10.1051/0004-6361/201219794, 1208.4947

Momose, M., Morita, A., Fukagawa, M., Muto, T., Takeuchi, T., Hashimoto, J., Honda, M., Kudo, T., Okamoto, Y.K., Kanagawa, K.D., Tanaka, H., Grady, C.A., Sitko, M.L., Akiyama, E., Currie, T., Follette, K.B., Mayama, S., Kusakabe, N., Abe, L., Brandner, W., Brandt, T.D., Carson, J.C., Egner, S., Feldt, M., Goto, M., Guyon, O., Hayano, Y., Hayashi, M., Hayashi, S.S., Henning, T., Hodapp, K.W., Ishii, M., Iye, M., Janson, M., Kandori, R., Knapp, G.R., Kuzuhara, M., Kwon, J., Matsuo, T., McElwain, M.W., Miyama, S., Morino, J.I., Moro-Martin, A., Nishimura, T., Pyo, T.S., Serabyn, E., Suenaga, T., Suto, H., Suzuki, R., Takahashi, Y.H., Takami, M., Takato, N., Terada, H., Thalmann, C., Tomono, D., Turner, E.L., Watanabe, M., Wisniewski, J., Yamada, T., Takami, H., Usuda, T., Tamura, M.: Detailed structure of the outer disk around HD 169142 with polarized light in H-band (2015). ArXiv e-prints 1505.04937

Muto, T., Grady, C.A., Hashimoto, J., Fukagawa, M., Hornbeck, J.B., Sitko, M., Russell, R., Werren, C., Curé, M., Currie, T., Ohashi, N., Okamoto, Y., Momose, M., Honda, M., Inutsuka, S., Takeuchi, T., Dong, R., Abe, L., Brandner, W., Brandt, T., Carson, J., Egner, S., Feldt, M.,

Fukue, T., Goto, M., Guyon, O., Hayano, Y., Hayashi, M., Hayashi, S., Henning, T., Hodapp, K.W., Ishii, M., Iye, M., Janson, M., Kandori, R., Knapp, G.R., Kudo, T., Kusakabe, N., Kuzuhara, M., Matsuo, T., Mayama, S., McElwain, M.W., Miyama, S., Morino, J.I., Moro-Martin, A., Nishimura, T., Pyo, T.S., Serabyn, E., Suto, H., Suzuki, R., Takami, M., Takato, N., Terada, H., Thalmann, C., Tomono, D., Turner, E.L., Watanabe, M., Wisniewski, J.P., Yamada, T., Takami, H., Usuda, T., Tamura, M.: Discovery of small-scale spiral structures in the disk of SAO 206462 (HD 135344B): implications for the physical state of the disk from spiral density wave theory. Astrophys. J. Lett. **748**, L22 (2012). doi:10.1088/2041-8205/748/2/L22, 1202.6139

Muzerolle, J., Allen, L.E., Megeath, S.T., Hernández, J., Gutermuth, R.A.: A spitzer census of transitional protoplanetary disks with AU-scale inner holes. Astrophys. J. **708**, 1107–1118 (2010). doi:10.1088/0004-637X/708/2/1107, 0911.2704

Ogilvie, G.I., Lubow, S.H.: On the wake generated by a planet in a disc. Mon. Not. R. Astron. Soc. **330**, 950–954 (2002). doi:10.1046/j.1365-8711.2002.05148.x, astro-ph/0111265

Ohta, Y., Fukagawa, M., Sitko, M.L., Muto, T., Kraus, S., Grady, C.A., Wisniewski, J.P., Swearingen, J.R., Shibai, H., Sumi, T., Hashimoto, J., Kudo, T., Kusakabe, N., Momose, M., Okamoto, Y., Kotani, T., Takami, M., Currie, T., Thalmann, C., Janson, M., Akiyama, E., Follette, K.B., Mayama, S., Abe, L., Brandner, W., Brandt, T.D., Carson, J.C., Egner, S.E., Feldt, M., Goto, M., Guyon, O., Hayano, Y., Hayashi, M., Hayashi, S.S., Henning, T., Hodapp, K.W., Ishii, M., Iye, M., Kandori, R., Knapp, G.R., Kuzuhara, M., Kwon, J., Matsuo, T., McElwain, M.W., Miyama, S., Morino, J.I., Moro-Martín, A., Nishimura, T., Pyo, T.S., Serabyn, E., Suenaga, T., Suto, H., Suzuki, R., Takahashi, Y.H., Takami, H., Takato, N., Terada, H., Tomono, D., Turner, E.L., Usuda, T., Watanabe, M., Yamada, T., Tamura, M.: Extreme asymmetry in the polarized disk of V1247 Orionis*. Publ. Astron. Soc. Jpn. **68**, 53 (2016). doi:10.1093/pasj/psw051, 1605.01453

Owen, J.E., Clarke, C.J., Ercolano, B.: On the theory of disc photoevaporation. Mon. Not. R. Astron. Soc. **422**, 1880–1901 (2012). doi:10.1111/j.1365-2966.2011.20337.x, 1112.1087

Paardekooper, S.J., Mellema, G.: Dust flow in gas disks in the presence of embedded planets. Astron. Astrophys. **453**, 1129–1140 (2006). doi:10.1051/0004-6361:20054449, astro-ph/0603132

Paardekooper, S.J., Lesur, G., Papaloizou, J.C.B.: Vortex migration in protoplanetary disks. Astrophys. J. **725**, 146–158 (2010). doi:10.1088/0004-637X/725/1/146, 1009.5197

Papaloizou, J.C.B., Pringle, J.E.: The dynamical stability of differentially rotating discs with constant specific angular momentum. Mon. Not. R. Astron. Soc. **208**, 721–750 (1984). doi:10.1093/mnras/208.4.721

Pérez, L.M., Isella, A., Carpenter, J.M., Chandler, C.J.: Large-scale asymmetries in the transitional disks of SAO 206462 and SR 21. Astrophys. J. Lett. **783**, L13 (2014). doi:10.1088/2041-8205/783/1/L13, 1402.0832

Perez, S., Casassus, S., Ménard, F., Roman, P., van der Plas, G., Cieza, L., Pinte, C., Christiaens, V., Hales, A.S.: CO gas inside the protoplanetary disk cavity in HD 142527: disk structure from ALMA. Astrophys. J. **798**, 85 (2015). doi:10.1088/0004-637X/798/2/85, 1410.8168

Perrin, M.D., Graham, J.R., Kalas, P., Lloyd, J.P., Max, C.E., Gavel, D.T., Pennington, D.M., Gates, E.L.: Laser guide star adaptive optics imaging polarimetry of Herbig Ae/Be stars. Science **303**, 1345–1348 (2004). doi:10.1126/science.1094602, astro-ph/0402615

Pinilla, P., Birnstiel, T., Ricci, L., Dullemond, C.P., Uribe, A.L., Testi, L., Natta, A.: Trapping dust particles in the outer regions of protoplanetary disks. Astron. Astrophys. **538**, A114 (2012). doi:10.1051/0004-6361/201118204, 1112.2349

Pinilla, P., de Boer, J., Benisty, M., Juhász, A., de Juan Ovelar, M., Dominik, C., Avenhaus, H., Birnstiel, T., Girard, J.H., Huelamo, N., Isella, A., Milli, J.: Variability and dust filtration in the transition disk J160421.7-213028 observed in optical scattered light. Astron. Astrophys. **584**, L4 (2015). doi:10.1051/0004-6361/201526981, 1510.00412

Quanz, S.P., Amara, A., Meyer, M.R., Kenworthy, M.A., Kasper, M., Girard, J.H.: A young protoplanet candidate embedded in the circumstellar disk of HD 100546. Astrophys. J. Lett. **766**, L1 (2013a). doi:10.1088/2041-8205/766/1/L1, 1302.7122

Quanz, S.P., Avenhaus, H., Buenzli, E., Garufi, A., Schmid, H.M., Wolf, S.: Gaps in the HD 169142 protoplanetary disk revealed by polarimetric imaging: signs of ongoing planet formation? Astrophys. J. Lett. **766**, L2 (2013b). doi:10.1088/2041-8205/766/1/L2, 1302.3029

Raettig, N., Klahr, H., Lyra, W.: Particle trapping and streaming instability in vortices in protoplanetary disks. Astrophys. J. **804**, 35 (2015). doi:10.1088/0004-637X/804/1/35

Rafikov, R.R.: Nonlinear propagation of planet-generated tidal waves. Astrophys. J. **569**, 997–1008 (2002). doi:10.1086/339399, astro-ph/0110496

Railton, A.D., Papaloizou, J.C.B.: On the local stability of vortices in differentially rotating discs. Mon. Not. R. Astron. Soc. **445**, 4409–4426 (2014). doi:10.1093/mnras/stu2060, 1410.1323

Rapson, V.A., Kastner, J.H., Millar-Blanchaer, M.A., Dong, R.: Peering into the Giant-planet-forming Region of the TW Hydrae Disk with the Gemini Planet Imager. Astrophys. J. Lett. **815**, L26 (2015). doi:10.1088/2041-8205/815/2/L26, 1512.01865

Reggiani, M., Quanz, S.P., Meyer, M.R., Pueyo, L., Absil, O., Amara, A., Anglada, G., Avenhaus, H., Girard, J.H., Carrasco Gonzalez, C., Graham, J., Mawet, D., Meru, F., Milli, J., Osorio, M., Wolff, S., Torrelles, J.M.: Discovery of a companion candidate in the HD 169142 transition disk and the possibility of multiple planet formation. Astrophys. J. Lett. **792**, L23 (2014). doi:10.1088/2041-8205/792/1/L23, 1408.0813

Rice, W.K.M., Armitage, P.J., Wood, K., Lodato, G.: Dust filtration at gap edges: implications for the spectral energy distributions of discs with embedded planets. Mon. Not. R. Astron. Soc. **373**, 1619–1626 (2006). doi:10.1111/j.1365-2966.2006.11113.x, astro-ph/0609808

Rosotti, G.P., Ercolano, B., Owen, J.E., Armitage, P.J.: The interplay between X-ray photoevaporation and planet formation. Mon. Not. R. Astron. Soc. **430**, 1392–1401 (2013). doi:10.1093/mnras/sts725, 1301.3015

Sallum, S., Follette, K.B., Eisner, J.A., Close, L.M., Hinz, P., Kratter, K., Males, J., Skemer, A., Macintosh, B., Tuthill, P., Bailey, V., Defrère, D., Morzinski, K., Rodigas, T., Spalding, E., Vaz, A., Weinberger, A.J.: Accreting protoplanets in the LkCa 15 transition disk. Nature **527**, 342–344 (2015). doi:10.1038/nature15761, 1511.07456

Stolker, T., Dominik, C., Min, M., Garufi, A., Mulders, G.D., Avenhaus, H.: Scattered light mapping of protoplanetary disks (2016). ArXiv e-prints 1609.09505

Stone, J.M., Gardiner, T.A., Teuben, P., Hawley, J.F., Simon, J.B.: Athena: a new code for astrophysical MHD. Astrophys. J. Suppl. Ser. **178**, 137–177 (2008). doi:10.1086/588755, 0804.0402

Takahashi, S.Z., Inutsuka, S.I.: Two-component secular gravitational instability in a protoplanetary disk: a possible mechanism for creating ring-like structures. Astrophys. J. **794**, 55 (2014). doi:10.1088/0004-637X/794/1/55, 1312.6870

Tamura, M.: Subaru strategic exploration of exoplanets and disks with HiCIAO/AO188 (SEEDS). In: Usuda, T., Tamura, M., Ishii, M. (eds.) American Institute of Physics Conference Series, American Institute of Physics Conference Series, vol. 1158, pp. 11–16 (2009). doi:10.1063/1.3215811

Tamura, M., Hodapp, K., Takami, H., Abe, L., Suto, H., Guyon, O., Jacobson, S., Kandori, R., Morino, J.I., Murakami, N., Stahlberger, V., Suzuki, R., Tavrov, A., Yamada, H., Nishikawa, J., Ukita, N., Hashimoto, J., Izumiura, H., Hayashi, M., Nakajima, T., Nishimura, T.: Concept and science of HiCIAO: high contrast instrument for the Subaru next generation adaptive optics. In: Society of Photo-Optical Instrumentation Engineers (SPIE) Conference Series, Proceedings of SPIE, vol. 6269, p. 62690V (2006). doi:10.1117/12.670742

Tanga, P., Babiano, A., Dubrulle, B., Provenzale, A.: Forming planetesimals in vortices. Icarus **121**, 158–170 (1996). doi:10.1006/icar.1996.0076

Thalmann, C., Grady, C.A., Goto, M., Wisniewski, J.P., Janson, M., Henning, T., Fukagawa, M., Honda, M., Mulders, G.D., Min, M., Moro-Martín, A., McElwain, M.W., Hodapp, K.W., Carson, J., Abe, L., Brandner, W., Egner, S., Feldt, M., Fukue, T., Golota, T., Guyon, O., Hashimoto, J., Hayano, Y., Hayashi, M., Hayashi, S., Ishii, M., Kandori, R., Knapp, G.R., Kudo, T., Kusakabe, N., Kuzuhara, M., Matsuo, T., Miyama, S., Morino, J.I., Nishimura, T., Pyo, T.S., Serabyn, E., Shibai, H., Suto, H., Suzuki, R., Takami, M., Takato, N., Terada, H., Tomono, D., Turner, E.L., Watanabe, M., Yamada, T., Takami, H., Usuda, T., Tamura, M.:

Imaging of a transitional disk gap in reflected light: indications of planet formation around the young solar analog LkCa 15. Astrophys. J. Lett. **718**, L87–L91 (2010). doi:10.1088/2041-8205/718/2/L87, 1005.5162

Tsukagoshi, T., Momose, M., Hashimoto, J., Kudo, T., Andrews, S., Saito, M., Kitamura, Y., Ohashi, N., Wilner, D., Kawabe, R., Abe, L., Akiyama, E., Brandner, W., Brandt, T.D., Carson, J., Currie, T., Egner, S.E., Goto, M., Grady, C., Guyon, O., Hayano, Y., Hayashi, M., Hayashi, S., Henning, T., Hodapp, K.W., Ishii, M., Iye, M., Janson, M., Kandori, R., Knapp, G.R., Kusakabe, N., Kuzuhara, M., Kwon, J., McElwain, M., Matsuo, T., Mayama, S., Miyama, S., Morino, J.-i., Moro-Martín, A., Nishimura, T., Pyo, T.S., Serabyn, E., Suenaga, T., Suto, H., Suzuki, R., Takahashi, Y., Takami, H., Takami, M., Takato, N., Terada, H., Thalmann, C., Tomono, D., Turner, E.L., Usuda, T., Watanabe, M., Wisniewski, J.P., Yamada, T., Tamura, M.: High-resolution submillimeter and near-infrared studies of the transition disk around Sz 91. Astrophys. J. **783**, 90 (2014). doi:10.1088/0004-637X/783/2/90, 1402.1538

van der Marel, N., van Dishoeck, E.F., Bruderer, S., Birnstiel, T., Pinilla, P., Dullemond, C.P., van Kempen, T.A., Schmalzl, M., Brown, J.M., Herczeg, G.J., Mathews, G.S., Geers, V.: A major asymmetric dust trap in a transition disk. Science **340**, 1199–1202 (2013). doi:10.1126/science.1236770, 1306.1768

van der Marel, N., Pinilla, P., Tobin, J., van Kempen, T., Andrews, S., Ricci, L., Birnstiel, T.: A concentration of centimeter-sized grains in the ophiuchus IRS 48 dust trap. Astrophys. J. Lett. **810**, L7 (2015a). doi:10.1088/2041-8205/810/1/L7, 1508.01003

van der Marel, N., van Dishoeck, E.F., Bruderer, S., Pérez, L., Isella, A.: Gas density drops inside dust cavities of transitional disks around young stars observed with ALMA. Astron. Astrophys. **579**, A106 (2015b). doi:10.1051/0004-6361/201525658, 1504.03927

van der Marel, N., van Dishoeck, E.F., Bruderer, S., Andrews, S.M., Pontoppidan, K.M., Herczeg, G.J., van Kempen, T., Miotello, A.: Resolved gas cavities in transitional disks inferred from CO isotopologs with ALMA. Astron. Astrophys. **585**, A58 (2016). doi:10.1051/0004-6361/201526988, 1511.07149

Wagner, K., Apai, D., Kasper, M., Robberto, M.: Discovery of a two-armed spiral structure in the gapped disk around Herbig Ae star HD 100453. Astrophys. J. Lett. **813**, L2 (2015). doi:10.1088/2041-8205/813/1/L2, 1510.02212

Whitney, B.A., Robitaille, T.P., Bjorkman, J.E., Dong, R., Wolff, M.J., Wood, K., Honor, J.: Three-dimensional radiation transfer in young stellar objects. Astrophys. J. Suppl. Ser. **207**, 30 (2013). doi:10.1088/0067-0049/207/2/30, 1307.0561

Williams, J.P., Cieza, L.A.: Protoplanetary disks and their evolution. Annu. Rev. Astron. Astrophys. **49**, 67–117 (2011). doi:10.1146/annurev-astro-081710-102548, 1103.0556

Yen, H.W., Liu, H.B., Gu, P.G., Hirano, N., Lee, C.F., Puspitaningrum, E., Takakuwa, S.: Gas gaps in the protoplanetary disk around the young protostar HL Tau. Astrophys. J. Lett. **820**, L25 (2016). doi:10.3847/2041-8205/820/2/L25, 1603.01378

Zhang, K., Isella, A., Carpenter, J.M., Blake, G.A.: Comparison of the dust and gas radial structure in the transition disk [PZ99] J160421.7-213028. Astrophys. J. **791**, 42 (2014). doi:10.1088/0004-637X/791/1/42, 1406.6974

Zhang, K., Blake, G.A., Bergin, E.A.: Evidence of fast pebble growth near condensation fronts in the HL Tau protoplanetary disk. Astrophys. J. Lett. **806**, L7 (2015). doi:10.1088/2041-8205/806/1/L7, 1505.00882

Zhu, Z., Baruteau, C.: Gas and dust hydrodynamical simulations of massive lopsided transition discs - I. Gas distribution. Mon. Not. R. Astron. Soc. **458**, 3918–3926 (2016). doi:10.1093/mnras/stw202, 1511.03497

Zhu, Z., Stone, J.M.: Dust trapping by vortices in transitional disks: evidence for non-ideal magnetohydrodynamic effects in protoplanetary disks. Astrophys. J. **795**, 53 (2014). doi:10.1088/0004-637X/795/1/53, 1405.2790

Zhu, Z., Nelson, R.P., Hartmann, L., Espaillat, C., Calvet, N.: Transitional and pre-transitional disks: gap opening by multiple planets? Astrophys. J. **729**, 47 (2011). doi:10.1088/0004-637X/729/1/47, 1012.4395

Zhu, Z., Nelson, R.P., Dong, R., Espaillat, C., Hartmann, L.: Dust filtration by planet-induced gap edges: implications for transitional disks. Astrophys. J. **755**, 6 (2012). doi:10.1088/0004-637X/755/1/6, 1205.5042

Zhu, Z., Stone, J.M., Rafikov, R.R., Bai, X.-n.: Particle concentration at planet-induced gap edges and vortices. I. Inviscid three-dimensional hydro disks. Astrophys. J. **785**, 122 (2014). doi:10.1088/0004-637X/785/2/122, 1308.0648

Zhu, Z., Dong, R., Stone, J.M., Rafikov, R.R.: The structure of spiral shocks excited by planetary-mass companions. Astrophys. J. **813**, 88 (2015). doi:10.1088/0004-637X/813/2/88, 1507.03599

Chapter 10
Super-Earths: Atmospheric Accretion, Thermal Evolution and Envelope Loss

Sivan Ginzburg, Niraj K. Inamdar, and Hilke E. Schlichting

Abstract Combined mass and radius observations have recently revealed many short-period planets a few times the size of Earth but with significantly lower densities. A natural explanation for the low density of these super Earths is a voluminous gas atmosphere that engulfs more compact rocky cores. Planets with such substantial gas atmospheres may be a missing link between smaller planets, that did not manage to obtain or keep an atmosphere, and larger planets, that accreted gas too quickly and became gas giants. In this chapter we review recent advancements in the understanding of low-density super-Earth formation and evolution. Specifically, we present a consistent picture of the various stages in the lives of these planets: gas accretion from the protoplanetary disk, possible atmosphere heating and evaporation mechanisms, collisions between planets, and finally, evolution up to the age at which the planets are observed.

10.1 Introduction

The *Kepler* mission discovered a large population of transiting planets a few times the radius of Earth, $R_\oplus$, in orbits of a few to a few dozen days. For a subset of these close-in super Earths we also have a mass measurement from radial velocity or transit timing variation (TTV) observations. Using these combined mass and radius measurements, we find that many of the short-period super Earths have low densities that rule out a purely rocky composition. The low density indicates

S. Ginzburg (✉)
The Hebrew University, Jerusalem 91904, Israel
e-mail: sivan.ginzburg@mail.huji.ac.il

N.K. Inamdar
MIT, Cambridge, MA 02139, USA
e-mail: inamdar@mit.edu

H.E. Schlichting
University of California, Los Angeles, Los Angeles, CA 90095, USA
MIT, Cambridge, MA 02139, USA
e-mail: hilke@epss.ucla.edu

M. Pessah, O. Gressel (eds.), *Formation, Evolution, and Dynamics of Young Solar Systems*, Astrophysics and Space Science Library 445,
DOI 10.1007/978-3-319-60609-5_10

either a water-rich composition or a rocky (or icy) core covered with a voluminous gas atmosphere, which is the only option for many extremely low-density planets (Lopez et al. 2012; Lissauer et al. 2013). Here we adopt the latter interpretation and study the atmospheric accretion onto planetary cores and their evolution. It seems, according to this interpretation, that super Earths with atmospheres of a few % in mass are among the most abundant planets found by *Kepler* (Wolfgang and Lopez 2015).

Rocky cores can gravitationally accrete gas from the gas-rich protoplanetary disk that surrounds young stars for their first few Myr (Mamajek 2009; Williams and Cieza 2011; Alexander et al. 2014). However, explaining the observed low-density super Earths by gas accretion from the surrounding nebula is not trivial. If the gas accretion rate is too fast, a rocky core can acquire an atmosphere comparable to its own mass (Lee et al. 2014). At this stage, the gas accretion rate increases, and the planet quickly evolves into a gas giant (instead of a super Earth) via runaway growth (Bodenheimer and Pollack 1986; Pollack et al. 1996; Piso and Youdin 2014; Piso et al. 2015). If, on the other hand, the accretion is too slow, planets may not obtain substantial atmospheres before the gas disk disperses Moreover, gas atmospheres can be lost due to evaporation (Rogers et al. 2011; Lopez et al. 2012; Owen and Jackson 2012; Lopez and Fortney 2013; Owen and Wu 2013, 2016) or collisions (Inamdar and Schlichting 2015), leaving a bare rocky core behind.

In the following sections we review these various aspects of gas accretion and loss and study the conditions required to form a low-density super Earth. We focus on highlighting the key physical processes that dictate gas accretion and loss and quantify our results in intuitive order of magnitude estimates. Specifically, Sect. 10.2 studies gas accretion from the nebula and Sect. 10.3 discusses mechanisms that can hamper the accretion. Section 10.4 is devoted to the evaporation of atmospheres once the gas nebula disperses. Section 10.5 focuses on the late evolution of low-density super Earths, which is relevant for interpreting the observations of $\sim$Gyr old planets. Section 10.6 discusses atmosphere loss due to giant impacts and the review is summarized in Sect. 10.7.

10.2 Gas Accretion

We assume a rocky core, of mass M_c and radius R_c, embedded inside a gas disk with ambient temperature T_d and density ρ_d. The outer edge, separating the planet's atmosphere from the surrounding nebula, is given by $R_{\rm out} = \min(R_{\rm H}, R_{\rm B})$, with $R_{\rm H}$ and $R_{\rm B}$ denoting the Hill and Bondi radii, respectively. We assume that the atmosphere's mass is $M_{\rm atm} \ll M_c$.

The gas accretion can be divided into two stages, as depicted schematically in Fig. 10.1. Initially, gas adiabatically contracts onto the core in a dynamical timescale. By integrating the adiabatic power-law density profile (Ginzburg et al. 2016), we find that the atmosphere reaches a mass $f \equiv M_{\rm atm}/M_c \propto \rho_d$, which

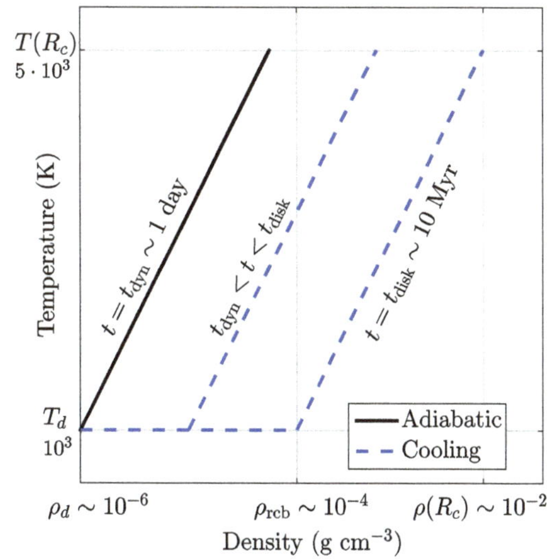

Fig. 10.1 Schematic temperature vs. density profiles (log. scale) of a super-Earth atmosphere during the nebular accretion phase. The initial adiabatic atmosphere (*solid black line*) is isentropic, while at later stages (two successive profiles are plotted) the cooling (and accreting) envelope is characterized by a nearly isothermal radiative outer layer, and a convective interior (*dashed blue lines*). Typical values of the density and temperature are provided. Figure after Ginzburg et al. (2016)

amounts to $f \sim 10^{-3}$, assuming that ρ_d is given by a minimum mass solar nebula (MMSN) model (Hayashi 1981), much lighter than observed envelopes.

Next, the atmosphere cools down and lowers its entropy. During this cooling process, the atmosphere develops an outer radiative, almost isothermal, envelope which is connected to a convective interior at the radiative–convective boundary (RCB), located at $R_{\rm rcb} \lesssim R_{\rm out}$ (Rafikov 2006). $R_{\rm rcb}$ is a good approximation for the planet's actual radius, because beyond it the density drops exponentially with a small scale height. As long as $R_{\rm rcb} \gg R_c$, the temperature profile remains roughly constant, so cooling is equivalent to an increase in density (see Fig. 10.1), and therefore in mass (the radius does not change significantly $R_{\rm rcb} \sim R_{\rm out}$, see Ginzburg et al. 2016). Thus, the planet's accretion rate is determined by the cooling timescale of the atmosphere (Lee et al. 2014; Piso and Youdin 2014). Quantitatively, we find the cooling time by dividing the atmosphere's energy $E_{\rm atm} \propto GM_cM_{\rm atm}$, with G denoting the gravitation constant, by the internal luminosity (calculated by combining the hydrostatic equilibrium and radiative diffusion equations)

$$L = \frac{\gamma - 1}{2} \frac{64\pi}{3} \frac{\sigma T_{\rm rcb}^4 R_{\rm B}}{\kappa \rho_{\rm rcb}}, \tag{10.1}$$

with σ denoting the Stephan-Boltzmann constant, $T_{\rm rcb} \sim T_d$ and $\rho_{\rm rcb}$ denoting the temperature and density at the RCB, and κ the opacity there. γ is the adiabatic index. Notice that the core does not contribute to the energy balance because the temperature on its surface is fixed at a constant $k_{\rm B}T(R_c) \sim GM_c\mu/R_c$ as long as $R_{\rm rcb} \gg R_c$, with $k_{\rm B}$ marking Boltzmann's constant and μ the molecular mass. Intuitively, the luminosity can be understood as $L \sim \sigma T_{\rm rcb}^4 R_{\rm rcb}^2/\tau$, with τ marking the optical depth at the RCB. The opacity increases mildly with the density (Freedman et al. 2008, 2014), implying, due to Eq. (10.1), that $L \propto \rho_{\mathrm{rcb}}^{-1} \propto M_{\mathrm{atm}}^{-1}$ (only the convective part of the atmosphere contributes significantly to the mass). Finally, we obtain the growth of the atmosphere with time $t = E_{\mathrm{atm}}/L \propto M_{\mathrm{atm}}^2(t)$ (Piso and Youdin 2014). Specifically, Ginzburg et al. (2016) find that by the time the disk disperses $t = t_{\mathrm{disk}}$, the atmosphere's mass fraction is given by

$$f \approx 0.02 \left(\frac{M_c}{M_{\oplus}}\right)^{0.8} \left(\frac{T_d}{10^3\ \mathrm{K}}\right)^{-0.25} \left(\frac{t_{\mathrm{disk}}}{1\ \mathrm{Myr}}\right)^{0.5}, \tag{10.2}$$

with $M_{\oplus}$ marking Earth's mass (see Lee and Chiang 2015, for a similar result).

We notice that while the mass of the initial adiabatic atmosphere is proportional to the nebula's density, Eq. (10.2) does not depend on ρ_d. More precisely, the dependence on ρ_d is logarithmic, and is therefore omitted (see Ginzburg et al. 2016). Intuitively, the radiative envelope decouples the density of the atmosphere $\propto \rho_{\mathrm{rcb}}$ from the outer boundary condition ρ_d (see Fig. 10.1). Thus, the bottleneck that typically determines the accretion rate is not the amount of available gas, but rather the rate at which this gas can radiate away its gravitational energy and settle onto the core. The logarithmic dependence on ρ_d implies that significant atmospheres can be accreted even in highly depleted disks (Inamdar and Schlichting 2015; Lee and Chiang 2016). Such depleted (and short lived, so $f \propto t_{\mathrm{disk}}^{1/2}$ is smaller) disks are relevant if atmosphere accretion is delayed until the assembly of the rocky core by collisions is over (see Sect. 10.6).

10.3 Atmosphere Heating

In Sect. 10.2 we demonstrated that atmosphere accretion is equivalent to cooling. Therefore, mechanisms that heat the atmosphere can hamper gas accretion. Such mechanisms have the potential to explain why super Earths did not continue to grow into gas giants, if accretion is somewhat more efficient than in Eq. (10.2), as suggested by Lee et al. (2014). In this section we focus on two examples: heating by planetesimals and by tides.

10.3.1 Heating by Planetesimals

The formation of the rocky cores of super-Earths and of gas giants can be divided into two phases (see, e.g., Goldreich et al. 2004). In the first phase, the cores grow by gravitationally attracting small building blocks (planetesimals) into their Hill sphere. The final mass in this phase is referred to as the "isolation mass". In the next phase, these isolation masses collide with each other due to orbital instability. This phase is referred to as the "giant impacts" phase.

The loss of gravitational energy of the impacting planetesimals dissipates heat inside the growing planet. The timescale for the core growth of giant planets in the outer disk is comparable to the gas disk lifetime, a few Myr (Pollack et al. 1996). Therefore, the accretion of solids and gas is simultaneous, and heating by planetesimals plays a crucial role in the cooling (equivalent to growth) of the gas atmosphere (Rafikov 2006, 2011). In the inner disk, however, the solid accretion time is shorter, implying that the core formation may be decoupled from the atmosphere growth (Lee et al. 2014). For this reason, many studies ignore planetesimal heating in the context of the close-in *Kepler* super Earths. However, a steady stream of new planetesimals may be supplied from larger semi-major axes as the planetesimal orbits decay due to their interaction with the gas disk. Even if the bulk of the planetesimals is consumed at an early stage (on a timescale much shorter than $t_{\rm disk}$), residual impacting debris can still interfere with the growth of the atmosphere.

How does the accretion of a solid mass ΔM_c during the disk's lifetime $t_{\rm disk}$ affect the gas accretion? We assume that the planetesimals impact at a constant rate and deposit their energy at the surface of the rocky core. In this case, the accretion luminosity is given by $L_{\rm acc} = GM_c\Delta M_c/(R_c t_{\rm disk})$. The internal luminosity at the end of accretion (when $t = t_{\rm disk}$), on the other hand, is given by $L = GM_c M_{\rm atm,0}/(R_c t_{\rm disk})$, assuming that the atmosphere's energy and mass are concentrated near R_c. This is relevant for the low values of γ, found by Lee et al. (2014) and Piso et al. (2015), due to hydrogen dissociation. $M_{\rm atm,0}$ marks the final atmosphere mass, if planetesimal impacts are ignored, which is calculated in Sect. 10.2. Since, according to Eq. (10.1), $L \propto M_{\rm atm}^{-1}(t)$, the internal luminosity reaches its minimum at the end of accretion ($t = t_{\rm disk}$). Therefore, if $\Delta M_c < M_{\rm atm,0}$ then $L_{\rm acc} < L$ during the accretion phase, and the heat generated by planetesimal impacts is evicted from the planet without affecting gas accretion.

For $\Delta M_c > M_{\rm atm,0}$, the internal luminosity L drops below the planetesimal-generated heat $L_{\rm acc}$ before the atmosphere reaches its final mass. Quantitatively, using $L \propto M_{\rm atm}^{-1}(t)$, the luminosities are equal when $M_{\rm atm}(t) = M_{\rm atm,0}^2/\Delta M_c$. At this stage gas accretion stops and the final atmosphere mass is therefore given by

$$\frac{M_{\rm atm}}{M_{\rm atm,0}} = \begin{cases} M_{\rm atm,0}/\Delta M_c & \Delta M_c > M_{\rm atm,0} \\ 1 & \Delta M_c < M_{\rm atm,0} \end{cases}. \tag{10.3}$$

Equation (10.3) shows that a mass in planetesimals comparable to the final atmospheric mass, $M_{\rm atm,0}$, can stop the gas accretion. This is interesting since the mass-radius relation for many close-in exoplanets suggests that they are enshrouded in gaseous envelopes containing a few percent of the planet's mass such that the accretion of planetesimals containing a few percent of an Earth's mass could lead to an early termination of the gas accretion. This is particularly interesting since the Earth is believed to have accreted about 1% of its total mass after the Moon-forming impact, suggesting that a planetesimal population containing a few % of an Earth mass did survive over time scales of 10^7–10^8 years in the inner solar system (Walker 2009; Schlichting et al. 2012). Equation (10.3), however, also demonstrates that heating by planetesimals cannot stop super Earths from evolving into gas giants by runaway gas accretion (see Sect. 10.1). The reason is that runaway accretion initiates when $M_{\mathrm{atm}} \sim M_c$ and the accreted gas significantly increases the planet's mass (and therefore its ability to attract more gas). In order to intervene with the accretion of such heavy atmospheres, Eq. (10.3) requires $\Delta M_c \sim M_c$, leading to faster gas accretion and runaway growth. Lee and Chiang (2015) reach the same conclusion by considering the non-linear relation between the atmosphere's mass and the planet's luminosity, which is more accurate for heavy, self-gravitating, atmospheres $M_{\mathrm{atm}} \sim M_c$.

We note that, in addition to their accretion heat, planetesimal impacts may also affect atmosphere growth by enhancing the atmosphere's heavy-element abundance, and therefore the opacity κ. We do not discuss this effect here.

In Sect. 10.6 we discuss the effects of the second phase of the core's assembly—giant impacts between the isolation masses.

10.3.2 Tidal Heating

Tidal heating has been proposed as a mechanism to inhibit the cooling of close-in gas giants (hot Jupiters), thus halting their contraction and explaining their puzzlingly large radii (Bodenheimer et al. 2001, 2003; Gu et al. 2003; Winn and Holman 2005; Jackson et al. 2008; Liu et al. 2008; Ibgui and Burrows 2009; Miller et al. 2009; Ibgui et al. 2010, 2011; Leconte et al. 2010). Can the same mechanism interfere with the cooling (and thereby accretion) of super-Earth atmospheres?

Ginzburg and Sari (2017) find the heat dissipation in the atmosphere due to circularizing tides

$$L_{\mathrm{circ}} = \frac{63\pi}{2} f \frac{e^2}{QP} \frac{GM_{\odot}^2}{R_{\mathrm{rcb}}} \left(\frac{R_{\mathrm{rcb}}}{a}\right)^6 \propto M_{\mathrm{atm}} R_{\mathrm{rcb}}^5, \tag{10.4}$$

with e denoting the orbital eccentricity, $Q \sim 10^5$ the tidal dissipation parameter of the atmosphere, $M_{\odot}$ the stellar mass, a the semi-major axis, and P the orbital period. During accretion, $R_{\mathrm{rcb}} \approx 0.5 R_{\mathrm{out}} = 0.5 \min(R_{\mathrm{H}}, R_{\mathrm{B}})$ (Ginzburg et al. 2016) is larger than the radius of Jupiter, leading to strong tidal heating, due to Eq. (10.4). The tidal power increases as $L_{\mathrm{circ}} \propto M_{\mathrm{atm}}(t)$, while the cooling luminosity decreases as $L \propto M_{\mathrm{atm}}^{-1}(t)$, according to Eq. (10.1). Therefore, atmosphere accretion will stop when $L = L_{\mathrm{circ}}$ and the tidal heat can no longer be evicted from the planet. If the tidal heating is strong enough (i.e., the planet is close to the star), then this condition is reached when $t < t_{\mathrm{disk}}$ and the atmosphere does not reach its full mass potential $M_{\mathrm{atm,0}}$, similar to Sect. 10.3.1. Quantitatively, from Ginzburg and Sari (2017):

$$\frac{M_{\mathrm{atm}}}{M_{\mathrm{atm,0}}} = \begin{cases} (P/P_{\mathrm{crit}})^{5/6} & P < P_{\mathrm{crit}} \\ 1 & P > P_{\mathrm{crit}} \end{cases}, \tag{10.5}$$

with $P_{\rm crit}$ marking the critical period, beyond which tides do not affect atmosphere accretion, given by

$$e \sim \left(Q\frac{P_{\rm crit}}{t_{\rm disk}}\right)^{1/2}\left(\frac{P_{\rm crit}}{t_{\rm dyn}}\right)^{1/3} \sim 0.2\left(\frac{P_{\rm crit}}{10\ {\rm day}}\right)^{5/6}, \tag{10.6}$$

for $t_{\rm disk} = 3$ Myr and with the core's dynamical time given by $t_{\rm dyn} \equiv (G\rho_c)^{-1/2} \approx 0.5$ h ($\rho_c \approx 5\,{\rm g\,cm^{-3}}$ is the rocky core's mean density). By combining Eqs. (10.2), (10.5), and (10.6) we obtain (see Ginzburg and Sari 2017, for details) the maximum gas mass fraction a rocky core may accrete when tides are taken into account, so that $f = \min(f_0 \equiv M_{\rm atm,0}/M_c, f_{\rm max})$ with

$$f_{\rm max} \approx \frac{2\%}{e}\left(\frac{M_c}{5M_\oplus}\right)^{0.8}\left(\frac{P}{10\ {\rm day}}\right)^{19/21}. \tag{10.7}$$

Equation (10.7) demonstrates that large eccentricities $e \gtrsim 0.2$ are necessary for tides to play a role in shaping super-Earth atmospheres. Due to the gas damping, such eccentricities are usually considered unlikely during the nebular phase. However, some studies suggest that planet–disk interactions may, under some circumstances, excite, rather than damp, the eccentricity (Goldreich and Sari 2003; Duffell and Chiang 2015; Teyssandier and Ogilvie 2016). Ginzburg and Sari (2017) find circularization timescales $t_{\rm circ} \sim 10^8$ yr $\ll$ Gyr (after the gas disk disperses). These relatively short timescales allow planets to cool, contract, and reach their observed radii after the tidal heating has ceased (see Sect. 10.5). In addition, the short timescales are consistent with the small observed eccentricities in Gyr-old systems.

Tidal heating has the potential to explain why super-Earths did not reach $f \sim 1$ and grow into Jupiters via runaway accretion. In addition, the dependence of $f_{\rm max}$ on the orbital period, as demonstrated in Eq. (10.7), might explain the scarcity of low-density super Earths in close proximity to the star (e.g. Youdin 2011). Nevertheless, in the following sections we ignore tides and focus on an alternative mechanism that may sculpt the observed super-Earth population—atmosphere evaporation.

10.4 Evaporation

After the gas disk disperses atmospheres can no longer grow in mass. At this stage, super-Earth evolution is governed by two processes: evaporation and cooling, which is now equivalent to contraction, rather than accretion (since the density increases with a constant $M_{\rm atm}$, see Ginzburg et al. 2016, for details).

Atmosphere evaporation can result from various mechanisms, with photoevaporation by high-energy stellar radiation commonly considered (Rogers et al. 2011; Lopez et al. 2012; Owen and Jackson 2012; Lopez and Fortney 2013; Owen and

Wu 2013; Lundkvist et al. 2016). The basic picture is that ionizing photons release energetic electrons which in turn heat the gas to high temperatures above the escape velocity. If the cooling of the gas is slow enough, the high-temperature gas escapes the planet's potential well. The widely used energy-limited model for photoevaporation linearly relates the gravitational energy of the escaping mass to the incident ionizing flux, so that the evaporation timescale of the atmosphere is given by $t_{\text{evap}} \propto M_{\text{atm}}/(\epsilon T_{\text{eq}}^4)$. T_{eq} marks the equilibrium temperature on the planet's surface and ϵ accounts for both the evaporation efficiency and the fraction of the ionizing radiation out of the total bolometric flux. ϵ is considered to be approximately constant for $t_{\text{UV}} \sim 100$ Myr, while the star is UV active, and then it decreases with time as $\epsilon \propto t^{-1.25}$ (Jackson et al. 2012; Lopez et al. 2012; Owen and Jackson 2012, and references therein). For $t > t_{\text{UV}}$ the ratio of evaporation timescale to age increases as $t_{\text{evap}}/t \propto t^{0.25}$. Therefore, if an atmosphere survived until t_{UV}, it will keep most of its mass in later times.

In addition to evaporation by external irradiation, mass from the loosely bound outer layers of the atmosphere can also be lost spontaneously, due to heat from the contracting inner layers of the atmosphere or from the underlying rocky core, combined with loss of pressure support from the vanishing gas-disk (Ikoma and Hori 2012; Owen and Wu 2016; Ginzburg et al. 2016). In fact, since the cooling luminosity of the planet is the energy source unbinding the outer atmosphere, the ratio between the cooling and evaporation timescales is simply the ratio between the atmosphere's energy (concentrated in its inner layers) and the binding energy of the outer layers

$$\frac{t_{\text{evap}}}{t_{\text{disk}}} = \frac{t_{\text{evap}}}{t_{\text{cool}}} = \frac{E_{\text{evap}}}{E_{\text{atm}}} = \left(\frac{R_{\text{rcb}}}{R_c}\right)^{-(3-2\gamma)/(\gamma-1)}, \tag{10.8}$$

where the last equality is derived for atmospheres that have their energy concentrated in the inner layers, while mass in the outside (as for the diatomic $\gamma = 7/5$. See Ginzburg et al. 2016, for derivation and other cases). Equation (10.8) shows that evaporation dominates super-Earth evolution (after disk dispersal) as long as $R_{\text{rcb}} \gg R_c$. Therefore, super Earths spontaneously shed their outer layers (dozens of percents in mass) and shrink to a radius comparable to R_c. Moreover, since $t_{\text{cool}} = t_{\text{disk}}$ by definition when the disk disperses and it can be shown that t_{cool} remains constant during the evaporation, the mass shedding occurs on a timescale comparable to t_{disk}.

10.4.1 Thin Atmosphere

After a time $\sim t_{\text{disk}}$ (a few Myr) atmospheres shrink from their initial size $\sim R_{\text{out}}$ to a radius comparable to the size of the rocky core R_c, as explained above. To study this regime, we redefine $R_{\text{rcb}} \equiv R - R_c$, where R is the radius of the RCB (essentially, the planet's radius). This definition coincides with the previous one for the thick atmosphere regime ($R_{\rm rcb} \gg R_c$). As in the thick regime, the thin regime is also characterized by a competition between cooling (i.e. contraction) and evaporation. While photoevaporation does not change conceptually in this phase, spontaneous evaporation may differ dramatically in the thin regime.

During the thick phase, the temperature at the base of the atmosphere, and therefore the temperature on the surface of the adjacent core remains constant $k_{\rm B}T(R_c) \sim GM_c\mu/R_c$. Therefore, assuming the core is convective (a plausible assumption for high temperatures for which the core is molten), it does not cool and its temperature profile remains constant. In the thin regime, however, an adiabatic atmosphere dictates the following temperature at its base:

$$k_{\rm B}T(R_c) = k_{\rm B}T_d + \frac{GM_c\mu}{R_c}\frac{R_{\rm rcb}}{R_c} \approx \frac{GM_c\mu}{R_c}\frac{R_{\rm rcb}}{R_c}, \tag{10.9}$$

where the last approximation assumes that the atmosphere is not ultra-thin $R_{\rm rcb}/R_c > R_c/R_{\rm B}$ (see Ginzburg et al. 2016). Equation (10.9) shows that once $R_{\rm rcb} < R_c$, the temperature on the surface of the rocky core becomes dependent on the thickness of the atmosphere $R_{\rm rcb}$. Consequently, as the atmosphere cools (and contracts), $T(R_c) \propto R_{\rm rcb}$ decreases, so the rocky core cools as well. Thus, while for $R_{\rm rcb} > R_c$ the rocky core does not play a role in the planet's cooling, for $R_{\rm rcb} < R_c$ we have to take into account its heat capacity. Quantitatively, the available energy for cooling in the thin regime is

$$E = gR_{\rm rcb}\left(\frac{\gamma}{2\gamma - 1}M_{\rm atm} + \frac{1}{\gamma}\frac{\gamma - 1}{\gamma_c - 1}\frac{\mu}{\mu_c}M_c\right), \tag{10.10}$$

with $g \equiv GM_c/R_c^2$ denoting the surface gravity and μ_c and γ_c marking the rocky core's molecular weight and adiabatic index, respectively. The first term in Eq. (10.10) represents the (gravitational and thermal) energy of the gaseous atmosphere and the second term accounts for the heat capacity of the rocky core (which is approximately incompressible).

In summary, during the thin phase we distinguish, following Eq. (10.10), between heavy atmospheres ($f \gtrsim \mu/\mu_c$), that regulate their own cooling, and light atmospheres ($f \lesssim \mu/\mu_c$), that are dominated by the heat capacity of the underlying rocky core. Assuming that the atmosphere is composed of hydrogen and helium and the core has an Earth-like composition, the above distinction is at a mass fraction of a few percent. The binding energy of the thin atmosphere is $E_{\rm evap} = GM_cM_{\rm atm}/R_c = gM_{\rm atm}R_c$. Therefore, heavy atmospheres cool and contract without spontaneous evaporation (since $R_{\rm rcb} < R_c$ and $E < E_{\rm evap}$). Light atmospheres, on the other hand, are lost completely because when they enter the thin regime $R_{\rm rcb} = R_c$ and $E > E_{\rm evap}$. Consequently, the atmosphere cannot cool and contract, and therefore $R_{\rm rcb} \propto E$ remains constant while $M_{\rm atm}$ evaporates (see Ginzburg et al. 2016, for details). One consideration that can save light atmospheres from complete loss is the loss timescale which can exceed the age of the system, in which case light atmospheres can survive to the present day not because there is insufficient energy to complete the loss but simply because their loss timescales exceed several Gyrs.

10.4.2 The Goldilocks Region

As explained in Sect. 10.1, acquiring and preserving atmospheres of a few percent from the protoplanetary disk is not trivial. If a rocky core is too massive or too cold (i.e. far away from the star), it will acquire an atmosphere which is too massive, according to Eq. (10.2), reaching $f \sim 1$ and then evolving via runaway accretion into a gas giant instead of a super-Earth. On the other hand, if the rocky core is too light or too close to the star, it will obtain a light atmosphere which is susceptible to evaporation, since $t_{\rm evap} \propto M_{\rm atm}/(\epsilon T_{\rm eq}^4)$ for photoevaporation, as explained above (spontaneous evaporation acts qualitatively similar).

In Fig. 10.2 we present the Goldilocks region in which rocky cores can accrete H/He envelopes at their current location, becoming low-density super Earths. Planets above the "Jupiter" line explode into Jupiters, while planets below the "UV" line lose their atmospheres due to UV photoevaporation. In the shaded area

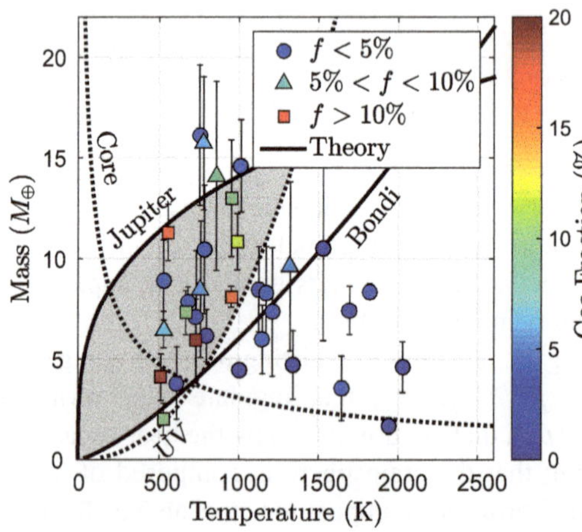

Fig. 10.2 Observed super-Earth population from Weiss and Marcy (2014). The planets are grouped according to their gas mass fraction f, estimated by Eq. (10.12), with low-density planets marked by *triangles* ($5\% < f < 10\%$) or *squares* ($f > 10\%$). The planet markers are also colour-coded according to f. The *top solid line* ("Jupiter") is according to Eq. (10.2) with $t_{\rm disk} = 10$ Myr and $f = 0.5$, while the *bottom dashed line* ("UV") represents the condition to survive photoevaporation $t_{\rm evap} \propto M_{\rm atm}/(\epsilon T_{\rm eq}^4) > t_{\rm UV}$. The other lines ("Core" and "Bondi") are relevant for spontaneous evaporation (see Ginzburg et al. 2016, for details). Inside the shaded area, planets manage to accrete and maintain gas envelopes without exploding into gas giants due to runaway accretion. Figure after Ginzburg et al. (2016)

between these lines we expect to find low-density super Earths. As seen in the figure, the observed low-density super-Earth population is indeed concentrated in the Goldilocks region, with mainly bare rocky cores outside it.

Migration can extend the Goldilocks region by separating the gas accretion from the UV evaporation, allowing planets to grow their atmospheres in more favourable conditions (see Ginzburg et al. 2016, for details).

10.5 Late Evolution

In Sect. 10.2 we discussed gas accretion from the protoplanetary disk, which vanishes after $t_{\mathrm{disk}} \sim 3\,\mathrm{Myr}$. In Sect. 10.4 we studied the evolution following disk dispersal. This consists of spontaneous evaporation for $\sim t_{\mathrm{disk}}$, in which the atmosphere shrinks to a thickness $R_{\mathrm{rcb}} \sim R_c$, followed by photoevaporation for $t_{\mathrm{UV}} \sim 100\,\mathrm{Myr}$, which evaporates the atmospheres of light or hot (i.e. close to the star) rocky cores. However, the planets that we observe are ~Gyrs old. In this section we focus on the late evolution of low-density super Earths, after photoevaporation has ceased to play a role.

At an age $t > t_{\mathrm{UV}}$, heavy envelopes ($f \gtrsim \mu/\mu_c$, see Sect. 10.4.1) cool and contract with a constant mass M_{atm}. Quantitatively, the atmosphere's mean density is determined by the adiabatic relation $\rho/\rho_{\mathrm{rcb}} \sim (T(R_c)/T_d)^{1/(\gamma-1)} \propto R_{\mathrm{rcb}}^{1/(\gamma-1)}$, with the last relation from Eq. (10.9). Since the atmosphere's mass is given by $M_{\mathrm{atm}} \sim \rho R_c^2 R_{\mathrm{rcb}}$, we find that $\rho_{\mathrm{rcb}} \propto R_{\mathrm{rcb}}^{-\gamma/(\gamma-1)}$. We combine this result with Eqs. (10.1) and (10.10) and derive the contraction of the atmosphere with time $t = E/L \propto R_{\mathrm{rcb}}^{-1/(\gamma-1)}$.

For the diatomic $\gamma = 7/5$, gas envelopes contract as $R_{\mathrm{rcb}} \propto t^{-2/5}$. However, gas envelopes do not compress indefinitely, and at some stage they reach the T_{eq} temperature floor, or the maximal gas density ceiling ρ_{max}, becoming liquid. Ginzburg et al. (2016) use the scaling relation above to show that super-Earth atmospheres reach ρ_{max} in a few Gyrs. Numerical evolution models (e.g. D'Angelo and Bodenheimer 2016) confirm that gas atmospheres stop contracting after ~10 Gyr, somewhat longer than the approximate estimate in Ginzburg et al. (2016). For simplicity, we assume here that atmospheres of observed planets are close to ρ_{max}, approximately consistent with the analytical and numerical estimates above. We evaluate this density using the equation of state of Nettelmann et al. (2008) as

$$\rho_{\mathrm{max}} \approx 0.5\,\mathrm{g\,cm^{-3}} \left(\frac{p}{\mathrm{Mbar}}\right)^{0.4}, \tag{10.11}$$

with $p \approx M_{\mathrm{atm}} g/(4\pi R^2)$ denoting the typical atmospheric pressure.

With the above estimate for the atmosphere's density, we can infer an observed planet's atmosphere mass fraction $f \equiv M_{\rm atm}/M_c$ from its mass $M \approx M_c$ and radius R by

$$f = \frac{\rho_{\max}}{\rho_c}\left[\left(\frac{R}{R_c}\right)^3 - 1\right], \tag{10.12}$$

with the rocky core's density and radius given by $\rho_c \propto R_c \propto M_c^{1/4}$, taking into account the mild gravitational compression (e.g. Valencia et al. 2006). Our estimate for the observed f is displayed in Fig. 10.3. This crude estimate is in agreement (approximately) with more elaborate numerical time-dependent estimates, e.g., Lopez et al. (2012).

It is important to mention that heating mechanisms can change our interpretation of the density, and therefore of f. Because cooling is equivalent to contraction in these late evolutionary stages, heating mechanisms can slow down or stop the contraction, in a similar manner to their effect on accretion in earlier stages (see Sect. 10.3). While tidal heating cannot stop contraction for long ($t_{\rm circ} \ll$ Gyr, see Sect. 10.3.2), Ohmic heating due to interaction of atmospheric winds with the planet's magnetic field might play an important role (see Valencia and Pu 2015). Heating mechanisms that delay contraction imply that the values of f given in Fig. 10.3 and in Lopez et al. (2012) might be overestimated, due to an overestimate of the density at a given age.

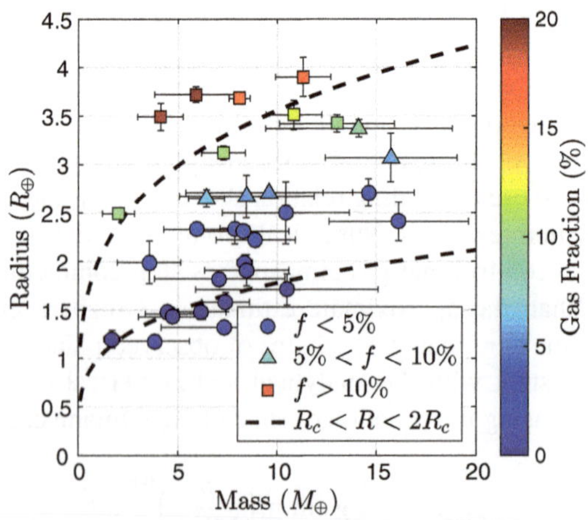

Fig. 10.3 Observed super-Earth population from Weiss and Marcy (2014). The planets are grouped according to their gas mass fraction f, estimated by Eq. (10.12), with low-density planets marked by *triangles* ($5\% < f < 10\%$) or *squares* ($f > 10\%$). The planet markers are also colour-coded according to f. The *two dashed black lines* mark the radius of the rocky core $R_c(M_c)$ and $2R_c(M_c)$. Planets with substantial atmospheres are expected to be found roughly between the two lines. Figure after Ginzburg et al. (2016)

10.6 Diversity of the Super-Earth Population and Giant Impacts

In the previous sections we showed that with simple physical arguments we can define a relatively narrow mass/temperature range in which planets are massive and cold enough to acquire and retain a significant atmosphere, while not too massive and cold to undergo runaway gas accretion and turn into Jupiters. Figure 10.2 shows that observed low-density super-Earths indeed reside in the predicted range of our model. However, the diversity among these super-Earths in terms of the total amount of gas they accreted is not well explained by our model, since, as we show in Eq. (10.2), the amount of accreted gas is primarily a function of core mass, temperature and disk lifetime. In addition, many observed exoplanetary systems that are in tightly packed orbital configurations also show significant diversity in their gas-mass fractions and bulk densities (see Fig. 10.4). Both of these observations are challenging to explain by gas accretion and subsequent sculpting by photo-evaporation alone. We suggest that the large observed range in exoplanet bulk densities maybe due to one or two giant impacts that occurred late in their evolution once the gas disk dissipated. We show below that giant impacts can modify the bulk composition of a super-Earth by factors of a few and in some cases lead to complete atmospheric loss (Inamdar and Schlichting 2016; Liu et al. 2015). Such late giant impacts are likely to be common because super-Earths must have formed in the presence of the gas disk and their dynamical interaction with the disk is expected to have resulted in migration and efficient eccentricity damping leading to densely packed planetary systems. As the gas disk dissipated mutual excitations lead to eccentricity growth culminating in one or two giant impacts ultimately resulting in planetary systems with long-term stability (Cossou et al. 2014).

10.6.1 Giant Impacts and Atmospheric Mass Loss

We use one-dimensional hydrodynamic simulations to calculate the envelope mass loss resulting from a giant impact. As the giant impact occurs it launches a strong shock that transverses the entire planet and that results in a global ground motion that in turn launches a shock into the atmosphere (see Fig. 10.5). We only model the adiabatic part of the atmosphere since the isothermal outer layer contains negligible mass. The propagation of the shock in the atmosphere is tracked by solving the hydrodynamic equations with a finite-difference Lagrangian scheme. If a fluid parcel is accelerated to velocities above its initial escape velocity from the planet, it is considered lost. The global mass loss is determined by integrating the local mass loss over the entire surface of the planet accounting for the global distribution of the different ground velocities (see Inamdar and Schlichting 2016, for details).

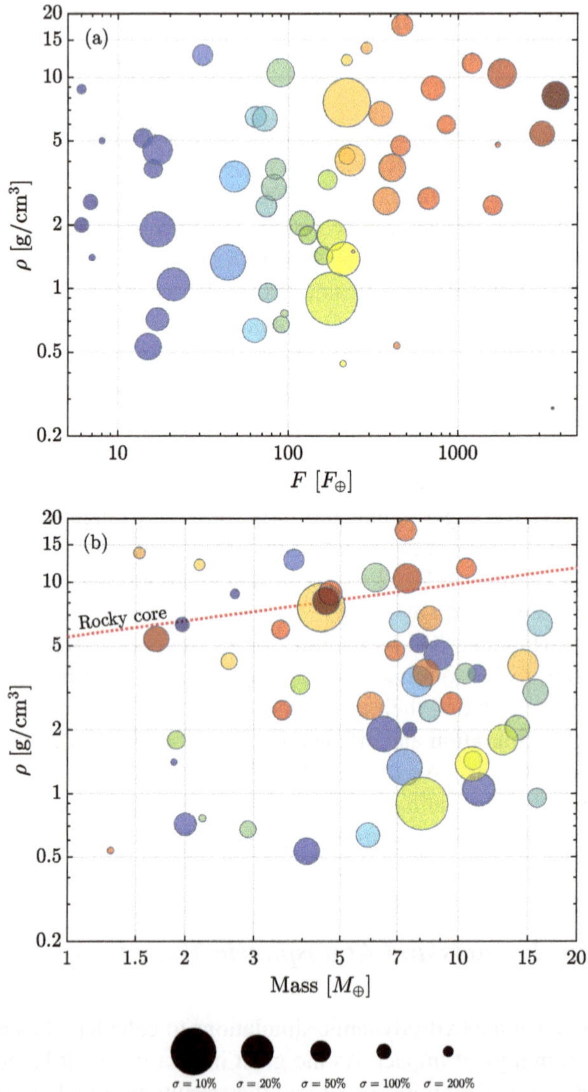

Fig. 10.4 Densities of exoplanets with $R < 4R_{\oplus}$. The surface area of each data point is inversely proportional to the 1σ error of the density estimate, such that the most secure density measurements correspond to the largest points. The normalization of the error bars is shown at the *bottom*. The colours of the points represent the amount of flux received from the host star. (**Panel a**) shows mean density as a function of flux, F, in units of the Earth flux, $F_{\oplus}$. (**Panel b**) displays exoplanet densities as a function of planet mass in units of Earth masses, $M_{\oplus}$. Most data are taken from Weiss and Marcy (2014) and the references therein. Additional data taken from Jontof-Hutter et al. (2015) and Barros et al. (2015). For reference, a mean density curve assuming a purely rocky planet (Seager et al. 2007) is shown with a *dotted red line*. Figure after Inamdar and Schlichting (2016)

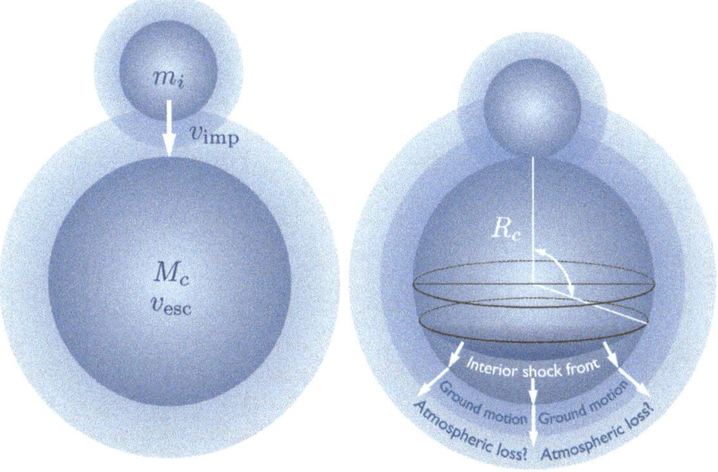

Fig. 10.5 The geometry of a giant impact. On the *left*, an impactor of mass m_i and impact velocity $v_{imp} \sim \sqrt{2}v_{esc}$ approaches a target of mass M_c and escape velocity v_{esc}. After the collision, a shock is generated through the target core. The shock—which propagates so as to conserve the linear momentum of the impactor—results in ground motion on the target body. This ground motion launches a shock into the overlying gas envelope, leading to hydrodynamic mass loss. For further details, see Inamdar and Schlichting (2015)

Figure 10.6 displays the global atmospheric loss as a function of the momentum of the impact. The results are displayed for atmospheric mass fraction of $f = 1\%$ and $f = 5\%$ spanning typical values for the mass fraction of super-Earth envelopes. By conservation of energy, the speed at which the two bodies collide with one another, v_{imp}, is given by $v_{imp}^2 = v_{esc}^2 + v_{\infty}^2$, where $v_{esc} = \sqrt{2G(M_c + m_i)/(R_c + r_i)}$ is the mutual escape velocity of the two bodies and v_{∞} the (relative) velocity dispersion of the two bodies. Here, M_c and m_i are the mass of the target and impactor, respectively and R_c and r_i are their corresponding radii. The largest protoplanets/planets will gravitationally stir the other bodies around them, exciting the velocity dispersion to roughly v_{esc}, so that $v_{imp} \sim \sqrt{2}v_{esc}$. Such that for equal mass impactors, the normalized impactor momentum is $\sim$0.7 implying that roughly half of the total envelope will be lost. This in turn will modify the bulk density of super-Earths by a factor of a few as shown in Inamdar and Schlichting (2016).

This atmospheric mass loss calculated in Fig. 10.6 is likely an underestimate of the total envelope loss because even the part of the envelope that was not immediately lost in the impact is susceptible to subsequent loss by photo-evaporation and Parker winds (Liu et al. 2015) because of its large inflated radius after the collision.

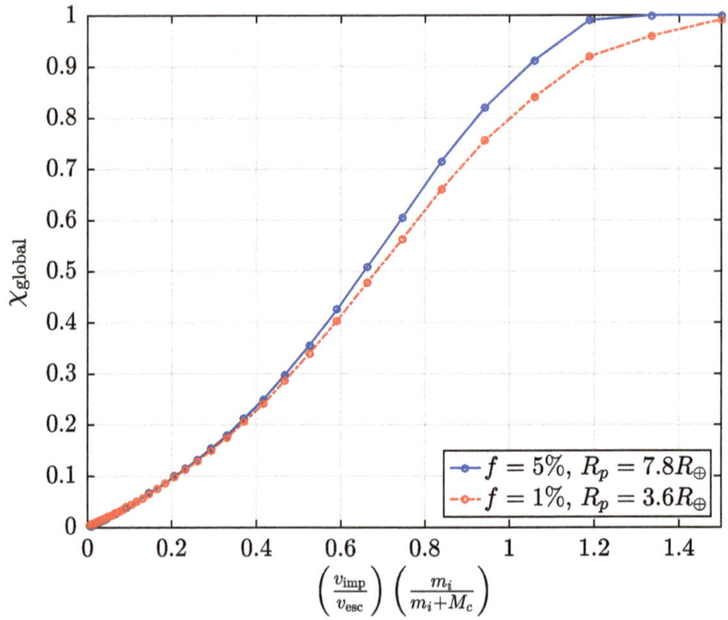

Fig. 10.6 Mass loss curve for late stage giant impacts. On the *vertical axis*, we show the envelope mass loss fraction due to a giant impact. On the *horizontal axis*, we show impactor linear momentum normalized to $M_c v_{esc}$. After the dissipation of the gas disk, the radial extent and structure of the envelope depends on the planet's cooling history. Here we assume $4M_\oplus$ core and envelope mass fractions of $f = 1\%$ (*red line*) and $f = 5\%$ (*blue line*). The total planet radii correspond to results from thermal evolution models after 50 Myr of cooling at 0.1 AU. A typical late stage merger between equal mass impactors (such that the normalized impactor momentum on the horizontal axis is $\sqrt{2}/2 \approx 0.7$) results in roughly half of the envelope mass being stripped away. Figure after Inamdar and Schlichting (2016)

10.7 Summary

In this chapter we studied the formation and evolution of voluminous gas atmospheres of short-period super Earths. Such gas envelopes, which constitute a few percent of the planet's mass, are a possible explanation to the low densities of many observed planets.

We showed that with simple physical arguments we can define a relatively narrow mass/temperature range in which planets are massive and cold enough to acquire and retain a significant atmosphere, while not too massive and cold to undergo runaway gas accretion and turn into Jupiters. Figure 10.2 shows that observed low-density super-Earths indeed reside in the predicted range of our model. However, the diversity among these super-Earths in terms of the total amount of gas they accreted is not well explained by gas accretion and subsequent sculpting by photo-evaporation alone.

We show that giant impacts can modify the bulk composition of a super-Earth by factors of a few and in some cases lead to complete atmospheric loss (Inamdar and Schlichting 2016; Liu et al. 2015) and suggest that giant impacts maybe responsible for the large observed range in exoplanet bulk densities, especially in tightly packed multiple planet systems.

In summary, explaining the low densities of close-in super Earths by gas accretion from the surrounding nebula has been the focus of intensive study for the last few years. While the different studies converge to a consistent understanding of the accretion itself, the importance of other mechanisms, such as heating (during or after gas accretion), migration (see D'Angelo and Bodenheimer 2016), and collisions are subject of active research.

Acknowledgements We thank Re'em Sari for discussions.

References

Alexander, R., Pascucci, I., Andrews, S., Armitage, P., Cieza, L.: The dispersal of protoplanetary disks. In: Protostars and Planets VI, pp. 475–496. University of Arizona Press, Tuscon (2014). 1311.1819

Barros, S.C.C., Almenara, J.M., Demangeon, O., Tsantaki, M., Santerne, A., Armstrong, D.J., Barrado, D., Brown, D., Deleuil, M., Lillo-Box, J., Osborn, H., Pollacco, D., Abe, L., Andre, P., Bendjoya, P., Boisse, I., Bonomo, A.S., Bouchy, F., Bruno, G., Cerda, J.R., Courcol, B., Díaz, R.F., Hébrard, G., Kirk, J., Lachurié, J.C., Lam, K.W.F., Martinez, P., McCormac, J., Moutou, C., Rajpurohit, A., Rivet, J.P., Spake, J., Suarez, O., Toublanc, D., Walker, S.R.: Photodynamical mass determination of the multiplanetary system K2-19. Mon. Not. R. Astron. Soc. **454**, 4267–4276 (2015). 1510.01047

Bodenheimer, P., Pollack, J.B.: Calculations of the accretion and evolution of giant planets The effects of solid cores. Icarus **67**, 391–408 (1986)

Bodenheimer, P., Lin, D.N.C., Mardling, R.A.: On the tidal inflation of short-period extrasolar planets. Astrophys. J. **548**, 466–472 (2001)

Bodenheimer, P., Laughlin, G., Lin, D.N.C.: On the radii of extrasolar giant planets. Astrophys. J. **592**, 555–563 (2003). astro-ph/0303541

Cossou, C., Raymond, S.N., Hersant, F., Pierens, A.: Hot super-Earths and giant planet cores from different migration histories. Astron. Astrophys. **569**, A56 (2014). 1407.6011

D'Angelo, G., Bodenheimer, P.: In Situ and Ex Situ formation models of Kepler 11 planets. Astrophys. J. **828**, 33 (2016). 1606.08088

Duffell, P.C., Chiang, E.: Eccentric Jupiters via disk-planet interactions. Astrophys. J. **812**, 94 (2015). 1507.08667

Freedman, R.S., Marley, M.S., Lodders, K.: Line and mean opacities for ultracool dwarfs and extrasolar planets. Astrophys. J. Suppl. Ser. **174**, 504–513 (2008). 0706.2374

Freedman, R.S., Lustig-Yaeger, J., Fortney, J.J., Lupu, R.E., Marley, M.S., Lodders, K.: Gaseous mean opacities for giant planet and ultracool dwarf atmospheres over a range of metallicities and temperatures. Astrophys. J. Suppl. Ser. **214**, 25 (2014). 1409.0026

Ginzburg, S., Sari, R.: Tidal heating of young super-Earth atmospheres Mon. Not. R. Astron. Soc. **464**, 3937–3944 (2017). https://arxiv.org/abs/1608.03718

Ginzburg, S., Schlichting, H.E., Sari, R.: Super-Earth atmospheres: self-consistent gas accretion and retention. Astrophys. J. **825**, 29 (2016). 1512.07925

Goldreich, P., Sari, R.: Eccentricity evolution for planets in gaseous disks. Astrophys. J. **585**, 1024–1037 (2003). astro-ph/0202462

Goldreich, P., Lithwick, Y., Sari, R.: Planet formation by coagulation: a focus on Uranus and Neptune. Annu. Rev. Astron. Astrophys. **42**, 549–601 (2004). astro-ph/0405215

Gu, P.G., Lin, D.N.C., Bodenheimer, P.H.: The effect of tidal inflation instability on the mass and dynamical evolution of extrasolar planets with ultrashort periods. Astrophys. J. **588**, 509–534 (2003). astro-ph/0303362

Hayashi, C.: Structure of the solar nebula, growth and decay of magnetic fields and effects of magnetic and turbulent viscosities on the nebula. Prog. Theor. Phys. Suppl. **70**, 35–53 (1981)

Ibgui, L., Burrows, A.: Coupled evolution with tides of the radius and orbit of transiting giant planets: general results. Astrophys. J. **700**, 1921–1932 (2009). 0902.3998

Ibgui, L., Burrows, A., Spiegel, D.S.: Tidal heating models for the radii of the inflated transiting giant planets WASP-4b, WASP-6b, WASP-12b, WASP-15b, and TrES-4. Astrophys. J. **713**, 751–763 (2010). 0910.4394

Ibgui, L., Spiegel, D.S., Burrows, A.: Explorations into the viability of coupled radius-orbit evolutionary models for inflated planets. Astrophys. J. **727**, 75 (2011). 0910.5928

Ikoma, M., Hori, Y.: In Situ accretion of hydrogen-rich atmospheres on short-period super-Earths: implications for the Kepler-11 planets. Astrophys. J. **753**, 66 (2012). 1204.5302

Inamdar, N.K., Schlichting, H.E.: The formation of super-Earths and mini-Neptunes with giant impacts. Mon. Not. R. Astron. Soc. **448**, 1751–1760 (2015). 1412.4440

Inamdar, N.K., Schlichting, H.E.: Stealing the gas: giant impacts and the large diversity in exoplanet densities. Astrophys. J. Lett. **817**, L13 (2016). 1510.02090

Jackson, B., Greenberg, R., Barnes, R.: Tidal heating of extrasolar planets. Astrophys. J. **681**, 1631–1638 (2008). 0803.0026

Jackson, A.P., Davis, T.A., Wheatley, P.J.: The coronal X-ray-age relation and its implications for the evaporation of exoplanets. Mon. Not. R. Astron. Soc. **422**, 2024–2043 (2012). 1111.0031

Jontof-Hutter, D., Rowe, J.F., Lissauer, J.J., Fabrycky, D.C., Ford, E.B.: The mass of the Mars-sized exoplanet Kepler-138 b from transit timing. Nature **522**, 321–323 (2015). 1506.07067

Leconte, J., Chabrier, G., Baraffe, I., Levrard, B.: Is tidal heating sufficient to explain bloated exoplanets? Consistent calculations accounting for finite initial eccentricity. Astron. Astrophys. **516**, A64 (2010). 1004.0463

Lee, E.J., Chiang, E.: To cool is to accrete: analytic scalings for nebular accretion of planetary atmospheres. Astrophys. J. **811**, 41 (2015). 1508.05096

Lee, E.J., Chiang, E.: Breeding super-Earths and birthing super-puffs in transitional disks. Astrophys. J. **817**, 90 (2016). 1510.08855

Lee, E.J., Chiang, E., Ormel, C.W.: Make super-Earths, not Jupiters: accreting nebular gas onto solid cores at 0.1 AU and beyond. Astrophys. J. **797**, 95 (2014). 1409.3578

Lissauer, J.J., Jontof-Hutter, D., Rowe, J.F., Fabrycky, D.C., Lopez, E.D., Agol, E., Marcy, G.W., Deck, K.M., Fischer, D.A., Fortney, J.J., Howell, S.B., Isaacson, H., Jenkins, J.M., Kolbl, R., Sasselov, D., Short, D.R., Welsh, W.F.: All six planets known to orbit Kepler-11 have low densities. Astrophys. J. **770**, 131 (2013). 1303.0227

Liu, X., Burrows, A., Ibgui, L.: Theoretical radii of extrasolar giant planets: the cases of TrES-4, XO-3b, and HAT-P-1b. Astrophys. J. **687**, 1191–1200 (2008). 0805.1733

Liu, S.F., Hori, Y., Lin, D.N.C., Asphaug, E.: Giant impact: an efficient mechanism for the devolatilization of super-Earths. Astrophys. J. **812**, 164 (2015). 1509.05772

Lopez, E.D., Fortney, J.J.: The role of core mass in controlling evaporation: the Kepler radius distribution and the Kepler-36 density dichotomy. Astrophys. J. **776**, 2 (2013). 1305.0269

Lopez, E.D., Fortney, J.J., Miller, N.: How thermal evolution and mass-loss sculpt populations of super-Earths and sub-Neptunes: application to the Kepler-11 system and beyond. Astrophys. J. **761**, 59 (2012). 1205.0010

Lundkvist, M.S., Kjeldsen, H., Albrecht, S., Davies, G.R., Basu, S., Huber, D., Justesen, A.B., Karoff, C., Silva Aguirre, V., Van Eylen, V., Vang, C., Arentoft, T., Barclay, T., Bedding, T.R., Campante, T.L., Chaplin, W.J., Christensen-Dalsgaard, J., Elsworth, Y.P., Gilliland, R.L., Handberg, R., Hekker, S., Kawaler, S.D., Lund, M.N., Metcalfe, T.S., Miglio, A., Rowe, J.F., Stello, D., Tingley, B., White, T.R.: Hot super-Earths stripped by their host stars. Nat. Commun. **7**, 11201 (2016). 1604.05220

Mamajek, E.E.: Initial conditions of planet formation: lifetimes of primordial disks. In: Usuda, T., Tamura, M., Ishii, M. (eds.) American Institute of Physics Conference Series, American Institute of Physics Conference Series, vol. 1158, pp. 3–10 (2009). 0906.5011

Miller, N., Fortney, J.J., Jackson, B.: Inflating and deflating hot Jupiters: coupled tidal and thermal evolution of known transiting planets. Astrophys. J. **702**, 1413–1427 (2009). 0907.1268

Nettelmann, N., Holst, B., Kietzmann, A., French, M., Redmer, R., Blaschke, D.: Ab initio Equation of state data for hydrogen, helium, and water and the internal structure of Jupiter. Astrophys. J. **683**, 1217–1228 (2008). 0712.1019

Owen, J.E., Jackson, A.P.: Planetary evaporation by UV X-ray radiation: basic hydrodynamics. Mon. Not. R. Astron. Soc. **425**, 2931–2947 (2012). 1206.2367

Owen, J.E., Wu, Y.: Kepler planets: a tale of evaporation. Astrophys. J. **775**, 105 (2013). 1303.3899

Owen, J.E., Wu, Y.: Atmospheres of low-mass planets: the "boil-off". Astrophys. J. **817**, 107 (2016). 1506.02049

Piso, A.M.A., Youdin, A.N.: On the minimum core mass for giant planet formation at wide separations. Astrophys. J. **786**, 21 (2014). 1311.0011

Piso, A.M.A., Youdin, A.N., Murray-Clay, R.A.: Minimum core masses for giant planet formation with realistic equations of state and opacities. Astrophys. J. **800**, 82 (2015). 1412.5185

Pollack, J.B., Hubickyj, O., Bodenheimer, P., Lissauer, J.J., Podolak, M., Greenzweig, Y.: Formation of the giant planets by concurrent accretion of solids and gas. Icarus **124**, 62–85 (1996)

Rafikov, R.R.: Atmospheres of protoplanetary cores: critical mass for nucleated instability. Astrophys. J. **648**, 666–682 (2006). astro-ph/0405507

Rafikov, R.R.: Constraint on the giant planet production by core accretion. Astrophys. J. **727**, 86 (2011). 1004.5139

Rogers, L.A., Bodenheimer, P., Lissauer, J.J., Seager, S.: Formation and structure of low-density exo-Neptunes. Astrophys. J. **738**, 59 (2011). 1106.2807

Schlichting, H.E., Warren, P.H., Yin, Q.Z.: The last stages of terrestrial planet formation: dynamical friction and the late veneer. Astrophys. J. **752**, 8 (2012). 1202.6372

Seager, S., Kuchner, M., Hier-Majumder, C.A., Militzer, B.: Mass-radius relationships for solid exoplanets. Astrophys. J. **669**, 1279–1297 (2007). 0707.2895

Teyssandier, J., Ogilvie, G.I.: Growth of eccentric modes in disc-planet interactions. Mon. Not. R. Astron. Soc. **458**, 3221–3247 (2016). 1603.00653

Valencia, D., Pu, M.: Ohmic dissipation in mini-Neptunes. In: AAS/Division for Extreme Solar Systems Abstracts, vol. 3, p. 30105 (2015)

Valencia, D., O'Connell, R.J., Sasselov, D.: Internal structure of massive terrestrial planets. Icarus **181**, 545–554 (2006). astro-ph/0511150

Walker, R.J.: Highly siderophile elements in the Earth, Moon and Mars: update and implications for planetary accretion and differentiation. Chem. Erde-Geochem. **69**, 101–125 (2009)

Weiss, L.M., Marcy, G.W.: The mass-radius relation for 65 exoplanets smaller than 4 earth radii. Astrophys. J. Lett. **783**, L6 (2014). 1312.0936

Williams, J.P., Cieza, L.A.: Protoplanetary disks and their evolution. Annu. Rev. Astron. Astrophys. **49**, 67–117 (2011). 1103.0556

Winn, J.N., Holman, M.J.: Obliquity tides on hot Jupiters. Astrophys. J. Lett. **628**, L159–L162 (2005). astro-ph/0506468

Wolfgang, A., Lopez, E.: How rocky are they? The composition distribution of Kepler's sub-Neptune planet candidates within 0.15 AU. Astrophys. J. **806**, 183 (2015). 1409.2982

Youdin, A.N.: The exoplanet census: a general method applied to Kepler. Astrophys. J. **742**, 38 (2011). 1105.1782

Chapter 11
Constraints from Planets in Binaries

Kaitlin M. Kratter

Abstract In this chapter I will discuss how planets place powerful constraints on the formation and early evolution of binary star systems. In addition, these systems demonstrate extreme modes of planet formation that can inform models of our own Solar System. I begin with a very brief overview of the theoretical mechanisms for forming binaries and their planets. I will present as case studies two triple-star systems comprising at least one planet orbiting one of the stars. Combining the information from these studies reveals that (1) many binaries with periods of 10–100 s of days form in their current orbital configurations as part of the star formation process (2) the frequency of tertiary companions in close binary systems may be indicative of three body instabilities other than the Kozai-Lidov mechanism (3) planet formation can proceed in highly perturbed, truncated disks, (4) a reservoir of material beyond the ice line may not be required to instigate planet formation.

11.1 Introduction

Heppenheimer (1974) considered models of planet formation in binaries many decades before their first discovery. Since the first detection of exoplanets in the 1990s, the diversity of known systems has increased tremendously. This increase is primarily due to the *Kepler* space mission, which has discovered over 3000 planet candidates to date (Coughlin et al. 2016). Planets in multi-star systems are now known to be relatively common, though much less frequent than around single stars (Wang et al. 2014, 2015; Kraus et al. 2016). Depending on the relative separations, planets can be found orbiting a single star in a multi-system (often referred to as S-type) or both stars in a close binary, so-called P-type orbits. The "S" refers to satellite and the "P" to planetary. In this review we use circumstellar and circumbinary for the two types, respectively, see Fig. 11.1. The frequency of planets in binaries as compared to single star systems is (unsurprisingly) a function of binary

K.M. Kratter (✉)
Department of Astronomy, University of Arizona, 933 N Cherry Avenue, Tucson, AZ 85721, USA
e-mail: kkratter@email.arizona.edu

M. Pessah, O. Gressel (eds.), *Formation, Evolution, and Dynamics of Young Solar Systems*, Astrophysics and Space Science Library 445,
DOI 10.1007/978-3-319-60609-5_11

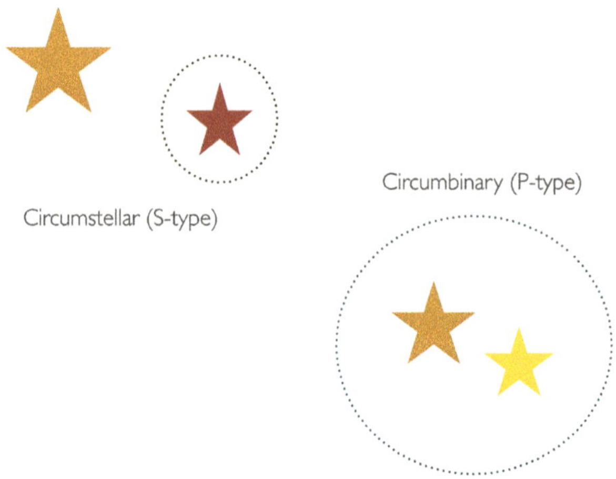

Fig. 11.1 Schematic of Satellite or S-type orbits as compared to Planetary, or P-type orbits. The latter are often referred to as circumbinary planets or even "Tatooines"

separation. At binary separations beyond 1000 AU, planet frequency is unaffected (Wang et al. 2014). At smaller separations of <100 AU, the planet occurrence rate is depressed by a factor of $\sim$3 (Kraus et al. 2016). These statistics apply only to circumstellar systems, whereas circumbinary occurrence rates are more poorly constrained (Martin et al. 2015; Li et al. 2016). In this chapter, we focus on planets in "dynamically interacting" binaries—those with relative separations that suggest present-day or formation-epoch interaction. Note that we use the term binary even though some of the systems are hierarchical triples or even quadruples. As of this writing, no planets are known to orbit more than two stars in any given system, however we shall see that the triples still play a crucial role.

The first planet discovered in a dynamically interacting binary was γ Cephei-b (Hatzes et al. 2003). This $\approx 2M_{\text{Jup}}$ planet has a semi-major axis of only 2 AU in a binary with semi-major axis of 20 AU, and remains among the tightest binaries with a planet in a circumstellar orbit. Since this discovery, a handful of other systems have been revealed with RV measurements (Eggenberger 2010), *Kepler* follow-up (Kraus et al. 2016; Dupuy et al. 2016), and direct imaging (Wagner et al. 2016). Additionally, *Kepler* revealed the first circumbinary planets (Doyle et al. 2011; Welsh et al. 2014). All of the circumbinary systems should be classified as dynamically interacting; not only are 11 of 12 known circumbinary planets within a factor of 2 of the instability limit due to proximity to the binary (see Sect. 11.4 below), but all are thought to have formed in the disk that surrounded and was shaped by the binary itself.

In Fig. 11.2, we illustrate the distribution of known planets in binaries. We show the planetary semi-major axis, a_p, versus the ratio of planet to stellar semi-major axis, a_p/a_*. In some cases, the stellar or planetary separation is simply the projected separation, as true orbits are unknown. We demarcate the rough stability boundaries

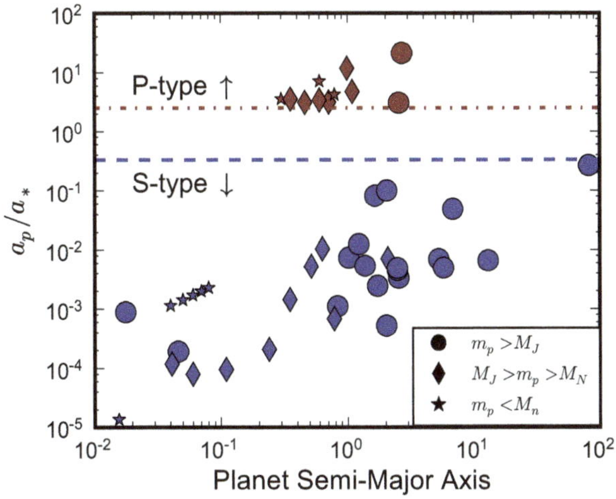

Fig. 11.2 Binary planet semi-major axis versus planet-star semi-major axis ratio for well-characterized S and P type systems. The data was complied from Wagner et al. (2016), Kostov et al. (2016) and the references therein. In some cases where orbits are poorly characterized, the semi-major axes are merely projected separations. Masses are approximate in some cases. The *dashed* and *dash-dotted lines* represent the approximate location of the instability boundary of each system type for a circular equal mass pair

for both systems. A few trends are immediately identifiable in this figure. First, the circumbinary systems are much more closely clustered towards the instability boundary. There remains debate about the cause of this apparent pile-up; it is possible that the observed distribution is consistent with a log-uniform semi-major axis distribution similar to single stars (Li et al. 2016). For the circumstellar systems, we see the clear outlier of HD131399 (Wagner et al. 2016) at the far right end of the plot. We discuss this system in detail in Sect. 11.5. This plot excludes many of the Kepler systems whose circumstellar status has only recently been discovered, but for which individual orbits are not yet available (Kraus et al. 2016).

What makes these systems so compelling is their ability to not only provide stringent constraints on the formation of the binaries that host the planets, but also inform our models of planet formation around single and binary stars alike. In order to better gauge the utility of these systems from a theoretical perspective, we begin with a brief overview of common binary formation models in Sect. 11.2. In Sect. 11.3 we describe how planet formation models are interwoven with binary formation. Next, in Sect. 11.4 we summarize the literature on planet stability in multi-star systems. In Sect. 11.5, we take two case studies of planetary systems that illuminate a key aspect of either binary formation or planet formation. We conclude in Sect. 11.6.

11.2 Binary Formation Models

It is not possible to provide an adequate review of all of binary formation in a short subsection. Here we endeavor to qualitatively describe the important aspects of the theories only. For more complete reviews, see Tohline (2002), Kratter (2011), and Duchêne and Kraus (2013).

11.2.1 Primordial Formation Models

We begin with primordial formation models: ones in which the binary forms and arrives in its final configuration by the end of the lifetime of the protoplanetary disk. We choose this special definition as we are concerned with planets that might form in such a disk—this is by no means a common distinction in the binary literature, although perhaps it should be. The two best supported theories for primordial binary formation are turbulent core fragmentation (Fisher 2004; Offner et al. 2010; Bate 2012) and disk fragmentation (Adams et al. 1989; Bonnell and Bate 1994; Kratter and Matzner 2006; Kratter et al. 2010).

11.2.1.1 Turbulent Core Fragmentation

The turbulent core fragmentation model of binary formation is consistent with the modern picture of star formation occurring in turbulent, filamentary structures in molecular clouds (Li et al. 2014). In this model, binaries appear in two ways; some cores or filaments develop multiple overdensities that are initially gravitationally bound, and remain so throughout the collapse process. In other cases, independent, initially unbound overdensities collapse, but become gravitationally bound either due to the background gravitationally potential of the cluster, or due to dissipation as the cores or protostars move through the gas. Note that while this process has been observed in a variety of numerical models cited above, there is not yet a predictive theory for how multiple star systems arise in this fashion.

There is also strong observational evidence for this mode of star formation at multiple evolutionary stages. At the very earliest stages, widely separated binaries have been observed in a variety of molecular lines and dust continuum (Tobin et al. 2015; Chen et al. 2013; Pineda et al. 2015). In addition, slightly evolved binaries (Class I–II) with protostellar disks have been resolved by ALMA. These systems show a misalignment between the inclinations of the two disks, indicating that at least one of the disks is misaligned with the binary orbital plane (Akeson and Jensen 2014; Williams et al. 2014). This misalignment indicates that the angular momentum reservoirs, and thus mass reservoirs, of the disks are likely distinct. With separations of order a few hundred AU, they are consistent with formation via turbulent fragmentation (Offner et al. 2016).

11.2.1.2 Disk Fragmentation

Binary formation via disk fragmentation occurs when protostellar disks grow sufficiently massive to become strongly self-gravitating. Disks respond to increasing self-gravity by first driving strong angular momentum transport via spiral arms. If this transport is insufficient to process the material infalling from the background core, the disk mass will rise and lead to runaway fragmentation (Kratter and Lodato 2016). This process is most likely to occur for more massive cores $\gtrsim$ a few $M_\odot$ (Krumholz et al. 2007; Kratter et al. 2008). The first compelling detection of this process in the act was recently presented by Tobin et al. (2016a), which revealed a young triple star system forming in a massive disk with spiral arms.

11.2.2 Secondary Formation Models

After the initial stage of cloud collapse, binary and multiple star systems may form via dynamical interaction, especially in dense clusters. One of the strongest indications that dynamical evolution plays a role in at least reshaping multi-system orbits, if not creating them entirely, is the preponderance of wide tertiary companions in close binaries (Tokovinin 2014; Moe and Di Stefano 2017). These extremely hierarchical orbits are consistent with either violent dynamical encounters shortly after birth (Bate 2012), or long timescale secular evolution via Kozai-Lidov cycles, followed by tidal circularization of the inner binary (Lidov 1962; Kozai 1962; Mazeh and Shaham 1979; Fabrycky and Tremaine 2007). Kozai-Lidov (hereafter KL) oscillations can arise in hierarchical triple systems under a relatively wide range of conditions, depending on the relative orbit inclinations, initial eccentricities, mass ratios, and separations of the three components (Naoz 2016). However in most cases, large relative inclination between the two orbits is required (Anderson et al. 2016). This secular oscillation exchanges eccentricity in the orbit of the inner pair for inclination between the two orbits. KL oscillations are a promising mechanism for generating close binaries because the inner pair is driven to high eccentricity, at which point the pericenter of the orbit can be sufficiently close for tidal forces to drive orbit circularization at small separations.

In the simplest quadrupole approximation, KL oscillations will occur when the relative orbit inclination $i_b > 39.2°$. In this limit, the perturbed bodies orbital angular momentum vector is conserved, which leads to a relation between the initial inclination and maximum eccentricity:

$$e_{\max} = \sqrt{1 - (5/3)\cos^2 i}, \tag{11.1}$$

where e_{max} is the eccentricity of the inner orbit. The timescale for KL oscillations scales with the orbital periods and mass ratios as:

$$\tau_{KL} = \frac{2P_{out}^2}{3\pi P_{in}} \frac{M_{tot}}{m_{pert}} (1 - e_{out}^2)^{3/2} \tag{11.2}$$

(Innanen et al. 1997; Kiseleva et al. 1998).

KL oscillations are suppressed if other forces cause the inner pair's pericenter to precess faster than the oscillation timescale above, because precession averages out the net effect of the inner pair's eccentricity. General Relativistic effects, stellar rotation, tides, and an extra body in the system can all act to shut off the oscillations. In the standard model for forming tight triple systems (and also misaligned Hot Jupiters), tidal forces dominate (Fabrycky and Tremaine 2007; Anderson et al. 2016). When tidal forces ultimately shut off the oscillations, the resultant system contains a very short period binary with a wide, inclined tertiary. The distribution of relative inclinations is likely somewhat bimodal, though this is a mass dependent statement (Anderson et al. 2016). KL oscillations are likely an important channel for binaries with orbital periods less than 10 days (Fabrycky and Tremaine 2007). Primordial formation mechanisms struggle to produce binaries with such short periods, as the radii of protostars on the Hayashi track exceed these separations (see, e.g., Krumholz and Thompson 2007). As we discuss below, these channels have important implications for circumbinary planets.

11.3 Constraints on Planets in Binaries from Formation Models

We have highlighted above the generic mechanisms for forming binaries. How does the planet formation process fit in to these theories? We break the process of planet formation into two critical stages. First, we discuss the constraints, both theoretical and observational, on the formation and evolution of the protoplanetary disks out of which planets form. Subsequently we describe how the onset of planet formation is modified in binary systems, and provide some speculation on possible solutions.

11.3.1 Disk Formation and Evolution

The presence of a binary companion can strongly impact the formation and evolution of protoplanetary disks. In the case of circumstellar disks in close binaries, tidal truncation is likely responsible for setting the outer edge of the disk, rather than cloud angular momentum as in single stars (Artymowicz and Lubow 1994; Goodman et al. 1993). Disks are also truncated in the circumbinary case near the inner Lindblad resonance. A key issue in understanding the origin of the planets is a careful accounting of when a disk will form, how it survives the star formation process, and how it persists to produce planets.

Surveys of young star forming regions like Taurus have discovered both circumstellar and circumbinary disks that might form planets (Harris et al. 2012). However, the sub-mm flux, which is a proxy for mass at fixed age and distance, is substantially smaller in disks within and surrounding binary stars systems. In addition, there is a deficit of disks around the closest binaries, which would be the progenitors of circumbinary systems (Harris et al. 2012). There is evidence that some of these disks may by escaping detection at young ages due to either short lifetimes or failure to identify all spectroscopic binaries in young clusters; debris disks do not seem to show as extreme a deficit (Trilling et al. 2007; Rodriguez et al. 2015). On-going RV planet searches in young clusters should help resolve this uncertainty (L. Prato, private communication).

From a theoretical perspective there are several threats to the survival of circumbinary and circumstellar disks. For the former case, the formation timescale of the binary is the most substantial problem. As noted above, one of the most promising mechanisms for creating closer binaries are the so-called secondary formation channels which rely on dynamical or secular evolution to rearrange higher order multiples into a close binary and a distant (sometimes ejected) tertiary (Reipurth and Clarke 2001). Both mechanisms are likely destructive to a pre-existing protoplanetary disk (see also Clarke and Pringle 1991).

The rearrangement of orbits in an unstable triple system usually proceeds over 10^2–10^4 orbital periods, and involves multiple close encounters between any of the three components before the orbits stabilize in a new configuration (Eggleton and Kiseleva 1995; Perets and Kratter 2012). Any disks present at the onset of the instability would likely be disrupted by the interaction. However, large scale star formation simulations suggest that if these interactions occur early enough in the star formation process, while there is still ample gas in the star-forming environment, a new disk of material might re-accumulate surrounding the newly formed tight binary (Throop and Bally 2008; Bate 2012; Muñoz et al. 2015). This model implies that circumbinary planet hosts underwent rapid orbit evolution in order to retain a substantial disk.

Circumbinary disks might also arise around close binaries that have undergone circumbinary disk-driven migration. This process has been studied extensively in the context of binary black-hole accretion disks (Lodato et al. 2009; Nixon et al. 2011; Lubow et al. 2015). In the protostellar case, it remains unclear over how many decades in radius disk-driven migration can take place for plausible disk parameters (Armitage and Bonnell 2002), and even outward migration can occur under some circumstances (Kratter et al. 2010). The major limitation on disk-driven migration derives from the inefficiency with which angular momentum is extracted from the orbit via Lindlblad resonances. As first discussed by Syer and Clarke (1995) (see also Lodato et al. 2009; Kocsis et al. 2012), once the disk mass becomes small compared to the binary mass, migration will be slow. The mass of the disk is limited by gravitational instability. Misalignment of the binary with the disk might aid in this process (Nixon et al. 2011), though such a configuration seems unlikely in the protostellar case.

As noted above, KL oscillations could in principle be responsible for generating circumbinary hosts at the centers of disks. But such disks would then be subject to strong perturbations from the eccentricity cycles of the inner pair; for disks with typical standoff radii from the binary (a factor of a few), the binary itself would drastically re-sculpt the inner edge of the disk as its eccentricity grows to $e > 0.9$. Thus it is difficult to imagine planet formation following this process, unless it occurs early when a disk can restabilize. The relative formation times suggests that instabilities arising for planets in binaries undergoing KL oscillations may not be relevant. This argument is consistent with the absence of circumbinary planets around binaries close enough to have undergone tidal circularization following KL cycles, although other arguments exist to destroy these planets (Muñoz and Lai 2015; Hamers et al. 2016; Martin et al. 2015). Note that KL during the pre-main sequence phase might result in wider tidally circularized orbits due to inflated stellar radii (Moe and Kratter 2017).

KL cycles may also be inhibited by the presence of a massive circumbinary disk required to form planets. Precession of the inner binary induced by the disk can overwhelm the KL oscillations. The precession timescale of the binary driven by the disk self-gravity is:

$$\varpi = \pi\phi(a_b/a_{\rm in}, \mu) n_b \frac{\Sigma_0 a_0^p a_b^3}{M_b a_{\rm in}^{1+p}} \tag{11.3}$$

(see the appendix of Silsbee and Rafikov 2015a), where ϕ is a function that attains values of order unity, a_b and $a_{\rm in}$ are the binary and inner disk semi-major axes respectively, a_0 is a reference distance of 1 AU, and Σ_0 is the disk surface density at that location. M_b is the total stellar mass, n_b is the binary mean motion, and p is the slope of the surface density profile. For typical disk parameters and binaries akin to the Kepler systems, this corresponds to a precession timescales of order a few 10^3 years (Silsbee and Rafikov 2015a).

For circumstellar disks, the formation issues for the binary itself are less restrictive. Both core and disk fragmentation (with some orbital evolution) can plausibly produce circumstellar planet progenitor systems. Since the orbits can arise without dynamical or secular instabilities between multiple stars, the disk survival is not threatened in the same manner. Nevertheless, binarity may reduce the typical size of the disk and thus the reservoir of material available for planet formation. For the typical binary separation, 50–100 AU (Raghavan et al. 2010; Moe and Di Stefano 2017), tidal truncation will shrink disks to 15–30 AU. Whether truncation impacts planet formation at much smaller radii (independent of the binary perturbations, which we discuss below) is an open question. Najita and Kenyon (2014) have pointed out that the typical dust mass reservoir in single-star disks is low compared to the typical masses of Kepler exoplanetary systems. The implications of such a finding are that (a) grain growth to beyond mm sizes begins early, hiding substantial amounts of dust, and (b) radial drift may resupply solids to the inner regions of disks from a reservoir of material at larger radii (Youdin and Chiang 2004). Clearly, truncation in binaries would limit the available reservoir of material. If such material is important, one would expect circumstellar planets to skew lower in mass as compared to single star planets. γ-cephei provides at least one massive counterexample. At present the statistics on circumstellar planets are sufficiently limited that such a trend is not yet identifiable (Wang et al. 2014). The circumstellar systems have a median mass above that of circumbinary systems, but selection effects due to detection via the RV method could be responsible.

Two other characteristics unique to disks in binaries are warping and tilt oscillations due to misalignment between the disk(s) and the binary orbit plane. Wide binary systems ($\sim$200 AU) have been identified by ALMA (Akeson and Jensen 2014; Williams et al. 2014) where one if not both of the disks are misaligned with the orbital plane. Relative inclination between the binary orbit and the disks can induce tilt oscillations, damped KL oscillations, disk warping, and in extreme circumstances disk breaking (Larwood et al. 1996; Bate 2000; Lubow and Ogilvie 2000; Batygin 2012; Martin et al. 2013; Fu et al. 2015; Lubow and Martin 2016). For the dynamically interacting systems on which we focus in the review, misalignment should damp relatively quickly. The precession rate relative to the disk's outer orbital frequency Ω_d is:

$$\omega_p/\Omega_d = K\cos\delta\frac{M_s}{M_p}\left(\frac{R}{D}\right)^3, \tag{11.4}$$

where K is an order unity factor depending on the surface density distribution in the disk, D is the binary separation, δ the inclination, and M_s, M_p, the secondary and primary stellar masses. The simplest estimate of the alignment timescale is:

$$t_{\text{align}} \sim \frac{R^2}{\nu}\left(\frac{H}{R}\right)^4\left(\frac{\Omega_d}{\omega_p}\right)^2, \tag{11.5}$$

where (R^2/ν) is the viscous timescale at radius, R for effective viscosity ν, and (H/R) is the disk aspect ratio (Bate 2000). However, parametric instabilities are likely to arise when δ exceeds the disk opening angle, set by H/R (Gammie et al. 2000). This drops the alignment timescale dramatically, to of order:

$$t_{\text{align}} \sim \omega_p^{-1}/\sin\delta. \tag{11.6}$$

Thus modestly misaligned systems (a few 10 s of degrees) have their inclinations damped on timescales comparable to the precession time, or of order 100–1000 s of orbital periods. Moreover, this extra source of dissipation may contribute to faster accretion rates, and potentially shorter disk lifetimes (Bate 2000).

In the case of even more extreme misalignment, disks undergo damped KL oscillations (Martin et al. 2014; Fu et al. 2015). These oscillations lead to eccentricity growth in the disk in addition to inclination variations. Over timescales of 10 s of orbits, the oscillations decay, leaving the disk at inclinations below the critical value for KL oscillations, but still eccentric (Martin et al. 2014). These likely revert to the case discussed above, where dissipation damps the inclination over somewhat longer timescales. Taken together, these studies suggest that planet formation in misaligned circumstellar disks where the binary dynamics are important may be rare. Nevertheless in the following section, we discuss how remnant misalignments might impact growth.

11.3.2 From Dust to Planets

Instigating the growth of planetesimals from microscopic dust grains remains one of the most challenging phases of planet formation for theorists, even around single stars (Youdin and Kenyon 2013). Micron and sub-micron-sized dust grains present in the ISM must clump together in the disk over many orders of magnitude to generate 1–500 km planetesimals. At the upper end of this size distribution, gravitational focusing takes over and allows planetary core growth to proceed somewhat efficiently. Early growth is hampered by several effects. First, once dust grains reach a cm in size, growth into pebbles 1–10 cm, and boulders 10–100 m is slow or impossible. Pebbles and boulders are too small to gravitationally attract each other, but electrostatic forces are insufficient to cause coagulation at speeds characteristic of particles in protoplanetary disks. Secondly, grains between 1 cm and 1 m are strongly affected by imperfect coupling with the gas in protoplanetary disks. Because the particles are a pressureless fluid, they orbit at the Keplerian speed, while the gas orbits at slightly sub-Keplerian speeds due to the outwardly decreasing pressure gradient. The resultant headwind felt by the particles causes an inward drift on timescales of 100–1000 s of orbits. Finally, the relative velocities of these size particles tend to increase, which contributes to more destructive rather than constructive, collisions. This threatens to remove intermediate size material from the disk before it can successfully accumulate into planets. See Armitage (2011) for a detailed review.

In binaries, the problem of growth up to the stage where gravitational focusing takes over is exacerbated by the potential for higher relative velocities between particles due to the forced eccentricity in the disk, or high mutual inclinations driven by a misaligned companion (Moriwaki and Nakagawa 2004; Marzari et al. 2009; Marzari and Nelson 2009; Batygin et al. 2011; Fragner et al. 2011). In a series of recent papers (Rafikov 2013a,b; Rafikov and Silsbee 2015a,b; Silsbee and Rafikov 2015a,b; Vartanyan et al. 2016), these authors review the mechanisms for excitation of planetesimal eccentricity in circumstellar and circumbinary orbits, pointing out that both the binary and the (typically) eccentric protoplanetary disk modify planetesimal orbits. They highlight conditions under which relative planetesimal velocities are suppressed to allow growth up to >100 km sizes, at which point even relatively high collisional velocities in eccentric disks are insufficient to result in destructive rather than constructive collisions. Previous work has shown that gas drag can aid in reducing collisional velocities between equal size bodies by forcing apsidal alignment (Marzari and Scholl 2000). However, this only reduces collisional velocities for equal sized bodies. They also find that disks which are sufficiently massive and thus self-gravitating are conducive to planetesimal formation, because the disks apsidal precession rate is dominated by disk self-gravity. The rapid apsidal precession of planetesimals driven by the disk reduces the maximum eccentricity to which they are excited, resulting in substantially lower collisional velocities. A careful accounting of the eccentricity evolution of planetesimals in binary systems suggests that only special regions of the disk will be conducive to planetesimal growth. Nevertheless, these constraints are not strong enough to rule out planet formation from relatively small planetesimals in the bulk of the disk.

A promising solution to the radial drift and fragmentation problem around single and binary stars is the particle-gas Streaming Instability (Youdin and Goodman 2005), which allows small pebbles to spontaneously collect into local overdensities as they undergo radial drift. These overdensities are large enough to become self-gravitating, and spontaneously collapse to form planetesimals ranging in size from 10s to 100s of km (Simon et al. 2016). Such large bodies maintain low enough collisional velocities across much of the parameter space to avoid destruction. It remains to be demonstrated that the streaming instability will operate successfully in eccentric disks, or even axisymmetric disks in the presence of a binary. There is no obvious reason, however, why the mechanism should break down under such circumstances. Ogilvie and Barker (2014) have developed a formalism for eccentric disks in the shearing box approximation, which offers a promising approach for validating the instability in the new regime.

For circumbinary planets, another substantial difference in formation may arise from the unique structure of circumbinary disks. As noted in Sect. 11.3.1, the inner boundary of such disks experiences a net outward torque. 1D viscous disk models suggest that this inner torque might produce an extended deadzone at a few times the binary semi-major axis (Martin et al. 2013). A large deadzone might provide a particularly favorable location for planetesimal agglomeration due to the decreased turbulence.

Subsequent to formation, planetary cores may of course undergo migration in their natal disk (Kley and Nelson 2012). For circumstellar systems, migration would likely proceed as around a single star, perhaps modified by disk eccentricity. For circumbinary systems, simulations suggest that inward migration remains the norm, and that the details of the disk structure impact the location at which migration stalls (Kley and Haghighipour 2014). The detailed outcome of planetary migration even in single star systems remains hotly debated. Though migration models will be required to fully understand binary planet origins and architecture, we can independently consider how their architecture is shaped by dynamical stability.

11.4 Binary Planet Dynamics

One of the most striking distinctions between planets around single stars as compared to binaries is the stringent limits on their orbital properties based on (a) stability in the presence of the stars in their current orbital configuration and (b) survival through the evolution that might be concurrent or subsequent to the process of planet formation. We first discuss the basic constraints on stability ignoring stellar orbit evolution, then review the more complicated issues of intertwined evolution.

11.4.1 Single Planet Stability

The simplest constraint to place on the existence of planets in binaries is based on long-term orbital stability. In P and circumstellar orbits, there are certain regions of phase space that planets are prohibited from occupying. This parameter space was broadly explored by Holman and Wiegert (1999), who carried out a series of numerical integrations of test particles in both classes of systems for a range of mass ratios and eccentricities. They find that in circumstellar systems, the binary companion semi-major axis must be $\sim$4 times the planet's semi-major axis for equal mass binaries. This factor increases for more disparate masses and high eccentricities. For circumbinary systems, the planet must have a semi-major axis more than twice that of the binary stars. For smaller mass ratios, the relative planet-star distances decrease, while at higher binary eccentricities they increase. This work, building on that of Rabl and Dvorak (1988) focused on co-planar, test particle orbits, and single planet systems.

The empirical result was followed up analytically by Mudryk and Wu (2006), who found that they could explain both the overall trend in the Holman and Wiegert (1999) results, and the deviations from it, via the destabilizing forces of overlapping mean-motion resonances. This work also revealed that the previously ignored free eccentricity of the planets had a minimal effect on the stability boundary.

11.4.2 Multi-Planet Systems

More recently, the dynamics of multi-planet circumbinary systems has been explored (Kratter and Shannon 2014; Smullen et al. 2016). These works have examined the required spacing between multi-planet systems around binaries for orbital stability as compared to that around single stars (Chambers et al. 1996; Smith and Lissauer 2009; Morrison and Kratter 2016). Kratter and Shannon (2014) found that the required spacing for stability is similar to that of > 5 planet systems around single stars, or $> 7R_{\mathrm{H,m}}$, where the mutual hill radius is defined as:

$$R_{\mathrm{H,m}} = \left(\frac{a_1 + a_2}{2}\right)\left(\frac{m_{p1} + m_{p2}}{3M_*}\right)^{1/3}, \tag{11.7}$$

where a_1 and a_2 are the semi-major axes of the planets. While the aforementioned study focused on well-spaced, circular planets, Smullen et al. (2016) and Sutherland and Fabrycky (2016) investigated the fates of more eccentric, scattering planets in binary systems. These works found that planet–planet and planet–star collisions were substantially reduced as compared to single star systems in favor of ejection from the planetary system. In initially overpacked systems, the binary has little observable influence on the surviving planetary system architecture when compared to single planet systems. This finding suggests that scattering is not responsible for differences in the observed mass or semi-major axis distribution of circumbinary planets, aside from removal from within the inner dynamically unstable region.

11.4.3 Secular Interactions in Systems with Mutual Inclination

Thus far we have discussed the KL mechanism in the context of the formation of the binary or the evolution of the disk. We have argued that typical disk alignment timescales are short for dynamically interacting circumstellar systems that might undergo KL oscillations (Bate 2000). For large enough binary separations, the disk alignment timescales become comparable to the dissipation timescales of a few Myr (Williams and Cieza 2011).

In circumbinary systems, it is possible that a wide tertiary might excite KL oscillations in a binary after the formation of planets. If KL oscillations in the binary were suppressed by the circumbinary disk at early times [see Eq. (11.3)], it is possible that the dissipation of the disk and formation of the planet would render the system susceptible to the oscillations. How might these late-time KL oscillations impact a circumbinary binary system?

The Kepler mission has only found CBPs around binaries with periods larger than 7 days. Such binaries are marginally too wide to have undergone efficient tidal circularization following eccentricity excitation, at least on the main sequence. Muñoz et al. (2015) first noted that these oscillations would be very destructive to circumbinary planets already in existence around the wider binaries prior to undergoing KL cycles. As summarized by Hamers et al. (2016), there are three regimes for CBPs in triple systems that are susceptible to KL cycles. If the mass of the planet is too high, then the precession induced in the binary orbit will shut off KL cycles entirely, just like the massive disk. When the mass of the planet is lower, two outcomes are possible. First, the planet may be destabilized by the excitation of eccentricity in the binary. Alternatively, the planet itself will participate in KL cycles and become strongly misaligned with the binary, and thus evade detection in transit surveys. Although Martin et al. (2015) have noted that even highly inclined planets will ultimately transit due to the precession of the planets longitude of ascending node, they are much more difficult to detect as the timing between subsequent transits can be much longer than the planets orbital period.

These dynamical arguments provide a compelling explanation for the lack of observed planets around close binaries, but it remains unclear whether dynamical instability removes them, or binary formation channels prevent them from existing. The proximity of the observed CBPs to the instability boundary of the binary provides somewhat compelling evidence that the stars did not undergo radical orbital evolution subsequent to the formation of the planet.

11.5 Triple System Case Studies

We have focused on the joint theory of binary formation and planet formation, but in many cases we have found that a third star is a crucial part of the picture. Although a theory of planets in triples remains undeveloped, we now consider two planetary systems as case studies. We highlight the theoretical implications of the existence of each system.

11.5.1 Kepler 444

Characteristics The star Kepler 444a hosts five, sub-Earth size planets in short period orbits ranging from 5–11 days. The primary is orbited by a pair of M-dwarfs at a projected separation of 66 AU. By combining astrometric constraints and RV monitoring, Dupuy et al. (2016) concluded that the M-dwarf's center of mass is most likely on a highly eccentric orbit around the primary with a pericenter approach of 5 AU, and $e = 0.86$. The orbital properties of the M-dwarf pair itself remain unconstrained, though the semi-major axis must be <0.3 AU. Like the transiting planets, which are by definition edge-on with respect to the observer, the M-dwarf pair also orbits in the plane of the sky. Dynamical modeling suggests that the planets and outer binary are likely co-planar as well. Misalignments of more than a few degrees between the two result in precession of the five planets so that concurrent transits of all five are rare. If the triple system arose prior to dispersal of the primary's protoplanetary disk, it would have been truncated at ~2 AU (Lubow and Artymowicz 1996).

Implications The present day system is dynamically stable (both the triple system and the planets). Interpretation of this system from a theoretical perspective depends on whether or not the triple orbit is primordial with respect to the planet. In other words, did planet formation occur in the presence of the eccentric M-dwarf pair, or did the orbit evolve subsequently? Dupuy et al. (2016) argue that the orbits are most likely primordial due to the coplanarity of the planets and the stars. One possible origin for the hierarchical triple is either turbulent core fragmentation followed by orbit migration or rapid dynamical instability. Although such fragmentation would not provide initially co-planar orbits, the primary's disk would align with the M-dwarf pair rapidly (Bate 2000), or might have reaccumulated following disruption during close dynamical encounters. Alternatively, the system could have all formed in a co-planar configuration via disk fragmentation, not unlike the forming hierarchical triple identified by Tobin et al. (2016a). Non-primordial mechanisms are harder to reconcile with the aligned, stable orbits. Late time arrival into this configuration either via dynamical instability of the triple or via KL oscillations would have likely disrupted a pre-existing planetary system. Moreover, the coincidence of co-planarity begs explanation. Without the dissipative gas disk, any misalignment should remain, causing persistent tilt oscillations.

A final argument for the disk fragmentation scenario stems from the need for a high mass disk within the truncation radius of $\sim$2 AU to provide ample material for the planetary systems. This mass requirement is exacerbated by the fact that Kepler 444A is metal poor, and thus one might suspect the disk to have had an unusually low dust-gas ratio, requiring a higher than average gas mass. For planet formation to proceed with $\sim$10% efficiency in the truncated disk, the initial surface density implies a disk that was gravitational unstable at 50–100 AU, consistent with the current location of the M-dwarf pair.

Independent of the formation method of the triple, the orbits suggest that the planet forming disk was truncated within the ice line. The lack of icy material suggests that planetesimal formation should commence even without a large population of sticky ice grains. The truncated disk scenario might favor inside-out planet formation models (Chatterjee and Tan 2015) where pebbles accumulate at the inner deadzone boundary, and form subsequent generations of planets, all tightly packed. The semi-major axes of the Kepler 444 planets are all within a factor of several of the predicted inner boundary of the deadzone (Mohanty et al. 2013).

11.5.2 HD131399

Characteristics The second triple star system we discuss highlights the opposite regime of planet formation and hierarchical triples. Wagner et al. (2016) discovered that HD131399A, a roughly 15 Myr old, $2M_\odot$ star, hosts a massive, $> 4M_{\rm Jup}$ planet at a projected separation of $\approx$80 AU. In many ways, the closest analogs to the planet are the HR 8799 companions in terms of the mass and separation (Marois et al. 2010). The orbital architecture is otherwise quite different. In HD131399, the primary and planet are orbited by a tight binary (HD131399B and C), at a projected separation of 315–320 AU. If both the planet and the BC pair are on coplanar, circular orbits with semi-major axis equivalent to their projected separation, the system is marginally dynamically stable. Thus unlike other multi-star, or star+planet systems, the orbits are not hierarchical. Substantial eccentricity in the BC pair can easily destabilize the planetary orbit, though due to the large planet-star mass ratio (the combined mass of the BC pair is $1.5M_\odot$), the planet's orbit has little impact on the stability of the triple. Because the planet and stellar orbits are so widely spaced, the astrometric orbit constraints are not nearly as powerful as for Kepler 444, and thus it is not possible to place meaningful constraints on the orbit parameters. Follow-up observations will be crucial for distinguishing the formation mechanisms, as detailed below.

Implications HD131399 holds the distinction of being the circumstellar system closest to the boundary of long-term dynamical instability based on the projected separations. This proximity is much more common amongst the circumbinary system (see Fig. 11.2). Because the system is also quite young, it is possible that the orbits are not long-term stable as noted by Wagner et al. (2016) and Veras et al. (2017). See Nielsen et al. (2017) for a non-planetary interpretation of HD131399b.

Consider first the formation of the hierarchical triple. As with Kepler 444, primordial formation via turbulent core or disk fragmentation seems likely. If HD131399A hosted a protoplanetary disk, it would have been truncated at the present day location of the giant planet. Following the models of Rafikov and Silsbee (2015b), formation in-situ via core accretion would be difficult due to the strong perturbations from the BC pair. An alternative in-situ formation scenario is the direct triggering of gravitational instability in the outer disk (Meru 2015). Previous work has suggested that the close passage of an external perturber to a disk typically heats the disk, suppressing fragmentation, but modern simulations with more sophisticated cooling models sometimes find that if GI is initially triggered in the outer regions of the disk, the interior may become susceptible as well. Recently, Fu et al. (2017) have argued that KL oscillations can potentially induce fragmentation via gravitational instability in disks otherwise stable to GI. This is an intriguing possibility for this system, although there is not yet an indication that the orbits are sufficiently inclined. Moreover, depending on the as yet unknown orbit of the BC component, KL oscillations might be suppressed entirely.

11.6 Discussion

In this chapter we have provided a qualitative overview of theoretical constraints related to the formation and survival of planets in dynamically interacting binaries. We argue that the existence of planets in binaries can place meaningful constraints on binary formation models. The existence of circumbinary planets around short period binaries argues for rapid formation of short period orbits. This is also consistent with the lack of planets around the shortest period systems (Muñoz and Lai 2015; Hamers et al. 2016; Martin et al. 2015), which might undergo KL oscillations on longer timescales. Close binary formation via fragmentation of the protostellar disks or cores with rapid dynamical orbit re-arrangement is plausible (Tobin et al. 2016b; Offner et al. 2016). The preponderance of circumstellar systems reveals that planet formation is likely efficient, and may occur even when the binary companions excites substantial eccentricity in the disk and planetesimals (Rafikov and Silsbee 2015b). This may indicate that planetesimal formation via the streaming instability is a robust mechanism to avoid the fragmentation barrier. Systems like Kepler 444 further reveal that planet formation does not rely on the existence of a reservoir of material beyond the iceline, because the disk was likely truncated interior to this region prior to the formation of the planets (Dupuy et al. 2016). Formation in a truncated disk lends credence to models of planet formation such as Chatterjee and Tan (2015), which favor episodic formation of planets at deadzone inner boundaries. Finally, HD 131399 provides an example of the opposite regime of multi-star planet formation (Wagner et al. 2016). HD131399 poses a unique challenge to formation models as it resembles neither a hierarchical, extreme mass ratio multi-star system, nor a typical planetary system.

Acknowledgements KMK is supported by the National Science Foundation under Grant No. AST-1410174.

References

Adams, F.C., Ruden, S.P., Shu, F.H.: Eccentric gravitational instabilities in nearly Keplerian disks. Astrophys. J. **347**, 959–976 (1989). doi:10.1086/168187

Akeson, R.L., Jensen, E.L.N.: Circumstellar disks around binary stars in Taurus. Astrophys. J. **784**, 62 (2014). doi:10.1088/0004-637X/784/1/62, 1402.5363

Anderson, K.R., Storch, N.I., Lai, D.: Formation and stellar spin-orbit misalignment of hot Jupiters from Lidov-Kozai oscillations in stellar binaries. Mon. Not. R. Astron. Soc. **456**, 3671–3701 (2016). doi:10.1093/mnras/stv2906, 1510.08918

Armitage, P.J.: Dynamics of protoplanetary disks. Annu. Rev. Astron. Astrophys. **49**, 195–236 (2011). doi:10.1146/annurev-astro-081710-102521, 1011.1496

Armitage, P.J., Bonnell, I.A.: The brown dwarf desert as a consequence of orbital migration. Mon. Not. R. Astron. Soc. **330**, L11–L14 (2002). doi:10.1046/j.1365-8711.2002.05213.x, astro-ph/0112001

Artymowicz, P., Lubow, S.H.: Dynamics of binary-disk interaction. 1: Resonances and disk gap sizes. Astrophys. J. **421**, 651–667 (1994). doi:10.1086/173679

Bate, M.R.: Predicting the properties of binary stellar systems: the evolution of accreting protobinary systems. Mon. Not. R. Astron. Soc. **314**, 33–53 (2000). doi:10.1046/j.1365-8711.2000.03333.x, astro-ph/0002143

Bate, M.R.: Stellar, brown dwarf and multiple star properties from a radiation hydrodynamical simulation of star cluster formation. Mon. Not. R. Astron. Soc. **419**, 3115–3146 (2012). doi:10.1111/j.1365-2966.2011.19955.x, 1110.1092

Batygin, K.: A primordial origin for misalignments between stellar spin axes and planetary orbits. Nature **491**, 418–420 (2012). doi:10.1038/nature11560

Batygin, K., Morbidelli, A., Tsiganis, K.: Formation and evolution of planetary systems in presence of highly inclined stellar perturbers. Astron. Astrophys. **533**, A7 (2011). doi:10.1051/0004-6361/201117193, 1106.4051

Bonnell, I.A., Bate, M.R.: The formation of close binary systems. Mon. Not. R. Astron. Soc. **271** 999–1004 (1994). doi:10.1093/mnras/271.4.999, astro-ph/9411081

Chambers, J.E., Wetherill, G.W., Boss, A.P.: The stability of multi-planet systems. Icarus **119**, 261–268 (1996). doi:10.1006/icar.1996.0019

Chatterjee, S., Tan, J.C.: Vulcan planets: inside-out formation of the innermost super-Earths. Astrophys. J. Lett. **798**, L32 (2015). doi:10.1088/2041-8205/798/2/L32, 1411.2629

Chen, X., Arce, H.G., Zhang, Q., Bourke, T.L., Launhardt, R., Jørgensen, J.K., Lee, C.F., Foster, J.B., Dunham, M.M., Pineda, J.E., Henning, T.: SMA observations of class 0 protostars: a high angular resolution survey of protostellar binary systems. Astrophys. J. **768**, 110 (2013). doi:10.1088/0004-637X/768/2/110, 1304.0436

Clarke, C.J., Pringle, J.E.: Star-disc interactions and binary star formation. Mon. Not. R. Astron. Soc. **249**, 584–587 (1991). doi:10.1093/mnras/249.4.584

Coughlin, J.L., Mullally, F., Thompson, S.E., Rowe, J.F., Burke, C.J., Latham, D.W., Batalha, N.M., Ofir, A., Quarles, B.L., Henze, C.E., Wolfgang, A., Caldwell, D.A., Bryson, S.T., Shporer, A., Catanzarite, J., Akeson, R., Barclay, T., Borucki, W.J., Boyajian, T.S., Campbell, J.R., Christiansen, J.L., Girouard, F.R., Haas, M.R., Howell, S.B., Huber, D., Jenkins, J.M., Li, J., Patil-Sabale, A., Quintana, E.V., Ramirez, S., Seader, S., Smith, J.C., Tenenbaum, P., Twicken, J.D., Zamudio, K.A.: Planetary candidates observed by Kepler. VII. The first fully uniform catalog based on the entire 48-month data set (Q1-Q17 DR24). Astrophys. J. Suppl. Ser. **224**, 12 (2016). doi:10.3847/0067-0049/224/1/12, 1512.06149

Doyle, L.R., Carter, J.A., Fabrycky, D.C., Slawson, R.W., Howell, S.B., Winn, J.N., Orosz, J.A., Prčsa, A., Welsh, W.F., Quinn, S.N., Latham, D., Torres, G., Buchhave, L.A., Marcy, G.W., Fortney, J.J., Shporer, A., Ford, E.B., Lissauer, J.J., Ragozzine, D., Rucker, M., Batalha, N., Jenkins, J.M., Borucki, W.J., Koch, D., Middour, C.K., Hall, J.R., McCauliff, S., Fanelli, M.N., Quintana, E.V., Holman, M.J., Caldwell, D.A., Still, M., Stefanik, R.P., Brown, W.R., Esquerdo, G.A., Tang, S., Furesz, G., Geary, J.C., Berlind, P., Calkins, M.L., Short, D.R., Steffen, J.H., Sasselov, D., Dunham, E.W., Cochran, W.D., Boss, A., Haas, M.R., Buzasi, D., Fischer, D.: Kepler-16: a transiting circumbinary planet. Science **333**, 1602 (2011). doi:10.1126/science.1210923, 1109.3432

Duchêne, G., Kraus, A.: Stellar multiplicity. Annu. Rev. Astron. Astrophys. **51**, 269–310 (2013). doi:10.1146/annurev-astro-081710-102602, 1303.3028

Dupuy, T.J., Kratter, K.M., Kraus, A.L., Isaacson, H., Mann, A.W., Ireland, M.J., Howard, A.W., Huber, D.: Orbital architectures of planet-hosting binaries. I. Forming five small planets in the truncated disk of Kepler-444A. Astrophys. J. **817**, 80 (2016). doi:10.3847/0004-637X/817/1/80, 1512.03428

Eggenberger, A.: Detection and characterization of planets in binary and multiple systems. In: Goździewski, K., Niedzielski, A., Schneider, J. (eds.). EAS Publ. Ser. **42**, 19–37 (2010). doi:10.1051/eas/1042002, 0910.3332

Eggleton, P., Kiseleva, L.: An empirical condition for stability of hierarchical triple systems. Astrophys. J. **455**, 640 (1995). doi:10.1086/176611

Fabrycky, D., Tremaine, S.: Shrinking binary and planetary orbits by kozai cycles with tidal friction. Astrophys. J. **669**, 1298–1315 (2007). doi:10.1086/521702, 0705.4285

Fisher, R.T.: A turbulent interstellar medium origin of the binary period distribution. Astrophys. J. **600**, 769–780 (2004). doi:10.1086/380111, astro-ph/0303280

Fragner, M.M., Nelson, R.P., Kley, W.: On the dynamics and collisional growth of planetesimals in misaligned binary systems. Astron. Astrophys. **528**, A40 (2011). doi:10.1051/0004-6361/201015378, 1104.1460

Fu, W., Lubow, S.H., Martin, R.G.: The kozai-lidov mechanism in hydrodynamical disks. III. Effects of disk mass and self-gravity. Astrophys. J. **813**, 105 (2015). doi:10.1088/0004-637X/813/2/105, 1509.01280

Fu, W., Lubow, S.H., Martin, R.G.: Fragmentation of Kozai-Lidov disks. Astrophys. J. Lett. **835**, L29 (2017). doi:10.3847/2041-8213/835/2/L29, 1612.07673

Gammie, C.F., Goodman, J., Ogilvie, G.I.: Linear and non-linear theory of a parametric instability of hydrodynamic warps in Keplerian discs. Mon. Not. R. Astron. Soc. **318**, 1005–1016 (2000). doi:10.1046/j.1365-8711.2000.03669.x, astro-ph/0001539

Goodman, A.A., Benson, P.J., Fuller, G.A., Myers, P.C.: Dense cores in dark clouds. VIII - velocity gradients. Astrophys. J. **406**, 528–547 (1993). doi:10.1086/172465

Hamers, A.S., Perets, H.B., Portegies Zwart, S.F.: A triple origin for the lack of tight coplanar circumbinary planets around short-period binaries. Mon. Not. R. Astron. Soc. **455**, 3180–3200 (2016). doi:10.1093/mnras/stv2447, 1506.02039

Harris, R.J., Andrews, S.M., Wilner, D.J., Kraus, A.L.: A resolved census of millimeter emission from Taurus multiple star systems. Astrophys. J. **751**, 115 (2012). doi:10.1088/0004-637X/751/2/115, 1203.6353

Hatzes, A.P., Cochran, W.D., Endl, M., McArthur, B., Paulson, D.B., Walker, G.A.H., Campbell, B., Yang, S.: A planetary companion to γ cephei A. Astrophys. J. **599**, 1383–1394 (2003). doi:10.1086/379281, astro-ph/0305110

Heppenheimer, T.A.: Outline of a theory of planet formation in binary systems. Icarus **22**, 436–447 (1974). doi:10.1016/0019-1035(74)90076-1
Holman, M.J., Wiegert, P.A.: Long-term stability of planets in binary systems. Astron. J. **117**, 621–628 (1999). doi:10.1086/300695, astro-ph/9809315
Innanen, K.A., Zheng, J.Q., Mikkola, S., Valtonen, M.J.: The kozai mechanism and the stability of planetary orbits in binary star systems. Astron. J. **113**, 1915 (1997). doi:10.1086/118405
Kiseleva, L.G., Eggleton, P.P., Mikkola, S.: Tidal friction in triple stars. Mon. Not. R. Astron. Soc. **300**, 292–302 (1998). doi:10.1046/j.1365-8711.1998.01903.x
Kley, W., Haghighipour, N.: Modeling circumbinary planets: the case of Kepler-38. Astron. Astrophys. **564**, A72 (2014). doi:10.1051/0004-6361/201323235, 1401.7648
Kley, W., Nelson, R.P.: Planet-disk interaction and orbital evolution. Annu. Rev. Astron. Astrophys. **50**, 211–249 (2012). doi:10.1146/annurev-astro-081811-125523, 1203.1184
Kocsis, B., Haiman, Z., Loeb, A.: Gas pile-up, gap overflow and Type 1.5 migration in circumbinary discs: application to supermassive black hole binaries. Mon. Not. R. Astron. Soc. **427**, 2680–2700 (2012). doi:10.1111/j.1365-2966.2012.22118.x, 1205.5268
Kostov, V.B., Orosz, J.A., Welsh, W.F., Doyle, L.R., Fabrycky, D.C., Haghighipour, N., Quarles, B., Short, D.R., Cochran, W.D., Endl, M., Ford, E.B., Gregorio, J., Hinse, T.C., Isaacson, H., Jenkins, J.M., Jensen, E.L.N., Kane, S., Kull, I., Latham, D.W., Lissauer, J.J., Marcy, G.W., Mazeh, T., Müller, T.W.A., Pepper, J., Quinn, S.N., Ragozzine, D., Shporer, A., Steffen, J.H., Torres, G., Windmiller, G., Borucki, W.J.: Kepler-1647b: the largest and longest-period kepler transiting circumbinary planet. Astrophys. J. **827**, 86 (2016). doi:10.3847/0004-637X/827/1/86, 1512.00189
Kozai, Y.: Secular perturbations of asteroids with high inclination and eccentricity. Astron. J. **67**, 591 (1962). doi:10.1086/108790
Kratter, K.M.: The formation of close binaries. In: Schmidtobreick, L., Schreiber, M.R., Tappert, C. (eds.) Evolution of Compact Binaries, Astronomical Society of the Pacific Conference Series, vol. 447, p. 47 (2011). 1109.3740
Kratter, K., Lodato, G.: Gravitational instabilities in circumstellar disks. Annu. Rev. Astron. Astrophys. **54**, 271–311 (2016). doi:10.1146/annurev-astro-081915-023307, 1603.01280
Kratter, K.M., Matzner, C.D.: Fragmentation of massive protostellar discs. Mon. Not. R. Astron. Soc., **373**, 1563–1576 (2006). doi:10.1111/j.1365-2966.2006.11103.x, astro-ph/0609692
Kratter, K.M., Shannon, A.: Planet packing in circumbinary systems. Mon. Not. R. Astron. Soc. **437**, 3727–3735 (2014). doi:10.1093/mnras/stt2179, 1311.2942
Kratter, K.M., Matzner, C.D., Krumholz, M.R.: Global models for the evolution of embedded, accreting protostellar disks. Astrophys. J. **681**, 375–390 (2008). doi:10.1086/587543, 0709.4252
Kratter, K.M., Matzner, C.D., Krumholz, M.R., Klein, R.I.: On the role of disks in the formation of stellar systems: a numerical parameter study of rapid accretion. Astrophys. J. **708**, 1585–1597 (2010). doi:10.1088/0004-637X/708/2/1585, 0907.3476
Kraus, A.L., Ireland, M.J., Huber, D., Mann, A.W., Dupuy, T.J.: The impact of stellar multiplicity on planetary systems. I. The ruinous influence of close binary companions. Astron. J. **152**, 8 (2016). doi:10.3847/0004-6256/152/1/8, 1604.05744
Krumholz, M.R., Thompson, T.A.: Mass transfer in close, rapidly accreting protobinaries: an origin for massive twins? Astrophys. J. **661**, 1034–1041 (2007). doi:10.1086/515566, astro-ph/0611822
Krumholz, M.R., Klein, R.I., McKee, C.F.: Radiation-Hydrodynamic Simulations of Collapse and Fragmentation in Massive Protostellar Cores. Astrophys. J. **656**, 959–979 (2007). doi:10.1086/510664, astro-ph/0609798
Larwood, J.D., Nelson, R.P., Papaloizou, J.C.B., Terquem, C.: The tidally induced warping, precession and truncation of accretion discs in binary systems: three-dimensional simulations. Mon. Not. R. Astron. Soc. **282**, 597–613 (1996). doi:10.1093/mnras/282.2.597, astro-ph/9604013
Li, Z.-Y., Banerjee, R., Pudritz, R.E., Jørgensen, J.K., Shang, H., Krasnopolsky, R., Maury, A.: The earliest stages of star and planet formation: core collapse, and the formation of disks and

outflows. In: Protostars and Planets VI, pp. 173–194. University of Arizona Press, Tucson, AZ (2014). doi:10.2458/azu_uapress_9780816531240-ch008, 1401.2219
Li, G., Holman, M.J., Tao, M.: Uncovering circumbinary planetary architectural properties from selection biases. Astrophys. J. **831**, 96 (2016). doi:10.3847/0004-637X/831/1/96, 1608.01768
Lidov, M.L.: The evolution of orbits of artificial satellites of planets under the action of gravitational perturbations of external bodies. Planet. Space Sci. **9**, 719–759 (1962). doi:10.1016/0032-0633(62)90129-0
Lodato, G., Nayakshin, S., King, A.R., Pringle, J.E.: Black hole mergers: can gas discs solve the 'final parsec' problem? Mon. Not. R. Astron. Soc. **398**, 1392–1402 (2009). doi:10.1111/j.1365-2966.2009.15179.x, 0906.0737
Lubow, S., Artymowicz, P.: Young binary star/disk interactions. In: AAS/Division of Dynamical Astronomy Meeting #27, Bulletin of the American Astronomical Society, vol. 28, p. 1182 (1996)
Lubow, S.H., Martin, R.G.: The evolution of planet-disk systems that are mildly inclined to the orbit of a binary companion. Astrophys. J. **817**, 30 (2016). doi:10.3847/0004-637X/817/1/30, 1512.02141
Lubow, S.H., Ogilvie, G.I.: On the tilting of protostellar disks by resonant tidal effects. Astrophys. J. **538**, 326–340 (2000). doi:10.1086/309101, astro-ph/0003028
Lubow, S.H., Martin, R.G., Nixon, C.: Tidal torques on misaligned disks in binary systems. Astrophys. J. **800**, 96 (2015). doi:10.1088/0004-637X/800/2/96, 1412.7741
Marois, C., Zuckerman, B., Konopacky, Q.M., Macintosh, B., Barman, T.: Images of a fourth planet orbiting HR 8799. Nature **468**, 1080–1083 (2010). doi:10.1038/nature09684, 1011.4918
Martin, R.G., Armitage, P.J., Alexander, R.D.: Formation of circumbinary planets in a dead zone. Astrophys. J. **773**, 74 (2013). doi:10.1088/0004-637X/773/1/74, 1306.5241
Martin, R.G., Nixon, C., Lubow, S.H., Armitage, P.J., Price, D.J., Doğan, S., King, A.: The kozai-lidov mechanism in hydrodynamical disks. Astrophys. J. Lett. **792**, L33 (2014). doi:10.1088/2041-8205/792/2/L33, 1409.1226
Martin, D.V., Mazeh, T., Fabrycky, D.C.: No circumbinary planets transiting the tightest Kepler binaries - a possible fingerprint of a third star. Mon. Not. R. Astron. Soc. **453**, 3554–3567 (2015). doi:10.1093/mnras/stv1870, 1505.05749
Marzari, F., Nelson, A.F.: Interaction of a giant planet in an inclined orbit with a circumstellar disk. Astrophys. J. **705**, 1575–1583 (2009). doi:10.1088/0004-637X/705/2/1575, 0909.4375
Marzari, F., Scholl, H.: Planetesimal accretion in binary star systems. Astrophys. J. **543**, 328–339 (2000). doi:10.1086/317091
Marzari, F., Thébault, P., Scholl, H.: Planet formation in highly inclined binaries. Astron. Astrophys. **507**, 505–511 (2009). doi:10.1051/0004-6361/200912379, 0908.0803
Mazeh, T., Shaham, J.: The orbital evolution of close triple systems - the binary eccentricity. Astron. Astrophys. **77**, 145–151 (1979)
Meru, F.: Triggered fragmentation in self-gravitating discs: forming fragments at small radii. Mon. Not. R. Astron. Soc. **454**, 2529–2538 (2015). doi:10.1093/mnras/stv2128, 1509.03635
Moe, M., Di Stefano, R.: Mind your Ps and Qs: the interrelation between Period (P) and Mass-ratio (Q) distributions of binary stars. Astrophys. J. Suppl. Ser. **230**, 15 (2017). doi:10.3847/1538-4365/aa6fb6, 1606.05347
Moe, M., Kratter, K.M.: Dynamical formation of close binaries during the pre-main-sequence phase. Astrophys. J. 21 pp. http://adsabs.harvard.edu/abs/2017arXiv170609894M
Mohanty, S., Ercolano, B., Turner, N.J.: Dead, undead, and zombie zones in protostellar disks as a function of stellar mass. Astrophys. J. **764**, 65 (2013). doi:10.1088/0004-637X/764/1/65, 1212.3798
Moriwaki, K., Nakagawa, Y.: A planetesimal accretion zone in a circumbinary disk. Astrophys. J. **609**, 1065–1070 (2004). doi:10.1086/421342
Morrison, S.J., Kratter, K.M.: Orbital stability of multi-planet systems: behavior at high masses. Astrophys. J. **823**, 118 (2016). doi:10.3847/0004-637X/823/2/118, 1604.01037
Muñoz, D.J., Lai, D.: Survival of planets around shrinking stellar binaries. Proc. Natl. Acad. Sci. **112**, 9264–9269 (2015). doi:10.1073/pnas.1505671112, 1505.05514

Muñoz, D.J., Kratter, K., Vogelsberger, M., Hernquist, L., Springel, V.: Stellar orbit evolution in close circumstellar disc encounters. Mon. Not. R. Astron. Soc. **446**, 2010–2029 (2015). doi:10.1093/mnras/stu2220, 1410.4561

Mudryk, L.R., Wu, Y.: Resonance overlap is responsible for ejecting planets in binary systems. Astrophys. J. **639**, 423–431 (2006). doi:10.1086/499347, astro-ph/0511710

Najita, J.R., Kenyon, S.J.: The mass budget of planet-forming discs: isolating the epoch of planetesimal formation. Mon. Not. R. Astron. Soc. **445**, 3315–3329 (2014). doi:10.1093/mnras/stu1994, 1409.7021

Naoz, S.: The eccentric kozai-lidov effect and its applications. Annu. Rev. Astron. Astrophys. **54**, 441–489 (2016). doi:10.1146/annurev-astro-081915-023315, 1601.07175

Nielsen, EL., De Rosa, RJ., Rameau, J., Wang, JJ., Esposito, T.M., Millar-Blanchaer, M.A., Marois, C., Vigan, A., Ammons, S.M., et al.: Evidence that the directly-imaged planet HD 131399 Ab is a background star. Astrophys. J. (2017). arXiv:1705.06851. http://adsabs.harvard.edu/abs/2017arXiv170506851N

Nixon, C.J., Cossins, P.J., King, A.R., Pringle, J.E.: Retrograde accretion and merging supermassive black holes. Mon. Not. R. Astron. Soc. **412**, 1591–1598 (2011). doi:10.1111/j.1365-2966.2010.17952.x, 1011.1914

Offner, S.S.R., Kratter, K.M., Matzner, C.D., Krumholz, M.R., Klein, R.I.: The formation of low-mass binary star systems via turbulent fragmentation. Astrophys. J. **725**, 1485–1494 (2010). doi:10.1088/0004-637X/725/2/1485, 1010.3702

Offner, S.S.R., Dunham, M.M., Lee, K.I., Arce, H.G., Fielding, D.B.: The turbulent origin of outflow and spin misalignment in multiple star systems. Astrophys. J. Lett. **827**, L11 (2016). doi:10.3847/2041-8205/827/1/L11, 1606.08445

Ogilvie, G.I., Barker, A.J.: Local and global dynamics of eccentric astrophysical discs. Mon. Not. R. Astron. Soc. **445**, 2621–2636 (2014). doi:10.1093/mnras/stu1795, 1409.6487

Perets, H.B., Kratter, K.M.: The triple evolution dynamical instability: stellar collisions in the field and the formation of exotic binaries. Astrophys. J. **760**, 99 (2012). doi:10.1088/0004-637X/760/2/99, 1203.2914

Pineda, J.E., Offner, S.S.R., Parker, R.J., Arce, H.G., Goodman, A.A., Caselli, P., Fuller, G.A., Bourke, T.L., Corder, S.A.: The formation of a quadruple star system with wide separation. Nature **518**, 213–215 (2015). doi:10.1038/nature14166

Rabl, G., Dvorak, R.: Satellite-type planetary orbits in double stars - a numerical approach. Astron. Astrophys. **191**, 385–391 (1988)

Rafikov, R.R.: Building tatooine: suppression of the direct secular excitation in kepler circumbinary planet formation. Astrophys. J. Lett. **764**, L16 (2013a). doi:10.1088/2041-8205/764/1/L16, 1212.2217

Rafikov, R.R.: Structure and evolution of circumbinary disks around supermassive black hole binaries. Astrophys. J. **774**, 144 (2013b). doi:10.1088/0004-637X/774/2/144, 1205.5017

Rafikov, R.R., Silsbee, K.: Planet formation in stellar binaries. I. Planetesimal dynamics in massive protoplanetary disks. Astrophys. J. **798**, 69 (2015a). doi:10.1088/0004-637X/798/2/69, 1405.7054

Rafikov, R.R., Silsbee, K.: Planet formation in stellar binaries. II. Overcoming the fragmentation barrier in α centauri and γ cephei-like systems. Astrophys. J. **798**, 70 (2015b). doi:10.1088/0004-637X/798/2/70, 1408.4819

Raghavan, D., McAlister, H.A., Henry, T.J., Latham, D.W., Marcy, G.W., Mason, B.D., Gies, D.R., White, R.J., ten Brummelaar, T.A.: A survey of stellar families: multiplicity of solar-type stars. Astrophys. J. Suppl. Ser. **190**, 1–42 (2010). doi:10.1088/0067-0049/190/1/1, 1007.0414

Reipurth, B., Clarke, C.: The formation of brown dwarfs as ejected stellar embryos. Astron. J. **122**, 432–439 (2001). doi:10.1086/321121, astro-ph/0103019

Rodriguez, D.R., Duchêne, G., Tom, H., Kennedy, G.M., Matthews, B., Greaves, J., Butner, H.: Stellar multiplicity and debris discs: an unbiased sample. Mon. Not. R. Astron. Soc. **449**, 3160–3170 (2015). doi:10.1093/mnras/stv483, 1503.01320

Silsbee, K., Rafikov, R.R.: Birth locations of the Kepler circumbinary planets. Astrophys. J. **808**, 58 (2015a). doi:10.1088/0004-637X/808/1/58, 1504.00460

Silsbee, K., Rafikov, R.R.: Planet formation in binaries: dynamics of planetesimals perturbed by the eccentric protoplanetary disk and the secondary. Astrophys. J. **798**, 71 (2015b). doi:10.1088/0004-637X/798/2/71, 1309.3290

Simon, J.B., Armitage, P.J., Li, R., Youdin, A.N.: The mass and size distribution of planetesimals formed by the streaming instability. I. The role of self-gravity. Astrophys. J. **822**, 55 (2016). doi:10.3847/0004-637X/822/1/55, 1512.00009

Smith, A.W., Lissauer, J.J.: Orbital stability of systems of closely-spaced planets. Icarus **201**, 381–394 (2009). doi:10.1016/j.icarus.2008.12.027

Smullen, R.A., Kratter, K.M., Shannon, A.: Planet scattering around binaries: ejections, not collisions. Mon. Not. R. Astron. Soc. **461**, 1288–1301 (2016). doi:10.1093/mnras/stw1347, 1604.03121

Sutherland, A.P., Fabrycky, D.C.: On the fate of unstable circumbinary planets: Tatooine's close encounters with a death star. Astrophys. J. **818**, 6 (2016). doi:10.3847/0004-637X/818/1/6, 1511.03274

Syer, D., Clarke, C.J.: Satellites in discs: regulating the accretion luminosity. Mon. Not. R. Astron. Soc. **277**, 758–766 (1995). doi:10.1093/mnras/277.3.758, astro-ph/9505021

Throop, H.B., Bally, J.: Tail-end bondi-hoyle accretion in young star clusters: implications for disks, planets, and stars. Astron. J. **135**, 2380–2397 (2008). doi:10.1088/0004-6256/135/6/2380, 0804.0438

Tobin, J.J., Looney, L.W., Wilner, D.J., Kwon, W., Chandler, C.J., Bourke, T.L., Loinard, L., Chiang, H.F., Schnee, S., Chen, X.: A sub-arcsecond survey toward class 0 protostars in perseus: searching for signatures of protostellar disks. Astrophys. J. **805**, 125 (2015). doi:10.1088/0004-637X/805/2/125, 1503.05189

Tobin, J.J., Kratter, K.M., Persson, M.V., Looney, L.W., Dunham, M.M., Segura-Cox, D., Li, Z.Y., Chandler, C.J., Sadavoy, S.I., Harris, R.J., Melis, C., Pérez, L.M.: A triple protostar system formed via fragmentation of a gravitationally unstable disk. Nature **538**, 483–486 (2016a). doi:10.1038/nature20094, 1610.08524

Tobin, J.J., Looney, L.W., Li, Z.Y., Chandler, C.J., Dunham, M.M., Segura-Cox, D., Sadavoy, S.I., Melis, C., Harris, R.J., Kratter, K., Perez, L.: The VLA nascent disk and multiplicity survey of perseus protostars (VANDAM). II. Multiplicity of protostars in the perseus molecular cloud. Astrophys. J. **818**, 73 (2016b). doi:10.3847/0004-637X/818/1/73, 1601.00692

Tohline, J.E.: The origin of binary stars. Annu. Rev. Astron. Astrophys. **40**, 349–385 (2002). doi:10.1146/annurev.astro.40.060401.093810

Tokovinin, A.: From binaries to multiples. II. Hierarchical multiplicity of F and G dwarfs. Astron. J. **147**, 87 (2014). doi:10.1088/0004-6256/147/4/87, 1401.6827

Trilling, D.E., Stansberry, J.A., Stapelfeldt, K.R., Rieke, G.H., Su, K.Y.L., Gray, R.O., Corbally, C.J., Bryden, G., Chen, C.H., Boden, A., Beichman, C.A.: Debris disks in main-sequence binary systems. Astrophys. J. **658**, 1289–1311 (2007). doi:10.1086/511668, astro-ph/0612029

Vartanyan, D., Garmilla, J.A., Rafikov, R.R.: Tatooine nurseries: structure and evolution of circumbinary protoplanetary disks. Astrophys. J. **816**, 94 (2016). doi:10.3847/0004-637X/816/2/94, 1509.07524

Veras, D., Mustill, A.J., Gänsicke, B.T.: The unstable fate of the planet orbiting the A star in the HD 131399 triple stellar system. Mon. Not. R. Astron. Soc. **465**, 1499–1504 (2017). doi:10.1093/mnras/stw2821, 1611.00007

Wagner, K., Apai, D., Kasper, M., Kratter, K., McClure, M., Robberto, M., Beuzit, J.L.: Direct imaging discovery of a Jovian exoplanet within a triple-star system. Science **353**, 673–678 (2016). doi:10.1126/science.aaf9671, 1607.02525

Wang, J., Xie, J.-W., Barclay, T., Fischer, D.A.: Influence of stellar multiplicity on planet formation. I. Evidence of suppressed planet formation due to stellar companions within 20 AU and validation of four planets from the Kepler multiple planet candidates. Astrophys. J. **783**, 4 2014. doi:10.1088/0004-637X/783/1/4, 1309.7097

Wang, J., Fischer, D.A., Horch, E.P., Xie, J.W.: Influence of stellar multiplicity on planet formation. III. Adaptive optics imaging of Kepler stars with gas giant planets. Astrophys. J. **806**, 248 (2015). doi:10.1088/0004-637X/806/2/248, 1505.05363

Welsh, W.F., Orosz, J.A., Carter, J.A., Fabrycky, D.C.: Recent Kepler results on circumbinary planets. In: Haghighipour, N. (ed.) Formation, Detection, and Characterization of Extrasolar Habitable Planets, IAU Symposium, vol. 293, pp. 125–132 (2014). doi:10.1017/S1743921313012684, 1308.6328

Williams, J.P., Cieza, L.A.: Protoplanetary disks and their evolution. Annu. Rev. Astron. Astrophys. **49**, 67–117 (2011). doi:10.1146/annurev-astro-081710-102548, 1103.0556

Williams, J.P., Mann, R.K., Di Francesco, J., Andrews, S.M., Hughes, A.M., Ricci, L., Bally, J., Johnstone, D., Matthews, B.: ALMA observations of a misaligned binary protoplanetary disk system in orion. Astrophys. J. **796**, 120 (2014). doi:10.1088/0004-637X/796/2/120, 1410.3570

Youdin, A.N., Chiang, E.I.: Particle pileups and planetesimal formation. Astrophys. J. **601**, 1109–1119 (2004). doi:10.1086/379368, astro-ph/0309247

Youdin, A.N., Goodman, J.: Streaming instabilities in protoplanetary disks. Astrophys. J. **620**, 459–469 (2005). doi:10.1086/426895, astro-ph/0409263

Youdin, A.N., Kenyon, S.J.: From Disks to Planets, p. 1. Springer, Dordrecht(2013). doi:10.1007/978-94-007-5606-9_1

Chapter 12
Planet Population Synthesis via Pebble Accretion

Bertram Bitsch and Anders Johansen

Abstract The aim of planet population synthesis is to incorporate all physical processes from planet formation theories to arrive at a synthetic population of planets that can be compared to observations. In this way, shortcomings in the theories can be shown. In particular, planet population synthesis has to incorporate theories and models regarding (1) core accretion, (2) gas accretion, (3) planet migration and (4) disk evolution. A general problem of the core accretion scenario is that the building time of a planetary core of a few Earth masses with just the accretion of planetesimals takes longer than the lifetime of the protoplanetary disk. This time-scale problem can be overcome when taking the accretion of pebbles into account. We will review here the differences of planetesimal and pebble accretion on the formation of planets that migrate through evolving protoplanetary disks.

12.1 Introduction

Since the discovery of the first exoplanet a bit more than 20 years ago (Mayor and Queloz 1995), the number of exoplanetary discoveries has risen dramatically. Today, more than 2000 exoplanets are discovered, by various detection methods. These findings included two classes of planets which are not harboured in our own solar system: hot Jupiters and super-Earths (Fig. 12.1). Hot Jupiters are Jupiter sized planets that orbit their host star in just up to a few days, while super-Earths are planets of above two Earth masses, presumably rocky, hence the name. Theories about planet formation have to explain the abundances and frequencies of these discovered planets. Of particular interest is here the population of cold gas giants ($r > 1$ AU) and of super-Earth systems, which are very abundant (Fressin et al. 2013). One way to test the theories is population synthesis, where the available

B. Bitsch (✉) • A. Johansen
Lund Observatory, Department of Astronomy and Theoretical Physics, Lund University, 22100 Lund, Sweden
e-mail: bert@astro.lu.se; anders@astro.lu.se

M. Pessah, O. Gressel (eds.), *Formation, Evolution, and Dynamics of Young Solar Systems*, Astrophysics and Space Science Library 445,
DOI 10.1007/978-3-319-60609-5_12

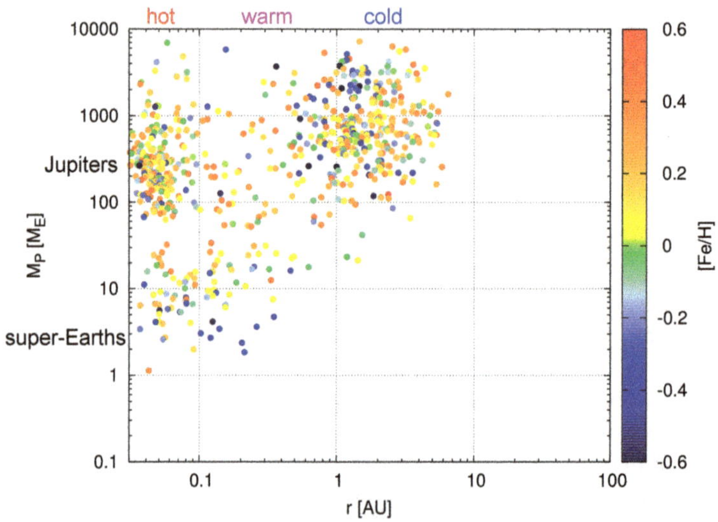

Fig. 12.1 Mass-orbital distance diagram of detected exoplanets either through transits or RV. The *colour coding* gives the metallicity of the planet's host star. Data taken from exoplanets.org

theories of planetary growth and migration as well as disk evolution are combined to synthesise a population of planets. This synthesised population should match the observations in order to confirm the theories of planet formation. If any aspects of the synthesised population does not match observations, theories have to be revised and improved.

The earliest population synthesis simulations pointed out the problem of planetary migration (Alibert et al. 2004; Ida and Lin 2004), namely that type-I migration of low mass planets is way too fast compared to the building time of planetary cores through the accretion of planetesimals. As the planetary core grows in the protoplanetary disk, it interacts with it and moves through it (Ward 1997), where the migration time-scale is much shorter than the lifetime of the protoplanetary disk (Tanaka et al. 2002). This means that planetary cores migrate into the central star before reaching runaway gas accretion, which is when they can potentially open up a gap in the disk that allows the transition into the slower type-II migration regime, potentially saving them from migrating inwards all the way to the star (Baruteau et al. 2014). In order to generate planetary cores that can grow to giant planets at a few AU via gas accretion, planetary type-I migration had to be slowed down by a factor of 100 (Ida and Lin 2004).

Another main problem in the core accretion scenario was the formation of the core itself. In order to form Jupiter's core at 5 AU, the surface density of planetesimals in the protoplanetary disk had to be increased by a factor of $\sim$6 in order to build big enough cores[1] (Pollack et al. 1996). However, the growth time-scale increases with orbital distance, so forming the cores of Saturn or of the ice giants requires an even larger increase of the solid component of the protoplanetary disk, which is unrealistic.

Here we will show that pebble accretion can help to solve both problems at the same time. Pebble accretion does not only accelerate the growth time-scales of protoplanetary cores, but also allows an efficient formation far away from the central star (Ormel and Klahr 2010; Lambrechts and Johansen 2012). Combining the shorter growth time-scales of pebble accretion-with unreduced planet migration rates, Bitsch et al. (2015b) showed that the formation of giant planets at large orbits is possible in evolving protoplanetary disks. We build on this model and show here the direct comparison with observational data.

This chapter is structured as follows. We first shortly summarise the disk evolution (Sect. 12.2), the planetary growth mechanism via pebble and planetesimal accretion (Sect. 12.3) as well as planet migration (Sect. 12.4). We then combine these things into our population synthesis approach involving pebble accretion and compare these results to core accretion solely by planetesimals (Sect. 12.5). Finally we give a summary in Sect. 12.6.

12.2 Disc Evolution

Inside the disk, dust grains collide and form pebbles (Zsom et al. 2010; Birnstiel et al. 2012; Ros and Johansen 2013) that migrate through the disk due to gas drag (Weidenschilling 1977; Brauer et al. 2008). Clouds of these pebbles can collapse under their own gravity and form planetesimals, aided by the streaming instability (Youdin and Goodman 2005; Johansen and Youdin 2007). These planetesimals can then accrete leftover pebbles and finally form the cores of giant planets (Lambrechts and Johansen 2012), which migrate through the disk (Ward 1997; Paardekooper and Mellema 2006; Kley and Crida 2008; Kley et al. 2009). Even though all these processes happen on different time and length scales, they all strongly depend on the underlying disk structure (temperature T, gas surface density Σ_g, aspect ratio H/r, viscosity ν), making the protoplanetary disk structure a key parameter for understanding planet formation. Every planet formation model is dependent on the underlying disk structure and the results apply only for the assumed disk structure.

One of the most used disk models is the Minimum Mass Solar Nebula (MMSN), which is essentially a power law in surface density and temperature (Weiden-

[1]The main problem here is the planetesimal isolation mass, at which the planet stops growing. It increases with the available amount of planetesimals, Eq. (12.8).

schilling 1977; Hayashi 1981), where the surface density profile is given by

$$\Sigma_{\rm g}(r) = \beta_\Sigma \left(\frac{r}{\rm AU}\right)^{-3/2} . \tag{12.1}$$

The aspect ratio of the disk follows

$$\frac{H}{r} = 0.033 \left(\frac{r}{\rm AU}\right)^{1/4} , \tag{12.2}$$

which indicates that there are no planet traps due to the entropy driven corotation torque, because those traps exist only when H/r decreases with orbital distance (Bitsch et al. 2014). The time evolution of the surface density can be parameterised through

$$\beta_\Sigma = \Sigma_0 \exp\left(\frac{-t}{\tau_{\rm disk}}\right) , \tag{12.3}$$

where $\tau_{\rm disk}$ is the disk's lifetime and $\Sigma_0 = 1700\,{\rm g/cm^3}$.

However, the temperature structure of a protoplanetary disk is much more complex than a simple power law. The temperature structure is given by a balance between viscous and stellar heating with radiative cooling. Viscous heating dominates in the dense inner parts of the protoplanetary disk, while stellar heating dominates the structure of the outer disk. As the disk evolves in time, it loses mass and thus viscous heating becomes less important in the later stages of the disk evolution.

The opacity of the disk $\kappa_{\rm R}$ is of great importance, as it sets the cooling rates of the disk, $D \propto 1/\kappa_{\rm R}$. The opacity is determined mainly by small dust grains, where the exact value of the opacity depends on the grain size and composition as well as on the dust-to-gas ratio (or metallicity $Z_{\rm dust}$). At the water ice line, water ice grains evaporate and can thus not contribute to the opacity any more, resulting in a jump in opacity. This then dramatically changes the disk structure and with it the radial gradients of surface density and temperature in the disk (Bitsch et al. 2014, 2015a; Baillié et al. 2015). These changes of the disk structure have important consequences for the formation of planetesimals and planets as well as on their migration properties, which also scales with the dust-to-gas ratio (Paardekooper et al. 2011; Bitsch et al. 2015b).

In this work, we will also relate on the disk structure model presented by Bitsch et al. (2015a), where also the change of the luminosity of the central star as it evolves is included. This leads to a decrease in stellar heating, resulting a colder disk in the outer regions of the protoplanetary disk as time evolves. We will show the differences in planet population synthesis studies between this disk model and the MMSN. For both disk models we use an α viscosity of 0.0054.

12.3 Planet Growth Mechanisms

After the formation of the planetary core, the planetary envelope can slowly contract until $M_{\rm env} \sim M_{\rm core}$ when runaway gas accretion sets in and the planet becomes a gas giant (Mizuno 1980). Contraction of the envelope and runaway gas accretion requires a solid core of at least a few Earth masses. However, the formation of a planetary core just by planetesimals or planetary embryos can take longer than the lifetime of the protoplanetary disk, if the metallicity is solar (Levison et al. 2010). In recent years new studies of pebble accretion showed that this time-scale problem can be evaded (Ormel and Klahr 2010; Lambrechts and Johansen 2012) and that the formation of planetary cores well within the lifetime of the protoplanetary disk is possible.

The reason why pebble accretion is much more efficient than planetesimal accretion is caused by the fact that pebbles feel gas drag in contrast to planetesimals. Inside the disk a radial gradient in pressure mimics a force that decreases gravity and causes the gas to orbit at a sub-Keplerian speed. The pebbles, on the other hand, do not feel the pressure gradient and want to orbit on a Keplerian speed, where the velocity difference between gas and pebbles acts as a headwind, causing inward drift of the pebbles, known as radial drift.

When planetesimals enter the Hill sphere of the growing planet, they are only accreted to a tiny fraction, while the rest of the planetesimals are scattered away. Pebbles, on the other hand, feel gas drag and lose angular momentum as they drift through the disk. When the pebbles then enter into the Hill sphere and their friction time is shorter than the time to cross the Hill radius, they can be accreted onto the planet. The planet can therefore accrete pebbles from within its entire Hill sphere, enhancing the planetary growth rate by a factor of 100–1000 (Lambrechts and Johansen 2012). Additionally this growth mechanism allows the formation of planetary cores very efficiently at large orbital distances, explaining the growth of the ice giants in our solar system (Lambrechts and Johansen 2014; Lambrechts et al. 2014).

The starting mass of the planetary seeds in our synthesis simulations is the pebble transition mass $M_{\rm t}$, where the bodies start to accrete in the efficient Hill accretion regime [Eq. (12.6)], which is given by

$$M_{\rm t} = \sqrt{\frac{1}{3}\frac{(\eta v_{\rm K})^3}{G\Omega_{\rm K}}}\,, \tag{12.4}$$

where G is the gravitational constant, $v_{\rm K} = \Omega_{\rm K} r$, and

$$\eta = -\frac{1}{2}\left(\frac{H}{r}\right)^2 \frac{\partial \ln P}{\partial \ln r}\,, \tag{12.5}$$

which determines the subkeplerianity of the gas. Here, H/r is the disk's aspect ratio and $\partial \ln P/\partial \ln r$ is the radial pressure gradient in the disk. We use this mass

[Eq. (12.4)] as starting mass also for simulations where the planetary core grows due to the accretion of planetesimals.

We would like to stress here that the pebble transition mass is larger than the typical mass of planetesimals inferred from the streaming instability, which is roughly 100 km (Johansen et al. 2015). These planetesimals can then grow to reach the pebble transition mass via the accretion of pebbles in the inefficient Bondi accretion regime (Lambrechts and Johansen 2012) or by accreting other leftover planetesimals. This growth phase can take up to 1–1.5 Myr, depending on the local disk properties and how many planetesimals formed inside the disk that can accelerate the initial growth process.

Planets at pebble transition mass accrete from the entire Hill sphere at a rate given by (Lambrechts et al. 2014)

$$\dot{M}_{\mathrm{c,\,2D}} = 2\left(\frac{\tau_{\mathrm{f}}}{0.1}\right)^{2/3} r_{\mathrm{H}} v_{\mathrm{H}} \Sigma_{\mathrm{peb}} \,, \tag{12.6}$$

where τ_{f} is the Stokesnumber of the pebbles, r_{H} is the Hill radius, v_{H} is the Hill speed $r_{\mathrm{H}}\Omega_K$, and Σ_{peb} the surface density of pebbles. However, if the planets are small, their Hill radius can be smaller than the pebble scale height of the disk ($H_{\mathrm{peb}} = H_{\mathrm{g}}\sqrt{\alpha/\tau_{\mathrm{f}}}$), which only allows accretion in the slower 3D regime (Morbidelli et al. 2015). Pebble accretion self-terminates when the planet reaches the so-called pebble isolation mass (Lambrechts et al. 2014)

$$M_{\mathrm{iso}} \approx 20\left(\frac{H/r}{0.05}\right)^{3} \mathrm{M}_{\mathrm{Earth}} \,. \tag{12.7}$$

At M_{iso} the planet is massive enough to carve a partial gap in the protoplanetary disk, which is caused by an exchange of angular momentum between the planet and the disk. The disk's material outside of the planetary orbit is accelerated and orbits at a super-Keplerian speed. This means that pebbles entering this region of the disk do not feel a headwind any more, but are accelerated and thus stop drifting inwards. The planet therefore shields itself from pebble accretion. More details on this can be found in Lambrechts et al. (2014) and Bitsch et al. (2015b).

Alternatively, planetary cores can grow via the accretion of planetesimals, which can form in the protoplanetary disk either by the streaming instability (Youdin and Goodman 2005; Johansen and Youdin 2007) or in vortices (Raettig et al. 2015).

In classical models, planets grow through the accretion of planetesimals, where the isolation mass is different compared to pebble accretion. It is given by (Kokubo and Ida 2002; Raymond et al. 2014)

$$M_{\mathrm{iso,pla}} = 0.16\left(\frac{b}{10R_{\mathrm{H}}}\right)^{3/2}\left(\frac{\Sigma_{\mathrm{pla}}}{10}\right)^{3/2}\left(\frac{r}{1\mathrm{AU}}\right)^{1.5(2-s_{\mathrm{pla}})}\left(\frac{M_\star}{M_\odot}\right)^{-0.5} \mathrm{M}_{\mathrm{E}} \,, \tag{12.8}$$

where b is the orbital separation of the growing embryos, which we set to $10R_{\mathrm{H}}$. Here, Σ_{pla} is the surface density in planetesimals, and s_{pla} the negative gradient of the surface density of planetesimals, which we set equal to the gradient in gas surface density.

The accretion rate of planetesimals onto a planetary embryo is set by two different regimes given by planetesimal dynamics: (1) dispersion-dominated growth (Safronov 1969) and (2) shear-dominated growth (Greenberg et al. 1991). In the dispersion dominated regime the scale height of the planetesimal swarm and the approach speed onto the planetary embryo are determined by random velocities v_{ran} that are much larger than the hill speed, $v_{\mathrm{H}} = \Omega_{\mathrm{K}} R_{\mathrm{H}}$. In the shear dominated case, the approach speed v_{ap} is set directly by the Hill speed. These two limits can be combined in the limit of strong gravitational focusing (where $v_{\mathrm{ran}} = v_{\mathrm{H}}$ and $v_{\mathrm{ap}} = v_{\mathrm{H}}$) in the following way

$$\dot{M}_{\mathrm{c,plan}} = \pi R^2 \Sigma_{\mathrm{pla}} \Omega_{\mathrm{K}} \frac{6p^{-1}}{\zeta^2} , \tag{12.9}$$

where $\zeta = v_{\mathrm{ran}}/v_{\mathrm{H}}$ determines the accretion regime and $p = R/R_{\mathrm{H}}$ is the size of the core relative to its Hill radius. We can rewrite this as

$$p = \frac{R}{R_{\mathrm{H}}} = \left(\frac{4\pi G \rho_{\bullet}}{9\Omega_{\mathrm{K}}^2} \right) , \tag{12.10}$$

where $\rho_{\bullet}$ is the density of the planetary core. As a conservative case we set $\zeta = 2$. We assume that all planetesimals form at the time when the disk forms, so that the initial planetesimal distribution in the disk does not change in time as the disk evolves and is therefore solely determined by the initial surface density of planetesimals and thus by the metallicity, where we use $\Sigma_{\mathrm{pla}} = Z\Sigma_{\mathrm{g}}$.

As the planet stops accreting solids (either planetesimals or pebbles), the planetary envelope can start to cool and contract on a time-scale of several 100 kyrs, as described in Piso and Youdin (2014) and outlined in Bitsch et al. (2015b). As soon as $M_{\mathrm{env}} \sim M_{\mathrm{core}}$, the envelope contracts rapidly and the planet undergoes runaway gas accretion to form gas giants, where our gas accretion rate is following the simulations of Machida et al. (2010), where the gas accretion rate is directly proportional to the disk's local surface density at the planet's location.

12.4 Planet Migration

Growing protoplanets interact with the surrounding gas and migrate through it. The migration rates are different for low mass planets that are still fully embedded in the disk (type-I migration) and giant planets that open gaps in the disk (type-II migration). The migration rates are obtained in high resolution hydro-simulations including radiation transport for planets in the type-I migration regime (Kley and Crida 2008; Kley et al. 2009; Bitsch and Kley 2011b; Lega et al. 2014, 2015).

However, these simulations are computationally very expensive, especially when long-term evolution of planet(s) are taken into account. Instead the migration rates for low-mass planets in N-body simulations or population synthesis studies are mimicked by fits to hydrodynamical simulations which give the torque acting on embedded planets (Paardekooper et al. 2011). Our simulations contain only single planets, where we assume that their orbit is circular and in mid-plane of the disk. This approach is justified as the disk damps eccentricity and inclination on time-scales much shorter than the migration time-scale for small planets (Bitsch and Kley 2010, 2011a).

The total torque $\Gamma_{\rm tot}$ acting on low mass planets consists of two parts. The Lindblad torque $\Gamma_{\rm L}$ and the corotation torque $\Gamma_{\rm C}$, which arises from the corotation region around the planetary orbit. The total torque is the sum of the two

$$\Gamma_{\rm tot} = \Gamma_{\rm L} + \Gamma_{\rm C} \,. \tag{12.11}$$

The Lindblad torque is caused by spiral waves that are launched by the planet itself. These waves carry away angular momentum from the planet and scale with the radial gradients of surface density, $\Sigma_{\rm g} \propto r^{-s}$, and temperature $T \propto r^{-\beta}$. The Lindblad torque is given in Paardekooper et al. (2011) by

$$\gamma \Gamma_{\rm L}/\Gamma_0 = -2.5 - 1.7\beta + 0.1s \quad \text{and} \quad \Gamma_0 = \left(\frac{q}{h}\right)^2 \Sigma_P r_p^4 \Omega_P^2 \,, \tag{12.12}$$

where q is the mass ratio between planet and star, Σ_P the gas surface density of the disk at the planet's location, and r_P the distance of the planet to the host star. Clearly for normal disk structures, the Lindblad torque is negative and can cause rapid inward migration of planets above one Earth mass, where the migration time-scale is much shorter than the disk's lifetime (Ward 1997; Tanaka et al. 2002).

The corotation torque arises when material that is inside the planetary orbit is deflected by the planet to an exterior orbit, which results in a transfer of angular momentum from the planet to the disk material. Similarly material that is on an exterior orbit of the planet gets deflected inwards and gives angular momentum to the planet. This is the barotropic part of the corotation torque, which strongly depends on the radial gradients in gas surface density.

If there is a radial gradient in entropy, $\xi = \beta - (\gamma - 1.0)s$, where $\gamma = 1.4$ is the adiabatic index in the disk, additional effects come into play. If the disk is hotter interior to the planet and material is deflected outwards by the planet, the material adiabatically expands when embedded in the lower temperature region outside of the planet's orbit. This expansion creates an under-density outside of the planetary orbit. Reciprocally, when material is deflected from the cold outer parts to the inner hot parts of the disk, it adiabatically compresses and generates an over density interior to the planet. The over- and under-density create an additional torque onto the planet, which is the entropy related corotation torque, given by

$$\gamma \Gamma_{\rm C,ent}/\Gamma_0 = 7.9 \frac{\xi}{\gamma} \,. \tag{12.13}$$

Outward migration can thus arise in those regions of the disk, where the radial gradients of temperature, surface density, and entropy are steep enough to cause large corotation torques that can overpower the negative Lindblad torque. This normally happens in regions of the disk, where H/r drops radially. In stellar irradiated disks this corresponds to the shadowed regions, which are caused by opacity transitions, for example at the water ice line (Bitsch et al. 2013, 2014; Baillié et al. 2015). Note here that the exact shape of the region of outward migration depends on the jump in opacity, which can change with the total metallicity in dust grains (Bitsch et al. 2015a) and with the chemical composition of the disk (Bitsch and Johansen 2016).

However, the corotation torque is prone to saturation. As the corotation region gives angular momentum to the planet, its own angular momentum decreases and has to be replenished to allow continuous outward migration. This replenishment of angular momentum happens through viscous diffusion of material. We can therefore identify three time-scales that are responsible for outward migration:

- the libration time, which is the time the material needs to make the whole horseshoe turn, given by $\tau_{\rm lib} = 8\pi r_{\rm P}/(3\Omega_{\rm P} x_{\rm s})$, with $x_{\rm s} = 1.16 r_{\rm P}\sqrt{(q/h)}$ being the width of the horseshoe/corotation region
- the U-turn time, which is the time the material needs to make a U-turn close to the planet, given by $\tau_{\text{U-turn}} = h \times \tau_{\rm lib}$
- the viscous time, which is the time the material needs to be replenished in the corotation region, given by $\tau_\nu = x_{\rm s}^2/\nu$, where ν is the viscosity inside the disc

In order to achieve outward migration, the U-turn time has to be smaller than the viscous time, because otherwise the effects of the material transport during the U-turn movement would be washed away. The viscous time also has to be shorter than the libration time, so that the material that is deflected at one side of the planet has replenished its angular momentum when it arrives on the other side of the planet. We can write this as

$$\tau_{\text{U-turn}} < \tau_\nu < \tau_{\rm lib} \; . \tag{12.14}$$

This comparison of time-scales also implies a dependence on planetary mass, which is illustrated in Fig. 12.2.

As the planet grows, it shoves away more and more material from its orbit and carves a gap in the disk, where the planet sits in the middle of it. When the gap becomes deep enough, the corotation region is depleted and the migration rate changes. As the planet feels the torque from both sides of the disk, it stays in the middle of the gap and cannot move with respect to the disk, but only with the disk as it accretes onto the central star. The migration therefore happens on the viscous time-scale of the disk's accretion onto the central star. However, if the planet is much more massive than the gas outside the gap, it will slow down the viscous accretion. This happens if $M_{\rm P} > 4\pi\Sigma_{\rm g} r_{\rm P}^2$, which leads to the migration time scale of

$$\tau_{\rm II} = \tau_\nu \times \max\left(1, \frac{M_{\rm P}}{4\pi\Sigma_{\rm g} r_{\rm P}^2}\right) , \tag{12.15}$$

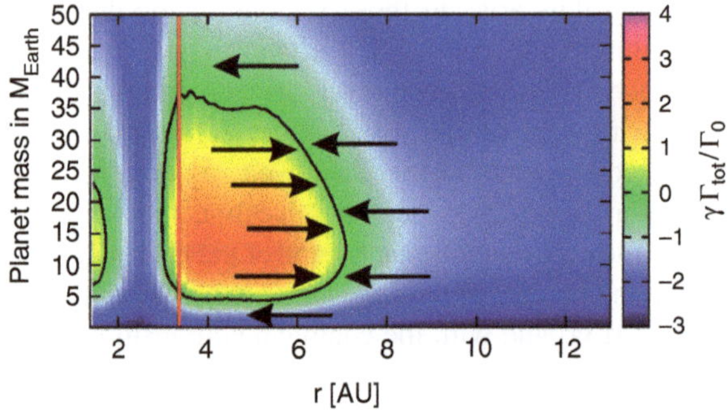

Fig. 12.2 Torque acting on planets as function of planetary mass and orbital distance in the disk model of Bitsch et al. (2015a). *Black contour lines* mark the transition between negative and positive torque, where a positive torque indicates outward migration. This is also indicated by the *thick black arrows*. The radial dependence of the region of outward migration is set by the local gradients in the disk, where the *vertical red line* marks the opacity transition at the water ice line. The upper and lower boundaries are set by torque saturation. Figure from Bitsch et al. (2015a), A&A, 575, A28, 2015, reproduced with permission © ESO

resulting in slower inward migration for massive planets (Baruteau et al. 2014). This is the classical type-II migration regime. However, in recent years this time-scale has been debated and new simulations suggest that type-II migration is not linked to the viscous evolution of the disk (Dürmann and Kley 2015). Additionally, the accretion of material during the migration phase of the planet influences the migration rate itself (Dürmann and Kley 2016; Crida and Bitsch 2016), but here we related back to the classical viscous migration rates for type-II migration [Eq. (12.15)].

12.5 Population Synthesis

All found exoplanets orbit around different types of stars, where not only the spectral type of the stars can be different, but also the metallicity of the host star. Additionally, the exact lifetime of the protoplanetary disk in which planets form and the exact birth time of the planetary seed are unknown. On top of that, the initial distribution of the planetary seeds is also not known. In order to get a homogeneous outcome of the simulations one has to assume a randomisation of certain parameters. In particular we will investigate the change of the following parameters on the evolution and growth of planets:

- The lifetime of the disk and the starting time of the planetary seeds, where we assume a median disk lifetime of 3 Myr, with a Gaussian distribution ($\sigma = 0.5$ Myr), where we fix the minimal and maximal lifetime of the disk to 1 and 5 Myr, respectively.

- The metallicity of the disk is also following a Gaussian distribution, where we put the median at [Fe/H] = −0.0625 ($\sigma = 0.23$), which corresponds to the metallicity of stars in the RV survey of Mayor et al. (2011) to which we compare our simulations.
- For the initial distribution of the planetary seed we assume a linear and a logarithmic distribution in semi-major axis in the regime of [0.1:50] AU.

In the following, we will discuss the influence of the disk's lifetime and of the disk's metallicity on planet growth. We will then put the results together and discuss the differences between the pebble accretion scenario and the growth via planetesimals.

12.5.1 Disc Lifetime

The lifetime of the protoplanetary disk is roughly a few Myr (Mamajek 2009), in which the evolution from planetary seeds to gas giants has to be completed. However, the exact lifetime of a protoplanetary disk is not exactly known and can vary due to external evaporation, cluster environments, and different viscosities inside the disk. Our mean lifetime of the protoplanetary disk is 3 Myr, with a σ of 0.5 Myr, where we limit the maximal lifetime to 5 Myr. Note that the evolution of the accretion rate in time is fixed and does not vary with the disk lifetime. This means that disks with shorter lifetime dissipate at higher accretion rates than disks with longer lifetime. A planetary seeds that forms in the disk, grows and migrates through it. The growth and migration rates are functions of the local disk properties [Eqs. (12.6) and (12.12)], indicating that the exact time when a planetary seed forms in the disk becomes very important.

In Fig. 12.3 we show the evolution of planetary seeds in so-called growth tracks, which show the planetary evolution in mass (via accretion of pebbles and gas) and semi-major axis. Here we have used $Z_{\text{dust}} = 0.15\%$ and $Z_{\text{peb}} = 1.35\%$, which corresponds to $Z_{\text{tot}} = 1.5\%$, corresponding to [Fe/H]= 0 and solar value. In Fig. 12.3 the planetary seeds start in a disk that is already 2 Myr old. If the disk's lifetime were only 3 Myr, their final position in the planetary mass-orbital distance diagram is indicated by the black circles, while their final position in a disk that were to live 5 Myr is indicated by the black square. Clearly a longer disk lifetime allows for further growth and migration of already formed cores in the disk. For example, the planetary seed forming at 5 AU only grows to become a super-Earth sized planet if the disk were to live 3 Myr, but can fully contract its envelope and go into runaway gas accretion if the disk lives for a longer time. A longer disk lifetime is also of crucial importance for planetary seeds forming in the outer parts of the protoplanetary disk, as the growth time-scales there are longer.

When comparing Fig. 3 to Fig. 2 of Bitsch et al. (2015b) some clear differences are visible. In particular, in Bitsch et al. (2015b) only one planet migrates all the way to the inner disk before the end of the disk's lifetime at 3 Myr and it is also a giant planet. This difference is due to the change in the disk structure due to the

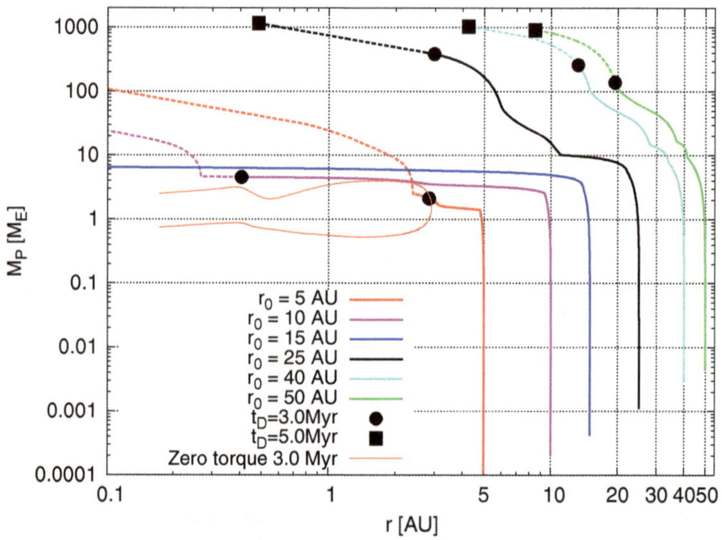

Fig. 12.3 Growth tracks for planets growing via pebble and gas accretion that migrate through the disk, where the planetary seeds start at a disk lifetime of 2 Myr. The *dots* mark the end of the disk's lifetime at 3 Myr, while the *squares* mark the end of the evolution if the disk were to live 5 Myr. Growth tracks that do not feature a circle or a dot reach the inner edge of the disk before the end of the disk's lifetime at 3 or 5 Myr even without being in the gas runaway phase. Note that the planets starting at 10 and 15 AU migrate inwards above the region of outward migration. Clearly a longer lifetime results in an enhanced growth and migration of the planet. Adapted from Bitsch et al. (2015b)

lower metallicity of micro metre sized dust grains, which results in a cooler disk with lower aspect ratio H/r. We assume here a low Z_{dust} as this gives a low opacity, resembling a more realistic grain size distribution, where most of the grains are large and thus have a small opacity. This results in a lower pebble isolation mass in Fig. 12.3 compared to Bitsch et al. (2015b), so the planetary core needs a longer time to accrete a gaseous envelope. Additionally, the lower metallicity in dust grains reduces the region of outward migration (Bitsch et al. 2015a), which allows planets to move inwards close to the central star even before they have reached the runaway gas accretion phase. Note here that the planet starting at 10 AU in Fig. 12.3 is just slightly too massive to be contained in the region of outward migration and therefore moves in slower compared to the planet starting at 15 AU, which reaches the inner edge of the disk even before the disk dissipates at 3 Myr.

Additionally, the amount of pebbles is higher here (1.35%) compared to Bitsch et al. (2015b), where $Z_{\text{pev}} = 1\%$ was used, which allows here a faster growth of the planetary seeds. In particular, the planetary seeds starting in the outer part of the protoplanetary disk ($r > 40$ AU) can form gas giants well within the lifetime of the protoplanetary disk. This is also aided by the smaller aspect ratio (smaller amount of dust grains result in colder disks), which concentrates the pebbles in the mid-plane of the disk and thus accelerates growth.

This clearly shows that the final planetary mass and orbital position is not only a function of the initial position of the planet, but also of the lifetime of the protoplanetary disk itself. Additionally this shows that the initial time when the planetary seed is formed in the disk is important, see also Fig. 4 in Bitsch et al. (2015b).

12.5.2 Metallicity

The metallicity in the protoplanetary disk gives the ratio of dust to gas, where a value of $Z_{tot} = Z_{dust} + Z_{peb} = 1.5\%$ corresponds to the solar value, [Fe/H]=0. As the planetary growth rate via pebble accretion rate scales directly with the surface density in pebbles, a change in metallicity has a direct influence on the growth rates of planets. In Fig. 12.4 the mass evolution of planetary cores that form in the same disk structure environment, but with different amounts of pebbles is shown. We keep the planets on fixed orbits for this simulations in order to show the effects of different Z_{peb} clearly, without any influence of planet migration, where the planet might migrate into regions where the accretion rates are different.

A larger metallicity results in a faster growth of the planet [Eq. (12.6)]. The planet therefore reaches pebble isolation mass faster (horizontal line in Fig. 12.4), when Z_{peb} is larger. As the disk evolves in time, H/r drops (Bitsch et al. 2015a) and with it the pebble isolation mass, which determines the final core mass. If the planet grows faster (due to high Z_{peb}), it can therefore reach a larger pebble isolation mass. This also means that the formation of gas giants is easier in disks with high metallicity, as

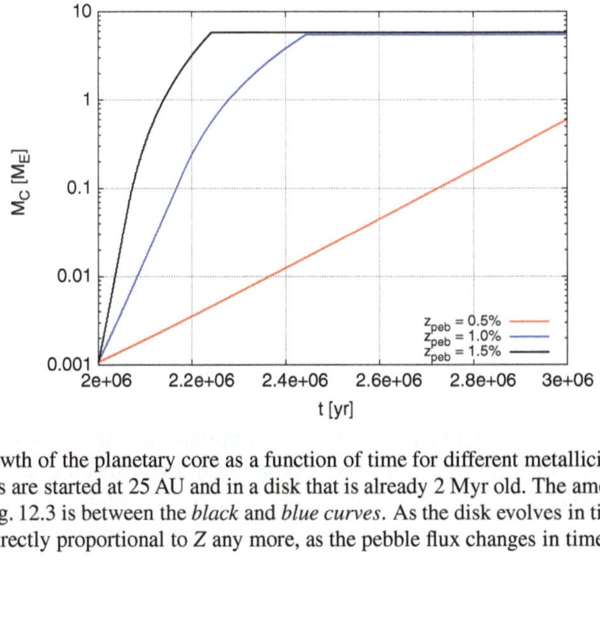

Fig. 12.4 Growth of the planetary core as a function of time for different metallicities in pebbles. Planetary seeds are started at 25 AU and in a disk that is already 2 Myr old. The amount of pebbles Z_{peb} used in Fig. 12.3 is between the *black* and *blue curves*. As the disk evolves in time, the growth rates are not directly proportional to Z any more, as the pebble flux changes in time

larger planetary cores are formed, which can contract their envelope on shorter time-scales and reach runaway gas accretion well within the lifetime of the protoplanetary disk. This is in agreement with observations (Fischer and Valenti 2005).

This comparison and the results of the previous section show that the metallicity is not only important for the growth of the planetary core, but also for the orbital evolution of the planet, indicating that the metallicity is a key parameter for the formation of planets.

12.5.3 Accretion via Pebbles

When using the (Bitsch et al. 2015a) disk model, we divide the metallicity into a metallicities for the dust grains $Z_{\rm dust}$ (for the disk structure, see Bitsch et al. 2015a) and for the pebbles $Z_{\rm peb}$ for planet growth (Bitsch et al. 2015b). We use in all the following simulations 10% of the total metallicity for the dust grains and 90% for the pebbles. This results in a cold inner disk, because the cooling rates are very high. This low aspect ratio in the inner disk results in a low pebble isolation mass and in a very small region of outward migration, where planetary cores can outgrow the region of outward migration before they reach runaway gas accretion (Bitsch et al. 2015b; Bitsch and Johansen 2016). The value of $Z_{\rm dust}$ is lower than in Bitsch et al. (2015b), because simulations regarding grain growth and fragmentation predict a size distribution where most of the mass of the dust is in larger particles (Birnstiel et al. 2011), which have a lower opacity than micro metre sized grains. When we use the MMSN disk model, we follow the H/r ratio as outlined above and use 90% of the metallicity in pebbles for planet growth and 10% for the dust. Note that the dust grains are also responsible for the cooling rates needed to calculate the migration of planets.

By varying the initial location of the planetary seed, the disk's lifetime and the metallicity as outlined above, we can provide the final mass and orbital distance of the grown planets and compare them to observations. As the planetary mass is crucial to compare, we rely on observational data that were obtained with the radial velocity method, as this method can constrain the planetary mass. In the RV survey of Mayor et al. (2011) 822 stars are observed, which show a planetary rate of 14% for planets with $M \sin(i) > 50 \rm M_E$ and for orbital periods smaller than 10 years (corresponding to 4.64 AU). We therefore simulate 822 growth tracks for each set of simulations to allow a direct comparison with observations. Each growth track is stopped when either the disk dissipates at the end of its lifetime, or when the planet reaches the inner edge of the disk at $r_{\rm inner} = 0.04$ AU or grows to 10,000 Earth masses.

In Fig. 12.5 we show our synthesised planet population in the MMSN disk, while Fig. 12.6 shows our synthesised planet population in the (Bitsch et al. 2015a) disk model. In the MMSN, the synthesised planets match very poorly with the RV data. Especially, the population of the hot giant planets is roughly a factor of 10 too massive and only a small amount of cold gas giants is produced. Additionally,

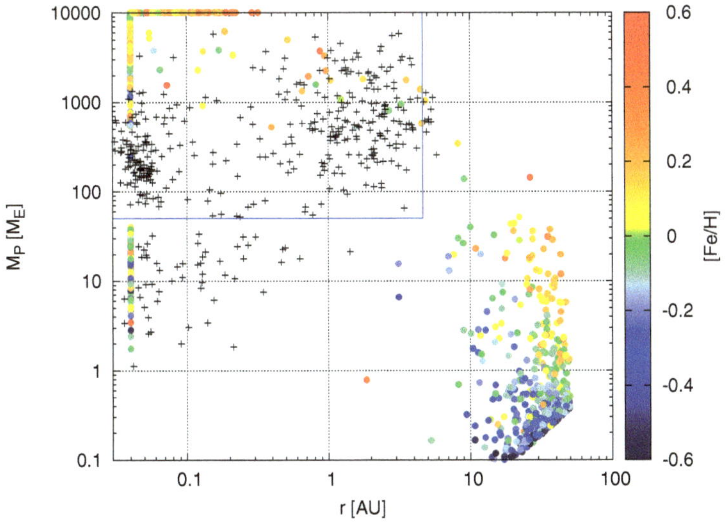

Fig. 12.5 Planet population via pebble accretion in the MMSN disk. The *black crosses* correspond to observations determined through RV measurements, while the *blue box* shows where the RV measurements are complete in the (Mayor et al. 2011) RV survey. The colour scale defines the total metallicity in units of [Fe/H], where 10% of the heavy elements are in micro metre sized dust particles and 90% in pebbles that can be accreted onto the planet

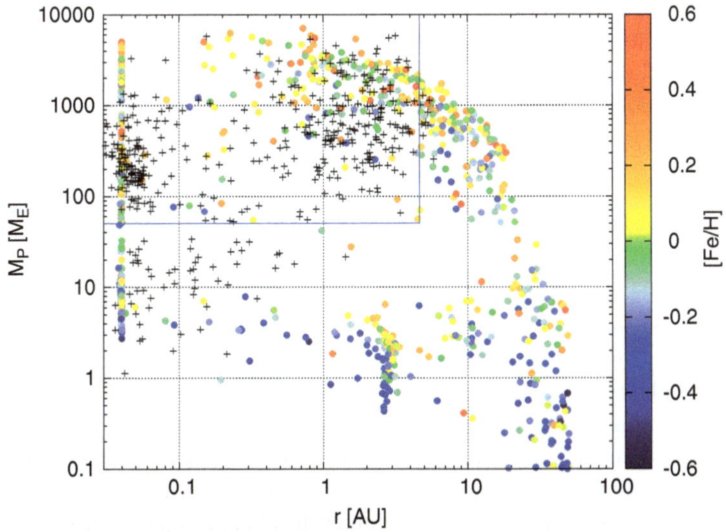

Fig. 12.6 Planet population via pebble accretion in the disk model of Bitsch et al. (2015a). The symbols are the same as in Fig. 12.5

basically no super Earth planets outside of 0.1 AU is formed. Additionally there are 230 planets in the RV box of completeness in the MMSN disk, which corresponds to 26.1%, about a factor of 2 higher than predicted by Mayor et al. (2011). On top of that, the final orbital positions and masses of those planets do not represent the observations at all. In contrast, the synthesis model using the (Bitsch et al. 2015a) disk model shows a better mass distribution for the giant planets and also the formation of super Earths is possible. However, the RV box of completeness shows 271 planets, corresponding to 30.7% which is also a factor of about 2 too high compared to the observations of Mayor et al. (2011).

The reason for the differences of both simulations is the underlying disk model, which significantly differs in the inner regions of the protoplanetary disk (Bitsch et al. 2015a). The MMSN power law disk follows a very steep gradient in surface density towards the inner disk ($\Sigma_g \propto r^{-3/2}$), while the (Bitsch et al. 2015a) model is much shallower ($\Sigma_g \propto r^{-1/2}$), which leads to completely different migration behaviour. In the MMSN disk, only inward migration is possible, while the planets in the (Bitsch et al. 2015a) model can stop their inward migration at the planet trap at the outer edge of the shadowed region caused by the water ice line. As the torque acting on planets also scales directly with the absolute value of the surface density [Eq. (12.12)], the MMSN model features much faster inward migration in the inner regions of the disk. This leads to the increased number of planets of a few to a few 10s of Earth masses at the inner edge of the disk.

As the MMSN disk model features no outward migration, basically no small planets for 1–10 AU are found in Fig. 12.5. The population of super-Earths sitting at a few AU in the (Bitsch et al. 2015a) model is caused by the planet trap at the water ice line. This stops the inward migration of small mass planets and the planets can stay there until the gas disk disappears, if the formation time of the planetary seed is late compared to the lifetime of the protoplanetary disk (see Fig. 12.3).

The gas accretion onto the planetary core is modelled following Machida et al. (2010), where the accretion rate is directly proportional to the gas surface density. This then leads to a much larger accretion rate in the inner disk in the MMSN model, allowing the giant planets to grow to much larger masses as they migrate in the inner disk (within a few AU). However, the masses of the hot gas giants in the inner disk are still larger than the masses of the observed hot Jupiters in the (Bitsch et al. 2015a) disk model.

In the very outer regions of the disk ($r > 10$ AU), the differences in the synthesised planet populations are quite small. This is caused by the fact that the initial growth time-scales via pebble accretion are quite similar in both disks even though the disk models differ in the outer disk. Additionally, all effects mentioned above multiply as the planet growth to become a gas giant, while the growth to an object of just a few Earth masses in the outer disk happens quite locally with only small amounts of planet migration (see Fig. 12.3).

Clearly the structure of the protoplanetary disk is of crucial importance for the outcome of planetary growth simulations, especially in the inner parts of the protoplanetary disk. The MMSN disk totally fails to reproduce either hot Jupiters (with the correct masses) or the super-Earth systems. The (Bitsch et al. 2015a) disk

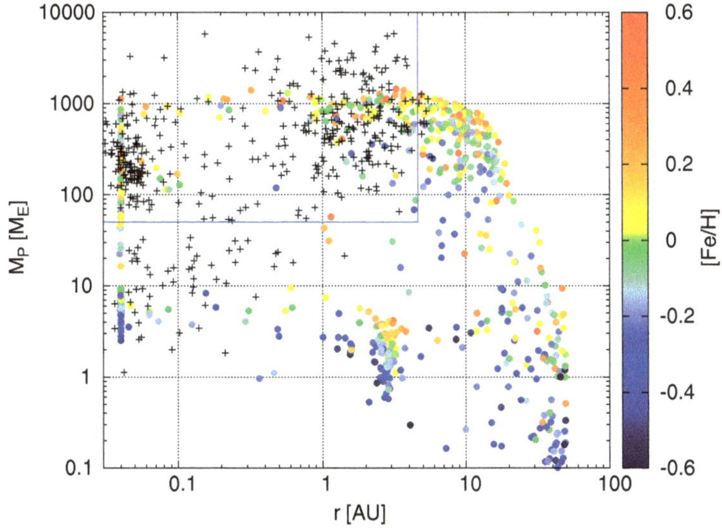

Fig. 12.7 Planet population via pebble accretion in the disk model of Bitsch et al. (2015a), but planetary seeds start only in disks that are already older than 1.5 Myr, otherwise the setup of the simulations is the same as in Fig. 12.6

model allows for a better match regarding giant planets than the MMSN, but even there the population of hot Jupiters is slightly too massive.

The planetary seeds in our simulations start at the pebble transition mass [Eq. (12.4)], which is actually a bit larger than the sizes of planetesimals formed by the streaming instability (Johansen et al. 2015). When these planetesimals are born, they can grow towards the pebble transition mass by accretion of other planetesimals or pebbles in the inefficient Bondi accretion branch (Lambrechts and Johansen 2012), which can take 1–1.5 Myr. In Fig. 12.7 we show the mass-orbital distance diagram, where planetary seeds start in disks that are at least 1.5 Myr old, with a minimal lifetime of 2 Myr. Otherwise the simulations have the same conditions as in Fig. 12.6.

The synthesised hot Jupiter planets match much better with the observations, as the final masses are reproduced very well. In general the final mass of the gas giants is lower, because the planets have less time to accrete a gaseous envelope in the disk. The late formation scenario also populates the parameter space that harbours hot super-Earths better, but inside the RV box of completeness the planetary rate is still about a factor of 1.5 too high (20.2%) compared to Mayor et al. (2011). However, our planets are in single systems (one-planet-per-disk) and we do not take scattering of planets into account, which can easily reduce this number. Also, scattering is an important process to explain the eccentricity distribution of giant planets.

As can be seen in Fig. 12.3, the initial starting position determines the final planetary mass and orbital distance. Therefore, the initial distribution of the planetary seeds in our population synthesis study determines the outcome of our

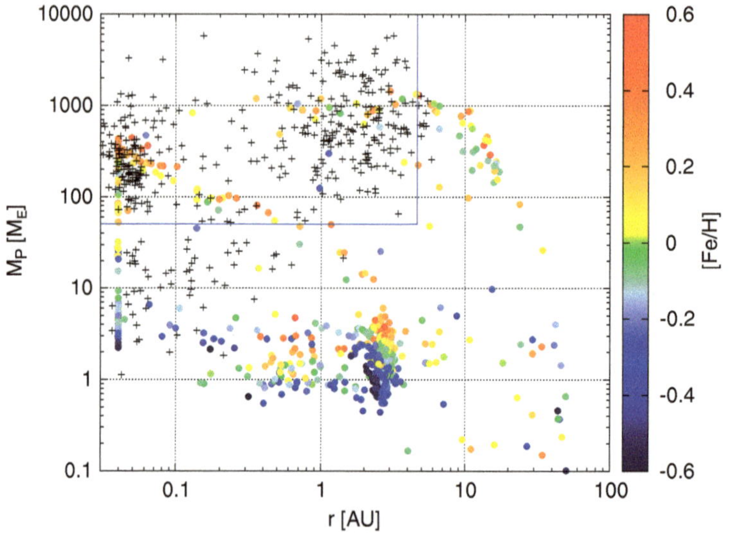

Fig. 12.8 Planet population via pebble accretion in the disk model of Bitsch et al. (2015a), but planetary seeds start only in disks that are older than 1.5 Myr with a logarithmic starting configuration

simulations. The linear starting distribution of planetary seeds as invoked above favours the formation of cold gas giants, as the many planetary seeds start at orbital distances of $r_{\rm init} > 15$ AU. If one were to invoke a logarithmic distribution of the starting location, many more planets form in the inner parts of the disk, where they do not have the chance to grow to become gas giants at wide orbits. We show in Fig. 12.8 the final planetary masses and orbital distances of planetary seeds that formed in a logarithmic starting configuration.

The logarithmic starting configuration of planetary seeds shows less giant planets compared to the linear starting configuration, because planets forming in the inner regions of the protoplanetary disk have a smaller pebble isolation mass and thus a smaller core mass, which makes it harder for those planets to reach the runaway gas accretion phase (Bitsch et al. 2015b). These failed gas giants are left as small rock-dominated planets of orbits of up to a few AU and are much more frequent compared to the results of the linear starting distribution of planetary seeds (Fig. 12.7), where more giant planets form, as giant planets form preferably in the outer disk. In this case, 11.0% of all planets end up in the RV box of completeness defined by Mayor et al. (2011). However, fitting the exact number of Mayor et al. (2011) is only meaningful if multiplicity and scattering are taken into account, which is lacking in our model.

The synthesised planets in Figs. 12.6, 12.7, and 12.8 all share a similar pattern, which is a lack of Neptune mass planets in the inner parts of the disk, which is actually where most of the observed planets in the (Mayor et al. 2011) survey are. The formation of gas giants in our model relies on the planetary seeds reaching pebble isolation mass so that they can contract their envelope and eventually go into runaway gas accretion. This is limited in the inner disk by the small pebble isolation mass. Small planetary cores therefore stay at the zero-torque position in the inner disk (Fig. 12.3) until they start to undergo runaway gas accretion. This set them all on a very specific type of growth track, as can be seen nicely in Fig. 12.8. This leaves a region of parameter space in planetary mass and orbital distance very void of synthesised planets (10–100 M_E between 0.1 and 1.0 AU). It is very hard to populate this mass bin with planets in simulations, because large planetary cores undergo rapid runaway gas accretion, allowing them to become even larger than 100 M_E within a very short timeframe. This was already noticed in the earliest planet population synthesis studies (Ida and Lin 2004) and identified as a planetary desert.

The formation of these super-Earths and Neptune planets is not entirely understood. In the literature many different ways have been proposed to form these planets (in-situ formation, migration, etc.), but none of the proposed mechanisms seems to satisfy the observational constraints (Ogihara et al. 2015). Additionally, multiplicity is probably a key ingredient needed to form those systems, which is beyond our one-planet-per-disk approach.

12.5.4 Accretion of Planetesimals

We investigate here the growth of planets, where the core builds up by accreting planetesimals. As planets migrate through he disk, they stir up the planetesimals and only a tiny fraction of those are actually accreted onto the planet (Tanaka and Ida 1999). Therefore our planetary seeds grow only until they reach the planetesimal isolation mass, in contrast to previous population synthesis simulations where the planetary seed can additionally accrete up to 100% of the planetesimals that it encounters during its migration (Dittkrist et al. 2014). As soon as the planets reach planetesimal isolation mass, gas accretion can start, which is modelled in the same way as for the pebble accretion simulations.

In Fig. 12.9 we show the orbital distance vs. planetary mass diagram of the growth simulations via planetesimals. Clearly, reaching final masses of above 10 M_{Earth} is very difficult and impossible at orbital distances larger than 1 AU. The reason why planet growth via planetesimals is not successful is twofold. First, the planetary growth rates are slow [Eq. (12.9)] compared to pebble accretion, giving less time to accrete gaseous envelopes after reaching isolation mass. In fact we also integrate 822 growth tracks as in the previous simulations featuring pebble accretion, but out of all these seeds, only ~250 reach a mass larger than 0.1M_E.

Secondly, the planetary cores that are forming suffer from inward type-I migration. As the growth rate is slow, they undergo strong inward type-I migration while still accreting some planetesimals at a low rate and thus cannot accrete a gaseous envelope. They reach the inner regions of the disk before they can start efficient runaway gas accretion and open a gap in the protoplanetary disk, where type-II migration could potentially save them from migrating all the way towards the inner

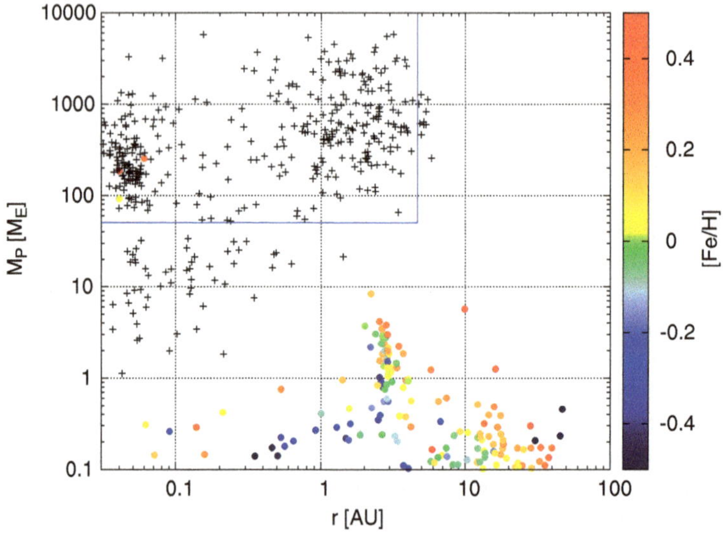

Fig. 12.9 Mass-orbital distance diagram of 822 planetary seeds growing by planetesimal accretion, where only 250 of these seeds grow above M_E. The *colour coding* shows the metallicity for each planet growth simulation. Forming gas giants outside of 0.1 AU is basically not possible

edge. This is consistent with the findings of Ida and Lin (2004) and Alibert et al. (2004) who suggested that planet migration is too fast to form giant planet cores in the planetesimal accretion scenario. The reason why recent population synthesis simulations are successful in producing giant planets is then caused by the different accretion scheme of solids, where the migrating planetary seeds can accrete up to 100% of all leftover planetesimals allowing faster growth (Dittkrist et al. 2014) and by using largely super-solar metallicities.

To overcome the problem of strong inward migration in the type-I migration regime, Coleman and Nelson (2016) invoked the formation of planetary seeds in zonal flows at distances of up to 30 AU from the central star, where the planets can grow without migrating and are then just released when they become massive enough to open up gaps in the disk. In this way the observed giant planet population can be matched. Another approach in overcoming the strong inward migration is to accelerate the growth rates of planets by increasing the amount of planetesimals in the disk, which is taken in recent population synthesis simulations.

We show the results of both approaches in Fig. 12.10, where the left panel shows the mass-orbital distance diagram of planetary seeds growing via planetesimal accretion, but with suppressing all planetary migration, while the right panel shows the mass-orbital distance diagram of planetary seeds growing via planetesimal accretion, where the planetesimal surface density Σ_{pla} has been increased by a factor of 10 and planetary migration is taken into account.

When suppressing planetary migration, the final mass of the planet is solely determined by the planetesimal isolation mass and by the time the planet has to

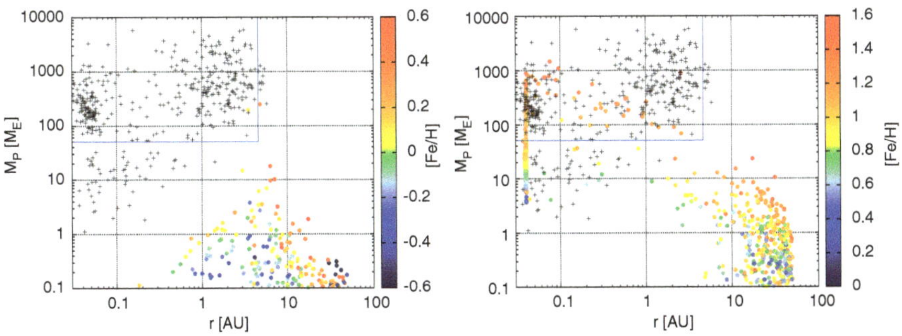

Fig. 12.10 Mass—orbital distance diagram of planetary seeds growing by planetesimal accretion, where planetary migration is completely suppressed *(left)* and in a disk where the metallicity of planetesimals is increased by a factor of 10 compared to the simulations before, but where migration is taken into account. This results in both cases in the growth of some cold gas giants compared to Fig. 12.9. Note here the different colour scales for the metallicities

grow. In the very inner regions of the disk the planetesimal isolation mass is very small, so that the planets stay below a few Earth masses. Outside of a few AU and inside of 10 AU, the planetesimal isolation mass is large enough to form cores of a few Earth masses than can accrete a gaseous envelope. However, at even larger orbital distances, the accretion rate becomes way too slow due to the low $\Sigma_{\rm pla}$, so that the planets do not grow big enough to attract a gaseous envelope.

When increasing the planetesimal surface density by a factor of 10, the growth time-scales of the planetary cores become short enough that the formation of planetary cores that can attract gaseous envelopes is possible at distances of up to a few AU. Additionally the planetesimal isolation mass increases [Eq. (12.8)], allowing the formation of larger planetary cores. Nevertheless, the formation of cold gas giants is only possible when [Fe/H]>1, which is not in agreement with observations (see Fig. 12.11 and the following section). In order to form cold gas giants in the planetesimal scenario, either planetary migration has to be ignored or the metallicity has to be increased to at least ten times the solar value, which does not reflect the giant planet occurrence rate in observations (Johnson et al. 2010; Buchhave et al. 2012). We therefore find both options to form giant planets in the planetesimal accretion scenario no realistic.

This is clearly different in the pebble accretion scenario, which allows fast growth of planetary cores even at large orbital distances at solar metallicity. The formation of planetary cores at these large orbital distances of 10–20 AU allows for a population of cold gas giants, which is impossible to form in the planetesimal accretion scenario at solar metallicity. The strength of the pebble accretion scenario therefore is that growth at large orbital distances is possible, so that forming planets do not move all the way into the central star before disk dissipation.

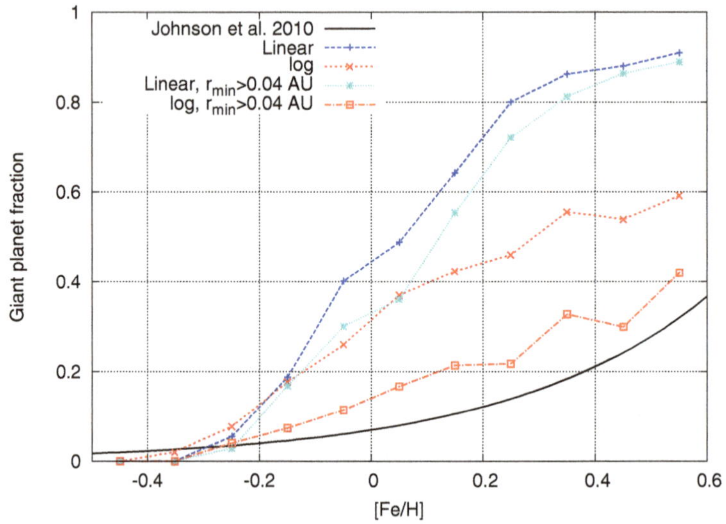

Fig. 12.11 Fraction of stars that harbour giant planets as a function of metallicity. The *black line* corresponds to Eq. (12.16) and reflects observations. The data of the simulations corresponds to the data shown in the mass-orbital distance diagrams of Figs. 12.7 and 12.8, so to planets formed in the late formation scenario. As in Johnson et al. (2010) we only take planets into account that are within 2.5 AU from the central star

12.5.5 Comparison to Observations

The strongest constraint for planet formation models is the comparison with observations. In this section we will discuss the influence of metallicity on the occurrence rate of giant planets, where we focus on the statistics obtained through RV surveys. Based on their RV measurements, Johnson et al. (2010) predict a giant planet occurrence rate around solar mass stars as a function of metallicity in the following form

$$f = 0.07 \times 10^{1.2[\mathrm{Fe/H}]}\,, \tag{12.16}$$

indicating that 7% of stars with solar metallicity ([Fe/H]=0) feature giant planets and an increasing giant planet occurrence rate with metallicity. Due to observational biases, Eq. (12.16) is only valid up to $r < 2.5$ AU, which is a different level of completeness compared to Mayor et al. (2011). We will therefore only compare our results up to this orbital distance. In Fig. 12.11 we show the fractions of stars that harbour giant planets for our simulations and compare this fraction to the observations of Johnson et al. (2010).

Our simulations follow the simplistic one-planet-per-disk approach, which only allows us to calculate the giant planet fraction by dividing the number of growth tracks that form giant planets within 2.5 AU by the total amount of planets within the same orbital distance. This approach has several biases, because we do not take scattering of bodies into account. This is not only important at the late stages of the disk where the final system architecture is determined, but also at the early stages, where small amounts of eccentricity can significantly reduce pebble accretion (Johansen et al. 2015) and might thus hinder planetary seeds to reach pebble isolation mass and become gas giants. Therefore the results of our simulations represent a maximum giant planet fraction per planetary seed per disk. So if a disk forms e.g. 10 planetary seeds at random initial positions r_0 in a disk with [Fe/H]=0, we predict a maximum of 5 giants.

Clearly, the formation of giant planets in the pebble accretion scenario scales with increasing metallicity. An increase of metallicity allows for faster accretion rates [Eq. (12.6)], which allows to reach the pebble isolation mass at an earlier stage (Fig. 12.4), giving more time to contract the envelope around the planet and therefore results in an earlier transition to runaway gas accretion. This results in a more efficient pathway to form giant planets. In fact, it seems that the planet formation frequency is too efficient in the pebble accretion scenario.

In addition to that, the starting position of the planetary seeds plays a crucial role, because the pebble isolation mass depends on the position in the disk, where larger pebble isolation masses can be reached in the outer parts of the protoplanetary disk. In the inner parts of the disk, the pebble isolation mass is low and the formation of giant planets is not possible. This explains why the fraction of giant planets for the logarithmic starting distribution is smaller than for the linear starting distribution.

In counting the fraction of planets for $r < 2.5$ AU, we also counted the planets that reached the inner edge of the simulation at 0.04 AU. When a planet reaches this distance we stop the simulation and do not take further evolution of the planet into account. Recent simulations have shown that the atmospheres of giant planets so close to the host star can be eroded away in time (Lanza 2013), leaving only the bare planetary core behind. These giant planets would then appear as small planets in observations. The light blue and orange line in Fig. 12.11 represent the same sample of simulations, but here we counted the giant planets that reached the inner edge of the disk at 0.04 AU as small planets due to the evaporation of their atmospheres. The logarithmic starting distribution now shows only a slightly larger giant planet fraction compared to the observations of Johnson et al. (2010), where this small excess can be additionally reduced due to scattering events, which we do not take into account in our simulations.

However, we want to note here that a comparison of a one-planet-per-disk approach in our simulations to observations is not necessarily correct, but we have to choose this approach due to our model. In reality, planetary growth most likely happens within a sea of planetesimals, where the planetesimals stir each other while accreting pebbles, so that only a few, but more than one, become dominant and form planetary systems (Levison et al. 2015). Additionally, observed planetary systems are also not systems with single planets, but contain multiple planets. Therefore, future planet population synthesis models have to take the effects of multiplicity into account, in order to achieve a better comparison with observations.

12.6 Summary

In this chapter we have combined results of planet growth via pebble and gas accretion with planetary migration and disk evolution to synthesise planet populations that we compared with observations of exoplanets. We have discussed the influence of the disk's lifetime and metallicity on the growth and migration of embedded planets and compared the pebble growth mechanism with growth of planetary cores via the accretion of only planetesimals in different disk models. We especially highlighted the interplay between planet growth and migration, which is the key to form long period gas giants.

A longer disk lifetime gives more time for the planet to grow in the disk, but also allows for more efficient planetary migration. In the inner disk, where the pebble isolation mass is low, a longer disk lifetime can trigger runaway gas accretion of cores (Fig. 12.3) and allow planets of a few Earth masses to accrete gaseous envelopes to become gas giants, if the cores are trapped in the region of outward migration. However, these planets then slowly migrate inwards in type-II migration and end up as hot Jupiters. Additionally planets that are slightly too massive to be contained in the region of outward migration migrate inwards as super-Earths. Planets that form in the outer parts of the disk, on the other hand, can then eventually reach their pebble isolation mass and become gas giants on wide orbits (Fig. 12.3).

The metallicity of the protoplanetary disk influences the growth of the planetary core by two factors. First, the growth speed accelerates with metallicity resulting in faster planetary growth, which is especially important at large orbital separations, where the growth rates might be too slow to reach pebble isolation mass for low pebble metallicities. Additionally a higher metallicity allows that the planetary core can reach a higher pebble isolation mass because the time evolving disk reduces its temperature in time. Nevertheless, pebble accretion is very robust and still allows growth at larger orbital distances, so that the formation of some cold gas giants in disks with sub solar metallicity is possible.

The final orbital position of the planets is determined by the migration rates in the disk. Low mass planets migrate generally inwards, except in regions where strong negative radial gradients of entropy are present. These strong gradients exist in the inner parts of the protoplanetary disk close to the water ice line in the disk model of Bitsch et al. (2015a), while they are absent in the MMSN disk. These local gradients are also dependent on the metallicity in dust grains in the disk, which determines the cooling rates, resulting in smaller regions of outward migration in disks with low metallicity (Bitsch et al. 2015a). Additionally, the migration speed in type-I migration scales with the surface density of the protoplanetary disk. As the MMSN disk has a much higher surface density in the inner regions of the disk, planets migrate faster in that disk. The result is that all the small mass planets migrate towards the inner edge of the disk in the MMSN disk (Fig. 12.5), in contrast the planets synthesised with the (Bitsch et al. 2015a) disk, where small mass planets can stay at a distance of a few AU.

The gas accretion rate is also a function of the local surface density, resulting in the fact that gas giants in the inner parts of the disk become more massive in the MMSN disk. In fact the planets become a factor of 10 or more too massive to resemble the observed mass distribution of hot Jupiters (Fig. 12.5). For planets formed in the (Bitsch et al. 2015a) disk model, the synthesised hot Jupiters are still too massive, but much less so compared to the MMSN disk model (Fig. 12.6). This is caused when the planetary seeds form early in the disk and can thus accrete for a longer time.

As the planetesimals born from the streaming instability are a significantly smaller than the pebble transition mass [Eq. (12.4)], the planetesimals first accrete in the inefficient Bondi accretion branch to reach pebble isolation mass within about 1–1.5 Myr (Lambrechts and Johansen 2012), which we do not take into account here. We therefore invoke that the planetary seeds (at pebble transition mass) start in disks that are at least 1.5 Myr old. As a consequence, the final mass of the giant planets is smaller than for planetary seeds that start early in the disk's lifetime (Fig. 12.7), which gives a much better match to the population of hot and cold Jupiters. Additionally slightly higher masses for the synthesised super-Earth population can be achieved, matching the observations quite well.

Planetesimal accretion, on the other hand, fails completely to produce gas giants at orbits outside of 1 AU. This is caused by the long growth time-scales of the planetary cores combined with the fast migration rates. Giant planets can only emerge in those disks, when migration is unrealistically turned off or when the metallicity is increased by a factor of 10 compared to solar value (Fig. 12.10). Both options seem unrealistic.

The pebble accretion scenario confirms the metallicities trend found in observations, where the fraction of giant planets increases with stellar metallicity (Fig. 12.11). In fact, the pebble accretion scenario is very efficient in the production of giant planets and overestimates the observations by Johnson et al. (2010). This can be due to several reasons: (1) we do not take multiplicity into account which can lead to competition between the different cores for the building blocks of the planet and scattering events that will then prevent planetary growth and (2) photoevaporation of planetary atmospheres close to the central star, which can transform hot Jupiters into planetary cores due to the complete loss of the planetary envelope (Lanza 2013).

In summary, pebble accretion can explain the formation of hot and cold gas giants, as well as of super-Earths in contrast to planetesimal accretion. However, in most of the simulations the final mass of the super-Earths is a bit too small compared to the RV data. Also considering the amount of super Earths planets found by the Kepler mission, the reproduction of super-Earths in our simulations can still be improved. This may be caused by the fact that we just evolve single planet systems, while hot super-Earth systems normally feature several planets, where mutual interactions play important roles. Future population synthesis simulations therefore have to combine pebble accretion with N-body dynamics in evolving protoplanetary disks in order to constrain planet formation theories even further.

Acknowledgements B.B. and A.J. thank the Knut and Alice Wallenberg Foundation for their financial support. A.J. was also supported by the European Research Council (ERC Starting Grant 278675-PEBBLE2PLANET) and the Swedish Research Council (grant 2014-5775).

References

Alibert, Y., Mordasini, C., Benz, W.: Migration and giant planet formation. Astron. Astrophys. **417**, L25–L28 (2004)

Baillié, K., Charnoz, S., Pantin, É.: Time evolution of snow regions and planet traps in an evolving protoplanetary disk (2015). astro-phEP p arXiv:1603.07674

Baruteau, C., Crida, A., Paardekooper, S.J., Masset, F., Guilet, J., Bitsch, B., Nelson, R., Kley, W., Papaloizou, J.C.B.: Planet-disc interactions and early evolution of planetary systems. In: Protostars and Planets VI. University of Arizona Press, Tuscon (2014). arXiv:1312.4293

Birnstiel, T., Ormel, C.W., Dullemond, C.P.: Dust size distributions in coagulation/fragmentation equilibrium: numerical solutions and analytical fits. Astron. Astrophys. **525**, A11 (2011)

Birnstiel, T., Klahr, H., Ercolano, B.: A simple model for the evolution of the dust population in protoplanetary disks. Astron. Astrophys. **539**, A148 (2012)

Bitsch, B., Johansen, A.: Influence of the water content in protoplanetary discs on planet migration and formation. Astron. Astrophys. **590**, A101 (2016)

Bitsch, B., Kley, W.: Orbital evolution of eccentric planets in radiative discs. Astron. Astrophys. **523**, A30 (2010). doi:10.1051/0004-6361, 1008.2656v1

Bitsch, B., Kley, W.: Evolution of inclined planets in three-dimensional radiative discs. Astron. Astrophys. **530**, A41 (2011a)

Bitsch, B., Kley, W.: Range of outward migration and influence of the disc's mass on the migration of giant planet cores. Astron. Astrophys. **536**, A77 (2011b)

Bitsch, B., Crida, A., Morbidelli, A., Kley, W., Dobbs-Dixon, I.: Stellar irradiated discs and implications on migration of embedded planets I: equilibrium discs. Astron. Astrophys. **549**, A124 (2013)

Bitsch, B., Morbidelli, A., Lega, E., Crida, A.: Stellar irradiated discs and implications on migration of embedded planets II: accreting-discs. Astron. Astrophys. **564**, A135 (2014)

Bitsch, B., Johansen, A., Lambrechts, M., Morbidelli, A.: The structure of protoplanetary discs around evolving young stars. Astron. Astrophys. **575**, A28 (2015a)

Bitsch, B., Lambrechts, M., Johansen, A.: The growth of planets by pebble accretion in evolving protoplanetary discs. Astron. Astrophys. **582**, A112 (2015b)

Brauer, F., Henning, T., Dullemond, C.P.: Planetesimal formation near the snow line in MRI-driven turbulent protoplanetary disks. Astron. Astrophys. **487**, L1–L4 (2008)

Buchhave, L.A., Latham, D.W., Johansen, A., Bizzarro, M., Torres, G., Rowe, J.F., Batalha, N.M., Borucki, W.J., Brugamyer, E., Cladwell, C., Bryson, S.T., Ciardi, D.R., Cochran, W.D., Endl, M., Esquerdo, G.A., Ford, E.B., Geary, J.C., Gilliland, R.L., Hansen, T., Isaacson, H., et al.: An abundance of small exoplanets around stars with a wide range of metallicities. Nature **486**, 375–377 (2012)

Coleman, G., Nelson, R.P.: Giant planet formation in radially structured protoplanetary discs (2016). astro-phEP p 1604.05191

Crida, A., Bitsch, B.: Runaway gas accretion and gap opening versus type I migration (2016). astro-phEP p arXiv:1610.05403

Dittkrist, K.M., Mordasini, C., Klahr, H., Alibert, Y., Henning, T.: Impacts of planet migration models on planetary populations. Effects of saturation, cooling and stellar irradiation. Astron. Astrophys. **567**, A121 (2014)

Dürmann, C., Kley, W.: Migration of massive planets in accreting disks. Astron. Astrophys. **574**, A52 (2015)

Dürmann, C., Kley, W.: The accretion of migrating giant planets (2016). astro-phEP p arXiv:1611.01070

Fischer, D.A., Valenti, J.: The planet-metallicity correlation. Astrophys. J. **622**, 1102–1117 (2005)

Fressin, F., Torres, G., Charbonneau, D., Bryson, S.T., Christiansen, J., Dressing, C.D., Jenkins, J.M., Walkowicz, L.M., Batalha, N.M.: The false positive rate of Kepler and the occurrence of planets. Astrophys. J. **766**, 81 (2013)

Greenberg, R., Bottke, W.F., Carusi, A., Valsecchi, G.B.: Planetary accretion rates - analytical derivation. Icarus **94**, 98–111 (1991)

Hayashi, C.: Structure of the solar nebula, growth and decay of magnetic fields and effects of magnetic and turbulent viscosities on the nebula. Prog. Theor. Phys. Suppl. **70**, 35–53 (1981)

Ida, S., Lin, D.N.C.: Toward a deterministic model of planetary formation. I. A desert in the mass and semimajor axis distributions of extrasolar planets. Astrophys. J. **604**, 388–413 (2004)

Johansen, A., Youdin, A.: Protoplanetary disk turbulence driven by the streaming instability: nonlinear saturation and particle concentration. Astrophys. J. **662**, 627–641 (2007)

Johansen, A., Mac Low, M.M., Lacerda, P., Bizzarro, M.: Growth of asteroids, planetary embryos and kuiper belt objects by chondrule accretion. Sci. Adv. **1**(3), 1500109 (2015)

Johnson, J.A., Aller, K.M., Howard, A.W., Crepp, J.R.: Giant planet occurrence in the stellar mass-metallicity plane. Publ. Astron. Soc. Pac. **122**(894), 905–915 (2010)

Kley, W., Crida, A.: Migration of protoplanets in radiative discs. Astron. Astrophys. **487**, L9–L12 (2008). doi:10.1051/0004-6361:200810033

Kley, W., Bitsch, B., Klahr, H.: Planet migration in three-dimensional radiative discs. Astron. Astrophys. **506**, 971–987 (2009). doi:10.1051/0004-6361, 0908.1863

Kokubo, E., Ida, S.: Formation of protoplanet systems and diversity of planetary systems. Astrophys. J. **581**, 666–680 (2002)

Lambrechts, M., Johansen, A.: Rapid growth of gas-giant cores by pebble accretion. Astron. Astrophys. **544**, A32 (2012)

Lambrechts, M., Johansen, A.: Forming the cores of giant planets from the radial pebble flux in protoplanetary discs. Astron. Astrophys. **572**, A107 (2014)

Lambrechts, M., Johansen, A., Morbidelli, A.: Separating gas-giant and ice-giant planets by halting pebble accretion. Astron. Astrophys. **572**, A35 (2014)

Lanza, A.F.: Star-planet magnetic interaction and evaporation of planetary atmospheres. Astron. Astrophys. **557**, A31 (2013)

Lega, E., Crida, A., Bitsch, B., Morbidelli, A.: Migration of Earth-size planets in 3D radiative discs. Mon. Not. R. Astron. Soc. **440**, 683–695 (2014)

Lega, E., Morbidelli, A., Bitsch, B., Crida, A., Szulágyi, J.: Outward Migration of planets in stellar irradiated 3D discs (2015). astro-phEP p 1506.07348

Levison, H.F., Thommes, E., Duncan, M.J.: Modeling the formation of giant planet cores. I. Evaluating key processes. Astron. J. **139**, 1297–1314 (2010)

Levison, H.F., Kretke, K., Duncan, M.J.: Growing the gas-giant planets by the gradual accumulation of pebbles. Nature **524**, 322–324 (2015)

Machida, M.N., Kokubo, E., Inutsuka, S.I., Matsumoto, T.: Gas accretion onto a protoplanet and formation of a gas giant planet. Mon. Not. R. Astron. Soc. **405**, 1227–1243 (2010)

Mamajek, E.E.: Initial conditions of planet formation: lifetimes of primordial disks. AIP Conf. Proc. **1158**, 3–10 (2009)

Mayor, M., Queloz, D.: A Jupiter-mass companion to a solar-type star. Nature **378**, 355–359 (1995)

Mayor, M., Marmier, M., Lovis, C., Udry, S., Segransan, D., Pepe, F., Benz, W., Bertaux, J.L., Bouchy, F., Dumusque, X., Lo Curto, G., Mordasini, C., Queloz, D., Santos, N.C.: The HARPS search for southern extra-solar planets XXXIV. Occurrence, mass distribution and orbital properties of super-Earths and Neptune-mass planets (2011). astro-phEP p arXiv:1109.2497

Mizuno, H.: Formation of the giant planets. Prog. Theor. Phys. **64**, 544–557 (1980)

Morbidelli, A., Lambrechts, M., Jacobson, S.A., Bitsch, B.: The great dichotomy of the Solar System: small terrestrial embryos and massive giant planet cores (2015). astro-phEP p arXiv:1506.01666

Ogihara, M., Morbidelli, A., Guillot, T.: A reassessment of the in situ formation of close-in super-Earths. Astron. Astrophys. **578**, A36 (2015)

Ormel, C.W., Klahr, H.H.: The effect of gas drag on the growth of protoplanets. Analytical expressions for the accretion of small bodies in laminar disks. Astron. Astrophys. **520**, A43 (2010)

Paardekooper, S.J., Mellema, G.: Halting type I planet migration in non-isothermal disks. Astron. Astrophys. **459**, L17–L20 (2006). doi:10.1051/0004-6361:20066304, arXiv:astro-ph/0608658

Paardekooper, S.J., Baruteau, C., Kley, W.: A torque formula for non-isothermal Type I planetary migration - II. Effects of diffusion. Mon. Not. R. Astron. Soc. **410**, 293–303 (2011)

Piso, A.M.A., Youdin, A.: On the minimum core mass for giant planet formation at wide separations. Astrophys. J. **786**, 21 (2014)

Pollack, J.B., Hubickyj, O., Bodenheimer, P., Lissauer, J.J., Podolak, M., Greenzweig, Y.: Formation of the giant planets by concurrent accretion of solids and gas. Icarus **124**, 62–85 (1996)

Raettig, N., Klahr, H.H., Lyra, W.: Particle trapping and streaming instability in vortices (2015). astro-phEP p 1501.05364

Raymond, S.N., Kokubo, E., Morbidelli, A., Morishima, R., Walsh, K.J.: Terrestrial planet formation at home and abroad. In: Protostars and Planets VI, pp. 595–618. University of Arizona Press, Tucson (2014)

Ros, K., Johansen, A.: Ice condensation as a planet formation mechanism. Astron. Astrophys. **552**, A137 (2013)

Safronov, V: Evoliutsiia doplanetnogo oblaka (1969)

Tanaka, H., Ida, S.: Growth of a migrating protoplanet. Icarus **139**, 350–366 (1999)

Tanaka, H., Takeuchi, T., Ward, W.R.: Three-dimensional interaction between a planet and an isothermal gaseous disk. I. Corotation and lindblad torques and planet migration. Astrophys. J. **565**, 1257–1274 (2002)

Ward, W.R.: Protoplanet migration by nebula tides. Icarus **126**, 261–281 (1997)

Weidenschilling, S.J.: Aerodynamics of solid bodies in the solar nebula. Mon. Not. R. Astron. Soc. **180**, 57–70 (1977)

Weidenschilling, S.J.: The distribution of mass in the planetary system and solar nebula. Astrophys. Space Sci. **51**, 153–158 (1977)

Youdin, A., Goodman, J.: Streaming instabilities in protoplanetary disks. Astrophys. J. **620**, 459–469 (2005)

Zsom, A., Ormel, C.W., Güttler, C., Blum, J., Dullemond, C.P.: The outcome of protoplanetary dust growth: pebbles, boulders, or planetesimals? II. Introducing the bouncing barrier. Astron. Astrophys. **513**, A57 (2010)

Index

M. Pessah, O. Gressel (eds.), *Formation, Evolution, and Dynamics of Young Solar Systems*, Astrophysics and Space Science Library 445,
DOI 10.1007/978-3-319-60609-5

Lightning Source UK Ltd.
Milton Keynes UK
UKHW02n1230130418
321005UK00001B/104/P

9 783319 606088